Western Europe's Natural Wonders

Western Europe's Natural Wonders is part of a series of books that provides a geologic guide to some of the most iconic, interesting, and geologically significant features of the Earth's surface. This volume of self-guided tours takes readers across three classic locations in Western Europe including the Mid-Atlantic Ridge in Iceland and classic locations in the development of geologic concepts in the Alps and Pyrenees. It examines plate tectonics from spreading centers to convergent margins and discusses the deformation of mountain ranges and sedimentary basins, reviews the importance of geology in a historical context, explains the role of geology in agriculture, mining, and energy development, and highlights mineral and fossil locations. Sections describe volcanism, glacial features, and the concepts of ice ages, thrust faulting, mountain building, and plate tectonics. Written in a conversational style, this is a lavishly illustrated set of field trips for students and teachers, professionals, and amateur Earth scientists to understand the key geologic aspects of local landscapes.

Features

- Explains the geology of regions with emphasis on landscape and scenery along a series of road transects.
- Includes numerous colorful maps and breathtaking photos of geological landscapes to illustrate various geological features.
- Provides background and insights that enhance the experience of the reader far beyond simple sightseeing.
- Easy to read and understood by the lay person and written by a career geologist with decades of experience in the field.

This book is for a variety of readers, from those with a deep knowledge and understanding of geology and earth sciences to those with a desire to learn more about how their natural surroundings and scenic parks came to be and what the scenery tells us about Earth's history. It is also an invaluable source of knowledge for professional Earth scientists who want to better understand their local geology, as well as tourism professionals, tourism boards, and agencies who wish to promote their scenic areas and parks.

Geologic Tours of the World

Series Editor
Gary Prost

Principal Geologist, G.L. Prost GeoConsulting of El Cerrito, California

This series of geological guidebooks provides everything needed to understand the landscape across the best-known natural wonders in the world. Each guidebook includes a geologic overview with maps and charts and describes the rocks and geologic history, plants and animals, and mineral and fossil locations. There are examples and photos of world-class scenery and geologic features, discussions of less known but equally interesting attractions, and information on accessing these sites. If you are walking along the sea cliffs of Great Britain, crossing the Rocky Mountains in Canada or the Italian Alps, gazing upon the Grand Canyon in the United States, exploring volcanoes in the Andes or glaciers in Iceland, what you are seeing is explained as well as how it came to be, and its significance. These self-guided tours are written for any readers interested in travel and geology and serve as an introduction to the geology of significant regions for educators, students, and professionals in Earth Science programs to develop their own explorations of their local geologic environment.

Published Titles

North America's Natural Wonders
Appalachians, Colorado Rockies, Austin-Big Bend Country, Sierra Madre
Gary Prost

North America's Natural Wonders
Canadian Rockies, California, The Southwest, Great Basin, Tetons-Yellowstone Country
Gary Prost

The United Kingdom's Natural Wonders
Scotland and Northern Ireland, Lake District and Yorkshire Dales, Wales and West Midlands, England
Gary Prost

South America's Natural Wonders
Patagonia, Neuquén Basin, Atacama Desert, Darwin's Footsteps (Mendoza-Valparaiso)
Gary Prost

Western Europe's Natural Wonders
Iceland, Pyrenees, and Western Alps
Gary Prost

Western Europe's Natural Wonders

Iceland, Pyrenees, and Western Alps

Gary L. Prost

CRC Press
Taylor & Francis Group
Boca Raton London New York

CRC Press is an imprint of the
Taylor & Francis Group, an **informa** business

Designed cover image: Photos by the author, Gary Prost. Front cover image - Clearing storm, Helbronner Peak, Mt Blanc Massif, Courmayeur, Italy. Back cover image - Alpine folding, Sorrosal Falls, Broto, Spain.

First edition published 2026
by CRC Press
2385 NW Executive Center Drive, Suite 320, Boca Raton FL 33431

and by CRC Press
4 Park Square, Milton Park, Abingdon, Oxon, OX14 4RN

CRC Press is an imprint of Taylor & Francis Group, LLC

© 2026 Gary Prost

ISBN: 978-1-032-56452-4 (hbk)
ISBN: 978-1-032-56451-7 (pbk)
ISBN: 978-1-003-43559-4 (ebk)

DOI: 10.1201/9781003435594

Typeset in Times
by Deanta Global Publishing Services, Chennai, India

To my wife Nancy, who lets me entertain my passion for geology.

Contents

Contents   xi

Preface

From the Mid-Atlantic Rift to mountains raised by the collision of Europe with Africa, Western Europe has so much to offer.

Most travelers aren't drawn solely by geology—they're after dramatic scenery, rich history, and unforgettable experiences. Iceland delivers. If you want to see volcanoes, glaciers, and hot springs, then Iceland is your kind of place. From steaming lava flows to massive ice caps, it is literally a land of fire and ice. The first tour in the European series delves into the volcanoes, hot springs, glaciers, geysers, and rift valleys of Iceland. Spectacular cliffs battered by the sea serve as rookeries for millions of gulls and puffins. As the only place where the Mid-Atlantic Ridge rises above sea level, this volcanic landscape is as dynamic as it is breathtaking—you never quite know when or where it might erupt next.

The southern margin of Europe is also geologically active. The magnificent Pyrenees owe their origin to the slow but relentless collision between the African and European tectonic plates. As part of the broader Alpine Orogeny, these mountains have been rising for some 50 million years. Their formation tells a tale of continental collision, fold-thrust belts, mobile salt layers, and the formation and reshaping of sedimentary basins. This geological journey loops through the central and eastern Pyrenees, crossing all of the major tectonic zones. We travel from the southern foreland basin into the central axial zone, and across to the northern foreland. Along the way, we explore deformation features related to both compression and salt tectonics. The dynamic relationship between mountain building (tectonics) and sedimentation –– from the early rifting stages to later compression and basin inversion – are revealed.

The towering Alps were born from the slow but relentless closure of the ancient Tethys Ocean, squeezed out of existence between the converging African and European continents. Over the last 90 million years, this powerful collision has shaped a mountain range that enchants visitors and intrigues scientists alike. The Alps have played a key role in shaping our understanding of glaciers, ice ages, continental collisions, and the architecture of fold-thrust belts. It's a region steeped in both geologic drama and human history – an ideal place for exploration. The Alps tour begins in the southern hinterland basin where vineyards are rooted in the fertile, geologically diverse soils near Turin. High-altitude rock panels reveal an astonishing secret: they were once ocean-floor sediments. Their presence atop mountains puzzled early geologists, but generations of study have revealed that these ancient seafloor layers were pushed skyward and thrust over younger rocks by nearly horizontal thrust faults. At the Matterhorn we find fragments of the African plate and from Europe's original southern margin that were shoved northwestward. Farther north and west, more slices of Europe were crumpled and thrust back over themselves. At the outer edges of the Alps, we reach the foreland Molasse Basin and the Central European Rift basins filled with sediments washed down from the rising mountains and, once again, ideal terrain for vineyards.

These trips can be run in either direction, but most start where there is easy access (a major town or airport). The transects as laid out can take up to two weeks, but these trips can easily be broken into smaller bits that can take as little time as a few hours. You don't have to do an entire tour in one go.

Times and distances are provided, but times are meant as a guide only and will change depending on traffic and weather conditions. Although described stops are provided, there are many more stops that can be made along these routes: don't feel constrained to use only those in the guidebook. Be aware that some routes, especially over the higher passes, may be closed in winter and even well into spring. All of these roads are accessible with a standard road car; four-wheel or all-wheel drive is not necessary for these trips.

A FEW WORDS ABOUT DRIVING IN WESTERN EUROPE

First, I strongly recommend getting a GPS navigation system such as Tom Tom, Google Maps, or Apple Maps. These were a lifesaver for me. Failing this, get a good navigator and have paper maps as backup.

It is always a good idea, before you leave, to look up and print out European road signs so that you are familiar with those that are uncommon in North America.

If you are renting a car and will be crossing international borders, AAA recommends that you get an International Driver's License. It costs a few bucks and does not require a test.

Be careful when pulling off the road, especially on narrow shoulders or curves in the road. Watch for traffic at all times. Safety first!

ICELAND

Most visitors either rent a car or join a tour. There are no trains in Iceland, so the best way to get around on your own is by car. Bus Passports allow you to visit some of the more popular tourist destinations. The more adventurous can hike or cycle around the country, but you better be prepared for rapidly changing weather. Route 1, the Ring Road, is the main road in Iceland and goes around the entire island except for the Westfjords. The Ring Road is close to the coast, and most of the tourist attractions in Iceland are near this road. The majority of the Ring Road has one lane in each direction. This paved road is 1,332 km (828 mi) long.

Maintain the speed limit, especially on gravel roads. The limit is 30–50 km/hr (18–30 mi/hr) in residential areas and 90 km/hr (54 mi/hr) on the main roads. The speed limit on gravel roads is 80 km/hr (48 mi/hr). Speed cameras are everywhere, and breaking traffic laws is punishable by large fines. Children must be in car safety seats or wear seat belts, depending on age. Driving under the influence of alcohol carries not only a very high fine, but also the loss of your driver's license. Only hands-free phones may be used while driving. Headlights should be on both day and night. Driving off-road is strictly forbidden due to the delicate environment.

Other things to be aware of include single-lane bridges and animals on the road. Warning signs indicate danger ahead, so be aware of them. F-roads are mountain roads: do not wander off them. These roads mostly require experienced drivers with high-clearance 4x4 vehicles, and it is suggested that two or more cars travel together in case one gets stuck. Most mountain roads are closed in the winter (Fun Iceland, https://www.funiceland.is/facts/travel-guide/driving-in-iceland/). All of the stops on this field trip are on paved or well-maintained gravel roads.

For information about road conditions, visit www.road.is (https://www.road.is/) or call 1777 or +354 522 1100. In case of emergencies, call 112.

The weather in Iceland is unpredictable and can change quickly from sunshine and blue sky to heavy rain and gale-force wind. It is therefore important to check the weather forecast and be prepared for all weather conditions. The official website for weather in Iceland is Road.is (https://www.road.is/).

Be sure to pull well off to the side of the road when stopping to examine the geology.

If you are going for a hike in the backcountry or near an active eruption, leave a note in your car window telling where you are going, who you are, how you can be reached, and when to expect you back. If you cannot be reached after your scheduled return time, they should call 112. Remember, weather conditions can change quickly. The Iceland Association for Search and Rescue is available to help.

SPAIN, FRANCE, AND ITALY

Speed cameras are almost everywhere, yet this does not deter most drivers from exceeding the speed limit. You will have to decide whether or not to respect the limits. Most native drivers do not respect personal space and will tailgate relentlessly unless you drive at least 10 km/h over the limit, and they *will* pass you at the first opportunity. Other than that, they tend to be polite drivers.

White panel vans seem to be the "pickup truck" of Europe. They are everywhere delivering whatever at high speeds to even the most remote locations.

Beware cyclists, especially on narrow, winding mountain roads. They are *everywhere* in France, less so elsewhere.

Get the smallest car possible, not only for better mileage, but so you can drive into and through old towns where there is barely a car's width between buildings and so you can enter and exit underground parking garages that can be exceedingly tight.

Regarding the high price of gasoline compared to North America, if you figure in the much better mileage in most European cars (mostly in the 40 to 50 miles/gallon range), and the exchange rate, the cost to drive a given distance does not vary much between the continents.

Toll roads are common in France and Italy, but the tolls are reasonable for the most part. You will have to decide whether you want to set your navigation to find the fastest or the shortest route to a location, and this may determine whether you pay tolls. You can also set most systems to avoid tolls.

I'll admit to feeling the usual sense of trepidation in preparing this volume. It's one thing to write about North America, where the landscapes are familiar to me. Venturing into a different continent – rich with history, dotted with Geoparks and World Heritage Sites – felt like a bold move for an outsider. The Alps, especially, hold a special place in the story of geology, being the cradle of thrust tectonics and the birthplace of the concept of Ice Ages. Still, I accepted the challenge, and I hope the result does justice to the subject.

One final note: naturally, there are countless other geological journeys across Europe that aren't covered in this volume. The possibilities are nearly endless, and I hope to explore more of them in future guides. I chose these three tours because they've long topped my personal wish list – for both their striking landscapes and their compelling geology. Hopefully, by the end of your journey – if not before – you'll agree they're worth the trip. After all, the goal of this book is simple: to make you say, *"I want to go there. I want to see that for myself."*

Charts are provided to show what rock layers exist in each area, how old the layers are, and what they are made of. Each region is placed in a plate-tectonic context, so that you know how and when mountains and basins formed. The geologic influences on the economy are discussed so that you know, for instance, the reasons that the Po Valley is a wine-growing region, and why much of the first evidence for glaciation came from the Alps.

Units are provided in metric first, then their Imperial equivalent.

Abbreviations are explained, then used. For example, millions of years is frequently abbreviated as Ma (for "mega-annum"), Ga (billions of years), Fm is used for Formation, Gp for Group, mbr for member, bbl for barrel, and billions of barrels of oil equivalent is BBOE.

Road descriptions, pullouts, travel times, mine and cave tours, contact names, websites and emails are current as of the time of writing. Contact a park or museum website for current hours and entry fees. All entrance fees and tour prices mentioned are current as of 2024 or 2025.

A simplified geologic time scale along with years before present is provided as a useful reference for those not familiar with the different geologic periods referred to in the trip descriptions.

Location of the three geologic tours in this volume.

EON	ERA	PERIOD	EPOCH		Ma
Phanerozoic	Cenozoic	Quaternary	Holocene		0.01
			Pleistocene	Late	0.8
				Early	1.8
		Tertiary (Neogene)	Pliocene	Late	3.6
				Early	5.3
			Miocene	Late	11.2
				Middle	16.4
				Early	23.7
		Tertiary (Paleogene)	Oligocene	Late	28.5
				Early	33.7
			Eocene	Late	41.3
				Middle	49.0
				Early	54.8
			Paleocene	Late	61.0
				Early	65.0
	Mesozoic	Cretaceous	Late		99.0
			Early		144
		Jurassic	Late		159
			Middle		180
			Early		206
		Triassic	Late		227
			Middle		242
			Early		248
	Paleozoic	Permian	Late		256
			Early		290
		Pennsylvanian			323
		Mississippian			354
		Devonian	Late		370
			Middle		391
			Early		417
		Silurian	Late		423
			Early		443
		Ordovician	Late		458
			Middle		470
			Early		490
		Cambrian	D		500
			C		512
			B		520
			A		543
Precambrian	Proterozoic	Late			900
		Middle			1600
		Early			2500
	Archean	Late			3000
		Middle			3400
		Early			3800?

Geologic time scale. Wiki Commons, diagram courtesy of the U.S. Geological Survey, https://commons.wiki media.org/wiki/File:Geologic_time_scale.jpg

Acknowledgements

The assistance and good company of Adam Prost (Chapter 1) and Guy Peasley (Chapter 2) were instrumental in allowing me to complete the field work on these tours and is gratefully acknowledged. Chapter specific acknowledgements follow. Thanks to Susan Sheets, travel agent extraordinaire, for arranging all of the details of the reconnaissance trips.

Chapter 1: The technological savvy and navigation skills of Adam Prost were instrumental in completing the field work on this tour and are gratefully acknowledged. He is the ideal travel companion – reliable behind the wheel, quick with a camera, and always up for a hike or adventure.

Chapter 2: Thanks to Samuel Eguiluz for suggesting, years ago, the TransPyrenean Geodrive that anchored this tour. Guy Peasley's assistance, companionship, and expert navigation were essential to completing the reconnaissance for this tour and are gratefully acknowledged.

Chapter 3: Thanks to my patient, calm, and competent travel agent, Susan Sheets, for bailing me out of a bad spot during a blizzard in Courmayeur.

About the Author

Gary L. Prost obtained his BSc in geology from Northern Arizona University in 1973 and an MSc (1975) and PhD (1986) in geology at Colorado School of Mines. Over the past 45 years, he has worked for Norandex (mineral exploration), Shell U.S.A. (petroleum exploration worldwide), the U.S. Geological Survey (geologic mapping, coal), the Superior Oil Company (mineral and oil exploration), Amoco Production Company (worldwide oil exploration, remote sensing, and structural geology), Gulf Canada (international new ventures), and ConocoPhillips Canada (Canadian Arctic exploration, gas field development, oil sands development, and reservoir characterization). He spent over 20 years working as a satellite image analyst in the search for hydrocarbons and minerals in more than 30 countries. During this time, he applied structural geology and remote sensing to exploration, development, and environmental projects. The second half of his career was spent working on regional studies, new ventures and frontier exploration, and oil and gas field development. Now retired, his most recent work has been in public outreach, leading field trips and

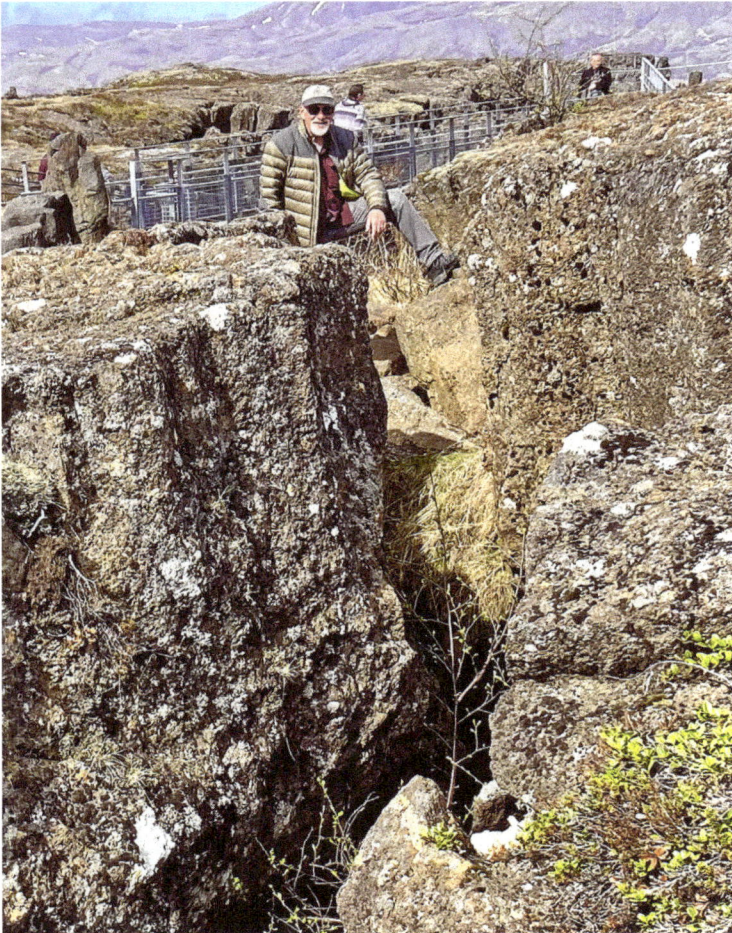

The author with one foot on the North American Plate, the other on the European Plate at the Mid-Atlantic Rift, Thingvellir, Iceland.

educating the public on topics of geological interest. He is the principal geologist for G.L. Prost GeoConsulting of El Cerrito, California, and he has been a registered professional geologist in Wyoming (United States) and in Alberta and the Northwest Territories (Canada). He has previously published eight books, including *Rocks and Riches – Exploring California's Stunning Geology* (Heyday, 2025), *Geologic Tours of the World – North America's Natural Wonders*, Volumes 1 and 2 (Taylor & Francis, 2020), *Geologic Tours of the World – the United Kingdom's Natural Wonders* (Taylor & Francis, 2023), *Geologic Tours of the World – South America's Natural Wonders* (Taylor & Francis, 2024), *The Geology Companion: Essentials for Understanding the Earth* (Taylor & Francis, 2018), *Remote Sensing for Geoscientists: Image Analysis and Integration* (fourth edition, Taylor & Francis, 2025), and the *English–Spanish and Spanish–English Glossary of Geoscience Terms* (Taylor & Francis, 1997).

1 Iceland

Gullfoss, Iceland's version of Niagara Falls. Photo courtesy of A. Tille, https://commons.wikimedia.org/wiki/File:GullfossOverview.jpg

OVERVIEW

Most people don't go touring for the geology alone; they want to see and experience stunning and scenic landscapes, exotic history, and charming attractions. Iceland has it all. And if you like volcanoes, glaciers, waterfalls, and hot springs, Iceland is your kind of place. It has Viking history and thermal spas, hothouse vegetables, and black sand beaches. The first tour in the European series explores the volcanoes, thermal springs, glaciers, waterfalls, and rift valleys of Iceland. The only subaerial exposure of the Mid-Atlantic Ridge, this presently active volcanic terrain is the source of both the "killer fog" of the 1780s and the 2010 shutdown of transatlantic air traffic. We examine the central rift of a mid-ocean ridge; discuss geothermal energy; visit basalt flows, columnar basalts at waterfalls and sea cliffs, volcanic cones and volcanic fissures; gaze upon endless glaciers; and examine what happens when eruptions occur beneath ice. Starting in Keflavik, we drive to the Blue Lagoon and the ongoing eruption just between there and Grindavík, and continue to the Seltún geothermal area of the Reykjanes Peninsula and active rift. We pass through the Rekjanes, Katla, and Saga UNESCO European Geoparks and two UNESCO World Heritage sites on Iceland, Vatnajökull National Park and Thingvellir National Park. We explore the site of the original geyser at Geysir, soak in several hot spring pools, and see one of the youngest islands in the world (and the only successful attempt to stop a volcanic flow) on Heimaey. You can stand with one foot on the North American tectonic plate and the other on the European plate at Thingvellir, and see the isolated yet dramatic volcanic edifice of the Snæfellsjökull Massif in Snæfellsjökull National Park and the spectacular sea cliffs of the Snæfellsnes Peninsula.

DOI: 10.1201/9781003435594-1

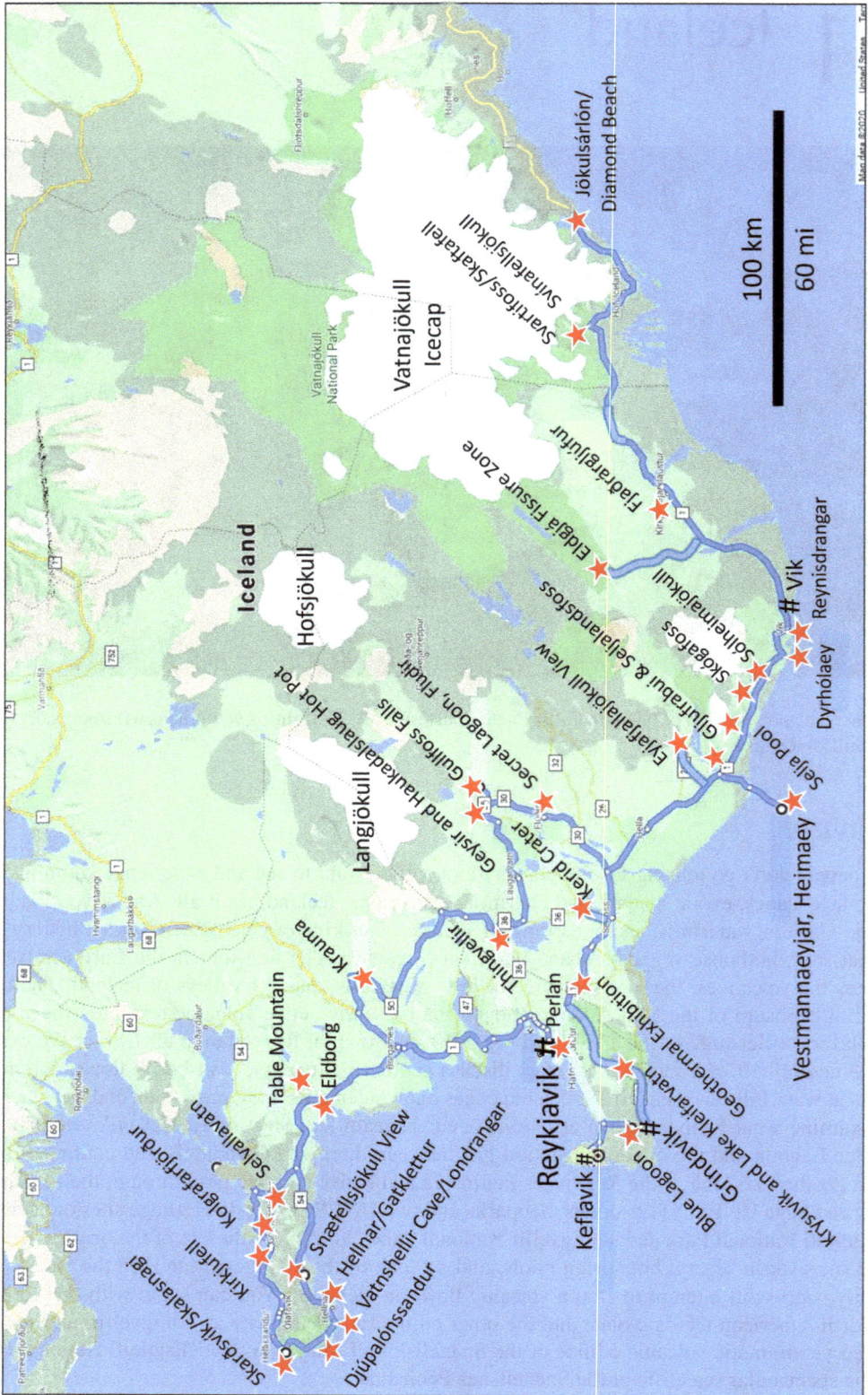

Major stops on the geologic driving tour of southern and western Iceland.

ITINERARY

Begin – Keflavík Airport
Stop 1 Reykjanes Peninsula
     1.1 Blue Lagoon and Sundhnúkur Eruptions
     1.2 Krýsuvík-Seltún Geothermal Field
     1.3 Lake Kleifarvatn Rift Valley
Stop 2 Geothermal Energy Exhibit
Stop 3 Kerid Crater
Katla Geopark
Stop 4 Eyjafjallajökull View
Stop 5 Seljalandsfoss and Gljúfrabúi
     5.1 Seljalandsfoss
     5.2 Gljúfrabúi
Stop 6 Selja Warm Pool
Stop 7 Skógafoss
Stop 8 Sólheimajökull Glacier
Stop 9 Dyrhólaey
Stop 10 Reynisdrangar
Side Trip 1 Eldgjá Fissure Zone and Ófærufoss
Stop 11 Fjaðrárgljúfur
Stop 12 Vatnajökull National Park
Stop 12.1 Svartifoss
Stop 12.2 Skaftafellsjökul Glacier
Stop 13 Svínafellsjökull Glacier
Stop 14 Breiðamerkurjökull Glacier
Stop 14.1 Glacier Lagoon/Jökulsárlón
Stop 14.2 Diamond Beach
Side Trip 2 Vestmannaeyjar, Heimaey
Stop 15 Secret Lagoon, Flúðir
Stop 16 Gullfoss Falls
Stop 17 Geysir and Strokkur
Stop 18 Thingvellir/Þingvellir National Park
     18.1 Visitor Center and Mid-Atlantic Rift
     18.2 Öxarárfoss
Stop 19 Krauma Hot Springs, Deildartunguhver
Stop 20 Snæfellsnes Peninsula
     20.1 Selvallavatn View
     20.2 Kolgrafarfjörður
     20.3 Kirkjufell Mountain and Falls
Snæfellsjökull National Park
     20.4 Skarðsvík Beach
     20.5 Skalasnagi Lighthouse
     20.6 Djúpalónssandur Beach
     20.7 Vatnshellir Cave
     20.8 Londrangar Viewpoint
     20.9 Hellnar Viewpoint
     20.10 Gatklettur and Cliff Overlooks
     20.11 Snæfellsjökull Viewpoint
Stop 21 Ljósufjöll Table Mountain
Stop 22 Eldborg á Mýrum

Stop 23 Reykjavík
    23.1 Perlan Natural History Museum
    23.2 Nauthólsvík Geothermal Beach, Reykjavík
End – Reykjavík

HISTORY

Iceland's position at the junction of warm southerly and cold northerly waters and air masses contributes to a unique climate that is unusually mild considering the high latitude of the island. The last Ice Age ended here about 9,000 years ago.

The island is thought to have been inhabited by a few Irish monks when Scandinavian Viking explorers began bumping into it in the 800s. The official story is that Iceland was "discovered" by Naddodd, a sailor looking for the Faroe Islands who got lost. He called the island *Snæland*, or "Snowland." The first Norseman who deliberately sailed to Iceland was Hrafna-Flóki Vilgerðarson. Flóki settled for the winter at Barðaströnd on the southwest coast. When summer arrived, and the island greened up, he realized that the place was indeed habitable. Flóki sailed back to Norway with the news. Perhaps the first permanent settler was the Norwegian Ingólfr Arnarson, who settled Reykjavík, or "Smoke Cove," in 874. The smoke probably referred to local steaming fumaroles. The settlers established the Althing, the oldest known parliamentary democracy, when they met to form a people's council at Thingvellir in 930. The settlers worshiped the Norse gods Odin, Thor, and Freyja, among others. By the 10th century, political pressure from Norwegian king Olaf Tryggvason led to increasing conversions to Christianity. In the year 1000, as a civil war between the religions seemed likely, the Althing appointed a local chieftain, Thorgeir Ljosvetningagodi, to settle the issue. He determined that the country should convert to Christianity, but that pagans would be allowed to continue to worship privately.

During this time, known as the Old Commonwealth, Icelanders began to document the nation's history in sagas, or historical stories. In the early 1200s, the age of the Sturlungs, a series of civil wars weakened Iceland, which eventually became subjugated to Norway through the Old Covenant (1262–1264). This effectively ended the commonwealth.

Climate change occurred around the time that Iceland became a colony of Norway. The Little Ice Age (roughly the years 1300 to 1900) caused shorter growing seasons and colder winters. Since Iceland had marginal agriculture at the best of times, the cooling climate caused extensive hardship and resulted in a kind of serfdom, the *vistarband*, in which landless peasants were bound to landowners for a year at a time. Farming barley became difficult, and longer winters meant more feed had to be laid-in for livestock. This led to trade with Europe where Icelandic cod was exchanged for grain. In 1319 Norway was united with Sweden, and in 1376 they were united with Denmark in the Kalmar Union. Upon the Union's dissolution in 1523, Iceland became a Danish protectorate.

By the middle of the 16th century, Christian III of Denmark began to impose Lutheranism on his subjects. Catholic opposition was strong, but ended in 1550 after the Battle of Sauðafell. Icelanders became Lutherans and it is still the dominant religion. Every community seems to have a scenic church.

In 1602, Iceland was forbidden to trade with countries (mainly England and Germany) other than Denmark. The Danish–Icelandic Trade Monopoly continued in effect until 1786 and effectively impoverished the country. This, together with the Little Ice Age and the eruption of Laki in 1783, led to severe suffering and depopulation. Even today the island has only 382,000 inhabitants.

A volcanic fissure in southern Iceland called Lakagigar, or Laki for short, erupted for 8 months starting in June 1783. The fissure, along with the adjacent Grímsvötn volcano, emitted an estimated 14 km^3 (3.4 mi^3) of basalt, 8 million tons of hydrofluoric acid vapor, and 120 million tons of sulfur dioxide gas, creating what became known as the "Laki haze." These gases and aerosols rose 15 km (10 mi) into the atmosphere and spread over Europe and the Middle East. On Iceland, the "móðuharðindin" (mist hardships), caused mainly by breathing sulfuric acid aerosols, killed 80% of the sheep, 50% of the cattle, and 50% of the horses on the island. As much as a quarter of the population died in the resulting famine. The deadly fogs extended from Ireland to Prague, causing crop failures and famine. The sun was blood colored; from his hotel in Paris Benjamin Franklin wrote of "a constant fog over all Europe, and a great part of North America." The erupted aerosols blocked the sun's rays, dropping global temperatures and causing the severe winter of 1783–84. Sulfur dioxide reacted in people's lungs to create sulfurous acid, causing them to slowly choke to death. An estimated 23,000 people were poisoned in Britain alone, while the extremely cold winter killed another 8,000. Weak monsoons related to the cooling caused droughts and famine in Egypt, Arabia, and India. In North America, the winter of 1784 was one of the coldest on record, with the Mississippi River freezing at New Orleans and ice on the Gulf of Mexico (Witze and Kanipe, 2014; Prost and Prost, 2018). And all this happened because of the distinctive composition of the fissure eruption and the fact that the ash and gas cloud made it to the stratosphere where winds could distribute them around the world.

An independence movement began in Iceland during the early 1800s. The Althing was restored in 1844, and, in 1874, a thousand years after the first settlement, Denmark granted Iceland a constitution and home rule. Iceland gained sovereignty in December 1918 and cut its last ties with Denmark in 1944.

Although Iceland was neutral in the Second World War, the United Kingdom invaded and peacefully occupied it in 1940 to prevent Nazi occupation. The Allies occupied the island until the end of the war.

Following World War II, Iceland became a founding member of NATO and joined the United Nations, although it remains outside the European Union. The economy is based largely on tourism, fishing, and aluminum smelting. Why aluminum smelting, you may ask. Aluminum smelting requires a lot of energy. Iceland imports aluminum ore and uses its abundant, inexpensive hydroelectric and geothermal energy to smelt this raw material into aluminum metal that it then exports. Following rapid financial growth, the 2008–11 Icelandic financial crisis devastated the economy, from which the island nation has slowly recovered (Wikipedia, History of Iceland).

GEOLOGY OF ICELAND

Iceland has been called the "land of fire and ice," and the description is fitting. The island is made up of volcanoes and lava flows, and it has more than its share of glaciers and ice caps. The distinction is that a glacier moves, whereas an ice cap is stationary (USGS, Glaciers and Icecaps).

Iceland is also a land of rifts. It is unique in that it is one of very few places in the world where a mid-ocean ridge spreading center can be examined on dry land. It lies on the Mid-Atlantic Ridge and separates the North American tectonic plate from the Eurasian plate. Prior to breakup, this area fell within the supercontinent Pangea, a continuous landmass that included all of the present continents (Sooflauschig, 2018). The rift that would become the Atlantic Ocean began in Triassic time, between 240 and 210 Ma (million years ago).

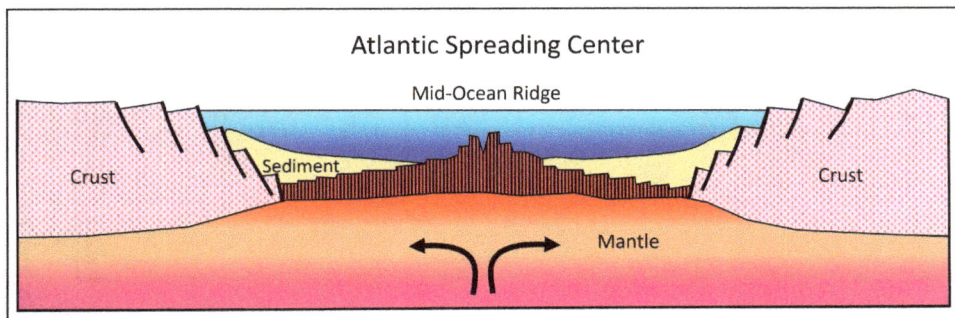

Schematic diagram of ocean floor spreading. The Mid-Ocean Ridge is composed mainly of near-vertical sheeted dikes and pillow basalts. Modified after https://commons.wikimedia.org/wiki/File:Ocean-birth_hg.png

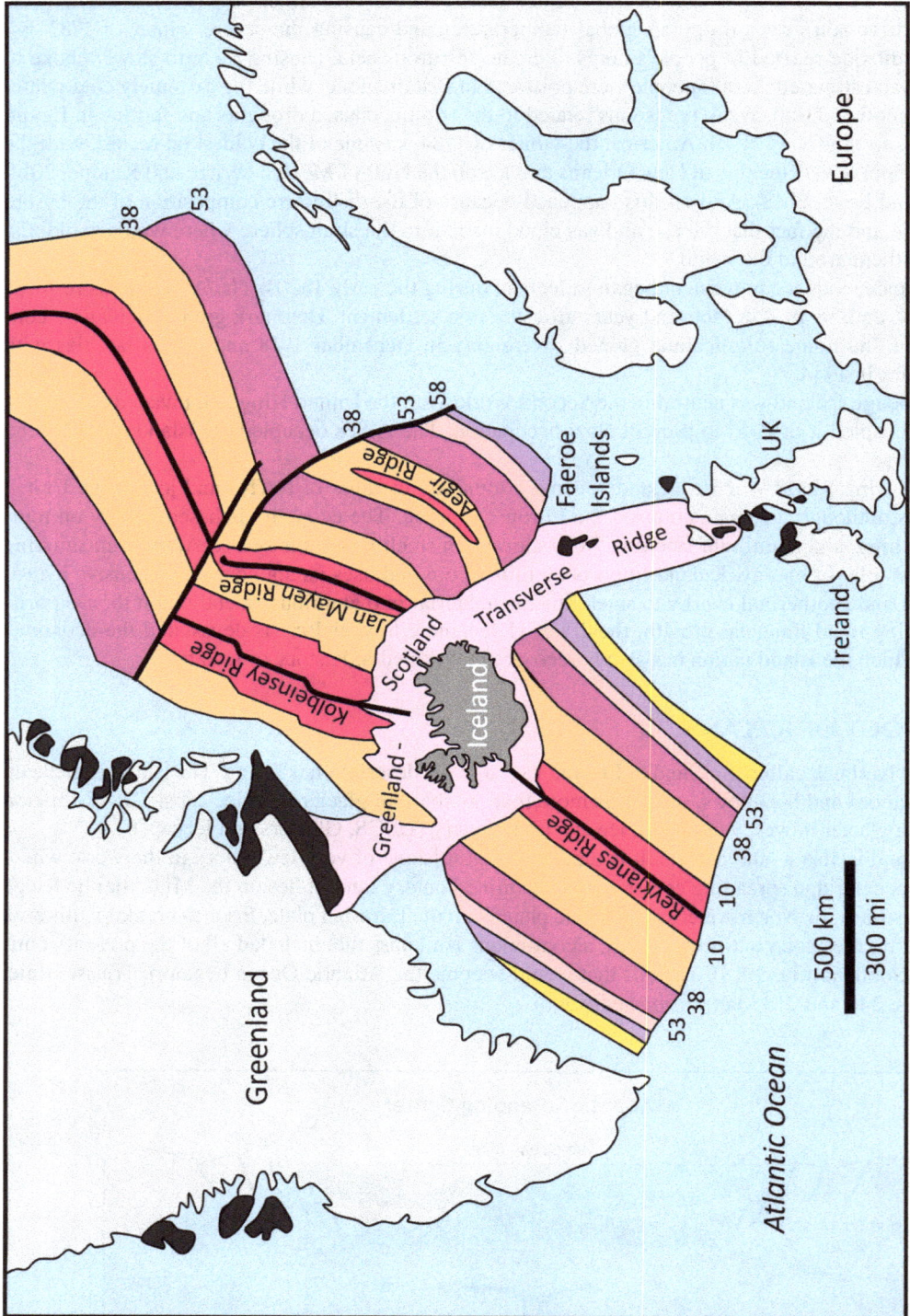

The Iceland hot spot, Mid-Atlantic Ridge segments, Greenland-Scotland Transverse Ridge, and ocean-bottom ages in millions of years. Black areas are flood basalts. Modified after Denk et al., 2011.

In this part of the Atlantic, Eurasia began to move away from North America in the early Paleogene, around 55 Ma (Denk et al., 2011). The actual combined spreading rate is around 2 cm/year (1 in/year), or 25 km (15.5 mi) in a million years (Thompson, 2010). Early rifting was accompanied by flood basalts (massive eruptions of highly fluid basaltic lava forming vast lava plains), some of which can be seen in eastern Greenland, others now in the Faeroe Islands, Northern Ireland (Giant's Causeway), and Scotland (at Ardnamurchan, Skye, Rhum, and Arran). At first much of the area was above sea level, a bulge known as the Greenland–Scotland Transverse Ridge, part of the North Atlantic Large Igneous Province. As a result of cooling and subsidence, today this bulge is mostly below sea level.

The center of rifting within Iceland appears to have shifted from western to central-eastern Iceland around 7 or 8 Ma (Ward, 1971).

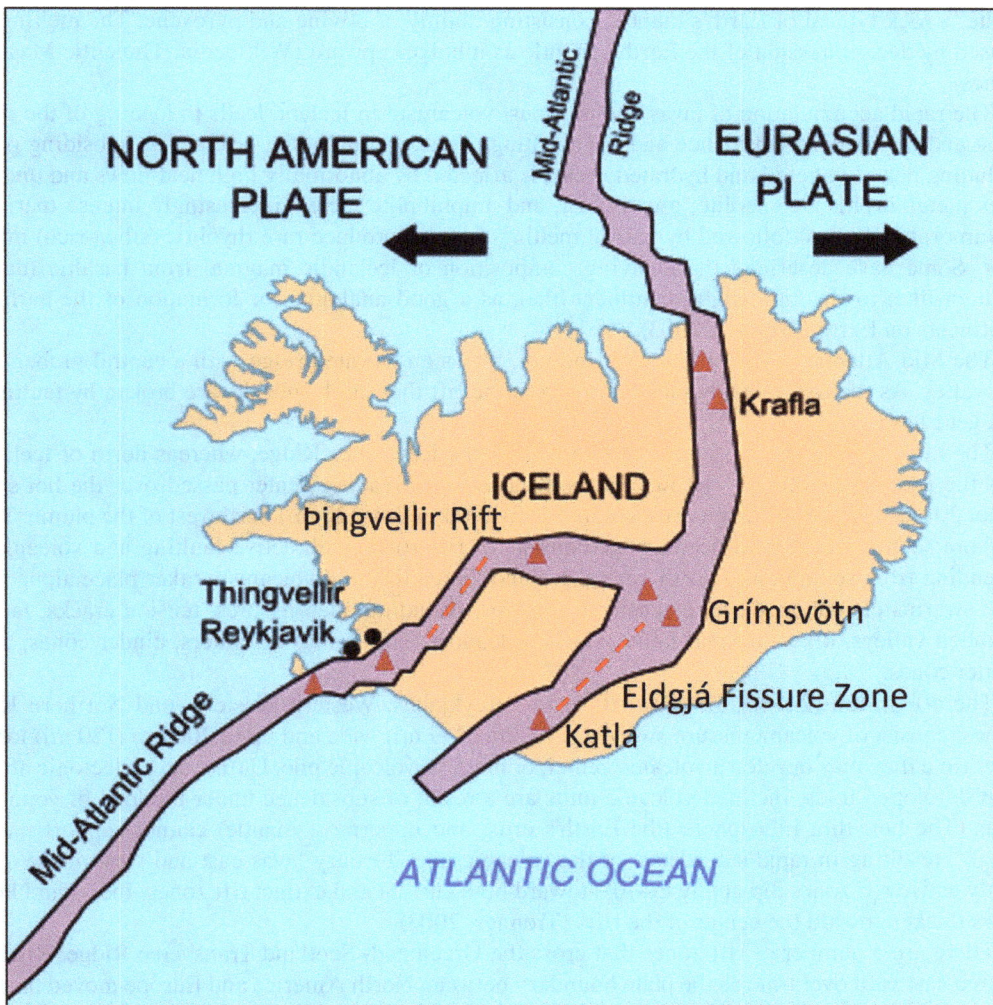

Major rift zones in Iceland. Red triangles are active volcanoes. The area encompassed by the western and eastern rifts in southern Iceland is the Hreppar Microplate. It is considered to be part of the Eurasian Plate. From https://en.wikipedia.org/wiki/File:Iceland_Mid-Atlantic_Ridge_map.svg Original source USGS, 2014, https://pubs.usgs.gov/gip/dynamic/understanding.html

In addition to being on the mid-oceanic ridge, it is thought by many geologists that there is also a mantle plume, or hot spot beneath Iceland today. The plume is caused by buoyant upwelling of Earth's mantle that leads to a crustal bulge, high heat flow, and extensive volcanism. The hot spot today lies under the Vatnajökull ice cap, where it extends downward 275–375 km (165–225 mi; Trønnes, 2003). Magma reservoirs are present at depths as shallow as 10 km (6 mi; Nature Iceland). The Iceland hotspot may be much older than the earliest rifting of what is now the northeast Atlantic mid-ocean ridge (Wikipedia, Geology of Iceland).

As with all oceanic spreading centers, the age of the ocean floor becomes older as one moves away from the central ridge. The youngest rock usually consists of sheeted dikes (accumulations of vertical sheets of black basaltic volcanic rock) and submarine volcanics such as pillow lavas. Pillow lavas solidified as pillow-shaped blobs that are characteristic of eruption under water. The volcanism is basaltic, mainly of the tholeiitic variety, which is typical of the ocean's crust. Tholeiitic rocks are the most common igneous rocks on Earth, and they are the only rocks that make up the surface of the Moon. Tholeiitic basaltic magmas are usually generated by partial melting of peridotite, a rock typical of Earth's mantle, consisting mainly of olivine and pyroxene. The melting is caused by decompression of the Earth's mantle as it bulges upward (Wikipedia, Tholeiitic Magma Series).

The rapid accumulation of lavas due to robust volcanism in Iceland leads to loading of the thin crust and consequent subsidence and slight tilting of the originally flat layers. The subsiding pile, including heavily altered and hydrated rocks, is affected by abnormally high heat flows and undergoes metamorphism to zeolite, greenschist, and amphibolite facies (increasingly intense thermal metamorphic grades) followed by partial melting that can produce rare rhyolitic (silica-rich) magmas. Some have described the evolving composition of Icelandic magma, from basaltic (dark, seafloor-like) to rhyolitic (light, continent-like), as a good analogue for formation of the earliest continents on Earth (Trønnes, 2003).

The Mid-Atlantic Rift, being young and hot, forms an elevated ridge with a central graben, or rift valley. As the rocks slowly move away from the rift they cool, subside, are broken by faulting, and get tilted.

The mid-ocean ridge south of Iceland is called the Reykjanes Ridge, whereas north of Iceland it is the Kolbeinsey Ridge. The Reykjanes–Kolbeinsey spreading center passed over the hot spot about 20 Ma ago and has since drifted about 150–200 km (90–120 mi) northwest of the plume. The onshore section of the mid-ocean ridge consists of rift zones with active faulting and volcanism extending from southwest to north-central Iceland. Spreading and volcanism takes place along fissure swarms characterized by extensional features, including open fissures, tension cracks, fault-bounded valleys, dikes, normal faults, shield volcanoes, and strings of craters, cinder cones, and spatter cones.

The 40–50 km (24–30 mi) wide rift zones (Reykjanes; Western, Eastern, and Northern Rift Zones) consist of volcanic fissure swarms 5–15 km (3–9 mi) wide and up to 200 km (120 mi) long. Over time they may develop a volcanic center, or massive volcanic pile. Unlike active tectonic areas with developed folds, inclined volcanic units are a result of subsidence under the load of younger lavas. The hot, thin lithosphere (the Earth's crust and uppermost mantle) cannot support much weight, resulting in rapid subsidence of the volcanic pile. Tertiary lavas east and west of the currently active rift zones dip gently (5–10°) toward both current and extinct rift zones. Individual lava flows thicken toward the center of the rifts (Trønnes, 2003).

There are a number of rift zones that cross the Greenland–Scotland Transverse Ridge. Rifting shifted eastward over time as the plate boundary between North America and Europe moved northwest relative to the hot spot (or the hot spot moved east relative to the mid-ocean ridge; Trønnes, 2003). The Iceland mantle plume that drives the hot spot has been active since 65–70 Ma, a time when Greenland was located above the plume (Sooflauschig, 2018). Between 15 and 24 Ma the spreading center was located in the Northwest Iceland Rift Zone, now off the northwest coast of Iceland. Around 15 Ma the Snæfellsnes–Húnaflói Rift Zone affected the Snæfellsnes Peninsula.

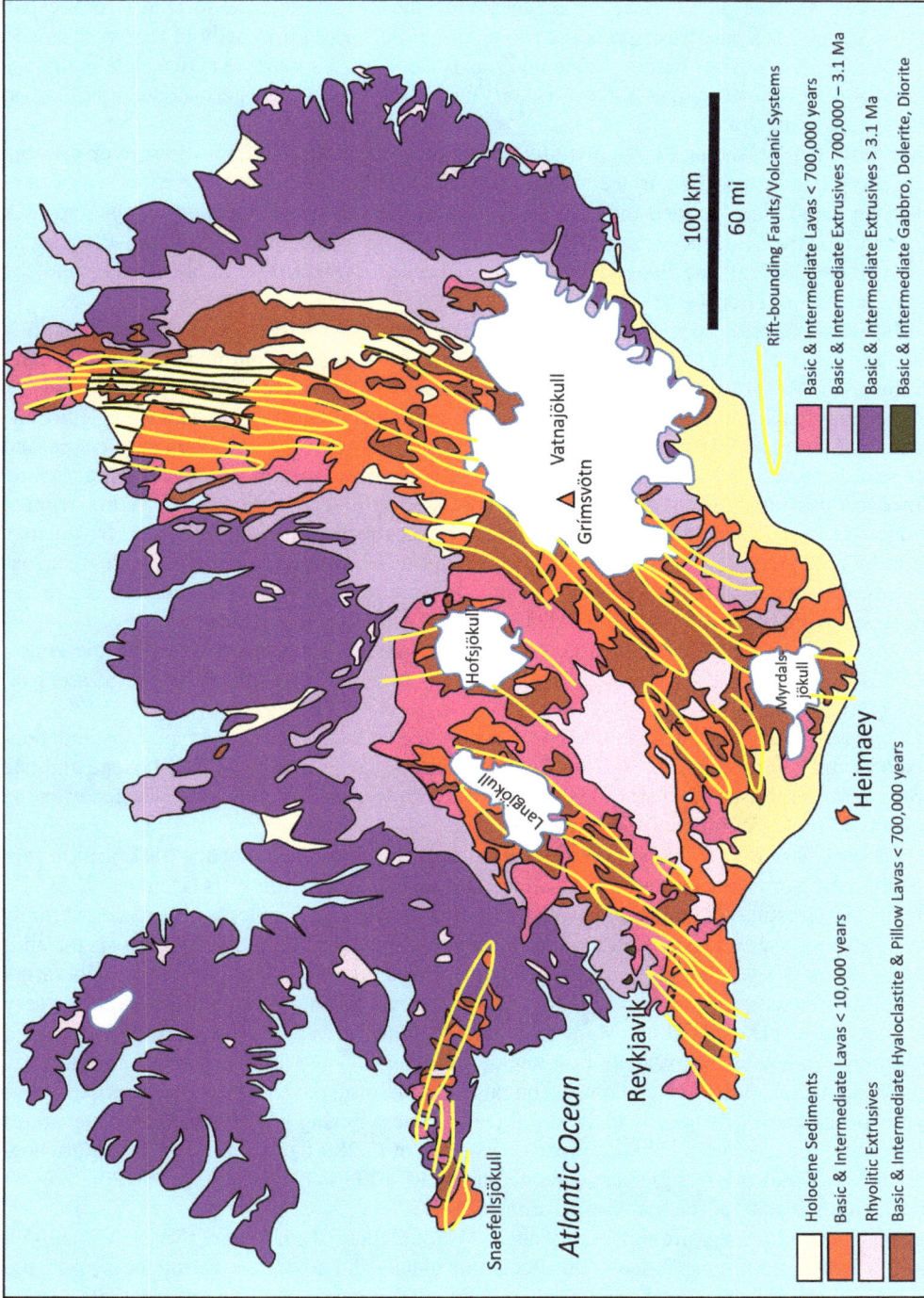

Geologic map of Iceland showing main rock units and volcanic systems. Ice caps are white. Modified after Jóhannesson, 2014.

Between 6 and 7 Ma the southern part of the Snæfellsnes–Húnaflói Rift Zone stopped spreading and the presently active Western Volcanic Zone (or Western Rift Zone) developed on the Reykjanes Peninsula. Two to three million years ago the northern part of the Snæfellsnes–Húnaflói Rift Zone became inactive and the presently active Eastern Volcanic Zone/Rift Zone formed (Denk et al., 2011). Since about 3 Ma east-west spreading in Iceland has occurred primarily in the Western and Eastern Volcanic Zones. The Western Volcanic Zone is the slowest spreading active plate boundary known, and some evidence suggests that it is starting to fail while the Eastern Zone is picking up the slack (Sinton et al., 2005).

The formation of Iceland as a large pile of volcanic rock began around 24 million years ago, but the oldest rocks exposed at the surface are only about 16 million years old. The rocks exposed at the surface in Iceland are divided into four general stratigraphic groups defined by age and rock type. The oldest is the Miocene–Pliocene Succession (16–3.1 Ma); next is the Pliocene–Pleistocene Succession (3.1–0.78 Ma); the upper Pleistocene Succession (780,000 to 11,500 years); and the Holocene Succession (11,500 years to Present).

The Miocene–Pliocene Succession of northwestern, western, northern, and eastern Iceland is mostly tholeiitic basalt (ocean floor–type) lava flows. Basaltic dikes are common and run vertically through the lavas. Most of the dikes are oriented northeast-southwest, the same as the rift zones.

The Pliocene–Pleistocene Succession occupies the area between the older Miocene–Pliocene Succession and the upper Pleistocene Succession. Volcanic rocks from the upper Pliocene and Pleistocene in Iceland differ from the older Miocene–Pliocene rocks mainly on the basis of texture. Lava flows are common during interglacial periods. During glacial periods the magma erupted beneath ice sheets and formed pillow lavas; explosive volcanism formed tephra (rock fragments). Basaltic glass cemented the tephra in the subglacial eruption creating hyaloclastite, a volcanic glass breccia (breccia is made of broken rock fragments). If the magma breached the surface of the ice, the hyaloclastites are capped by lava flows that form steep-sided table mountains.

The upper Pleistocene Succession is essentially the same as the present-day volcanic zones. Rocks of this period are mostly interglacial basaltic lava flows and subglacial pillow lavas and hyaloclastites (Denk et al., 2011).

There are rare examples of sedimentary rocks in Iceland. A sequence of marine and non-marine sediments is present on the Tjörnes Peninsula in northern Iceland. These Pliocene and late Pleistocene deposits consist of siltstone and sandstone, with fossils of marine mollusks and even the low-grade coal lignite (Wikipedia, Geology of Iceland).

The Holocene Succession is composed of recent lava flows and pyroclastics (tuff, ash), minor marine sediments, some glacial sediments (till, moraines, loess), and thin soils (Denk et al., 2011).

There is an interesting relationship between volcanism and glaciation. Iceland has been variably covered by large ice sheets and smaller ice caps for the past three million years. Viscosity is the ability to flow … think of low viscosity warm maple syrup compared to high viscosity room temperature honey. The low viscosity of the mantle plume makes magma generation by pressure-release very sensitive to rapid glacial unloading. Many have noted a link between deglaciation and increased volcanic activity in Iceland, in particular on Reykjanes Peninsula and around Snæfellsjökull.

The maximum elevation of volcanic vents on table mountains (so, vents that were exposed above the glaciers or ice caps) provides an estimate of ice thickness during glaciations. These elevations give a minimum ice thickness value of more than 1,000 m (3,280 ft) in central Iceland. Offshore terminal moraines indicate that glaciers extended up to 50–100 km (30–60 mi) beyond the present coastline during the peak of the Ice Ages (Trønnes, 2003).

Today glaciers and ice caps presently cover about 11% of Iceland; the largest of these is Vatnajökull (a "jökull" is a glacier or ice cap in Icelandic). Icelandic glaciers have been retreating for the past 100 years; Vatnajökull has lost as much as 10% of its volume (Wikipedia, Geology of Iceland).

As many glaciers blanket active volcanoes, subglacial eruptions pose the hazard of sudden flooding caused by glacial meltwater. These floods are known as jökulhlaups. *Do not* drive across an

Lithology Unit Biozone Member Fm Gp

Höskuldsvík Lava Group

450 m — 400 m — 300 m — 200 m — 100 m — 0 m

Units: 25, J, 24, I, H, G, 23, 22, 21, 20, 19, 18, 17, 16, 15, 14, 13, 12, 11, 10, 9, F, E, 8, 7, D, 6, C, 5, 4, B, 3, 2

Biozones: Serripes Zone, Mactra Zone, Tapes Zone

Members: Stapi, Nór, Tófugjá, Kambsgja, Halldjarnarstaðakambur, Tungukambur, Ytri Svarthamar, Brunngil, Skeifárbás, Hvift, Flekatá, Rekárloss, Stonsur, Stonsubakki, Hásbakki, Egilsgöta, Hústorta, Tungugerðiskambur

Fm: Stórhöfði, Tungai, Skeifá, Hringver, Reká, Tungugerði, Grænhöfði

Gp: Barmur Group

Kaldakvísl Group

Legend:
- Conglomerate
- Sandstone
- Mudstone
- Lignite
- Lava flow
- Tuff

Map: Iceland, # Reykjavik, Atlantic Ocean

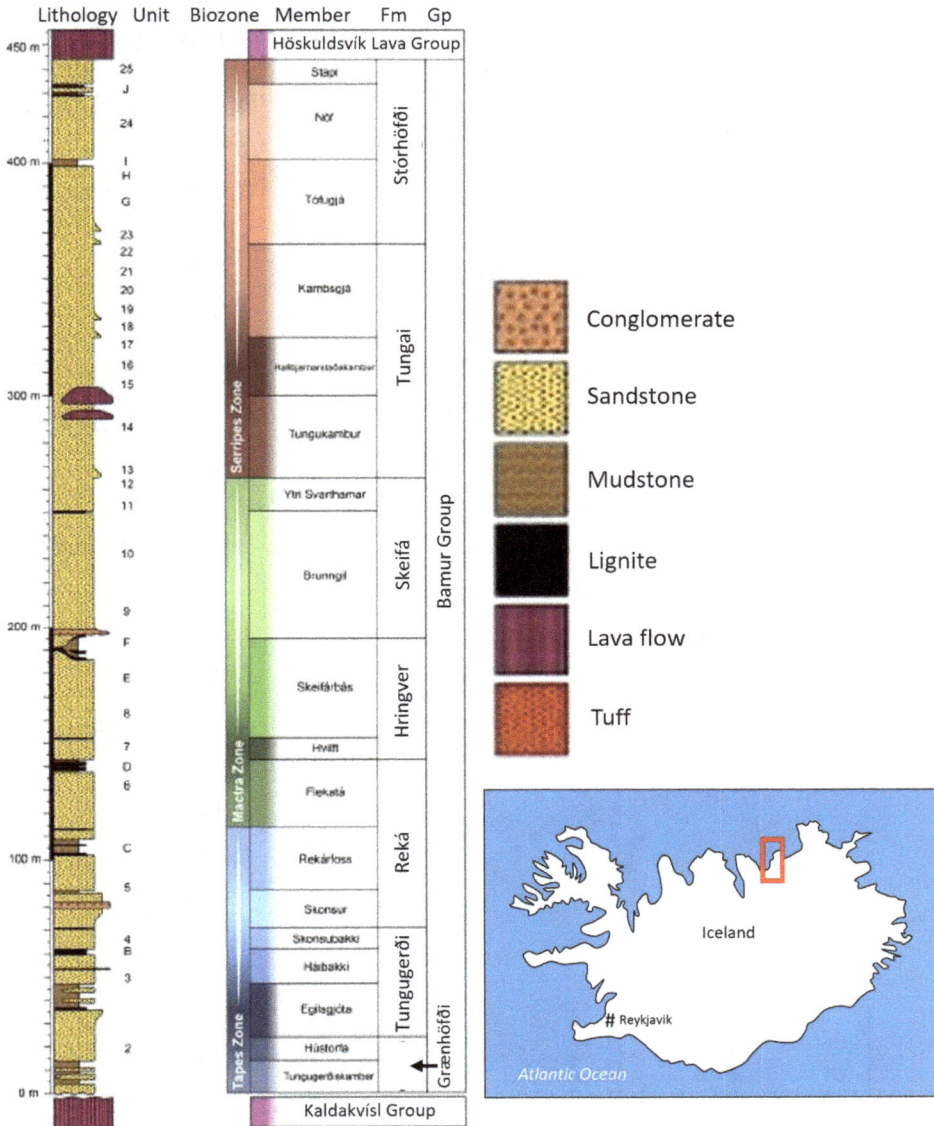

Iceland's stratigraphy is mostly volcanic. This 455 m (1,493 ft) section, measured in northeast Iceland, consists of stacked volcanic sandstones with interbedded lava flows, conglomerate, mudstone, lignite, and tuff. The Barmur Group (informally Tjörnes beds) is a sedimentary succession of latest Miocene to early Pliocene age (Allison et al, 2025).

outwash plain if an eruption is imminent. You can see the remains of old roadbeds that have been abandoned and steel bridges that have been twisted like pretzels by glacial meltwater floods.

BEGIN – KEFLAVÍK AIRPORT

Keflavík Airport to Blue Lagoon: *Follow Komur to Reykjanesbraut/Route 41; drive east on R-41 to R-43; turn right (south) on R-43 and drive to 426/Bláalónsvegur; turn right on 426 and drive to* ***Stop 1.1, Blue Lagoon*** *(63.881277, -22.453425), for a total of 23.2 km (14.4 mi; 25 min).*

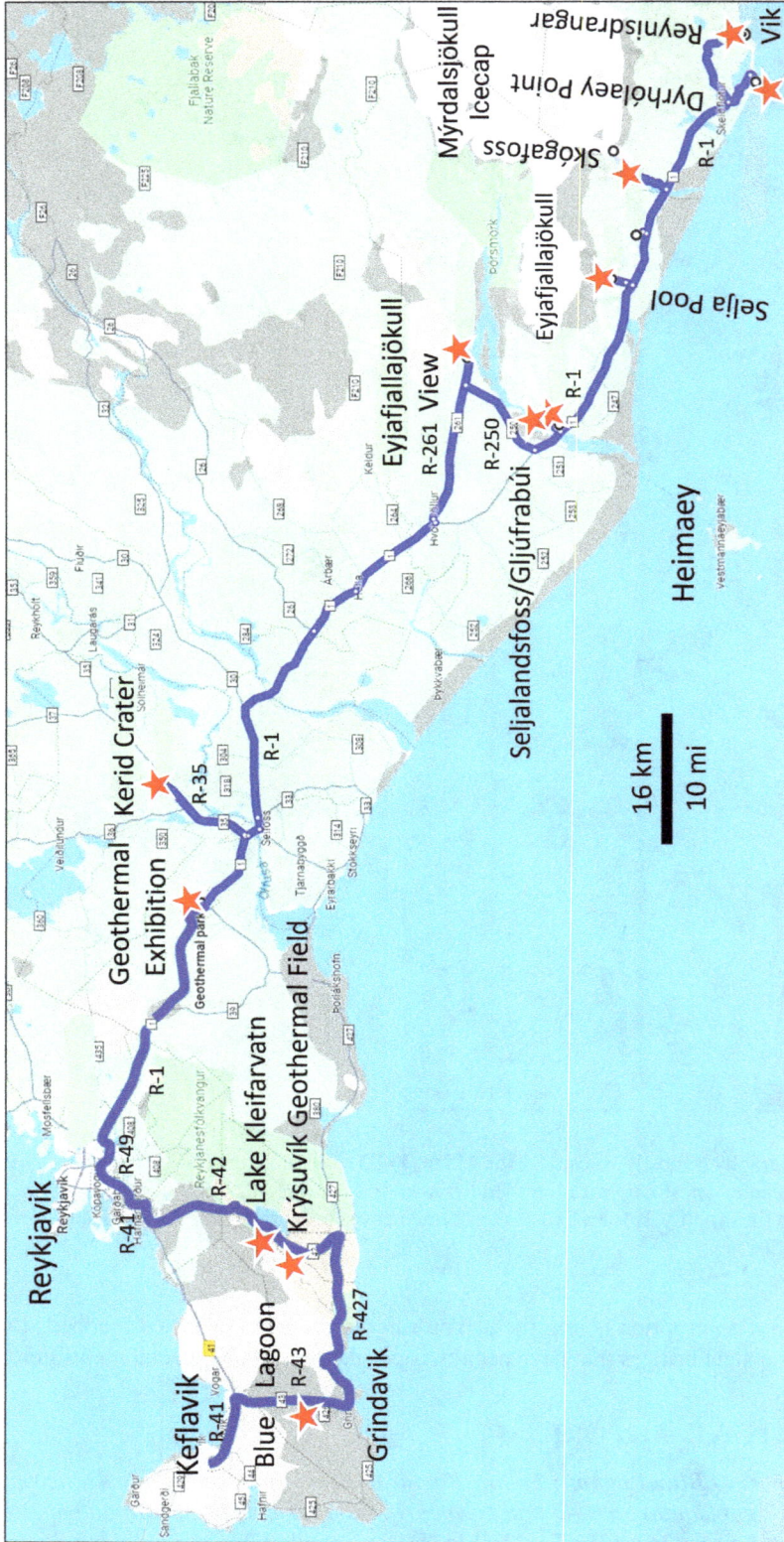

STOP 1 REYKJANES PENINSULA

The first thing I noticed when leaving the airport is the total lack of trees. You will see some, eventually, tucked into folds in the hills or around farms, but the land is for the most part tundra. The reindeer moss covering everything gives a soft, rounded look to the otherwise lava-covered landscape. Occasionally there are black basalt outcrops speckled with white lichen. Most of the architecture is functional, industrial, and concrete and steel, as if preparing for the next ashfall. Which, I suppose, is the point.

Google Maps image showing rifted, northeast-elongated faults and fissures where the Mid-Atlantic Ridge comes onshore between Grindavik and Lake Kleifarvatn.

This is where the Mid-Atlantic Ridge, specifically the Reykjanes Ridge, comes onshore. The area from Keflavík to Grindavík to Hellisheiðarvirkjun is characterized by northeast-elongated rifted and down-dropped valleys (grabens) and ridges formed by fissure eruptions along normal faults. Normal faults are essentially vertical breaks where one side moved up and the other side moved down. These are clearly seen on satellite images.

High-temperature geothermal and volcanic systems are located along these rifts. These are known collectively as the Reykjanes-Langjökull volcanic belt, and it has been active for the past 5 or 6 million years. There are five volcanic centers: from west to east they are the Reykjanes, Svartsengi, Krýsuvik, Brennisteinsfjöll, and Hengill. Each system has fissure swarms aligned with the Mid-Atlantic Ridge spreading center. All of these erupt basalt, except the Hengill. Hengill volcano is distinguished by palagonite tuff, an ash made up largely of altered basaltic glass. Periods of activity on the peninsula typically occur every thousand years or so, and eruptions can last for several hundred years. Volcanism is mainly fissure eruptions with accompanying cinder and spatter cones.

The Krýsuvík system is about 8 km (4.8 mi) wide and 50 km (30 mi) long. It is characterized by low tuff ridges, vents, lava flats, and geothermal activity. The last eruption on the Krýsuvík system was during the 1100s and is known as the Krýsuvík Fires. Extensive lava flows reached the sea on both the north and the south coasts of the peninsula.

STOP 1.1 BLUE LAGOON AND SUNDHNÚKUR ERUPTIONS

The 2023–25 Sundhnúkur eruptions are a series of volcanic eruptions in the Reykjanes Peninsula, just north of the coastal town of Grindavík. There have been more than a half dozen eruptions between December 2023 and May 2024, each following an earthquake swarm. The seismic and volcanic activity caused repeated evacuations of the Blue Lagoon and the town of Grindavík.

An intense earthquake swarm in the Eldvörp–Svartsengi volcanic system began on 24 October 2023, caused by a magmatic intrusion beneath the area. An evacuation was ordered in Grindavík, and large-scale subsidence in and around the town caused significant damage. The first eruption started on 18 December 2023, at the Sundhnúksgígar crater row north of Grindavík. Lava flowed laterally from both sides of a 4 km-long (2.5 mi- long) fissure. Indications are that magma accumulated in a chamber 4 to 5 km (2.5 to 3 mi) under Svartsengi. The eruption was relatively short, with the eruption ending on 21 December.

A second fissure erupted on 14 January 2024. A few hours later, a smaller fissure opened less than 100 meters (330 ft) from the town inside the protective berms, with its lava reaching and burning three homes. One individual was reported missing and presumed to have fallen into a crevice and died. A third eruption began on 8 February 2024 in the same area as the December eruption, with a fissure about 3 km (1.9 mi) long erupting lava on top of earlier flows. The lava flowed over a road and caused significant damage. A fourth eruption occurred on 16 March 2024, originating from a fissure approximately 3–4 km (1.9–2.5 mi) in length. This event was considered the most powerful eruption on the peninsula since 2021. This eruption lasted 54 days. A fifth eruption began on 29 May 2024. The sixth eruption began on 22 August 2024 and lasted 14 days. It covered an area of 15.8 km^2 (6.1 mi^2). Despite being the largest eruption in the series, it did not damage any infrastructure. The seventh eruption began on 20 November 2024 and lasted 18 days, engulfing the Blue Lagoon parking lot and threatening the area. An eighth eruption, on 1 April 2025, lasted only seven hours, making it the shortest and least intense eruption as of this writing (mid-2025).

About 1 km (0.6 mi) before the turnoff to Blue Lagoon, R-43/Grindavikurvegur has a stretch of graded gravel that crosses the third of the Sundhnúkur lava flows. Signs tell you not to walk on the still steaming basalt flows. The road is closed to traffic south of the turnoff to the Blue Lagoon as of June 2024.

New gravel road over steaming, two-week-old lava. Road to the Blue Lagoon, view north.

About 0.6 km (2,100 ft) after the turnoff to Blue Lagoon you begin to see blue water in ponds on the south side of the road. The water's milky blue shade is due to its high silica content. The silica forms a soft white mud on the bottom of the ponds that bathers rub on themselves as a healing mask. The water also contains salts and sulfur. Very few organisms live in the water apart from some blue-green algae.

The Blue Lagoon spa.

The Blue Lagoon spa – the largest outdoor spa (5,000 m², or 1.24 ac) and one of the most visited attractions in Iceland – is an artificial lagoon fed by water derived from the nearby Svartsengi geothermal power plant. The geothermal waters originate roughly 2,000 m (6,500 ft) below the surface, where freshwater and seawater combine at extreme temperatures. Superheated water vented from the ground near a lava flow is used to run turbines that generate electricity. After going through the turbines, the steam and hot water passes through a heat exchanger to provide heat for a municipal water heating system. Then the water is fed into the lagoon at a controlled temperature of 37–39°C (98–102°F). The entire pool is replenished every two days. Because of its mineral concentration, water cannot be recycled and must be disposed in the nearby landscape, a permeable lava field. Minerals in the water are soon deposited and render the ground impermeable, so the plant needs to continuously dig new ponds in the nearby lava field (Nakajima, 2014; Wikipedia, Blue Lagoon).

Visit

Address: Nordurljosavegur 9, Grindavík 240 Iceland
Phone: +354-420-8800
Website: https://www.bluelagoon.com/day-visit/the-blue-lagoon

Hours vary by season and day; it is most crowded between 10:00 and 14:00. 2024 hours are:
22 June – 20 August, 7:00–23:00
21 August – 21 June, 8:00–22:00

Entrance fee (2024): Comfort package from US\$71; Premium from US\$92; Signiture from US\$117 (2024 prices). Price varies depending on time of day; children under 13 are free.

Guests are required to shower prior to using the pool. The communal showers are separated by gender. Children age 8 and under are only allowed entry with the use of arm floaters, provided free of charge. The lagoon is not suitable for children under the age of 2 years.

The lagoon is accessible for wheelchair users with a ramp that extends into the water and a shower chair. There is also a private changing room available for those with special needs. There are multiple restaurants and a spa in the area.

Blue Lagoon to Krýsuvík-Seltún Geothermal Field: *Take R-426/Bláa lónið south to Grindavík and turn left (north) on R-425/Nesvegur; drive to R-43/Grindavíkurvegur and turn right (south); drive to Hópbraut and turn left (east); drive to R-427 and turn left (east); drive to R-42/Suðurstrandarvegur; turn left (north) on R-42 and drive to* **Stop 1.2, Krýsuvík-Seltún Geothermal Field** *parking area (63.895661, -22.051944) on the left for a total of 35.2 km (21.8 mi; 30 min). The parking area holds 50–60 vehicles. Restrooms are available.*

STOP 1.2 KRÝSUVÍK-SELTÚN GEOTHERMAL FIELD

The area we are driving through, around the town of Grindavík, is the landward continuation of the Reykjanes Ridge. The axis of the rift is offset a bit to the east on the Reykjanes Peninsula. Like most of Iceland, the peninsula is basalt: the topography consists of an older Pleistocene basalt with interglacial flows and subglacial table mountains and ridges and a younger unit consisting of post-glacial (less than about 9,000 years) lavas from fissures, crater rows, and shield volcanoes (a shield volcano is a broad, gently sloping volcano, resembling a shield, formed by the eruption of highly fluid basaltic lava). The faults and fissures pretty much all trend northeast and have post-glacial, that is recent offsets (Eiriksson and Bjornsson, 1974).

Because we are on the Atlantic spreading center, the heat flow is extremely high. Seltún is one of several geothermal fields in the area where groundwater is superheated by near-surface magma and flows to the surface as hot springs and geysers. The rocks, originally hyaloclastites and basalts, range from bright red (oxidized iron minerals) to bleached (clay minerals) due to alteration of their mineral components by hot magmatic fluids.

Krýsuvík lies in a northeast-elongated valley on the volcanic axis of the Reykjanes Peninsula. The most significant structural element in the Krýsuvík area is a northeast-orientated fissure swarm, which controls volcanism and forms the regional structural framework for the area. The Krýsuvík Fissure Swarm is a large, *en echelon* structural feature. *En echelon* describes a series of parallel faults or fissures that are offset in a staggered pattern. The swarm extends from the north coast near Reykjavík to the south coast. Lakes with straight shorelines suggest that their origin is structurally controlled. The Krýsuvík Valley and its surroundings are characterized by northeast-trending eruptive fissures and faults that created the uplifted and down-dropped topography of the region (Morales, 1992).

The Krýsuvík Geothermal Field covers about 40 km² (15.4 mi²). Fumaroles and steaming grounds; mud pits; hot, warm, and cold springs; and hot and warm grounds are located along northeast-oriented faults that serve as conduits for groundwater. The basaltic rock in the area is lumped into three groups: at the base, and most common, is upper Pleistocene hyaloclastite that erupted beneath an ice cap. This unit comprises pillow lavas, pillow breccias, and hyaloclastites. The middle group contains patches of lava, of late interglacial or early Holocene age, that cap the hyaloclastite.

The youngest group contains scoriacious tephra ejected from early Holocene explosive craters. Several explosion craters (maars) are located in the Krýsuvík field. Maars form when lava encounters groundwater, causing a steam explosion.

The Krýsuvík Fissure Swarm crosses the active plate boundary, a zone approximately 2–5 km (1.2–3.0 mi) wide trending about N76E. Most of the seismicity on the peninsula is confined to this zone and indicates predominantly right-lateral strike-slip faulting, that is, faults where the opposite side moves to the right.

Seltún Geothermal Field, Krýsuvík.

Two kinds of hydrothermal alteration are found in the field: slight alteration and extensive clay alteration. Slight alteration is indicated by red-brown and yellow-brown bedrock with opaline silica and zeolite minerals. Clay alteration is distinguished by completely altered and bleached bedrock (Khubaeva, 2007). Light-colored rock in the thermal areas is a function of kaolinite clay replacing original minerals, mostly feldspar, in the rock, whereas red clay is a result of the oxidation (rusting) of the iron mineral pyrite to limonite.

The first drilling in this area was by the naturalist Eggert Olafsson and Director of Health Bjarni Palsson in 1755. They were looking for sulfur, much in demand during times of war in Europe. At a depth of 10 m (33 ft) the boreholes erupted and drilling was discontinued. Sulfur was mined from Krýsuvík in the 1750s and 1760s, with as much as 72.5 tonnes (80 tons) being exported during the 18th century.

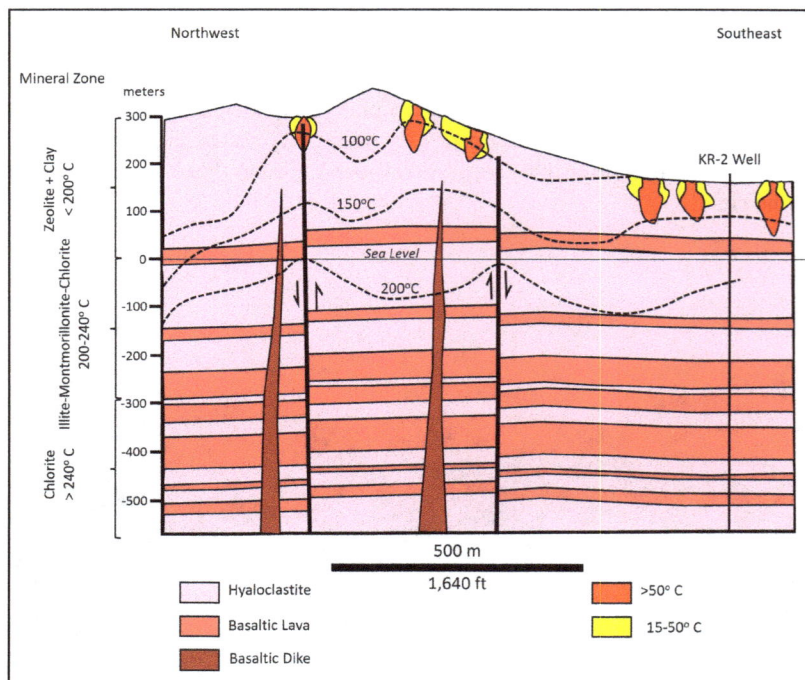

Cross section of the Seltún - Krýsuvík Geolthermal Area. Stratigraphy is based on the KR-2 well. Modified after Mawejje, 2007.

The first drilling for geothermal energy took place in 1941. The hot water was used to heat homes and generate electricity. In 1947 a borehole reached 230 m (750 ft) and erupted high-pressure steam. Drillling was stopped just before an explosion blew out a crater 43 m (140 ft) in diameter. Gravel and clay were thrown 700 m (2,300 ft) north of the vent.

The greatest heat appears to be at depths of around 300 m (1,000 ft). There are plans to exploit the geothermal heat to generate electricity.

*Krýsuvík to Lake Kleifarvatn Overlook: Return to R-42 and turn left (north); drive 3.0 km (1.8 mi; 2 min) to **Stop 1.3, Lake Kleifarvatn** turnout (63.925941, -21.997650) on the right. There is room for about six cars. No facilities are available.*

STOP 1.3 LAKE KLEIFARVATN RIFT VALLEY

As you drive to this stop you can pull over on the right (east) side of the road above the black sand beach. Be careful: if you drive onto the beach itself you will likely get stuck in the sand. Walk onto the beach to observe the falling lake levels as revealed by the old beach terraces/shorelines.

This stop has great views of the lake and you can get up close with the hyaloclastite. There is room for about six cars in the pullout.

Lake Kleifarvatn sits in a fault-bounded graben. Interestingly, this lake has only the smallest of stream inflow, and no outflow, indicating that it is (1) filled almost entirely by rainfall, and (2) in an actively subsiding basin. The lake is 5 km long and 2 km wide (3.2 by 1.3 mi). It is 97 m (318 ft) deep at the deepest point, but, curiously, that is changing. The lake has been slowly draining into the subsurface since earthquakes rocked the area in the year 2000, and in the 20 years since it has lost almost 20% of its surface area (Wikipedia, Kleifarvatn).

Rift-controlled Lake Kleifarvatn, view northeast. Kistufell, the table mountain is center-right; Sveifluháls, the hyaloclastite ridge, is center-left.

Normal faults and associated fissure zones influenced the shape of the valley and lake, but another factor may have contributed to lake morphology. Some have speculated that maar volcanism is suggested by elliptical depressions on the lake bottom (Morales, 1992). Maar craters form when magma comes into contact with groundwater and erupts explosively.

Sveifluháls, west of the lake, is a 22 km (13 mi) long basaltic hyaloclastite ridge erupted beneath an ice cap along a northeast-trending fissure. Pillow lavas at the base are covered by hyaloclastite tephra.

Close-up of the hyaloclastite. Looking northeast toward Lake Kleifarvatn.

Kistufell is a table mountain east of Kleifarvatn Lake. Like Sveifluháls, upper Pleistocene basalts erupted beneath an ice cap, but here the lavas broke through and flows formed at the top above pillow lavas and hyaloclastites. The flows now hold up the top of the mountain and give it the flat top. In addition to being a lava pile over a fissure vent, it has been suggested that some table mountains may be horsts, or uplifted fault blocks (Morales, 1992).

You can see water-laid hyaloclastite at this stop. It appears to be light to medium gray crossbedded fine-grained to pebbly conglomerate.

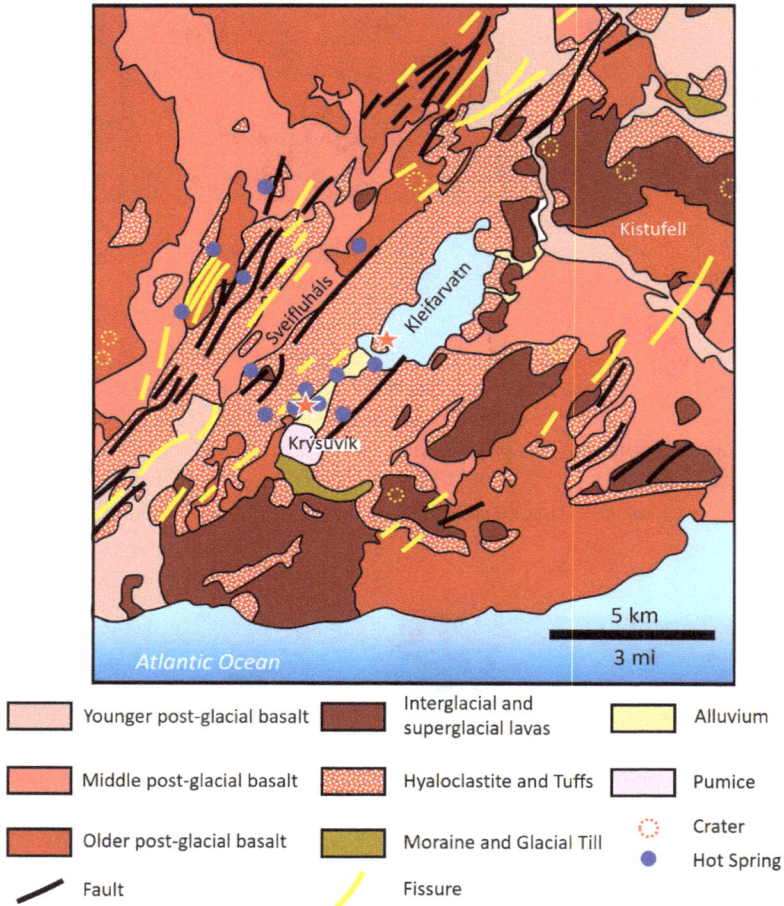

Geologic map of the Lake Kleifarvatn–Sveifluháls Ridge area. Stars are the Krýsuvík and Kleifarvatn stops. Modified after Abdelghafoor, 2007.

*Lake Kleifarvatn to Geothermal Energy Exhibition: Return north on R-42 to R-417/Bláfjallavegur; turn right (east) on R-417 and drive to Bláfjallavegur-1; turn left (north) on Bláfjallavegur-1 and drive to R-1/ Suðurlandsvegur/Þjóðvegur 1; turn right (east) on R-1 and drive to turnoff for Hellisheiðarvirkjun/Geothermal Energy Exhibition on the left. This is **Stop 2, Geothermal Energy Exhibition** (64.037667, -21.401177), for a total of 48.8 km (30.3 mi; 1 hr). There is a large parking area and facilities here.*

GEOTHERMAL ENERGY IN ICELAND

Until the early 1950s Iceland imported coal to generate electricity. By 2014, roughly 85% of primary energy use in Iceland came from local renewable resources, with geothermal accounting for 66% of Iceland's primary energy use. How did that happen? Someone was thinking ahead.

Iceland sits on an active volcanic system with extremely high heat flow. More than 200 volcanoes are located within the active volcanic zone stretching southwest to northeast across the island, and at least 30 of them have erupted in the past millennium. In the active volcanic zones, areas can have geothermal gradients over 150°C/km (98°F/1,000 ft). For comparison, the average continental heat gradient is 25°–30°C/km (72°–87°F/mi /mi). There are at least 20 high-temperature areas containing steam fields with underground temperatures reaching 250°C within 1,000 m depth (482°F within 3,000 ft). About 250 separate low-temperature areas with temperatures not exceeding 150°C (302°F) in the uppermost 1,000 m are found mostly in areas flanking the active zones. Over 600 hot springs (temperature over 20°C, or 68°F) are known.

The abundant heat is used in a number of ways. It is used directly to provide hot water to homes and melt ice on sidewalks and heat homes by circulating hot water. About 9 out of 10 households are heated with geothermal energy. Geothermal energy is used to heat 138 of the 169 swimming pools in Iceland, in addition to the popular recreational hot springs swimming areas like the Blue Lagoon. Indirectly, hot water under pressure flashes to steam with a drop in pressure. The steam is used to drive turbines that generate electricity.

In 2013 the total use of geothermal energy was 46.7 PJ (1,2972 GWh), with space heating accounting for 45%. Geothermal power plants produced 5.245 GWh, or 29% of the country's total electricity production in 2013. Because aluminum smelting depends on cheap, reliable energy, the industry began moving to Iceland in the 1990s. By 2013 the aluminum industry in Iceland used about 70% of produced electricity.

The Iceland Deep Drilling Project (IDDP) is a collaborative effort between Icelandic power companies, the Icelandic government, U.S. National Science Foundation, and Alcoa. IDDP was formed to determine the feasibility of using supercritical geothermal fluids for power production. A supercritical fluid in this case is water well above its boiling temperature, but still a fluid because of extreme pressure. Wells will be drilled to a depth of about 5 km (16,400 ft) to tap water at temperatures between 450°C to ~600°C (840° to 1,100° F). A 2003 study suggested supercritical fluid could produce ten times more power than conventional geothermal from 2.5 km (8,200 ft) deep geothermal wells. A conventional geothermal well yields about 5 MWe (megawatts of electricity). An IDDP well tapping a supercritical reservoir at temperatures above 450°C and at pressures of 23–26 MPa (3,336–3,771 psi) might yield ~50 MWe (National Energy Authority of Iceland, Iceland Deep Drilling Project; Worland, 2017; Wikipedia, Geothermal Power in Iceland).

Geothermal Energy Exhibition and plant.

STOP 2 GEOTHERMAL ENERGY EXHIBIT AND HELLISHEIÐI POWER PLANT

Located on Hengill volcano in the South of Iceland, the Hellisheidarvirkjun (or Hellisheidi) heat and power plant is the largest power station in Iceland. Hengill volcano, located in the middle of the western volcanic rift zone, consists of palagonite tuff (palagonite is altered basaltic volcanic glass). The area is one of the most extensive geothermal areas in the world, with many hot springs at the surface and a large magma chamber close to the surface. The last eruption occurred around 2,000 years ago, and the volcano is considered still active.

The uppermost 500 m (1,640 ft) of strata at the plant are tuffs; below that basalts are dominant. Wells are drilled into high-temperature aquifers (water-bearing layers) connected to faults and fissures that can often be seen at the surface.

The plant has provided electricity to Reykjavík since 2006. At that time two high-pressure turbines began producing 45 Mw of electricity. A year later an additional low-pressure turbine added 33 Mw. By 2008 the plant had added two more 45 Mw turbines, and this capacity was combined with steam generation at Stora-Skardsmýrarfjall to bring total electricity production to 213 Mw.

In 2020 the Hellisheidi power plant produced about 303 Mw of electricity and up to 400 Mw of thermal (hot water) energy, ranking it as the largest geothermal power station in the world by installed capacity. The plant uses 500 kg/s (132 gallons/sec) of 180°C (356° F) steam extracted from 30 wells between 2,000 and 3,000 m (6,560–10,000 ft) deep to generate electricity (Extreme Iceland, 2016).

The geothermal exhibit at Hellisheidi/Hellisheiði Geothermal Power Plant explains how geothermal power is generated and used in Iceland. Displays show the geology, technology, and history of geothermal energy in Iceland. The power plant provides an example of how geothermal energy can provide sustainable power well into the future (National Energy Authority Iceland, Geothermal Energy Exhibition).

In front of this facility, across a field, are Orca and Mammoth, two working prototype testbeds for direct air carbon capture of CO_2. As of 2024, the Mammoth plant is the largest direct air carbon capture and storage plant in the world: it can capture up to 36,000 mt (metric tons) of CO_2 from the air annually, equivalent to the emissions from 7,517 gasoline-powered passenger vehicles driven for one year. Climeworks, the plant operator, says that twelve of 72 collector containers have been installed onsite. Captured CO_2 is pumped into the Earth and permanently stored in deep rock layers. Mammoth is Climeworks's second commercial direct air capture facility and is about ten times bigger than their first plant, Orca.

The plant uses geothermal-generated electricity to power the direct air capture process. Filters absorb the CO_2, which is then released by heating and pumped into basalt, where it combines with calcium in the rock and is converted naturally over time to calcium carbonate (limestone). Project backers see this as the wave of the future. Critics say it works only because electricity in Iceland is cheap.

Visit

Individual self-guided audio tours, guided tours, and group tours are available. Most are around 1 hour.
Hours: The exhibition is open every day at 9:00–17:00. Please direct inquiries to syning@on.is
Entrance fees (2024) are 2115 kr (about USD $15.00) for adults and ~990 kr ($7.00) for seniors and teens. Children under 12 are free but must be accompanied by an adult.

Address: Orka náttúrunnar, Bæjarháls 1, 110 Reykjavík, Kt. 471119-0830
Phone: 591-2880.
Website: https://www.on.is/en/geothermal-exhibition/

The Service Center is open 8:30–16:30 weekdays. The phone number is 591-2700.
A coffee shop and souvenir shop are on site.

Geothermal Energy Exhibition to Kerid (Kerið) Crater: *Return to R-1 and turn left (east); continue driving east on R-1 to R-35/ Biskupstungnabraut; turn left (north) on R-35 and drive to Kerið Car Park on the right. This is **Stop 3, Kerið Crater** (64.041836, -20.886233), for a total of 39.4 km (24.5 mi; 31 min).*

Stop 3 Kerid/Kerið Crater

This scenic crater, located in the Western Volcanic Zone of Iceland, is 55 m (180 ft) deep and 170 by 270 m (560 by 890 ft) across. Kerid lies at the north end of a row of craters known as Tjarnarholar. Originally thought to be an explosion crater, there are no ash deposits associated with it. It is now thought that this volcanic caldera formed about 6,500 years ago after the magma chamber emptied during an eruption and collapsed in on itself. The aquamarine lake in the bottom is in stark contrast to the bright red volcanic material it sits in. The color of the water is a result of minerals leached out of the surrounding rock. The water itself, 7–14 m (23–46 ft) deep, is at the same level as the local water table. Swimming in the lake is not recommended: this is not geothermally heated water, but rather it is just a few degrees above freezing.

Kerid Crater.

The crater is on private property, and there is a modest 500 Ikr (~US $3.50 in 2024) entrance fee.

KATLA GEOPARK

Katla was designated a UNESCO Geopark in 2011. The park is located on the south-central coast of Iceland. At 9,542 km² (3,684 mi²), the park covers about 9% of Iceland. It is part of both the European Geoparks Network and the Global Network of National Geoparks. The purpose of a Geopark is to preserve geologic features of global significance. Among these features are the Katla, Eyjafjallajökull, Grímsvötn, Lakagígar, and Eldgjá volcanoes and fissure zone, and the Eyjafjallajökull, Mýrdalsjökull, and western Vatnajökull ice caps (Wikipedia, Katla Geopark).

Geologically, Katla Global Geopark is set in the Eastern Volcanic Zone (EVZ), an area that has had over 150 eruptions since the 9th century (VisitSouthIceland). The park is characterized by volcanoes, northeast-trending eruption craters, fissures, hyaloclastite ridges, ice-capped volcanoes, flow-capped table mountains, vast lava flows, and black sand beaches. In addition you will find glacial outwash, moraine- and ice-dammed lakes, glacial rivers, and outburst flood (jökulhlaup) deposits.

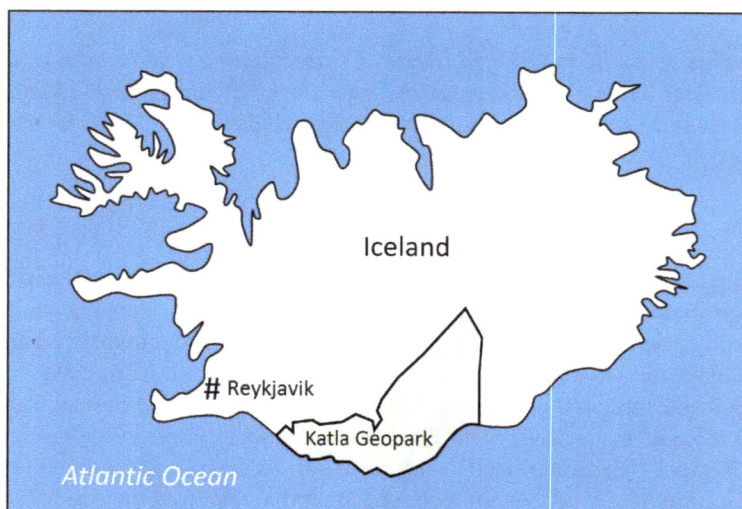

Map showing the extent of Katla Geopark. Modified after Katla Geopark Project Geological Report, https://www.katlageopark.com/media/39154/Geological-report.pdf

Four volcanic systems occur in the EVZ of Katla Geopark (Katla Geopark Project, Geological Report). Volcanic systems include both fissure swarms and their central volcanoes. Key features of the park include:

(1) Katla, one of the largest central volcanoes in Iceland, is largely covered by the Mýrdalsjökull ice sheet. This volcanic system includes the 78 km (47 mi) long Eldgjá fissure swarm. Katla caldera is located under the ice sheet which covers about 100 km^2 (39 mi^2) and is 700 m (2,300 ft) thick. The cone consists of both subglacial explosive volcanics and subaerial flood basalts from fissure eruptions. Rocks in the west and in the Þórsmörk area are mostly hyaloclastites younger than 55,000 years. This volcano has erupted roughly twice per century. The last eruption of Katla was in 1918 (Lonely Planet, Katla Geopark).

(2) The eruption of Eldgjá in 934–940 produced the largest basaltic flood lava in historical times. The fissure zone is about 50 km long, extending from Mýrdalsjökull ice cap in the southwest to the Vatnajökull ice cap in the northeast. It contains a graben, eruptive fissures, and a line of craters.

(3) Outwash from six jökulhlaups (glacial floods) flowed from under the western Mýrdalsjökull ice sheet, inundating Thorsmörk and areas downstream. The youngest of these is believed to have washed out a birch forest in Drumbabót. Remains of the forest contain *in situ* tree trunks standing 20–60 cm (8–24 in) up from the outwash plain. Since 1955 the Skaftá River has had floods, Skaftárhlaups, every one or two years. These jökulhlaups are associated with a geothermal area under the Vatnajökull ice sheet.

(4) The Tindfjöll volcanic system lies between Hekla and Eyjafjallajökull. Tindfjöll is probably the oldest continuously active central volcano in the EVZ. It is covered by a small glacier, Tindfjallajökull. Its caldera, 7 km (4.2 mi) in diameter, produced basalts and rhyolites. There have been no historical eruptions.

(5) The widespread Thorsmörk ignimbrite originated from an explosive eruption in the Tindfjöll system around 55,000 years ago. The ignimbrite is dominantly rhyolitic with a minor basaltic component.

(6) Grímsvötn is a 1,749 m (5,738 ft) high central volcano below the Vatnajökull ice sheet. A large (35 km^2; 13.5 mi^2) caldera and associated geothermal area sits below the Vatnajökull ice cap. The Grímsvötn volcanic system is tholeiitic (basaltic). Grímsvötn has erupted at least 40 times since 1598, most recently in 2004. The Laki Fissure, discussed previously, is located within the Grímsvötn fissure swarm. This area has been preserved since 1975 as a natural monument, and it is now within both Vatnajökull National Park and the Katla Geopark.

(7) The easternmost part of the Katla Geopark contains the oldest rocks in the area, hyaloclastite and lava flows dated at 2.5 Ma. The outcrops are in the lower part of Lómagnúpur sea cliff, the highest sea cliff in Iceland at 671 m (2,200 ft).

(8) Grænalón is the largest ice-dammed lake in Iceland. It is dammed by Skeiðarárjökull Glacier flowing south from the Vatnajökull ice sheet.

(9) Hyaloclastite ridges are unique to Iceland. They form during subglacial fissure eruptions. Several ridges extending 30–44 km (18–26 mi) can be seen between the Mýrdalsjökull and Vatnajökull ice caps. The dominant rock types in the ridges are tuff and breccia, although some include pillow lavas. All trend northeast, the same as fissures in the area. There is nothing like them anywhere else in the world.

(10) Black sand beaches and columnar basalt cliffs at Reynesfjara.

Visit

Over 40 information panels at 24 sites explain the landscape (UNESCO, 2017). There is no park office, but the geopark website provides information about the geology, geohazards, and history, as well as educational materials and maps.

Phone: +354-487-5757
Website: https://www.katlageopark.com/about-katla/

Kerid Crater to Eyjafjallajökull View: *Return south on R-35 to R-1; turn left (east) on R-1 and drive to R-261/Hlíðarvegur; at the roundabout take the 2nd exit onto R-261 and drive to (63.714308, -19.822158) on Fljótshlíðarvegur Road. This is **Stop 4, Eyjafjallajökull View**, for a total of 84.6 km (52.5 mi; 1 hr 8 min).*

STOP 4 EYJAFJALLAJÖKULL VIEW

Approximately 14 km (8.5 mi) southeast of this stop is Eyjafjallajökull sitting beneath its ice cap. Eyjafjallajökull is infamous for its eruption over six days in 2010.

This stratovolcano, whose name means "island mountain's glacier," rises to 1,651 m (5,417 ft) and is covered by one of the smaller ice caps on Iceland. A stratovolcano is a steep, classic cone-shaped volcano consisting of interbedded tuffs and flows of basaltic and andesitic lava. Picture Mount Fuji. A 2.5 km (1.5 mi) wide caldera lies beneath the ice cap, which covers the mountain above 1,000 m (3280 ft). Most of the eruptions have been explosive, but flows also emanate from fissure vents, particularly on the west side of the mountain. The system has been active for at least the past 800,000 years. Although historical eruptions are relatively rare, eruptions were recorded in 920, 1612, 1821–23, and 2010.

View southeast to Eyjafjallajökull and ice cap.

Unusually active seismicity and rapid bulging of the surface began in February 2010. In March 2010 almost 3,000 small earthquakes were recorded with epicenters at 7–10 km (4–6 mi) deep, indicating magma movement under the mountain. Fissure eruptions occurred away from the ice cap on 20 March; explosive eruptions from the central vent beneath the 200 m (656 ft) thick ice cap began on 14 April. This caused catastrophic melting of ice and jökulhlaups in rivers draining the volcano (Wikipedia, Eyjafjallajökull). Some meltwater flowed back into the erupting volcano, instantly vaporized, and supercharged the eruption, spewing clouds of glass-rich ash directly into the jet stream (Klescheva, 2014). This second eruption led to a shutdown of air traffic over Europe and the North Atlantic. More than 107,000 flights were canceled, costing the airline industry over $1.7 billion and stranding millions of passengers (Prost and Prost, 2018).

From this spot there is also a great view up the valley to the Mýrdalsjökull ice cap to the east.

Mýrdalsjökull ice cap, view east.

Eyjafjallajökull View to Seljalandsfoss: *Return west on R-261 to R-250/Dímonarvegur and turn left (south); drive south on R-250 to R-1 and turn left (southeast); drive to turnoff on unnamed road to Seljalandsfoss/Hamragardar and turn left (north); follow the road to **Stop 5.1, Seljalandsfoss** (63.615933, -19.992469) parking area on the right (east) for a total of 20.8 km (12.9 mi; 25 min).*

STOP 5 SELJALANDSFOSS AND GLJÚFRABÚI

If you haven't seen any yet, get ready for some spectacular waterfalls ("foss" in Icelandic). Iceland is well known for its spectacular falls, and these are just the first of many we will encounter. Why so many waterfalls? Heavy rainfall along with seasonal snowmelt and gradual melting of the ice caps provides the water. Lava flows, resistant to erosion, provide the cliffs. Together they make some awesome falls. Watch for rainbows in the mist if the sun is shining.

STOP 5.1 SELJALANDSFOSS

Seljalandsfoss is known not only for its beauty, but also for being able to walk behind the falls. The Seljalands River, which flows from the Eyjafjallajökull ice cap and volcano, drops 60 m (200 ft) at Seljalandsfoss. The cliff consists of highly resistant Hamragarðahraun lava over softer tillite (poorly sorted glacial outwash sediments) and, at the base, hyaloclastite (glassy volcanic breccia), which has been eroded out behind the falls.

Raincoats are recommended as the mist from the falls will soak everything near it. Parking costs 800 Ikr (2024; ~US $6) and is good for the day.

Seljalandsfoss.

Seljalandsfoss to Gljufrabui: *Take the walking path 600 m (2,000 ft) north to* ***Stop 5.2, Gljúfrabúi*** *(63.620976, -19.986994).*

Stop 5.2 Gljúfrabúi

Be prepared to get wet: to reach the falls you have to wade through the river, and mist from the falls covers everything. This is the same cliff as at Seljalandsfoss, but here it is known as Franskanef, "the French Nose." You can walk into the slot canyon cut by the falls and look up from inside the cliff.

Gljúfrabúi to Seljavallalaug Warm Pool: *Return south to Highway 1/Ring Road and turn left; drive 20.5 km (12.7 mi) to R242 on the left; turn left onto R242 and drive 2.8 km (1.7 mi) to the parking area at the end of the road. There is a 1.2 km (4,000 ft; 20 min) walk (one way) on an essentially level path to the pool. This is* ***Stop 6, Seljavallalaug Warm Pool*** *(63.558699, -19.622512), for a total of 23.9 km (14.8 mi; 24 min).*

Stop 6 Seljavallalaug Warm Pool

This is one of the oldest hot spring–heated swimming pools in Iceland (the oldest is at Flúðir). Built in 1923, the pool has a bare-bones changing room that is in some disrepair. There are no toilets or other facilities. The pool is 9 m (30 ft) long and 5 m (16 ft) wide. Hot water is piped in from nearby hot springs. The water in this outdoor pool is between 20°C and 30°C (68°–86° F), depending on the weather. The pool sits in an Alpine valley surrounded on three sides by imposing cliffs and pounding waterfalls. Look for the travertine buildup where one of the local hot springs enters the pool.

Gljúfrabúi Falls.

Seljavallalaug Warm Pool.

*Seljavallalaug Warm Pool to Skógafoss: Return south to R1/Ring Road; turn left (east) and drive to the turnoff to Skógafoss; turn left (northwest) onto Skógafoss and drive to the large pay parking area. A 280 m (920 ft) path leads to the base of the falls. This is **Stop 7, Skógafoss** (63.529469, -19.512802) for a total of 9.9 km (6.2 mi; 10 min). From the parking area it is 215 m (710 ft) to the lower falls viewing area. From the lower falls it is a ~5.6 km (3.4 mi) round-trip walk to the upper falls.*

STOP 7 SKÓGAFOSS

Skógafoss translates as "The Forest Waterfall," although there is no forest here. This 62 m (203 ft) high falls is the last and arguably the best of over 30 falls on the Skógá River. Massive amounts of water pour over the cliff edge, creating a mist that catches and prisms the sunlight. Cliff birds dart in and out of the dark recesses. The river is born in the Eyjafjallajökull and Mýrdalsjökull ice caps. The present-day cliffs were once at the coast, but sedimentation has deposited enough material to shift the coastline 5 km (3 mi) south.

Legend has it that Thrasi Thórólfsson, the first settler in this area, hid a gold-filled chest beneath the falls. Much later three men attempted to retrieve the treasure chest. They managed to grab one of the handles, but it snapped off as they were dragging the chest to the riverbank, and it sank once again beneath the water. That handle, once on the door of Skógar Church, can be seen today on the entrance door at Skógar Folk Museum just east of the falls. It is claimed that, when the sun shines on Skógafoss, you can still glimpse the gold at the base of the rainbow stretching across the spray from the falls.

A trail leads to the top of the waterfall and can be followed upstream to see many more cascades. There is a 5.6 km (3.4 mi) round-trip walk to the upper falls.

Skógafoss.

Skógafoss to Sólheimajökull Glacier: Return south to R-1 and turn left (east); drive east on R-1 to R-221/ Sólheimajökulsvegur; turn left (north) on R-221 and drive to the parking lot at the end of the road. This is **Stop 8, Sólheimajökull** *(63.529523, -19.369209), for a total of 11.7 km (7.2 mi; 14 min).*

STOP 8 SÓLHEIMAJÖKULL GLACIER

Sólheimajökull is an outlet glacier, a tongue of ice flowing down from the Mýrdalsjökull ice cap. The glacier appears to be melting rapidly and receding as a result of warmer annual temperatures (Wikipedia, Sólheimajökull). You can walk right onto the glacier and view lateral moraine material close at hand.

Sólheimajökull Glacier.

Sólheimajökull to Dyrhólaey Viewpoint: Return south to R-1 and turn left (east); drive east on R-1 to R-218/ Dyrhólavegur; turn right (south) on R-218 and drive to the parking area at the end of the road. This is **Stop 9, Dyrhólaey Viewpoint** *(63.402263, -19.130348), for a total of 24.3 km (15.1 mi; 26 min).*

STOP 9 DYRHÓLAEY

Dyrhólaey, meaning Door Hill Island, is a spectacular natural arch developed in sea cliffs. The roughly 50 m (165 ft) high arch fits easily into the 100 m (330 ft) high cliff. For comparison, the Arc de Triomphe in Paris is 50 m high and would fit inside this gap. The small islands, or sea stacks, are erosional remnants of the former cliff. The cliff, product of a submarine eruption around 100,000 years ago, consists of basalt flows on the east and tuffs to the west (Katla Geopark, Dyrhólaey).

There are fine views of black sand beaches stretching both east and west from this viewpoint.

This area is a nature reserve and is closed during bird nesting season, which extends from the 8th of May till the 25th of June between the hours of 9:00 and 17:00. Birds that have been spotted here include fulmars, guillemots, razorbills, gannets, puffins, and gulls.

Double natural arches at Dyrhólaey Point. View east.

Black sand beach looking west from Dyrhólaey viewpoint.

Dyrhólaey Viewpoint to Reynisdrangar: *Return north on R-218 to R-1; turn right (east) on R-1 and drive to Vik; turn right (south) on Víkurbraut and drive to the Vik sports field; turn right (south) and drive to black sand beach parking. This is **Stop 10, Reynisdrangar** (63.414300, -19.014423), for a total of 19.6 km (12.2 mi; 24 min).*

STOP 10 REYNISDRANGAR

This is an area of black beaches and scenic sea stacks. The sea cliffs to the east rise 66 m (216 ft) from the waves and consist of layers of columnar basalt, pillow lavas, and tuff. Pillow lavas tell us that those lavas were erupted under water. Sea stacks are tall, freestanding offshore pillars of rock formed by wave erosion along a coastline. They are the erosional remnants of sea cliffs that have eroded back to their current positions. The stacks show how far the lava flows once reached before the relentless action of the waves washed them away. Tuffs are volcanic ash that fell on land.

We are going to see a lot of columnar basalt as we tour across Iceland. These amazing columns are similar to those seen at the Giant's Causeway in Ireland and Scotland, Devils Tower in Wyoming, Devils Postpile in California, and others. You might wonder how they formed. Well, they are much like mud cracks in that they form by shrinkage, but in this case as the rock cools. Mechanically, the least amount of work is required to break the rock at 120° angles, which usually leads to perfectly hexagonal crack patterns. Occasionally you will find pentagonal cracks. It is these shrinkage cracks in the cooling basalt flow that formed the columns seen along this beach.

Or you can believe the Icelandic folk tales that tell how the columnar basalt and offshore stacks came to be. Two trolls were dragging a three-masted ship to land, while the crew struggled to keep the ship off the beach. Alas, at dawn the ship's bottom scraped against the rocks, but as the rising sun's rays struck the trolls they instantly turned to columns of stone and needles of rock (Klescheva, 2014; Wikipedia, Reynisdrangar).

Another legend recalls that a local man's wife was taken by two trolls during the night and frozen into pillars of stone. Although she was the love of his life, he had been unable to contain her free spirit, so she found her fate with the trolls, rocks, and the sea at Reynisfjara.

Vík í Mýrdal is famous as the southernmost village in Iceland. In 1991 the black sand beach here was ranked one of the ten most beautiful non-tropical beaches in the world (Wikipedia, Reynisdrangar).

Reynisdrangar beach, sea cliffs, and stacks.

Chunks of debris flow form talus at the base of the cliffs, Reynisdrangar black sand beach.

Reynisdrangar to Side Trip 1, Eldgjá Fissure Zone: *Return to R-1 and turn left (east); drive 48 km (29.8 mi) from Vik to the turnoff to Landmannalauger; turn left (north) on F-209; this road is only paved part way. The graded gravel road eventually becomes a rough track not suitable for low-clearance vehicles; bear right onto F-208 and drive to Gjátindur; turn right (east) on Gjátindur and drive 4.8 km (3.0 mi) to the end of the road. Park and walk 600 m (2,000 ft) to the west to overlook the fissure zone. This is* **Side Trip 1, the Eldgjá Fissure Zone** *(63.965663, -18.603173), for a total of 93.7 km (58.3 mi; 1 hr 43 min). Looking across the valley to the west is Ófærufoss plunging into the rift valley. When we tried to get here in 2024 there was a barrier closing the road at (63.772288, -18.507235),~17 km from R-1. There was still a decent view north to the rift valley.*

If you prefer to skip this side trip:

Reynisdrangar to Fjaðrárgljúfur: *Return to R-1 and turn left (east); drive to the turnoff to Hunkubakkar; turn left (north) on R-206/Holtsvegur and drive to* **Stop 11, Fjaðrárgljúfur** *(63.771200, -18.171646) for a total of 69.4 km (43.1 mi; 54 min).*

SIDE TRIP 1 ELDGJÁ FISSURE ZONE AND ÓFÆRUFOSS

Vikings first settled Iceland in about 870 CE. The period of settlement lasted for 60 years, by which time the population may have been as high as 30,000. Iceland was much more forested and veg-etated than it is now. The Vikings settled on farmsteads, but the soils were often thin and quickly eroded. Farming became marginal and the population capacity of Iceland was reached quickly. The eruption of Eldgjá disrupted farming on the temperate south coast, both by covering large areas in ash and flood deposits and by cooling temperatures. Some have argued that the Eldgja eruption lies at the heart of the saga of Ragnarök – twilight of the gods (Zijlstra, 2016).

The Eldgjá ("Fire Canyon") eruptions occurred in the East Volcanic Zone. The lava flows are basaltic, but there was also a great deal of ash and tephra. There are over 30 distinct layers of tephra related to the Eldgjá eruptions. The eruptions began around 935 CE and lasted up to eight years. They were among the largest flood basalt eruptions in historical times. The fissure is at least 50 km (30 mi) long, extending from the Mýrdalsjökull ice cap in the southwest (Katla) to the Vatnajökull

ice cap in the northeast. The most dramatic section of the fissure zone is an 8 km (5 mi) long section in the southwest where one fissure is 600 m (1970 ft) wide and 270 m (886 ft) deep. This graben (down-dropped block, or rift valley) is considered the largest volcanic canyon in the world (Wikipedia, Eldgja). The fissure zone contains grabens, individual eruptive fissures, and a row of spatter cones. In the southernmost part of the zone the eruption was subglacial and caused a glacial outburst flood. The activity in this rift zone provides evidence that the Mid-Atlantic Rift spreading center is shifting east from Thingvellir to this area.

The Eldgjá eruption was a catastrophic eruption with worldwide impact. Eldgjá tephra is widespread in the Northern Hemisphere. Tree ring studies indicate that the summer of 940 was one of the coolest in 1,500 years, perhaps 2°C (4°F) cooler than normal in Europe, central Asia, and even Canada.

Total magma volume produced in the Eldgjá eruption has been estimated to be around 19 km^3 (4.6 mi^3); it covered about 800 km^2 (309 mi^2). Lava flowed south along the Skaftá River and reached the Atlantic between Álftaver and Mýrdalssandur (Katla Geopark, Eldgjá; Oregon State University, Eldgja Fissure System).

The famous Laki Fissure, which erupted for several months in 1783, lies in the northeast section of the Eldgjá Fissure Zone about 20 km (13 mi) from where we are.

Ófærufoss, across the valley to the west, is a two-stepped waterfall on the Nydri-Ófæru River that drops into the Eldgjá graben. A natural stone arch once spanned the lower falls, but in 1993 it collapsed during the spring thaw (Katla Geopark, Eldgjá).

The northern part of Eldgjá, including Ófærufoss, and surrounding areas have been a part of Vatnajökull National Park since 2011 (Wikipedia, Eldgjá).

Side Trip 1, Eldgjá Fissure Zone to Fjaðrárgljúfur: *Return south to R-1 and turn left (east); drive on R-1 to turnoff to Hunkubakkar or Fjaðrárgljúfur; turn left (north) on R-206/Holtsvegur and drive to **Stop 11, Fjaðrárgljúfur** (63.771200, -18.171646), for a total of 65.8 km (40.9 mi; 1 hr 21 min).*

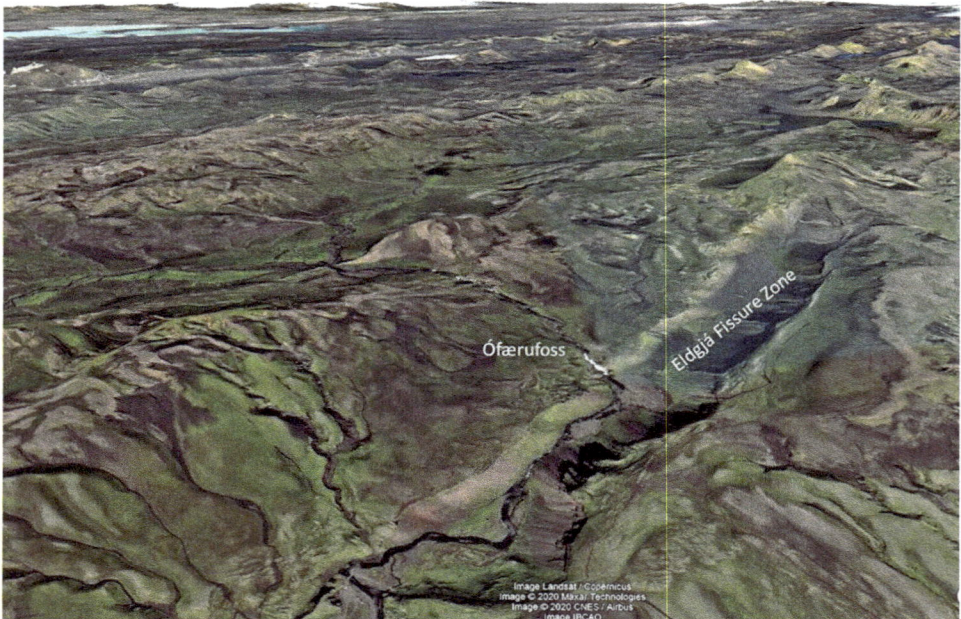

Google Earth oblique view north over the Eldgjá Fissure Zone and Ófærufoss. ©2020 Google Earth, Landsat/Copernicus, Maxar Technologies, CNES/Airbus, Image IBCAO.

View north to the rift valley (center) from the road barrier at 63.772288, -18.507235. Road F-208, approximately 17 km (10.6 mi) from the turnoff on Highway 1.

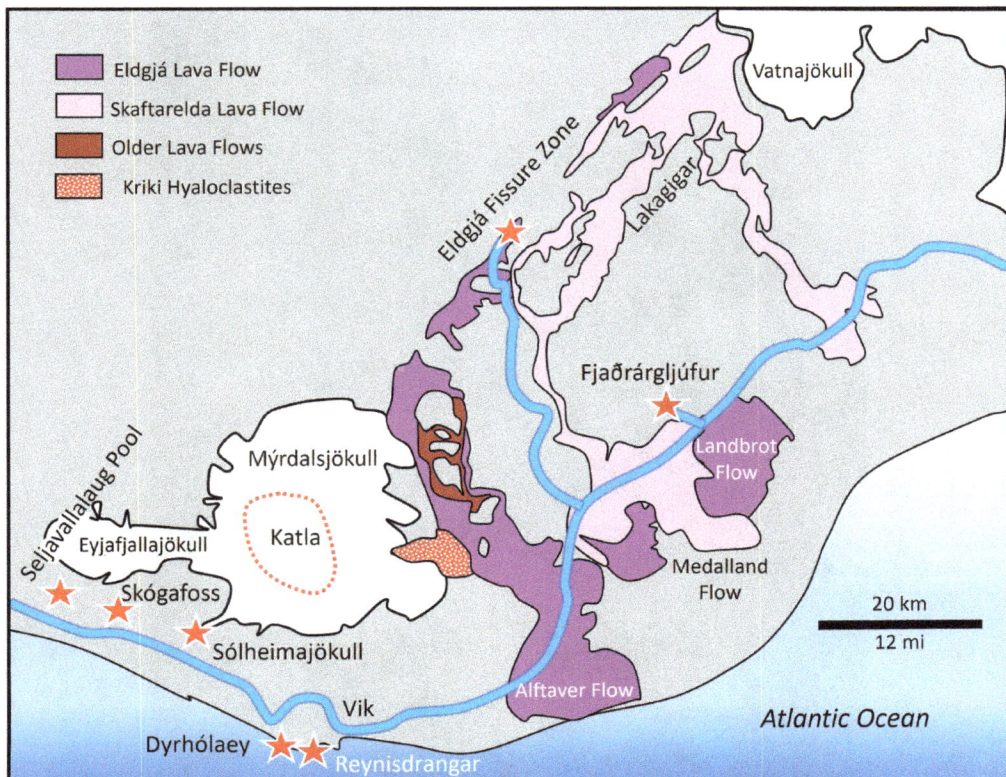

Map showing flows from the Eldgja fissure eruptions. Modified after Zijlstra, 2016.

Stop 11 Fjaðrárgljúfur

Virtually unknown before 2015, Fjaðrárgljúfur Canyon was put on the map by Justin Bieber's music video "I'll Show You." By 2018, the annual number of visitors had reached 300,000.

Fjaðrárgljúfur is a sheer canyon, about 100 m (330 ft) deep and 2 km (1 mi) long, carved by the Fjaðrá River as it flows to the Skaftá River. The canyon has vertical walls, and the shallow river meanders at its base. The bedrock in Fjaðrárgljúfur is a 2 million-year-old palagonite, altered basaltic glass that looks like a welded tuff.

Fjaðrárgljúfur Canyon formed about 9,000 years ago, at the end of the last Ice Age. As the glaciers retreated, a lake formed above the present canyon and filled with sediments. When the outlet to the lake was breached, the river began to cut down into the lake sediments. Sediment terraces on both sides of the valley reveal the original location of the lake.

It costs 10 kr to park (2024), and you'll need a smartphone to pay. There is parking for 20 to 30 cars. Washrooms are available. A short and easy trail starts at the parking lot and heads up along the canyon rim. Be careful, as walking near cliff edges can be dangerous, especially in winter. Due to important breeding grounds, hiking up the canyon itself and in the river is now forbidden (Katla Geopark, Fjaðrárgljúfur; Cars Iceland.com, Fjaðrárgljúfur).

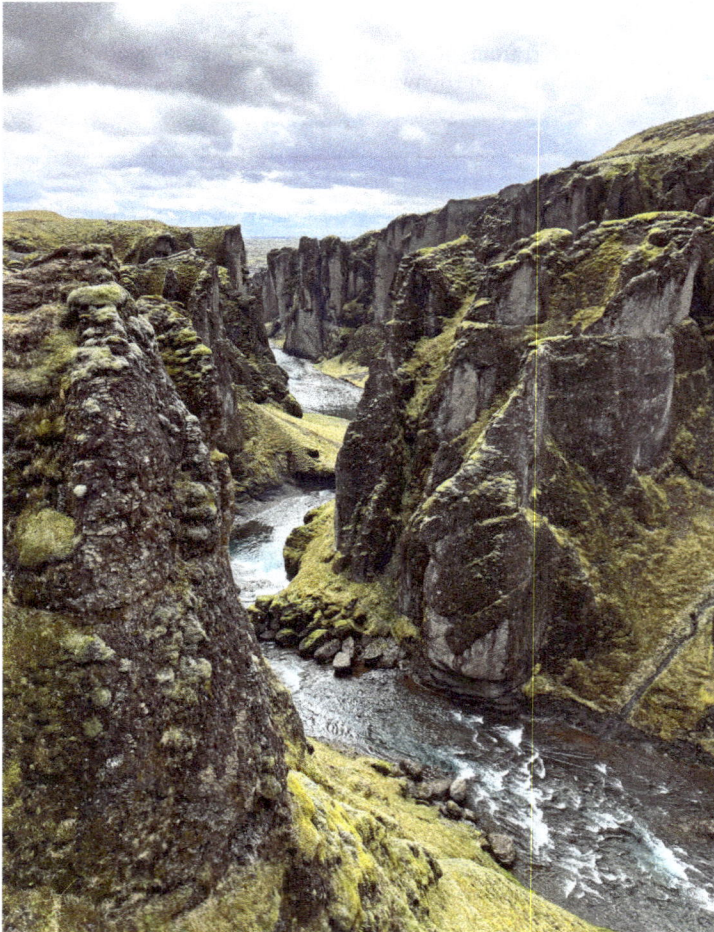

Fjaðrárgljúfur Canyon looking south.

Vatnajökull National Park

Vatnajökull National Park covers 1,400,000 ha (3.5 million ac), almost 14% of Iceland. It contains Vatnajökull ice cap, the largest glacier in Europe at 8,100 km^2 (3,127 mi^2) and 400 to 600 m (1,312 to 1,970 ft) thick. Within the park are 10 volcanoes, eight of which are beneath the ice cap. Two of these, Bárðarbunga and Grímsvötn, are among the four most active volcanoes in Iceland (Buz, 2014).

The park is volcanically active because it sits above the spreading Mid-Atlantic Ridge, expressed here as two intersecting rift zones. The Eastern Rift Zone extends northeast from Heimaey to Grímsvötn and Bárðarbunga; the North Rift Zone extends northward from these volcanoes.

Most of the rock in the park is basalt ranging in age from about 10 Ma to Recent (last eruption was in 2015). Shield volcanoes sit alongside fissure zones, linear spatter cone arrays, table mountains, and vast flood lavas. About 85% of the area is considered wilderness, and, other than a few historic farms, no development has occurred.

The Vatnajökull ice cap extends from a maximum elevation of 2,000 m (6,562 ft) to 300 m (984 ft) below sea level (Wikipedia, Vatnajökull National Park). It has, on average, been retreating since it reached its maximum extent in the late 1700s. The retreat has exposed both terminal and lateral moraines as well as glacial outwash plains and glacial flood plains. Interestingly, lakes beneath the ice cap are being studied for their unique single-celled life that is considered analogous to life on the early Earth and potential life in seas beneath the icy surfaces of the moons of Jupiter and Saturn (UNESCO, 2019).

The area became a national park in 2008; it was designated a UNESCO World Heritage Site in 2019.

For information about visiting the park and permits, see the park web site (English and Icelandic) at https://www.vatnajokulsthjodgardur.is/.

Fjaðrárgljúfur to Svartifoss: Return to R-1 and turn left (east); drive east on R-1 to R-998/ Skaftafellsvegur; turn left (north) on R-998 and drive to the Svartifoss parking area; from there a trail leads 130 m (420 ft) to Hundafoss, and another leads 940 m (3,100 ft) to Svartifoss. This is **Stop 12.1, Svartifoss** *(64.026900, -16.974590) for a total of 78.6 km (48.8 mi; 1 h 4 min).*

Stop 12.1 Svartifoss

At Svartifoss, the "Black Waterfall," a basalt flow characterized by polygonal cooling joints can actually be viewed from below. We first saw columnar basalt in the beach cliff at Reynisdrangar. Now add a waterfall, a plunge pool, and a rainbow in the mist and you have Svartifoss.

Svartifoss. The columnar basalt frames a spectacular cascade.

Svartifoss to Skaftafell Glacier: *Return south 1.9 km (1.2 mi; 5 min) to the Skaftafell Glacier parking area. Take the trail 1.4 km (4,470 ft) to the glacier overlook. This is* **Stop 12.2, Skaftafell Glacier** *(64.022911, -16.944098).*

STOP 12.2 SKAFTAFELLSJÖKULL GLACIER

About 4 km (2.4 mi) before this stop is the Skeiðará Bridge Monument on the right (east) side of the highway at (63.984697, -16.960106). This is all that is left of a bridge hit by a jökulhlaup. In November 1996 the road here was washed away by a glacial flood following an eruption below the Vatnajökull glacier. The floodwater rose over a period of 15 hours to a rate of 50,000 m3/s and carried with it huge blocks of ice. The 376 m long Gígjukvísl Bridge was completely washed away; the Skeiðará Bridge was badly damaged. These bent pieces of steel are a monument to that flood.

Skaftafellsjökull is a typical active temperate glacier with moraines and local development of kame and kettle topography (mounds and depressions in glacial outwash) that developed from a gradually melting debris-rich glacier. The area downstream from this piedmont (mountainfront) glacier, with closely spaced moraines and outwash fans, is typical of the south coast of Iceland (Evans et al., 2017).

The oldest rocks in this area are 5 Ma basalts. The stratigraphic section indicates frequent shifts between glacial and interglacial periods. The oldest tillite, at 4 Ma, shows the area has been subject to glaciation for at least that long (Wikipedia, Skaftafell). In contrast, Europe's Ice Age began about 2.6 Ma.

Detailed study of the rock layers in this area indicates 16 glacial and interglacial periods in the past 5 Ma. Glacial intervals are indicated by pillow basalts (underwater eruptions) and hyaloclastite ridges and breccias (subglacial eruptions). Interglacial periods are characterized by glacial tills and lava flows showing signs of erosion (Buz, 2014).

View north up Skaftafellsjökull. Notice that the central moraine, where two branches of the glacier come together, becomes a lateral moraine toward the end of the glacier. This is probably because the west side has more flowing ice than the east side.

There is an extensive glacial outwash plain below Skaftafell Glacier. View south to the coast.

This glacier was originally part of Skaftafell National Park, established in 1967. In 2008 it became part of the newly formed Vatnajökull National Park.

The Skaftafell Visitor Center provides information on the park and its geology, natural history, trails, and human history. Displays describe the intertwined history of the land and its people. Park rangers give guided walks and lectures, and children's activities are available during the summer season.

This area has a variety of hiking trails, and tent and car camping are available from 1 May to 30 September.

Skaftafellsjökull to Svínafellsjökull: *Return south on R-998 to R-1; turn left (east) on R-1 and drive 800 m (0.5 mi) to Svinafellajokulsvegur; turn left (northeast) on Svinafellajokulsvegur and drive to* **Stop 13, Svínafellsjökull** *(64.008325, -16.879489), for a total of 5.1 km (3.2 mi; 14 min). Walk 295 m (965 ft) to the closest overlook, or 550 m (1,800 ft) to the second overlook.*

Stop 13 Svínafellsjökull Glacier

The rocks at Svínafells, as at nearby Skaftafell, are largely derived from the subglacial Öræfajökull volcano. They are slightly altered tholeiitic basalt characterized by clinopyroxene phenocrysts (large black to green elongated crystals that indicate slowly cooling magma before being erupted at the surface). Silica-rich volcanics, in particular dikes, do exist along with some areas of extensive hydrothermal (hot magmatic fluids) alteration. Zeolites and other alteration minerals found in the older rocks are thought to be related to intense thermal activity associated with the Öræfajökull volcano (Prestvik, 1979).

Svínafellsjökull. Photo courtesy of Debivort, https://commons.wikimedia.org/wiki/File:Sv%C3%ADnafellsj%C3%B6kull.jpg

Glacial landforms in the area consist of sequences of recessional moraines deposited from the mid-Holocene to as recently as the last 100 years. There is abundant evidence of glacial floods (jökulhlaups) related to eruptions of Öraefajökull in 1362 and 1727. Accelerated retreat of Svínafellsjökull glacier since 2005 has left behind ice-cored eskers, kettle holes, ice caverns, and an ice-floored lake (Everest et al., 2017).

*Svínafellsjökull to Jökulsárlón/Glacier Lagoon: Return to R-1 and turn left (east); drive east on R-1 to the sign for **Stop 14.1, Jökulsárlón/Glacier Lagoon** (64.052586, -16.177827), for a total of 57 km (35.4 mi; 52 min).*

Stop 14 Breiðamerkurjökull Glacier

When the first settlers arrived in Iceland around 870 CE, the edge of Breiðamerkurjökull Glacier was about 20 km (12 mi) north of its present position. During the Little Ice Age, which ended in the late 1800s, the glacier had extended to within 1 km (0.6 mi) of the coast. Maps of the area from the 1700s show the glacier drained directly into the sea through a series of anastamosing channels. Breiðamerkurjökull Glacier has retreated rapidly since then, leaving a lake in the depression gouged out by the advancing glacier. Lateral moraines can be seen on both sides of the lake (Wikipedia, Jökulsárlón).

An ice shelf extends from the glacier into the lagoon. Since water temperature rarely rises above 1°C, ice above the water melts faster than that below the water, causing the shelf (Buz, 2014).

Stop 14.1 Glacier Lagoon/Jökulsárlón

Jökulsárlón (Glacier Lagoon) is a lake at the mouth of Breiðamerkurjökull Glacier, a tongue of the larger Vatnajökull ice cap, which can be seen along the skyline to the north. The lagoon covers about 18 km² (6.9 mi²). It is considered to be the deepest lake in Iceland, over 284 m (932 ft) at its deepest point. The size of the lake has quadrupled since the 1970s because of the retreating glacier.

Jökulsárlón looking northwest to the Vatnajökull ice cap.

From the lagoon you get spectacular views of the Vatnajökull ice cap, a vast icefield that towers 910 m (3,000 ft) above the landscape. Icebergs calve from the end of the retreating glacier and move across the lagoon with the wind and tidal currents. The bergs range in color from milky white to translucent aquamarine, depending on the sediment and air trapped in the ice.

Fish wander in from the sea along with the tides. Seals often gather at the outlet to feed on the fish. Seabirds, including Arctic terns, Arctic skua, gannets, and puffins, gather to catch herring, trout, salmon, and krill that are found in the lagoon.

Jökulsárlón has been a setting for four Hollywood movies: *A View to a Kill*, *Die Another Day*, *Lara Croft: Tomb Raider*, and *Batman Begins*, as well as the reality TV series *The Amazing Race* (Wikipedia, Jökulsárlón).

Jökulsárlón to Diamond Beach: *Walk to R-1 and cross it to the parking area at the beach. This is* ***Stop 14.2, Diamond Beach*** *(64.044630, -16.177415), for a total of 1.0 km (0.6 mi; 6 min).*

STOP 14.2 DIAMOND BEACH

Aquamarine icebergs set in black sand beaches. There is geology here, but this is primarily a scenic stop.

The beaches on both sides of the lagoon's outlet are spits of black volcanic sand that enclose Jökulsárlón. The "diamonds" on Diamond Beach are the bergy bits that beach themselves here after breaking free from the Breiðamerkurjökull Glacier, transiting the lagoon, and washing out to sea. They are clear, sparkling chunks of pure freshwater ice that change appearance based on how they capture the sun's rays at any given time of day. The spits are formed of sand moved about by the waves and longshore currents.

Iceberg "diamonds" on Diamond Beach.

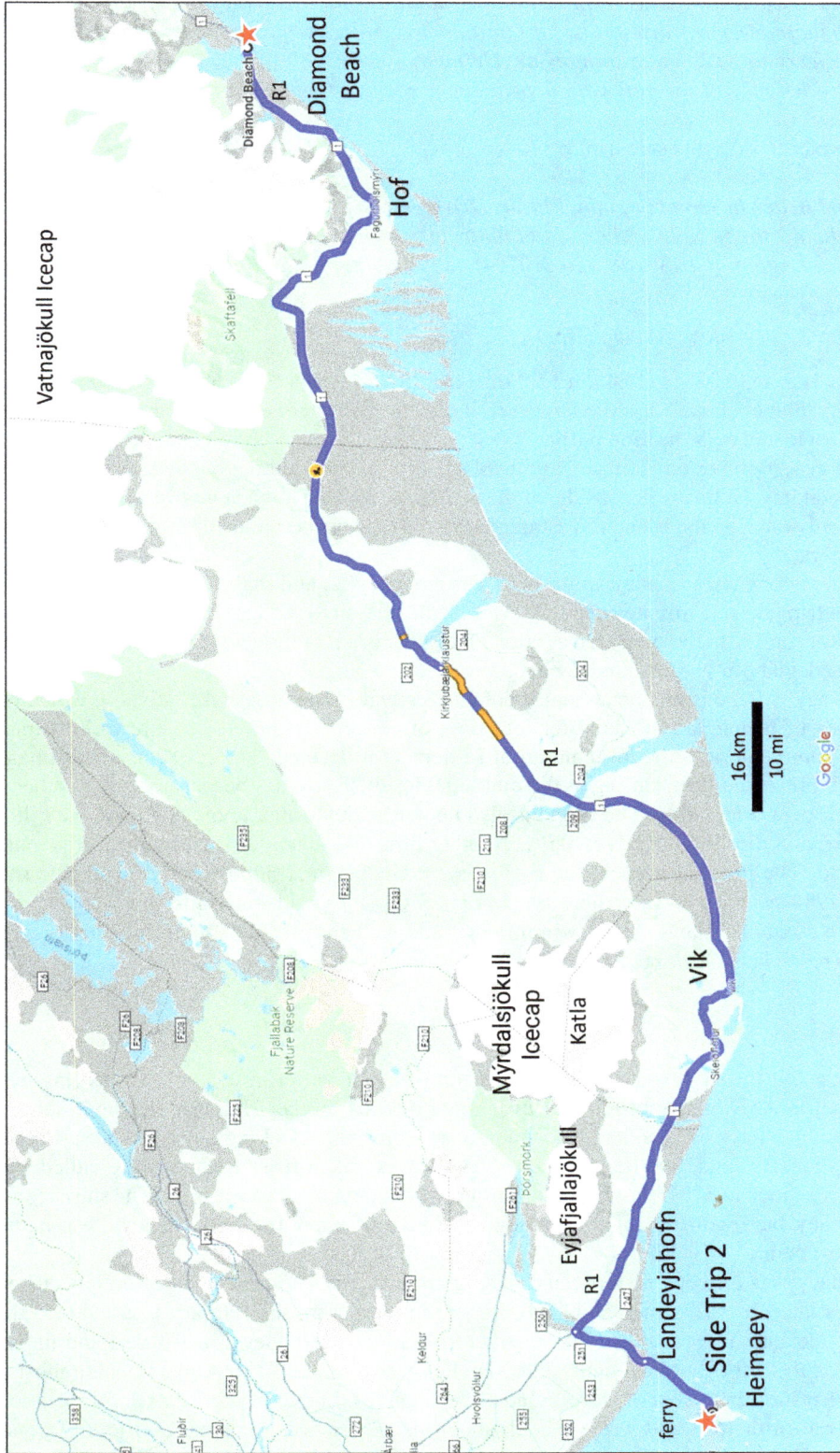

Diamond Beach to Side Trip 2 – Vestmannaeyjar, Heimaey: *Drive west on R-1 to Landeyjahafnarvegur; turn left (south) on Landeyjahafnarvegur and drive to the ferry terminal; take the ferry to* **ST2, Vestmannaeyjar, Heimaey** *(63.443489, -20.273537) for a total of 279 km (173 mi; 4 h 4 min). Time may vary depending on the ferry schedule.*

If you prefer to skip this side trip:

Diamond Beach to Secret Lagoon, Flúðir: *Drive west on R-1 to R-30/Skeiða- og Hrunamannavegur; turn right (north) on R-30 and drive to Flúðir; turn right (east) on Hvammsvegur and drive to* **Stop 15, Secret Lagoon** *(64.137708, -20.310724) on the right for a total of 338 km (210 mi; 4 hr 8 min).*

Side Trip 2, Vestmannaeyjar, Heimaey

Heimaey is a small island (13.4 km^2; 5.2 mi^2) that lies 7.4 km (4.6 mi) off the south coast of Iceland. Meaning "home island," it is the largest island in the Vestmannaeyjar archipelago and is home to 4,500 people and eight million puffins.

Most people living on Haimaey are involved with fishing, although tourism is a significant part of the economy. Heimaey is considered the most productive fishing center in the country.

The only town on the island is Vestmannaeyjar. The town is connected to the mainland by a ferry and airport.

The ferry connects Landeyjahöfn with Vestmannaeyjar, and makes five to six round-trips each day. The trip takes 35 minutes one way (Wikipedia, Heimaey; Claus, 2020). If you plan to visit, get your tickets online before you arrive (https://herjolfur.is/en/ or https://visitwestmanislands.com/tour /ferry-landeyjahofn-to-vestmannaeyjar/).

Heimaey is famous for the annual Þjóðhátíð festival ("The National Festival"), which brings in thousands of people to the island for four days of concerts, bonfires, and fireworks. First held in 1874, it commemorates the millennium of settlement in Iceland. The residents of Vestmannaeyjar had been prevented from sailing to the mainland for the festivities because of bad weather, so they held their own celebration. The island is also known as the former home of Keiko, the killer whale that starred in the 1993 film *Free Willy*. Another film, *The Deep* (2012), was filmed on and around the island. The film is based on the true story of Guðlaugur Friðþórsson, an Icelandic fisherman who in 1984 survived an incredible six hours in 5°C (41°F) cold water after his vessel capsized off the south coast of Iceland. After swimming to land he then walked another three hours across lava fields to reach help (Wikipedia, Heimaey).

History

The Sagas tell the story that Ingólfur Arnarson, the first settler in Iceland, was sailing west from Ingólfshöfði. At Hjörleifshöfði, Ingólfur learned that his brother Hjörleifr Hróðmarsson had been murdered by his slaves, who had subsequently gone missing. Looking out to sea he saw boats heading for a small island, and he gave chase. The slaves, taken from Ireland, were called Vestmenn (west men), since in 875 Ireland was the westernmost land known at the time. The slaves took refuge on Heimaey, but Ingólfur hunted them down and killed them in revenge. The town was named after these west men.

Heimaey was a sleepy farming and fishing community. The next event of note occurred many centuries later. In 1627, three shiploads of Arab pirates from the Barbary Coast raided the south coast of Iceland. They attacked Grindavík and Heimaey. Heimaey was isolated and unprotected. Known as the Turkish Abductions, 234 islanders were hauled off to Algiers, where most of them spent the rest of their lives as slaves. One of the captives, Lutheran minister Ólafur Egilsson, managed to return in 1628 and wrote a book about his experience. Another, Guðríður Símonardóttir, known as Tyrkja-Gudda (Turkish-Gudda), bought her way back to Iceland through Tunisia, Italy, and Denmark. In 1636, ransom was paid for 34 of the captives, and most of them returned to Iceland.

In the 20th century the island was known for its volcanic eruptions: the 1963 eruption, which created the new island of Surtsey off the south coast, and the Eldfell eruption of 1973, which created a 200 m (656 ft) high volcanic cone in what had been a meadow.

GEOLOGY

The Vestmannaeyjar archipelago is geologically very young. The islands were formed by eruptions over just the past 40,000 years. The archipelago contains 70–80 volcanoes, both above and below the sea.

There are 15 islands and about 30 stacks.

On 14 November 1963 an eruption lasting four years created the new island of Surtsey (Wikipedia, Heimaey). The oldest rocks on Vestmannaeyjar, on the north side of the island, formed about 40,000 years ago. Initially there were two separate islands. Eventually they were connected by younger volcanics. The rest of Heimeay didn't emerge until a series of eruptions about 6,000 years ago. Stórhöfði, the headland to the south, originally formed as a third island. When Helgafell erupted, all three islands merged into the present island (Claus, 2020).

The eruption that put Heimaey on the map began on 23 January 1973 at Eldfell. Fissures, accompanied by earthquakes, opened and grew to 1,600 m (5,200 ft) in length, and lava began to pour out. When the fissures eventually closed, the eruption focused on a single lava flow heading toward the town and harbor. Half a million m^3 (654,000 yd^3) of ash settled over the town. During the night, all 4,500 inhabitants were hastily evacuated by fishing boats.

The encroaching lava flow threatened to close and fill the harbor. The townspeople constantly sprayed the lava with cold seawater, causing some to solidify and much to be diverted, thus saving the harbor (McPhee, 1989). For almost five months, seawater was pumped through fire hoses onto the advancing lava to cool and harden it. The combination of slow-moving lava and an inexhaustible supply of water probably saved the town (Prost and Prost, 2018). The eruption ended on 3 July. It destroyed around 400 homes, but 80% of the town was saved along with the harbor. The eruption increased the area of Heimaey from 11.2 km^2 (4.3 mi^2) to 13.44 km^2 (5.19 mi^2)

Houses in Vestmannaeyjar buried by ash fall from the 1973 eruption of Eldfell, Iceland. Photo courtesy of the U.S. Geological Survey, https://pubs.usgs.gov/gip/heimaey/.

The ships Sandey and Lóðsinn pumped sea water onto the leading edge of the lava at the breakwater. Credit: Sigurgeir Jónasson, and U.S. Geological Survey. https://pubs.usgs.gov/of/1997/of97-724/lavaoperations.html

Eldfell means Hill of Fire. The islanders are using heat from the still cooling lava flows to provide hot water and to generate electricity. They also used some of the extensive tephra as landfill on which 200 new houses were built (Wikipedia, Heimaey).

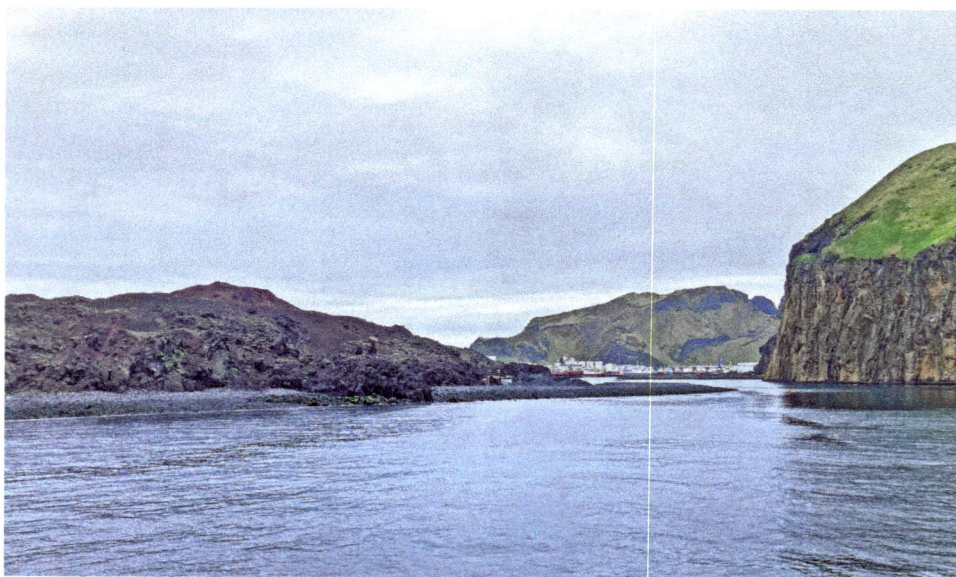

A lava flow from Eldfell (dark rock, left) almost blocks the entrance to Heimaey harbor. View is west toward Vestmannaeyjar and the harbor.

SIGHTS/WALKS

Most of Heimaey is accessible by walking. The main regret voiced by visitors is that they did not schedule enough time to see everything of interest.

The Eldheimar Museum is built over houses that were buried by the 1973 Eldfell eruption and have since been excavated. The first floor is dedicated to the eruption of Eldfell and its impact on the town of Vestmannaeyjar. The second level highlights the Surtsey nature reserve.

Eldheimar Museum on the flank of Eldfell.

Ruins of a home buried in ash from the Eldfell eruption, Eldheimar Museum.

Hours: Open daily 11:00 to 17:00.
Entry Fee (2024): 3,200 kr (~$22.90) adults; 2,700 kr ($19.30) seniors; under 10 years free.
Address: Suðurvegur / Gerðisbraut 10, 900 Vestmannaeyjar
Phone: +354-488-2700
Email: eldheimar@vestmannaeyjar.is
Website: https://eldheimar.is/en/

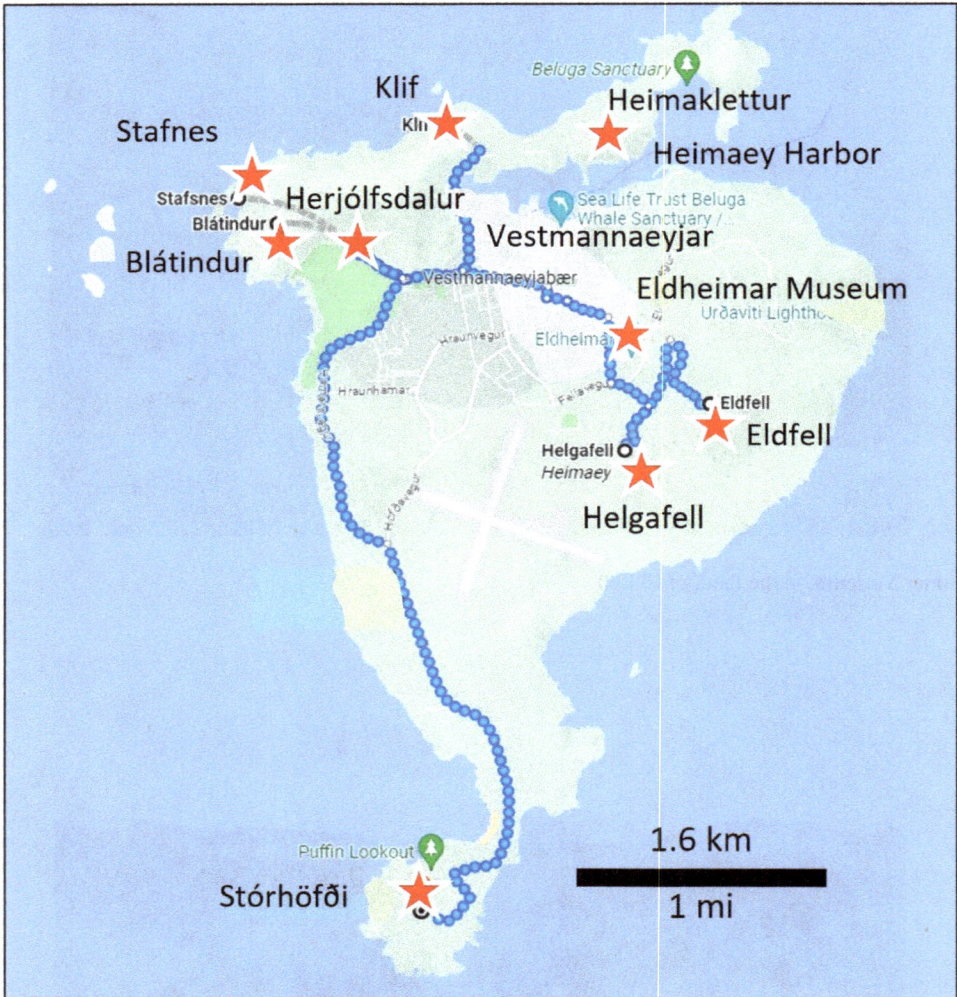

Walk to Heimaklettur: Heimaklettur is on the northern peninsula of Heimaey and the highest hill on the island at 283 m (928 ft). It is the iconic "Home Rock" of Vestmannaeyjar. There are ladders and chains to help you up the path to the top (Claus, 2020).

Heimaklettur (center) viewed from Eldfell. Photo courtesy of Szilas, https://commons.wikimedia.org/wiki/
File:View_from_the_Eldfell_towards_the_crater.jpg

Walk to Eldfell: The walk to the top of Eldfell is one almost every visitor to Heimaey wants to do.
This is the volcano that erupted out of a meadow in 1973 and almost destroyed the town. The climb
to the summit is considered relatively easy, as the ascent is gradual. On a clear day you can see
most of Heimaey as well as the new island of Surtsey, a World Heritage Site, 20 km (12 mi) south of
Heimaey. The parking lot is at about 90 m (290 ft); the summit is about 190 m (620 ft).

Eldfell Volcano and Vestmannaeyjar as seen from summit of Helgafell.

Walk to and climb the 227 m (745 ft) high **Helgafell** volcano for panoramic views of the island. This cone is often overlooked by tourists and can be a lot less crowded than its neighbor, Eldfell. The central vent is 90 m (290 ft) across. The last eruption of Helgafell was in 3950 BCE, give or take a few hundred years.

Helgafell from Vestmannaeyjar harbor.

Walk to Herjólfsdalur, a giant natural amphitheater with cliffs rising up on all sides. Sheltered from the wind, grassy Herjólfsdalur was the home of Herjólfur Barðursson, Vestmannaeyjar's first settler. Herjólf's valley is now the setting for the annual Þjóðhátíð festival each August.

Herjólfsdalur (right) and Blátindur (center left). Photo courtesy of H. Krapf, https://commons.wikimedia.org /wiki/File:2016-06-15_14-27-30_17.0_Iceland_-_Vestmannaeyjum_Vestmannaeyjar_4h_150%C2%B0.JPG

Blátindur is the imposing peak rising 273 m (895 ft) above Herjólfsdalur. It takes some effort to scramble to the summit, but the views are worth it.

Walk to Klif: At the west end of the northern peninsula (and on the way to Heimaklettur), this is part walk, part scramble. Klif is the name given two flat-topped table mountains, Stóra-Klif and Litla-Klif ("Big Cliff" and "Little Cliff"). Stóra-Klif is northeast of Litla-Klif. It's possible to hike up Stóra-Klif, but the route is challenging. Ropes and chains attached to the rock help you up. It is worth the effort, as the views are stunning.

Both mountains are tuyas, or table mountains. As mentioned previously, a tuya is a flat-topped, steep-sided volcanic hill formed when lava erupts through a glacier or ice sheet. The lava cools quickly and thus cannot travel far, piling up as a steep-sided hill. Lava flows make the flat top of the hill. At the base you often find pillow lavas; explosive volcanism forms tephra (rock fragments). Basaltic glass in the tephra forms hyaloclastite, a volcanic glass breccia.

Stóra-Klif (right) and Litla-Klif (left) from Vestmannaeyjar harbor. Photo courtesy of D. Delso, https://commons.wikimedia.org/wiki/File:Puerto_de_Vestmannaeyjar,_Heimaey,_Islas_Vestman,_Su%C3%B0urland,_Islandia,_2014-08-17,_DD_088.JPG

Walk to Stafsnes: This secluded beach is nestled in a cove on the northwest point of the island, below the imposing cliffs of Blátindur. It can be sheltered from the wind, and it is usually the warmest place on the island. Stafsnes is considered one of the most beautiful places on Heimaey, and the view to the southern islands is magnificent. Eider, large black and white sea ducks, have their nests up on the rock walls and along the rocky shores of Stafsnes. Stafsnes is not the most accessible place on Heimaey, but it is worth the effort. To get there you walk from Herjólfsdalur up to Dalfjall and then take a steep walk down to Stafsnes beach. The climb up is difficult because the slopes are often near-vertical (Heimaslóð, Stafsnes, https://www.heimaslod.is/index.php/Stafsnes).

Stórhöfði is the southernmost point on Heimaey and home to the largest puffin colony in Iceland. The name means Great Cape. This is considered the easiest hike, as the terrain is essentially flat. At Stórhöfði you find not only a puffin colony but also the geologically interesting Klauf beach. The cape is linked to the main island by a narrow isthmus formed 5,000 years ago by a lava flow from Helgafell. A small bird watching hut for viewing puffins is located about half way up the hill. June to August is considered the best for puffin viewing (Lonely Planet, Stórhöfði).

Stórhöfði is considered the windiest place in Europe, and it is the location of one of the oldest lighthouses in Iceland, having operated since 1906. Weather observations began at the lighthouse in 1921 (Wikipedia, Heimaey).

Stórhöfði Peninsula, view south. Photo courtesy of Ó. Sigurðsson, https://commons.wikimedia.org/wiki/File
:St%C3%B3rh%C3%B6f%C3%B0i_(7469276784).jpg

Side Trip 2, Heimaey to Secret Lagoon, Flúðir: *Get on the ferry from Heimaey to the ferry ter-*
minal at Landeyjahafnarvegur and drive to R-1; turn left (west) and take R-1 to R-30/Skeiða- og
Hrunamannavegur; turn right (north) on R-30 and drive to Flúðir; turn right (east) on Hvammsvegur
*and drive to **Stop 15, Secret Lagoon** (64.137708, -20.310724) on the right for a total of 110 km*
(68.4 mi; 2 hr 9 min). Times may vary due to ferry schedule.

STOP 15 SECRET LAGOON, FLÚÐIR

Secret Lagoon is an area of natural hot springs and a man-made pool near the village called Flúðir. Icelanders know it as "the old pool" or "gamla laugin." The water stays at 38°–40°C (~100°F) all year. The pool's bottom is all gravel and pebbles. A path by the pool allows you to explore this area. A little geysir nearby erupts every five minutes. During winter, the northern lights often put on a light show.

The pool was built in 1891 and is the oldest in Iceland. It was rebuilt in 1947, and the establishment was renovated again in 2014. New facilities next to the hot spring include showers, a bar, and an eating area. Food is not served daily, but can be arranged in advance. Though not as busy as the Blue Lagoon, it still attracts hundreds of people each day and requires a reservation in the summer.

This area is known for its thermal activity, which is used for heating greenhouses, allowing Iceland to produce fresh food year-round.

Secret Lagoon, Flúðir.

Geologically, the area consists of Plio-Pleistocene (roughly 1.5 to 2 Ma) hyaloclastites, basalt flows, and breccias with interbedded tillites. The surface is covered by a veneer of Holocene marine and river sediments. Structurally, the area has been broken into northwest-dipping tilted fault blocks by northeast-trending normal faults, mostly down to the southeast. The layers here are inclined about 20° west. Fractures with a similar northeast trend appear to control most of the geothermal springs. Low-temperature zeolite minerals in surface rocks indicate low-temperature and low-pressure metamorphism, that the rocks were buried at least 300–400 m (984 to 1,312 ft) before erosion (Tsokonombwe, 2012). Some marine sediments containing seashells have been found above the volcanics here, indicating that this area was below sea level during and shortly after the last glaciation (~10,000 years ago). There is a small geyser that erupts a plume of boiling water up to 2 m (6 ft). Multiple hot springs flow into the pool and in the surrounding fields. The chemistry of the fluids indicates a temperature of 155°C (311°F) in the hottest part of the system. Geothermal heat supports a thriving greenhouse industry for growing vegetables, notably tomatoes. If you get a chance, try the tomato soup, a local favorite, at the Hot Pot restaurant and greenhouse next door.

VISIT

The Secret Lagoon is compared favorably to the Blue Lagoon because it is less expensive and less busy. While true, the Blue Lagoon has facilities like saunas and steam rooms that are not available here. And unlike the Blue Lagoon, you won't have to wash silt out of your hair. Enjoy the setting with the lightly sulfuric steam rising from the springs.

Entry fees (2024):	Adults (15 years and older) 3,600 Isk; Seniors (67+) 2,500 Isk; Children (14 years and younger) free.
Hours:	From 1 June 2020 to 30 September Secret Lagoon is open every day from 10:00 to 20:00; 1 October to 31 May from 10:00 to 19:00.
Address:	Hvammsvegur – 845 Flúðir
Reception:	+354-555-3351
Tourist Information:	+354-853-3033
Email:	Use the "Contact Form" found on the website.
Website:	https://secretlagoon.is/

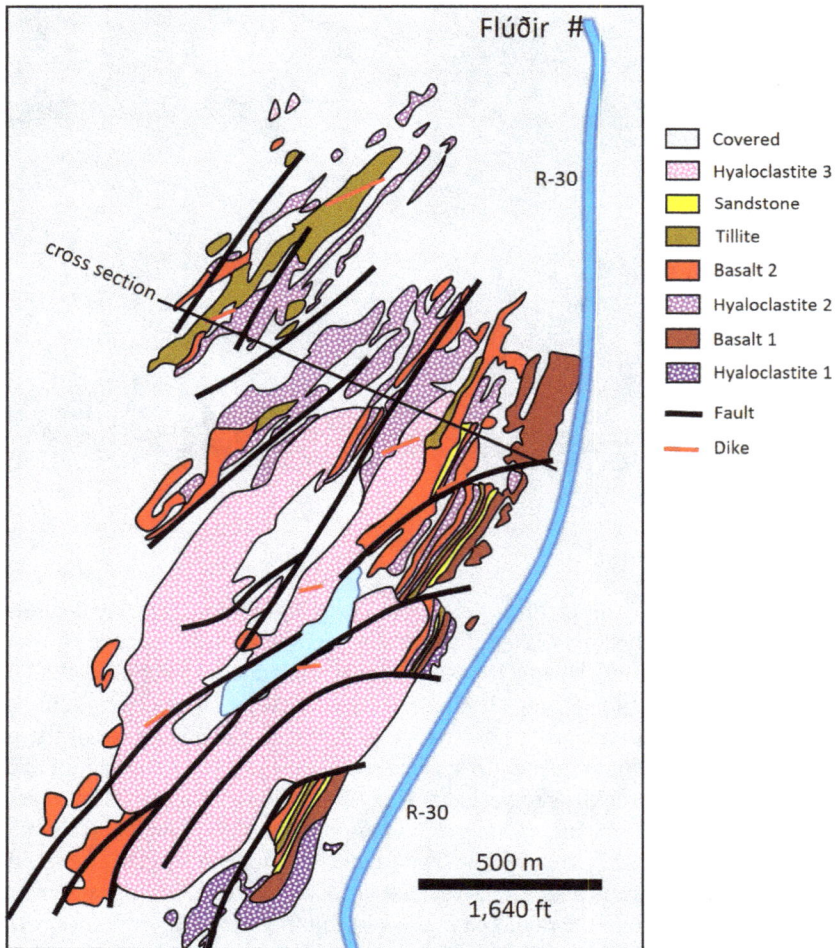

Geologic map of the Midfell-Flúðir geothermal area south of Flúðir. The hot springs pool is approximately 1 km (0.6 mi) northeast of Flúðir. Modified after Tsokonombwe, 2012.

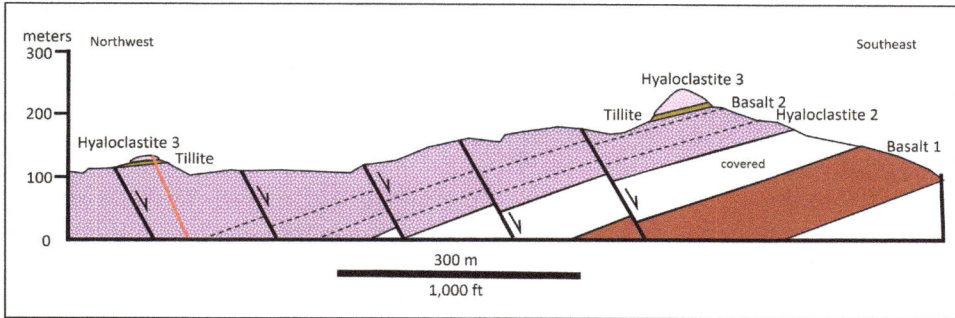

Cross-section through the Midfell-Flúðir geothermal area south of Flúðir. Location is shown on the previous figure. Modified after Tsokonombwe, 2012.

Secret Lagoon to Gulfoss: *Return west to Flúðir and turn right (north) on R-30/Skeiða- og Hrunamannavegur; after a few km turn right (northeast) to stay on R-30; drive to R-35; turn right on R-35 and drive to the turnoff to **Stop 16, Gulfoss Falls** (64.325109, -20.124962) on the right for a total of 31.9 km (19.8 mi; 28 min). Walk 270 m (890 ft) to the falls.*

STOP 16 GULLFOSS FALLS

Set in upland tundra, with views of what could be Iceland's Alps, the Hvítá River plunges over a two-tiered waterfall often dubbed Iceland's answer to Niagara Falls. Gullfoss (Golden Waterfall) on the Hvítá (White River) is Iceland's most beloved waterfall. The Hvítá is a glacial river that rises in the Icelandic highlands and emerges from Hvítárvatn Lake. After flowing 40 km (24 mi), the glacial silt-laden river reaches the double falls. At this point the river is 120 m (390 ft) wide and pushes an average of 141 m³ (5,000 ft³) over the falls every second. That is enough water to fill an Olympic swimming pool every 18 seconds. The upper drop is 11 m (36 ft); the lower falls are 22 m (72 ft) high. Below the falls the river has cut a gorge 40–70 m (130–230 ft) deep and 3 km (1.8 mi) long. At the lower lookout you can see lava flows over glacial till consisting of breccias and conglomerates.

Gullfoss Falls.

The river plunges over a resistant caprock of basaltic lava. Below the lava is glacial tillite, much easier to erode. Geologists have suggested that Gullfossgjúfur, the gorge below the falls, may have been formed during a jökulhlaup (glacial outburst) near the end of the last Ice Age. Gullfoss and Gullfossgjúfur formed where the river found it easy to erode along a fissure in the hard lava (Aabech, Gullfoss). The falls erode upstream at the rate of about 25 cm/yr (10 in/yr; https://justfunfacts.com /interesting-facts-about-gullfoss/).

Gullfoss is considered Europe's most powerful waterfall. The average flow is about 110 m³/s (4,000 ft³/s); during the spring snowmelt the flow reaches around 1,800 to 2,000 m³/s (63,000-71,000 ft³/s) (Hit Iceland, Gullfoss; Wikipedia, Gullfoss).

Notice that the two falls are linear breaks at 110° to one another, and that the lower falls is essentially parallel to Gullfossgjúfur. That is because they are controlled by joints and faults that are more readily eroded than the surrounding rock.

Glacial till consisting of breccia and conglomerate is interbedded between lava flows, Gullfoss Falls.

The story goes that at the turn of the 20th century, efforts began to acquire the rights to waterfalls in Iceland for hydroelectric development. In 1907 an Englishman sought to purchase Gullfoss. He offered to pay a price of ISK 50,000, nearly 50 times the value of the farmhouse of Tómas, the farmer who owned the water rights. His response: "I won't sell, friend."

In the autumn of 1907, a law was passed to ensure that only Iceland residents could acquire the waterfalls without seeking special permission. Be that as it may, speculators managed to gain control of many of the largest falls. When Gullfoss came into the possession of agents for foreign companies, Sigríður Tómasdóttir, the original farmer's daughter, became so upset that she threatened to throw herself into the falls if the contract was not withdrawn. She went from one official to another in Reykjavík, to no avail. A court handed down a judgment against her. Sveinn Björnsson, a young lawyer who later became Iceland's first president, joined the struggle. Together they managed to save the waterfalls. In 1928, when the agreed rental for the falls was not paid, the contract was voided. Einar Guðmundsson purchased the Brattholt property from Sigríður Tómasdóttir in 1939. In 1975 he wrote the Icelandic Nature Conservation Council a letter offering to make them a gift of all the land surrounding the falls, which the Council wished to declare a nature reserve. One hydroelectric proposal from 1977 would have produced 2,444 gigawatt hours of electricity annually and doubled Iceland's production of hydroelectric power at the time (https://www.vulkaner.no/n/gullfoss/egullfoss.html).

Einar turned over part of the Brattholt property to the Nature Conservation Council, stipulating only that the land thenceforth be used as provided for by the Nature Conservation Act. In 1978, Sigríður's struggle was commemorated by a plaque at the falls. The Ministry of Culture and Education signed an agreement creating a nature reserve around Gullfoss on 9 March 1979 (https://www.vulkaner.no/n/gullfoss/egullfoss.html). Some say that although it is popular, the story that Sigríður saved the waterfall from exploitation is apocryphal (Wikipedia, Gullfoss). True or not, it serves as a rallying point for Iceland's environmentalists.

Visit

There is no admission fee, and the parking lot is also free.

Gullfoss is usually crowded during the summer season. Facilities, including food and restrooms, are capable of handling thousands of visitors every day. The paths are paved for the most part, but they can be slippery when wet and icy in winter (https://www.hiticeland.com/)

If you are lucky enough to visit on a sunny day, the mist rising from the falls will be filled with rainbows.

Gullfoss to Geysir: *Return south and west on R-35 to F-333/Haukadalsvegur and turn right (north) to* **Stop 17, Geysir** *(64.313432, -20.297292), for a total of 9.7 km (6.0 mi; 9 min).*

Stop 17 Geysir and Strokkur

The word "geyser" comes from here. Literally. Geysir, from the Icelandic "to gush," was first mentioned in 1647 by Bishop Brynjólfur Sveinsson (Bischoff, 2014). The hot spring lies on the slopes of Laugarfjall hill, and it is near other erupting springs like Strokkur, about 50 m (160 ft) south (Wikipedia, Geysir). Research investigating the hot springs deposits (sinter) indicates that Geysir has been active, on and off, for the last 10,000 years.

It appears that the activity of the geyser is controlled by earthquakes. During tectonically quiet times the plumbing system gets plugged with mineral precipitates. Earthquakes reinvigorate the system by unplugging the fractures and pores in the rocks. It was Robert Bunsen, of Bunsen burner fame, who explained how geysers work in 1846. He explained that mixing of cool surface water with heated water below is difficult due to the small fractures in the subsurface. This prevents cooling of the deep water. The heat coming from magma in the subsurface heats the deep groundwater until it reaches the boiling point, at which point gas bubbles develop. Eventually some bubbles reach the surface and reduce the pressure at depth, much like boiling water lifts the lid of a pot to release pressure. Reducing pressure on water that is *above* the boiling point results in instant vaporization of the superheated water and the geyser erupts violently. The longer the fracture system, the harder it is for the water to mix and the higher the water pressure at depth. Higher pressure raises the boiling temperature of water at depth, making eruptions further apart. On the other hand, hotter rocks raise water to the boiling point faster, leading to shorter intervals between eruptions (Bischoff, 2014).

By the late 1800s Geysir had become dormant, plugged by mineral deposits. An earthquake in 1896 rejuvenated the system. In 1910, it was active every 30 minutes; five years later eruptions were up to six hours apart, and, by 1916, they had all but ceased. An earthquake in 2000 stimulated the geyser again, probably by opening deep fissures. Eruptions up to 122 m (400 ft) high occurred for two days, an all-time record for geysers. Eruptions were taking place an average of eight times a day, yet by 2003 it was erupting only three times a day. In fact, the nearby Strokkur geyser now erupts quite a bit more than the original Geysir, erupting to 30 m (98 ft) high every few minutes. Strokkur's activity has also been affected by earthquakes, though to a lesser degree than Geysir. Which is why most photos that purport to show Geysir erupting actually show Strokkur (Wikipedia, Geysir).

Strokkur erupting. When I was there (2024) the geyser erupted for 5 to 10 seconds once every 10 minutes or so. Photo courtesy of Adam Prost.

Geysir to Thingvellir National Park: *Drive south and west on R-35; continue straight onto R-37/Laugarvatnsvegur and drive to Laugarvatn; at the roundabout west of town take the first exit to R-365; continue straight on R-36/Þingvallavegur and drive to the Þingvellir turnoff on the left; this takes you to* **Stop 18.1, Þingvellir/Thingvellir National Park Visitor Center and Rift** *(64.255705, -21.131824), for a total of 60.6 km (37.7 mi; 49 min).*

Stop 18 Thingvellir/Þingvellir National Park

Thingvellir is both the cultural and the geologic heart of Iceland. Iceland was first settled around 874, and by 930 the population had grown to the point where a general assembly was needed. This location was chosen as the "assembly fields," where the Althing (parliament) would meet. It would continue to meet there until 1798. Thingvellir National Park was established in 1933 to preserve the cultural heritage of the oldest parliament in the world. The park was later expanded and designated a UNESCO World Heritage Site in 2004 (Wikipedia, Þingvellir).

Why a World Heritage Site? In addition to the cultural heritage, the earth here is literally tearing itself apart. The park lies atop the Mid-Atlantic Rift, where the North American tectonic plate is

Eruption of Geysir as depicted in *Sulphur in Iceland*, by Charles Carter Blake, 1873. https://commons.wikim edia.org/wiki/File:BLAKE(1873)_p06_The_Great_Geysir.jpg Original held and digitised by the British Library.

slowly moving west away from the European plate. This rifting, about 3–7 mm/yr (0.12–0.28 in/yr), is 20%–30% of the total spreading between the plates (Slotznick, 2014). The balance of the spreading occurs in the eastern volcanic rift zone that extends from Heimaey through the Eldgjá Fissure Zone and on to the northern coast near Kópasker.

STOP 18.1 VISITOR CENTER AND MID-ATLANTIC RIFT

Stop at the Visitor Center either before or after visiting the rift valley, which is just outside the gift shop. The rift can be clearly seen in the fault-bounded grabens (down-dropped valleys) and fissures. At its narrowest, you can walk into the Mid-Atlantic spreading center and, stretching your arms on either side, touch both the North American and the European tectonic plates.

Thingvellir is smack-dab in the heart of the Western Volcanic Zone, which was the main axis of spreading between 6 and 2 Ma. The Eastern Volcanic Zone formed about 2 Ma and appears to have slightly more active spreading at this time. The basalts at Thingvellir are about 10,000 years old, and we can tell that faulting is even younger because those lavas have been offset vertically by up

Thingvellir as depicted in *Letters from High Latitudes*, by Frederick Temple Blackwood, Marquis of Dufferin and Ava, 1867. https://commons.wikimedia.org/wiki/File:Letters_from_high_latitudes_-_being_some _account_of_a_voyage,_in_1856,_in_the_schooner_yacht_%22Foam%22,_to_Iceland,_Jan_Mayen,_and _Spitzbergen_(1867)_(14759120846).jpg

to 40 m (131 ft) along the Almannagjá normal fault. The Thingvellir graben is the deepest graben, with the largest fault offsets in Iceland. The deep graben, lack of recent volcanics, and slow spreading rate have led some to suggest the center of spreading has moved to the Eastern Volcanic Zone. Regardless, this zone is still spreading, and it is still seismically active (Slotznick, 2014).

View east across the entire Thingvellir rift valley. The valley is about 700 m (2,300 ft) wide here.

Western margin of the active Thingvellir rift.

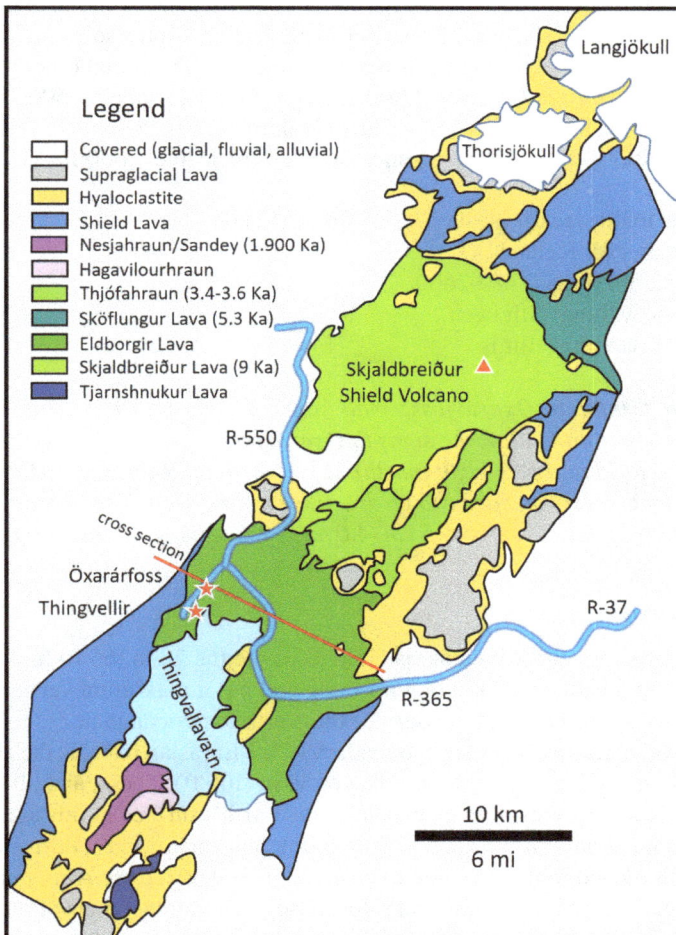

Geologic Map of the Thingvellir area. Faulting is too dense to show. Modified after Samundsson, 1992.

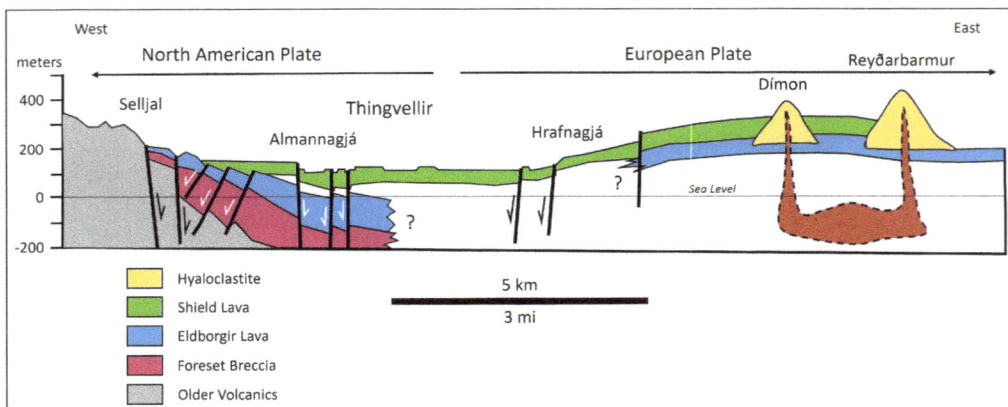

Cross Section of the Thingvellir area. Location is shown on the previous figure. Modified after Sæmundsson, 1992.

VISIT

The Thingvellir Visitor Center has interpretive displays and a short film that explain the historical and geological significance of the area. There is a small cafeteria and a souvenir shop. The park has camping, horseback riding, and hiking trails that lead to the main rift and other points of interest.

Entrance to the exhibits (2024) costs 1,000 Ikr (US $7.23) for adults, 500 Ikr (US $3.62) for seniors (67 and older) and students; and it is free to children under 18 years.

Parking costs 750 Ikr (US $5.42) for a single family vehicle and 1,000 Ikr (US $7.23) for a 6–8 passenger vehicle.

Hours: 10:00 to 16:00 weekdays; 10:00 to 17:00 weekends
Address: Bláskógabyggð, Iceland
Phone: +354-482-3613; (354) 482-2660
Website: https://www.thingvellir.is/en
Email: thingvellir@thingvellir.is

Thingvellir Visitor Center to Öxarárfoss: *Walk north 1.4 km (0.9 mi; 20 min) along the path behind the Visitor Center and follow the signs to* ***Öxarárfoss.***

Or return west to R-36 and turn right (north); drive north on R-36 to R-361/Öxarárfoss turnoff on the right; turn right (south) and drive to Öxarárfoss parking area. This is ***Stop 18.2, Öxarárfoss*** *(64.265002, -21.113985), for a total of 6.5 km (4.0 mi; 9 min). Walk 210 m (700 ft) west to the falls.*

STOP 18.2 ÖXARÁRFOSS

The Öxará River flows over the Almannagjá Fault to create the 20 m (66 ft) high Öxarárfoss (Axe Falls). The Almannagjá Fault is 7.7 km (4.8 mi) long and has a maximum vertical offset of 30–40 m (100–130 ft). It marks the eastern boundary of the North American plate.

Across the graben, marking the western boundary of the Eurasian plate, is the Hrafnagjá Fault. It is 11 km (6.8 mi) long and has a maximum offset of 30 m (100 ft). These, and all the fissures in the Western Volcanic Zone, are a result of extension of the Earth's surface in this area.

Some say this is not a natural falls. According to the Sturlunga Sagas (12th and 13th centuries), the Öxará River was moved to this location to channel the water into the Almannagjá Valley in the 9th century. The purpose was to provide water for those attending meetings of the Icelandic parliament, the Althing.

Scenic Öxarárfoss.

The name supposedly comes from the axe that killed a notorious female troll who terrorized the inhabitants and visitors to the area (Cohen, 2018). The dark history of the lovely falls continued in later centuries. In the 16th and 17th centuries, the calm, deep waters at the bridge over Öxará were used by Icelandic courts to drown women convicted of adultery, and the pool is still called Drekkingarhylur, the Deep Drowning Pool. Men were hanged here, and women were drowned by putting them in a sack and holding them under the water until they stopped moving. It is said that 18 women were drowned here (Hit Iceland, Öxarárfoss).

The western margin of the Thingvellir rift is occupied by the scenic and energetic Öxarár River for some distance below the falls.

Proposed Saga Geopark

A new UNESCO Global Geopark is being proposed near Reykholt, about 48 km (30 mi) north of Thingvellir. The proposed Saga Geopark includes 2,270 km^2 (876 mi^2) of lowlands to highlands at the edge of the Southwest Iceland volcanic rifts. The area is characterized by subglacial and intra-glacial volcanism, extensive low-temperature geothermal activity, and includes the world's largest boiling hot water spring at Deildartunguhver. The geopark has historical importance as the site of the Reykholt farm where Snorri Sturluson, the most famous saga writer in Iceland, lived and worked in the 13th century (Saga Geopark Project, 2015).

The southeast part of the proposed geopark is dominated by subglacial and intra-glacial volcanism less than 800,000 years old. Several post-glacial lava flows less than 11,000 years old are found in the area. The youngest and largest is Hallmundarhraun, which erupted around 940 CE. It contains some of the largest lava tube caves in the world. The best known is Surtshellir (Black's Cave, "hellir" being the Icelandic word for cave), about 2 km (1.3 mi) long. The area is home to a number of large shield volcanoes and table mountains composed of olivine-rich basalts, indicating a deep mantle source. The shield volcanoes were created during interglacial times; as mentioned earlier, eruptions under ice create stapis, or table mountains.

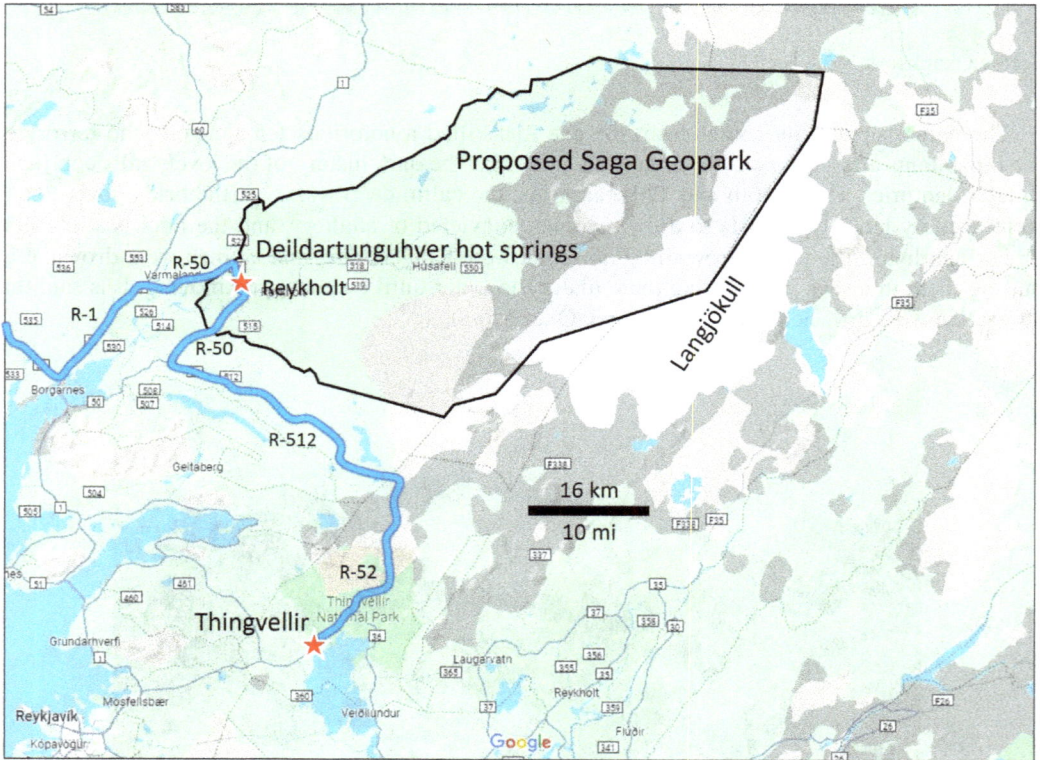

Proposed Saga Geopark.

The northwest part of the geopark is dominated by very young Quaternary lavas, 800,000 to 3,000 years old. These are closest to the active rift zone. Tertiary basalts older than 3 Ma are found farther west and north.

The main geologic features in the park are the extinct Húsafell and the active Prestahnjúkur central volcanoes. Prestahnjúkur is the only active volcanic system in the geopark. Earthquake swarms occur, on average, once a year, indicating rifting events along north-northeast trends parallel to the faults and fissures of the Western Volcanic/Rift Zone.

The oldest space heating system in Iceland was discovered during archaeological excavations at Reykholt farm in Borgarfjördur and date from the 13th century. The hot pool of the saga writer Snorri Sturluson still exists. In the early 1980s the largest hot spring in the world, Deildartunguhver ("hver" is a boiling hot spring), was piped about 70 km (44 mi) for use as a municipal heating system for the towns of Akranes and Borgarnes. Most farmhouses in this part of Iceland are heated by natural hot springs.

The Reykholt area is the largest low-temperature geothermal area in Iceland. Scientists began investigating geothermal activity in the 18th and 19th centuries. The most famous of these was the chemist Robert Bunsen, who spent the summer of 1846 in Iceland collecting geothermal water and rock samples. He wrote several articles about the nature of geothermal fields in Iceland, work which earned him the title "father of geothermal research."

Glaciers are an important part of the proposed Saga Geopark. Langjökull ("Long Glacier") is the second largest glacier in Iceland after Vatnajökull. It is a major ice cap with numerous outlet glaciers (Saga Geopark Project, 2015).

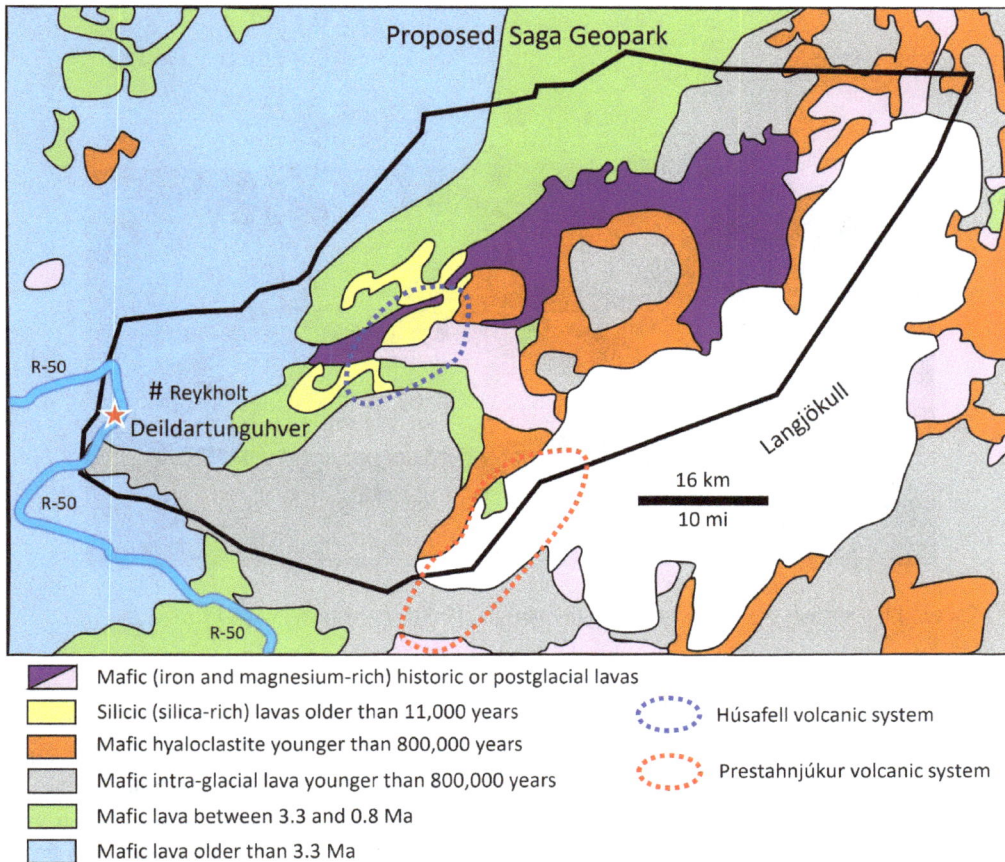

Geologic map of the proposed Saga Geopark. Modified after Saga Geopark Project, 2015.

Öxarárfoss to Krauma – Deildartunguhver Hot Springs: Drive north on R-361 to R-550; continue straight on R-550; bear right onto Uxahryggjavegur; turn left (northwest) on R-52; drive on R-52 to R-50/Borgarfjarðarbraut; turn right (northeast) on R-50 and drive to turnoff to Krauma/555 Deildartunguvegur on the left. This is Stop 19 Krauma Hot Springs (64.663387, -21.410085) for a total of 78 km (48.4 mi; 1 hr 13 min). The large parking area is free and holds 50 to 80 vehicles.

STOP 19 KRAUMA HOT SPRINGS, DEILDARTUNGUHVER

The Krauma hot springs at Deildartunguhver are considered the highest-flow hot springs in Europe at 180 l/s (2,853 gal/min). The literally boiling hot water issues from the springs at 100°C (212° F) and is diverted to five pools of varying temperatures. The different temperatures are achieved by mixing with cold water from the Rauðsgil River, which originates at the Ok Glacier (the smallest glacier in Iceland). There is also a sixth, cold water plunge pool that is 5°–8°C (41°–46° F) for those with no dermal nerve endings. In addition to the pools there is a relaxation room where you can settle by a fireplace and listen to soft music.

Deildartunga belonged to the Sigurbjörg family for 200 years until it was expropriated by the government of Iceland. In addition to the hot spring pools, about 1941 a greenhouse was built to utilize heat from the springs to raise vegetables.

One of several hot springs pools at Krauma resort and spa. Photo courtesy of Adam Prost.

VISIT

It is recommended that guests book the resort in advance, but you can buy tickets in person, subject to availability. They request that groups of 10 or more people to send an email to krauma@krauma .is. You can also contact krauma@krauma.is to reserve a table and for special offers if your group is larger than 15 people. Beverages, including wine, beer, and non-alcoholic beverages, can be purchased at the pools. Customers can also rent towels, bathing suits, and bathrobes.

Address:	Krauma is located at Deildartunguhver, 320 Reykholt on Route 50
Phone:	+354-555-6066
Website:	https://www.krauma.is/en or https://www.krauma.is/en/deildartunguhver
Email:	krauma@krauma.is.
Phone:	+354 5556066
Hours:	
	Spa: 11:00 to 21:00 daily
	Restaurant: 11:00 to 21:00 daily
Entrance (2024):	6,800 kr adult (17 years and older; ~$48.57)
	4,990 kr Seniors (over 67; ~$35.64)
	3,400 kr teens (13–16 years; ~$24.28)
	450 kr children (0–12 years; ~$3.21)
Rentals:	1,200 kr for bath towels (~$8.57)
	1,200 kr swimwear (~$8.57)
	1,600 kr bathrobe (~$11.43)

STOP 20 SNÆFELLSNES PENINSULA

The Snæfellsnes Peninsula is an area 80 km (48 mi) long and 10–20 km (6–12 mi) wide extending in an east-west orientation at the far western end of Iceland. About 4,000 people live in five fishing villages and on farms. In order to promote sustainable development of economic opportunities, preserve the environment, and maintain the local population, the Snæfellsnes Regional Park was established in 2014 (Andi Snæfellsness, Snæfellsnes Regional Park).

Geologically, the Snæfellsnes Peninsula corresponds to a 120 km (72 mi) long zone of magmatism that lies about 150 km (90 mi) west of Iceland's Western Volcanic Zone. The peninsula consists of upper Pleistocene (less than 2 Ma) and Holocene volcanic units that lie uncomfortably over Neogene tholeiitic flood basalts erupted from the now dormant Snæfellsnes–Húnaflói Rift Zone. Before about 6 Ma, the Mid-Atlantic spreading center went through the Snæfellsnes Peninsula. This Snæfellsnes–Húnaflói Rift Zone was abandoned when the rift migrated eastward to its current position in the Western Volcanic Zone. For that reason, the Snæfellsnes Peninsula no longer has significant active rifting and crustal extension, and it is generally considered aseismic (no earthquakes). Right-lateral strike-slip faults along the length of the peninsula are thought to be a result of unequal spreading between the Western, Eastern, and Northern Volcanic Zones (Maruszczak, 2018).

The lavas of the Snæfellsnes Peninsula are largely alkali basalts. Alkaline rocks have an excess of alkali content (i.e., Na_2O and K_2O) over silica (SiO_2). They are less common than the tholeiitic basalts that make up most of Iceland (Sykora, 2016).

There are two main kinds of volcanism on the peninsula: basaltic eruptions at scattered cinder cones and large central volcanoes that align roughly west-northwest along the trace of the Snæfellsnes–Húnaflói Rift Zone. The younger volcanics have been divided into three distinct systems, each categorized by their central volcano. From east to west these are Ljósufjöll, Lýsuskard, and Snæfellsjökull.

The Ljósofjöll Volcanic System is the farthest east and spans an area of the Snæfellsnes Peninsula greater than the Snæfellsjökull and Lysuskard systems combined. It has a distinct east-southeast alignment. The Ljósufjöll central volcano has produced the greatest volume of silicic eruptions on the Snæfellsnes Peninsula. The oldest silicic material is Pleistocene (~680 Ka). Most of the eruptions in the system are alalkali basalts. Post-glacial (younger than 13 Ka) eruptions are cinder cone basalts. Lýsuskard and Ljósufjöll do not produce fissure eruptions, as seen in the other volcanic zones, but rather produce eruptions around a large central volcano (Maruszczak, 2018).

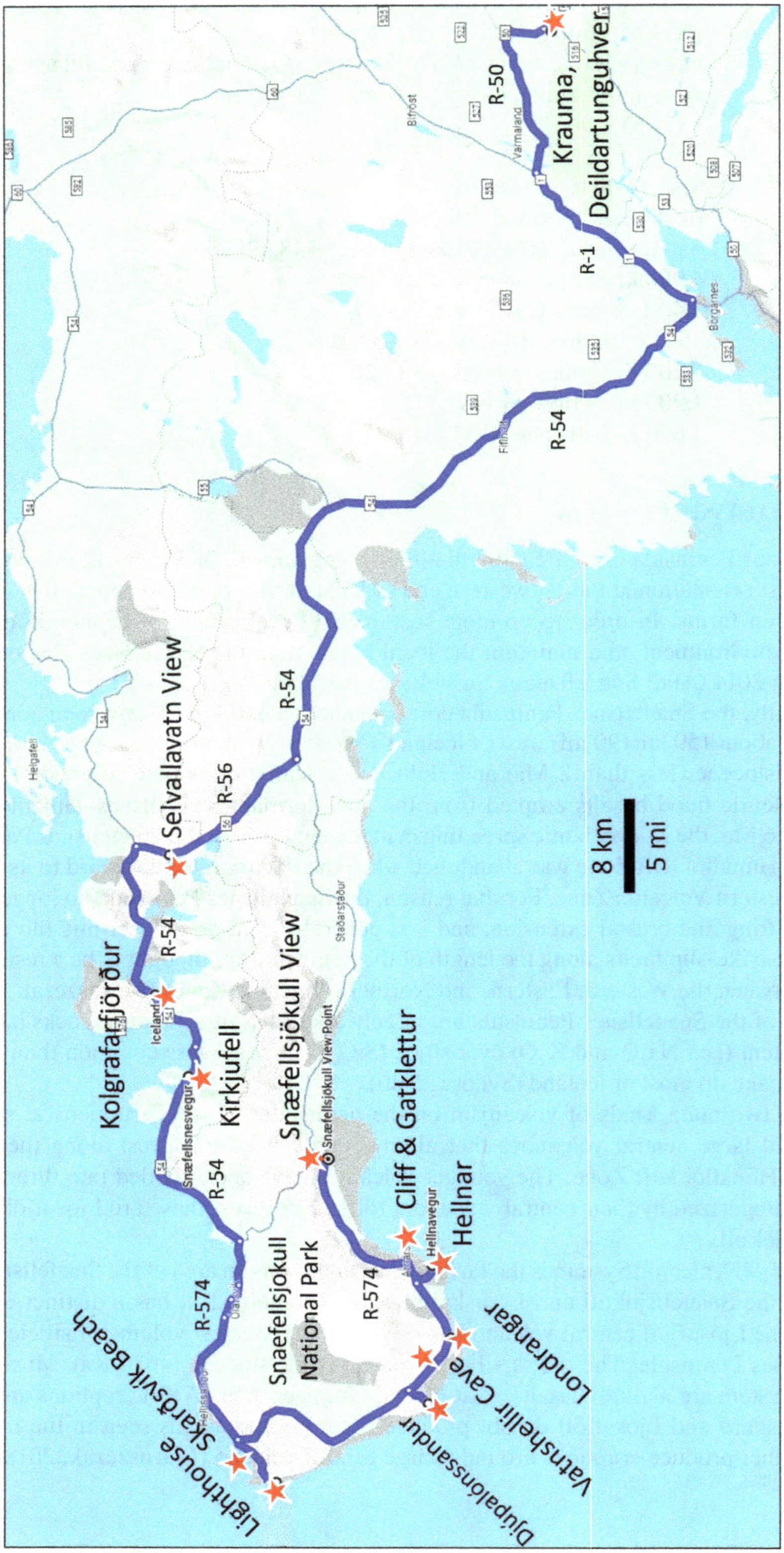
Krauma, Deildartunguhver
R-50
R-1
R-54
R-54
R-56
Selvallavatn View
R-54
Kolgrafarfjörður
Kirkjufell
R-54
Snæfellsjökull View
Cliff & Gatklettur
Hellnar
Snaefellsjokull National Park
R-574
R-574
Londrangar
Vatnshellir Cave
Djúpalónssandur
Skarðsvik Beach
Lighthouse
8 km
5 mi

The west-northwest Lýsuskard System in the central peninsula is ~30 km (18 mi) long and consists primarily of late Pliocene flood basalts up to 2 Ma with pockets of rhyolitic eruptions and post-glacial basalts. The system is now essentially inactive.

In the westernmost part of the peninsula, the Snæfellsjökull system is characterized by the large Snæfellsjökull central volcano. This stratovolcano is 1,446 m (4,744 ft) high and is capped by the Snæfellsjökull glacier. It gained fame as the volcano featured in Jules Verne's *Journey to the Center of the Earth.*

Snæfellsjökull is a long-lived volcanic center on the peninsula that has erupted magma ranging from basalt flows to rhyolite tephra (Sykora, 2016). Most of the system is covered by post-glacial lava flows that erupted from fissures and cinder cones around the flank of the mountain. The oldest dated rocks in this system are around 842 Ka (Maruszczak, 2018).

Krauma to Selvallavatn View: Return to R-50 and turn left (north); drive on R-50 to R-1/Þjóðvegur; turn left (southwest) on R-1 and drive to R-54/Snæfellsnesvegur at Borgarnes; turn right (west and north) on R-54 and drive to R-56/Vatnaleið; turn right (north) on R-56 and drive to **Stop 20.1** ***Selvallavatn View** (64.94167, -22.90440) pullout on the left for a total of 112 km (69.3 mi; 1 hr 20 min). There is free parking for about 12 cars, and there are picnic tables but no washrooms.*

STOP 20.1 SELVALLAVATN VIEW

Lava flowed into Selvallavatn Lake from cinder cones to the north. This is all part of the Ljósufjöll volcanic system at the east end of the Snaefellsnes Peninsula. The Ljósufjöll system comprises a group of olivine basaltic cinder cones and lava flows along short west-northwest fissures that extend over roughly 90 km (54 mi). The volcanic field contains the largest outcrops of silica-rich rhyolite and trachyte in the Snæfellsnes volcanic zone (trachyte is a rock composed mainly of sodium and potassium feldspars where the crystals are aligned due to magma flow). These rocks erupted from the middle to late Pleistocene continuing into the Holocene. The latest eruption took place in the year 960 (Smithsonian, Global Volcanism Program, https://volcano.si.edu/volcano.cfm?vn=370030).

Satellite images show the Gráakúla scoria cone, which erupted about 4,000 years ago, and its lava flow that entered Selvallavatn Lake.

View west to Selvallavatn from the picnic area. Lava flowed into the lake from the north.

Satellite image of Selvallavatn Lake and lava flow, as well as prominent cinder cones. ©2024 Google Earth; Maxar Technologies.

Selvallavatn to Kolgrafarfjörður: *Continue north on R-56 to R-54; turn left (west) on R-54 and drive to* ***Stop 20.2, Kolgrafarfjörður*** *(64.95167, -23.15868) pullout on the left for a total of 17.7 km (11 mi; 14 min). There is free parking for about eight vehicles, but no facilities.*

Stop 20.2 Kolgrafarfjörður

Kolgrafarfjörður is a fjord on the north coast of Snæfellsnes. A fjord is a long, narrow, generally steep-walled arm of the sea, usually along mountainous coasts, caused by rising sea levels drowning Ice Age valleys. This fjord is only about 40 m (130 ft) deep at its deepest point, and there are many shallows. It is best known for the herring that winter here and the orcas that feed on them.

In December 2012, there was a massive herring die-off in the fjord. An estimated 30,000 tons of fish died due to lack of oxygen, due either to calm weather and lack of turnover in the water column or to an overabundance of biota. A similar herring die-off occurred in the fjord during February 2013, with an estimated 22,000 tons of dead fish. I can only begin to imagine what that smelled like.

Panoramic view southeast to the inland part of Kolgrafarfjörður.

Kolgrafarfjörður to Kirkjufell Mountain: Continue west on R-54/Snæfellsnesvegur through the village of Kirkjufell to **Stop 20.3 Kirkjufell Mountain** *(64.928034, -23.313187) parking area on the left side of the road for a total of 9.7 km (6.0 mi; 11 min). This is a large pay parking lot, with room for about 30 cars and 8 buses, but there are no facilities. Walk 240 m (800 ft) to the falls.*

Stop 20.3 Kirkjufell Mountain and Kirkjufellsfoss

Kirkjufell mountain rises 463 m (1519 ft) above the small fishing town of Grundarfjörður. While it is the most photographed mountain in Iceland, it is mostly known as "Arrow Head Mountain" on the TV series *Game of Thrones* (Share, 2018; Renyard, 2019; Wikipedia, Kirkjufell).

Kirkjufell is a unique glacial landform, a natural pyramid peak. Geologists call it a "nunatak," the summit or ridge of a mountain that protruded above an ice field or glacier. The shape is a result of glaciers grinding away at the rock from both sides. Nunatak is the aboriginal Inuit word for a high rock sticking out of glaciers.

The layers of Kirkjufell are the result of several volcanic eruptions (flows and hyaloclastites) interlayered with glacial tills, all younger than 2 million years.

Kirkjufell is a challenging climb, but fish fossils have been found at the top. The fossils indicate uplift as a result of crustal rebound after melting of glaciers. To get to the top you need to book an experienced guide, because climbing this peak has proved fatal on several occasions.

You can easily walk to the nearby Kirkjufellsfoss waterfall that drops 16.4 m (54 ft) over three falls. If the sun is out, look for the iridescent rainbows in the mist.

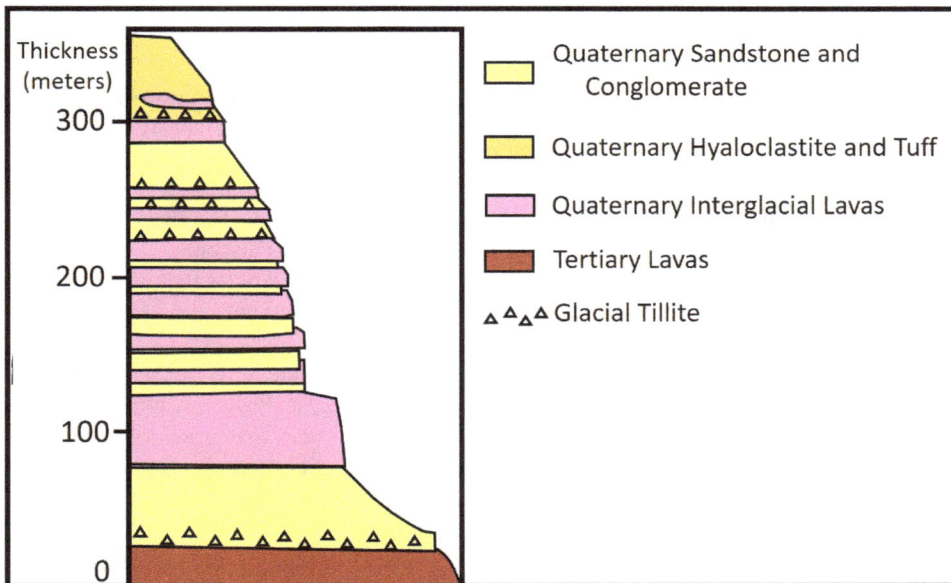

Stratigraphy of Kirkjufell Mountain. From a roadside geology display.

Kirkjufell Mountain and Kirkjufellsfoss.

SNÆFELLSJÖKULL NATIONAL PARK

Snæfellsjökull National Park was established on 28 June 2001 to protect and conserve the area's landscape, plants, and animals as well as important historical relics. The park covers 170 km² (65 mi²). The Búðahraun, Arnarstapi, and Hellnar nature reserves and the Bárðarlaug natural monument all fall under the jurisdiction of the national park. Búðahraun lava field in the southern and eastern Snæfellsnes Peninsula was designated a nature reserve in 1977. The coastal area around Arnarstapi and Hellnar was designated a nature reserve in 1979.

The park is dominated by Snæfellsjökull peak and its glacial mantle. The peak rises to 1,446 m (4,744 ft); it was first climbed in 1754 by Eggert Ólafsson and Bjarni Pálsson.

The Saga of Bárður Snæfellsás, dating to the early 14th century, reports that after some serious family issues Bárður gave up on human companionship and vanished onto the ice cap, after which he became the guardian spirit of Snæfell.

Jules Verne knew of this mountain and used its abundant lava caves as the setting for *Journey to the Center of the Earth*. First published in 1864, the protagonist, Professor Otto Lidenbrock, is an eccentric German geologist who believes there are volcanic tubes that reach to the center of the earth. He, his nephew Axel, and their Icelandic guide Hans descend into Iceland's Snæfellsjökull Volcano, where they have many adventures and encounter dangers, including cave-ins, subpolar tornadoes, an underground ocean, and Mesozoic creatures.

As an aside, the Icelandic word for cave, *hellir*, is probably related to the Old English word for the underworld, *hel* or *helle*, from which we get our word "hell."

The main vegetation in the park is the thick moss that covers most rocks, while small birch and mountain ash grow in lava depressions where they are protected from the cold wind. Wild berries are plentiful in late summer. Eleven of the 16 species of fern found in Iceland grow in the shadow of the mountain.

Snæfellsjökull National Park has outstanding whale-watching. Killer whales, minke whales, and porpoises are commonly spotted off the Snæfellsnes coast. Foxes are common, and mink can be seen near the coast where food is abundant.

Birding is excellent. Eiderduck is the most common species of duck. Large colonies of Arctic tern nest near Arnarstapi, Rif, and Öndverðarnes, with the colony at Rif being one of the largest in Europe. The tern spends the winter along the Antarctic ice shelf, flying up to 40,000 km (25,000 mi) per year. The most common birds include guillemot, razorbill, fulmar, kittiwake, and shag. Shag can be seen at Arnarstapi, where you can also see nesting kittiwakes. Black guillemots are most often spotted at Malarrif and Lóndrangar. The great black-backed gull, herring gull, and lesser black-backed gull are common. Rednecked phalaropes are often spotted on the ponds above Pumpa in Arnarstapi. Bird cliffs are accessible at Þúfubjarg and Saxhólsbjarg, but be careful not to disturb the birds, and mind your footing.

GEOLOGY

The area contains geologic formations from almost every era of Iceland's past. Most formations in and around the National Park are geologically "modern." The mountain is an active stratovolcano, built up over the last 800,000 years. The crater under the summit is 200 m (656 ft) deep and buried under ice. The glacier has been shrinking in recent years and it is now approximately 11 km^2 (4.25 mi^2). The last eruption was nearly 1,800 years ago, and it covered the northern part of Snæfellsnes and the Westfjords (northwestern Iceland) in ash.

Snæfellsjökull is the central volcano in a 30 km (18 mi) long volcanic system that extends from Öndverðarnes in the west to Mælifell in the east. The hills north of the glacier, around Bárðarkista, are palagonite tuff (altered basaltic ash), formed during eruptions under a glacier.

Snæfellsjökull (center right) looking southwest from along the north coast near Búlandshöfði View Point.

As suggested in Verne's book, the area has many lava caves. Visitors are advised not to enter the caves unless accompanied by someone familiar with them. Búðaklettur, an 88 m (290 ft) high cone that sourced the Búðahraun lavas some 5,000–8,000 years ago, has a few caves, the best known of which is Búðahellir. It was originally thought that Búðahellir was bottomless, and that there was a tunnel connecting it to the sea. The Búðahraun lava flows are known for their prominent mineral crystals. The yellow-green mineral is olivine (the common variety of the gem peridot), the white mineral is plagioclase feldspar, and the black mineral is pyroxene. The beach south of Búðir is almost pure green olivine sand.

VISIT

Entrance is free, as with all national parks and nature reserves in Iceland. Guided walks and tours are available at various spots. There are no campgrounds within the national park, but hikers and cyclists can pitch tents at the Arnarstapi campground. Cycling is permitted on roads and marked tracks, and horse riding is permitted on designated riding trails. Fires are not allowed, and dogs and other pets must be kept on a leash.

Visitor centers are located in Malarrif and Ólafsvík.

Gestastofa Visitor Center
Address: Malarrif, Snæfellsbær
Hours: 11:00 to 16:00 daily
Phone: +354-436-6888
Website: https://www.ust.is/english/snaefellsjokull-national-park/
Winter Hours, Visitor Center at Malarrif: 12:00 to 15:00 Monday – Friday; 12:00 to 16:00 Saturday and Sunday
Phone: National Park, office: +354-436-6860
Information: +354-436-6860
Email: snaefellsjokull@ust.is

Ólafsvík Visitor Center
Address: Kirkjutún 2, 355 Ólafsvík
Hours: 9:00 – 17:00 Monday–Friday, June–August, reduced hours rest of year
Phone: +354-433-6929
Email: atthagastofa@snb.is

Kirkjufell to Skarðsvík Beach: *Continue driving west on R-54/Snæfellsnesvegur to R-574/ Útnesvegur; continue straight (west) on R-574 to turnoff to Skarðsvík Beach on the right; turn right and drive to **Stop 20.4, Skarðsvík Beach** (64.880792, -23.985674), for a total of 38.8 km (24.1 mi; 36 min). There is free parking for 12 to 15 cars on both sides of the road; picnic tables, but no facilities.*

STOP 20.4 SKARÐSVÍK BEACH

You are less than 3 km (2 mi) from the westernmost point on the Snæfellsnes Peninsula. This is one of the few golden sand beaches in Iceland, where most beaches are black volcanic sand or pebbles. I have not studied the sand, but suspect it is composed mainly of shell fragments and/or is derived from more quartz and feldspar-rich lavas. Skarðsvík Beach is also known for its black basalt cliffs and sea cave. Lava tube caves are common throughout this area. Turn right (northeast) at the water and walk to where the sand ends. There is a small cave where, in 1962, a burial mound dating to the 10th century was discovered. Along with bones of an 18–25-year-old male there was a sword, spearhead, broken knife, and shield boss. These items can now be seen in the National Museum.

Skarðsvík Beach with rare golden sand and lava flows extending to the sea.

An explorer stands next to a lava tube in the sea cliffs, Skarðsvík Beach.

Skarðsvík Beach to Skalasnagi Lighthouse: *Continue driving southwest on a graded gravel road through broken lava flows; at the sea cliffs turn left (south) and drive to **Stop 20.5, Skalasnagi Lighthouse** (64.86398, -24.03874) for a total of 3.9 km (2.4 mi; 12 min). There is free parking for 5 to 10 cars; there are picnic tables but no washrooms.*

STOP 20.5 SKALASNAGI LIGHTHOUSE

Scenic orange Skalasnagi Lighthouse sits atop shear sea cliffs favored by nesting seabirds of all kinds. There is no access to the beach – there is no beach. This is mostly a birdwatching site. A large natural arch punctuates the cliffs. The drive reveals a rugged topography formed by the collapse of subsurface lava tubes within the basaltic flows.

Sea cliffs and natural arch at Skalasnagi Lighthouse.

Skalasnagi Lighthouse to Djúpalónssandur Beach: *Return north and east past Skarðsvík Beach to R-574/ Útnesvegur; turn right (south) onto R-574; take R-574 south to R-572 and turn right (southwest); drive to* **Stop 20.6, Djúpalónssandur Beach** *parking area (64.753351, -23.895213), for a total of 25.6 km (15.9 mi; 32 min). There is free parking for about 20 cars, picnic tables, and washrooms.*

STOP 20.6 DJÚPALÓNSSANDUR BEACH

Djúpalónssandur is a black pebble beach that was once a thriving fishing village. Today it is a ghost town. The dominant rocks in the cliffs are mugearite (an oligoclase-bearing basalt consisting of olivine, apatite, and various oxides) and benmoreite (a sodium-rich and silica-poor trachyandesite). Flow banding is evident in the dark gray lava. The black rocks signify lava that cooled quickly and is glassy, whereas red lava represents lava that came into contact with water and oxidized, or rusted.

The beach also contains the remains of the fishing trawler *Epine*, shipwrecked off the coast here on 13 March 1948. The wreck is maintained as a memorial to the 14 men who died that day (Wikipedia, Djúpalónssandur).

Djúpalónssandur black sand beach.

The largest rock on the beach is a large lava block called Söngklettur, the Singing Rock, perhaps because the wind whistles through the cracks. The red rock is said to be the church of the elves. At the water's edge at the south end of the beach is the Kerling, a petrified troll woman carrying a load of fish on her back.

Walk to the tidal lagoon called Svörtulón, the Black Lagoon, at the beach. On the edge of the lagoon is Gatklettur, a window rock hollowed by weathering and erosion.

About 1 km (0.6 mi) west of Djúpalónssandur beach is Dritvík Bay. From the 16th century until the mid-19th century Dritvík was the largest seasonal fishing station in Iceland. There are still some ruins there (Wikipedia, Djúpalónssandur; Fun Iceland, Djúpalónssandur). From early April to mid-May hundreds of men and women would row out to sea to fish the fertile waters. The name Dritvik means "excrement bay," probably referring to the abundant bird droppings.

There are four "lifting stones" on the beach that were used by the Dritvik fishermen to qualify men for work on fishing boats. The Fullsterkur ("full strength") weighs 154 kg (340 lb), Hálfsterkur ("half strength") at 100 kg (220 lb), hálfdrættingur ("weakling") at 54 kg (119 lb), and Amlóði ("Useless") 23 kg (51 lb). The hálfdrættingur was the minimum weight a man would have to lift to hip height to qualify. You are welcome to try to lift the stones.

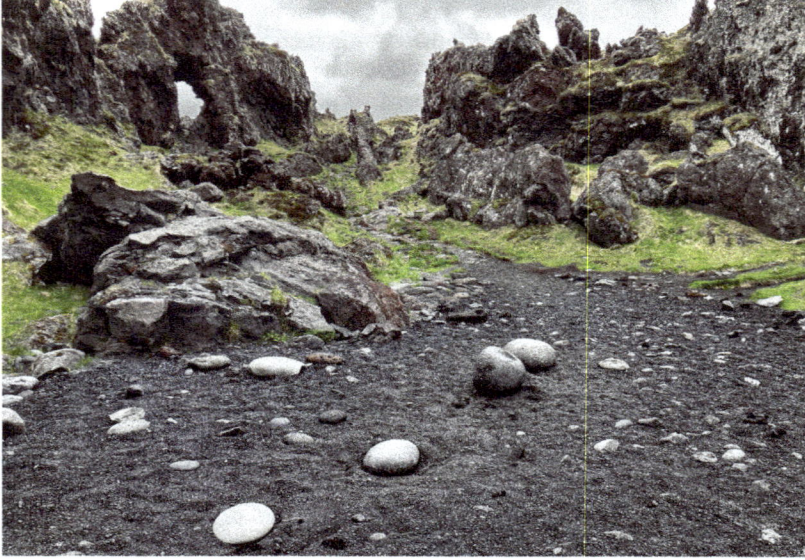

Djúpalónssandur, window rock, and lifting stones.

Djúpalónssandur Beach to Vatnshellir Cave: *Return to R-574 and turn right (southeast); drive to parking area on the left for **Stop 20.7, Vatnshellir Cave** (64.747823, -23.818036), for a total of 5.7 km (3.5 mi; 6 min). There is a large free parking area, but no washrooms.*

STOP 20.7 VATNSHELLIR CAVE

The area around Snæfellsnesjökull glacier contains hundreds of lava tubes. Local lore says that trolls lived in many of these caves.

Vatnshellir Cave is an 8,000-year-old lava tube created by the eruption of a nearby crater in the Purkhólar Group. Lava tubes form as molten lava moves downhill, cooling on the surface and creating a crust at the top of the flow. After the eruption stops, the lava continues to drain toward the sea, leaving behind an empty chamber.

The multihued walls contain red and ocher iron-rich rocks, yellow sulfur minerals, and green copper oxides (Guide to Iceland, Vatnshellir Cave).

Vatnshellir ("Water Cave") is a 205 m (670 ft) long lava tube with three levels. The uppermost (first level) has a partially collapsed roof, is just below the surface, and is 35 m (115 ft) long. The lower parts are on two levels and are in good condition. This part of the cave is called the Undirheimar (Underworld). The middle level is 12–20 m (39–66 ft) below the surface and is about 100 m (330 ft) long. At the southern end of the middle level there is a 12 m (39 ft) vertical lavafall that drops into the lowermost level. This part of the cave is about 32 m (105 ft) below the surface, almost horizontal and just over 70 m (230 ft) long. The Undirheimar was opened to the public in 2010 as "the first 20 vertical metres of the route to the centre of the earth" (Stefánsson, 2010).

VISIT

Vatnshellir is a protected site, as it is within the Snæfellsjökull National Park. For that reason Summit Adventure Guides, the tour operator, is the only group with permission to take you into Vatnshellir Cave. They give 45-minute guided tours hourly during the summer and three times a day during the fall and winter. You are advised to dress warm, since the temperature in the cave is just above freezing year-round. Hiking shoes/boots are recommended, but sneakers are acceptable.

Gloves are recommended for your safety and comfort. The tour is available all year. You can buy tickets at the cave, but to guarantee a space you should book ahead of time. The operator supplies helmets and flashlights; headlamps are not allowed. The cave is accessible for most people who are able to walk unsupported on uneven ground and are comfortable with the two spiral staircases in the cave. Note that the closest restrooms are at Malarrif Visitor Center 3 km (1.8 mi; 4 min) away (Summit Adventure Guides, Vatnshellir Cave). There is a large parking area, but no washrooms.

Prices (2024): Adults Isk 5,400 ($39.00); Seniors/Students Isk 4,200 ($30.00); Teenagers (12–17) Isk 2,400 ($17.00); Children (5–11) Free. Children under five years are not allowed on the tour.
Address: Gufuskalar, 360 Snaefellsbaer, Iceland Hours: 9:00 to 16:00 weekdays.
Email: info@summitguides.is
Phone: +354-787-0001
Website: https://www.summitguides.is/vatnshellir-cave-op1r6

Entrance to Vatnshellir Cave, a series of conjoined lava tubes.

Vatnshellir Cave to Lóndrangar Viewpoint: *Turn left (east) on R-574/Útnesvegur and drive 2.4 km (1.5 mi; 2 min) to **Stop 20.8, Lóndrangar Viewpoint** (64.737472, -23.775518). A 250 m (825 ft) stroll takes you to the cliff overlook. There is free parking for about 20 cars and picnic tables, but no washrooms.*

STOP 20.8 LÓNDRANGAR VIEWPOINT

Lóndrangar are two basalt pinnacles along the sea cliffs east of Malarrif. They are thought to be eroded remains of a volcanic plug and crater whose softer rocks have mostly eroded away. A volcanic plug, or neck, is the solid magma remaining in a volcano's conduit or plumbing system after volcanic activity has ended. The taller rock is about 75 m (246 ft) high and the shorter one is about 61 m (200 ft) high. Ásgrímur Bergþórsson climbed the higher rock in 1735; the shorter pinnacle wasn't climbed till 1938.

The cliffs at this stop are columnar basalt at the base, overlain by a light brown volcanic sediment, overlain, in turn, by more basalt flows.

Cliffs at Lóndrangar. Dark basalt flows occupy the lower parts of the sea cliffs, whereas light-colored volcanic sediments occupy the upper cliffs.

The cliffs around Lóndrangar are home to many seabirds including the puffin (during the warm months), northern fulmar, common murre, and black-legged kittiwake. Puffins nest on the higher cliffs.

The area around the cliffs is said to belong to elves, and for that reason the farmers in the area have never used the fields on the slopes here (Fun Iceland, Lóndrangar; Wikipedia, Lóndrangar; Guide to Iceland, Lóndrangar).

This area offers some of the best views of Snæfellsjökull Glacier on the Snæfellsnes Peninsula.

Lóndrangar to Hellnar Viewpoint: *Return to R-574/Útnesvegur turn right (east); drive to R-5730/ Hellnavegur and turn right (south); drive to **Stop 20.9, Hellnar Viewpoint** (64.750606, -23.646041) at the sea cliffs for a total of 7.7 km (4.8 mi; 7 min). There is free parking for about 15 cars, picnic tables, but no toilets.*

STOP 20.9 HELLNAR VIEWPOINT

As you turn onto R-5730 there is a low scoria ridge on the right (scoria is a basaltic rock ejected from a volcanic vent that cooled quickly and is full of gas bubbles). This is Bárðarlaug, an ancient lake-filled volcanic crater that has been scoured by a glacier.

Hellnar is an old fishing village on the southwest Snaefellsnes Peninsula. At one time it had been one of the largest fishing stations on the peninsula, dating back at least to 1560 (Guide to Iceland, Hellnar).

The rock at Hellnar, as at Djúpalónssandur, consists of mugearite and benmoreite. Again, you can see flow banding in the lava (Sykora, 2016).

The seacliff, Valasnös, is known for its columnar basalt and sea caves that probably started as lava tubes. The cliff's appearance is sensitive to changing light conditions throughout the day.

Large colonies of birds nest in the area, although puffins do not.

On a clear day there is a good view of Snæfellsjökull from this stop.

Sea cliff with columnar basalt and a sea cave, Hellnar.

Flow banding around an old lava tube in the cliff at Hellnar.

*Hellnar Viewpoint to Gatklettur and Cliff Viewpoints: Return north on R-5730 to R-574/Útnesvegur and turn right (east); drive to Arnarstapavegur and turn right (east); drive into Arnastapi to **Stop 20.10, Gatklettur/Cliff Viewpoints** (64.766301, -23.627578) parking area on the right for a total of 4.8 km (3.0 mi; 6 min). Gatklettur is a 260 m (850 ft) walk east of the parking area. There is a large free parking lot by two restaurants, but no other facilities.*

Walk to Cliff Viewpoint (64.764400, -23.624830), about 270 m (890 ft) south from the parking area.

STOP 20.10 GATKLETTUR AND CLIFF OVERLOOKS

A gatklettur is a natural arch formed by wave erosion of the basalt cliffs over the eons. This is a land of natural arches; there are several of them in various stages of erosion along the sea cliffs. There are also sea caves and columnar basalt. Interestingly, some of the columnar basalt is horizontal instead of vertical, forming interesting polygonal patterns in the cliffs that are used by nesting seabirds. There is a 2 km (a bit over one mile) path that is one of the nicest coastal walks along the sea cliffs that I have been on.

Gatklettur at Arnastapi.

A perilous natural window at Arnastapi. Probably not long for this world. Photo courtesy of Adam Prost.

More-or-less horizontal columnar jointing in cliffs at Cliff.

Enlargement of the central portion of the previous photos shows a cliffside gull rookery in columnar jointed basalt.

Gatklettur and Cliff Viewpoints to Snæfellsjökull Viewpoint: Return to R-574/Útnesvegur and turn right (north and east); drive 14.1 km (8.8 mi; 12 min) to *Stop 20.11, Snæfellsjökull Viewpoint* (64.826702, -23.454962) pulloff on the right. There is room for about three cars; no facilities.

Leaving Arnastappi you drive over part of the Hellnahraun lava field. This flow just north of Arnarstapi is around 4,000 years old. It flowed from a crater that is currently under the glacier.

STOP 20.11 SNÆFELLSJÖKULL VIEWPOINT

This stop provides excellent views to Snæfellsjökull volcano and the ice cap that mantles it.

There is also a plaque commemorating Axlar-Björn, the only known serial killer in Icelandic history. Born Björn Pétursson in 1555, he lived at Öxl west of Búðir. His story is the stuff of folklore and legend. Accounts differ, but either nine or 18 victims are claimed. The victims were travelers and farmhands that came to Öxl looking for work. Some stories say he killed them with an axe; others that he drowned them. Suspicions grew as people disappeared in the area while Björn's horses and other possessions increased. Axlar-Björn was arrested and eventually confessed to nine murders. Even more bodies were found when they searched his farm. In 1596 Axlar-Björn was sentenced to die by hanging, to be followed by breaking on the wheel, which seems a bit over the top even for those more violent times (Wikipedia, Axlar-Björn).

View west to Snæfellsjökull, the bright ice-capped shield volcano in the center.

Snæfellsjökull Viewpoint to Ljósufjöll Table Mountain: *Continue driving east on R-54 to to* **Stop 21, Ljósufjöll** *(64.815568, -22.264170) pullout on either side of the road for a total of 65.3 km (40.6 mi; 45 min).*

STOP 21 LJÓSUFJÖLL TABLE MOUNTAIN

The Ljósufjöll ("Mountains of the Light") volcano fissure system initiated about 700,000 years ago, in mid- to late Pleistocene time. The last eruption was about 1,000 years ago, so it is probably still active. The layers are inclined to the east, indicating loading there by a volcanic pile at least 760 m (2,500 ft) thick (Share, 2018).

Rhyolites from the Ljósufjöll central volcano suggest that silicic magma is generated by partial melting of basaltic crust followed by concentration of the lower-melting point feldspar minerals (Martin and Sigmarsson, 2007). As befits their high feldspar content, these rocks are lighter in color than the basalts around them.

The Ljósufjöll volcanic system at the east end of the Snaefellsnes Peninsula consists of alkali olivine basaltic cinder cones and lava flows along short fissures. The volcanic field is about 90 km (56 mi) long, 20 km (12 mi) wide at the east end, and narrows to about 10 km (6 mi) in the west. The volcanic field contains the most outcrops of silicic rhyolitic and trachytic rocks in the Snaefellsnes volcanic zone (Smithsonian Institution, Global Volcanism Program, https://volcano.si.edu/volcano .cfm?vn=370030).

Tilted strata of Ljósufjöll Table Mountain are up to 760 m (2,500 ft) thick. Only the lower 300 m (1,000 ft) can be seen in this photo due to the low cloud deck when we were there.

Ljósufjöll to Eldborg á Mýrum: *Continue south on R-54 and turn right (west) at the sign for Eldborg and Snorrastadir; drive to* **Stop 22, Eldborg á Mýrum** *(64.774165, -22.302144) to the large free parking area on the left (before crossing the river) for a total of 5.8 km (3.6 mi; 6 min). A 3.8 km (2.4 mi) walk (one way) takes you to the crater rim. There are no facilities, but there is a guesthouse. Walk west across the river to the trailhead.*

STOP 22 ELDBORG Á MÝRUM

Eldborg, meaning "Castle of Fire," is a slightly elliptical spatter cone 60 m (197 ft) high and 200 m (656 ft) wide (Thordarson and Höskuldsson, 2008). It is the largest of a linear chain of spatter cones that erupted from a 1 km (0.6 mi) long fissure around 5,000 years ago. It sits within the largest lava field in the Ljósufjöll volcano and fissure vent system (Share, 2018).

You can climb the crater using a trail that begins on the south side at Snorrastadir, on Route 5610. The trail passes through moss-covered lava fields (Another Iceland, Eldborg Crater). There is a small cinder cone on the right (east) as you near the main crater. Views from the top are amazing: on a clear day it is easy to see Snaefellsnes 50 km (31 mi) west of here.

View northwest from the trailhead to Eldborg á Mýrum explosion crater.

Panorama of the Eldborg á Mýrum crater from the crater rim.

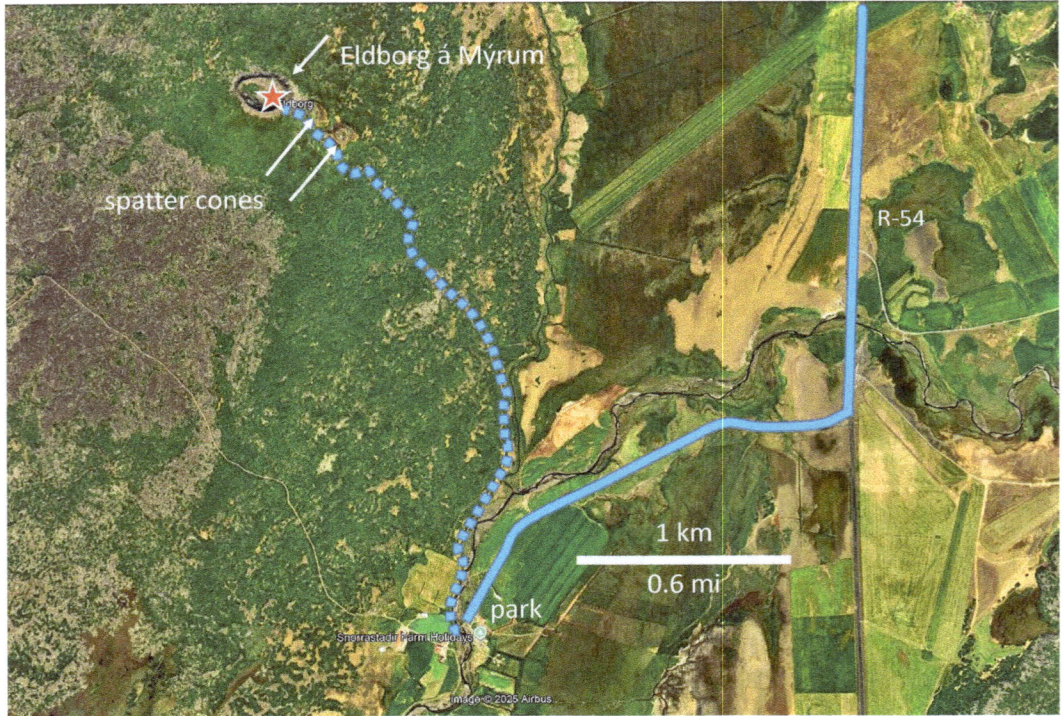

Google Earth image of Eldborg á Mýrum and nearby spatter cones. ©2025 Google Earth, Airbus.

Eldborg á Mýrum to Perlan Natural History Museum, Reykjavík: Return to R-54 and turn right (south); drive to R-1 and turn right (south); continue straight (west) on R-49 on the outskirts of Reykjavík; turn left (south) onto R-40/ Kringlumýrarbraut; take the exit toward Rvk-Fossvogur/ Rvk-Öskjuhilió and merge onto Bústaðavegur; turn left (south) to stay on Bústaðavegur; at the roundabout, take the 1st exit onto Varmahlíð and drive to Stop 23.1, Perlan Natural History Museum (64.129022, -21.916550), for a total of 110 km (68.3 mi; 1 hr 27 min). There is a large free parking lot.

STOP 23 REYKJAVÍK

As mentioned previously, this picturesque city has been continuously occupied since 874, the time of the first settlement on Iceland. It is the capital and largest city on the island. At latitude 64°08' N, it is the world's northernmost federal capital. The population of the metropolis in 2020 was around 131,000, and it is the cultural, economic, and governmental center of the country. Reykjavík is considered among the cleanest, greenest, and safest cities in the world.

During the last Ice Age (prior to about 12,000 years ago) most of the city area was covered by an ice sheet, while other parts of the city were below the sea. After the ice melted and the weight of the ice was removed much of the area rose above the sea, as indicated by clam-bearing sediments as much as 43 m (141 ft) above present sea level. The hills of Öskjuhlíð and Skólavörðuholt are likely the remains of former shield volcanoes that were active during interglacial periods.

The oldest rocks in the area are roughly 700,000-year-old basalt flows, but most of the rocks are post-glacial (younger than 10,000 years). As recently as 4,500 years ago basaltic lava flowed down the valley of the Elliðaá River and reached the sea at Elliðavogur Bay. The area is part of the Reykjanes volcanic belt and rift system and is seismically active (Wikipedia, Reykjavík).

Reykjavík, looking north from the Perlan Museum. Prominent church is Hallsgrimskirkja.

Stop 23.1 Perlan Natural History Museum

Perlan is a must-see stop in Reykjavík, an iconic building that hosts a one-of-a-kind interactive nature exploratorium. It is a learning laboratory where visitors can explore Icelandic nature through science and human perception. The observation deck provides a 360° view of the city and surrounding countryside. Perlan also has a restaurant, bar, café, ice cream parlour, and gift shop for guests to enjoy.

Current exhibitions (2024) include the Volcano Show, the Ice Cave and Glaciers Exhibit, the Northern Lights Exhibit, Látrabjarg Seabird Cliff Exhibit, Water in Iceland Exhibit, Forces of Nature, the Virtual Fish Tank, Iceland timeline, Mývatn, Ok, and more.

Parking is free. Entry is 5,390 kr (~$38.50) for an adult, 3,390 kr (~$24) for seniors.

Entrance to the Perlan Natural History Museum, Reykjavík.

Ice cave, Perlan Museum.

Aurora show, Perlan Museum.

Visit

Perlan is situated on Reykjavík's highest hill, close to the center of town. Getting to Perlan is quite easy, as many paths lead up to the building. You can get there by car, bus, taxi, walking, or cycling. Walking from the city center of Reykjavík takes about 30 minutes. A taxi takes about five minutes from the center. The local bus has a regular schedule every day and buses number 18 and 13 stop closest to Perlan.

Address: 105 Reykjavík, Iceland
Hours: 9:00 to 22:00.
Phone: +354-566-9000
Website: https://perlan.is/

Natural History Museum to Nauthólsvík Geothermal Beach: *Head east on Varmahlíð; at the roundabout, take the 2nd exit onto Bústaðavegur; turn left (northwest) to stay on Bústaðavegur; turn left (west) onto Flugvallarvegur; turn left (south) onto Nauthólsvegur; at the roundabout, take the 1st exit and stay on Nauthólsvegur; turn right (south) into the parking area for **Stop 23.2, Nauthólsvík Geothermal Beach** (64.122000, -21.928253), for a total of 3.0 km (1.9 mi; 6 min).*

Stop 23.2 Nauthólsvík Geothermal Beach, Reykjavík

Considered the Ibiza of the North Atlantic, Nauthólsvík Geothermal Beach is for people who love splashing around in the ocean and sunbathing on golden sand. The beach has been a popular place to swim since 1949. The hot tubs and man-made lagoon opened in 2000. Hot water is pumped into a man-made lagoon to create a swimming area that averages between 15° to 19°C (59° to 66° F). A hot tub sits in the middle of the lagoon. You can swim from the warm cove to the cold bay and back. This beach is open only in the summer. The beach is next to the aquatic center where sailing, kayaking, and other watersports are available.

Visit

The beach is equipped with changing rooms, toilets, showers, steam baths, and hot tubs. There is a large parking area that holds up to 150 cars.

Hours (Summer):   10:00 to 19:00.
Entry fee: 890 kr (2024)
Address: at the end of Nauthólsvegur Road (64.122000, -21.928253)
Phone: +354-511-6630
Website: https://nautholsvik.is/en/#top

Iceland is truly a land of contrasts: volcanoes and glaciers; massive waterfalls and spouting geysers; hot spring pools and ice cold lakes, rivers, and beaches. You never know what you're going to see here: this island is alive. A week earlier we were at the Blue Lagoon and there was no volcanic activity. On our way to the airport at the end of our visit the Sýlingarfell volcano, located near the Blue Lagoon, could be seen erupting beyond the mossy tundra 11 km (7 mi) south of the highway.

1 km
0.6 mi
Reykjavík
Perlan
Perlan
Museum
Nauthólsvík Geothermal Beach
6 min
3.1 km
Harpa Concert Hall and Conference Centre
The Icelandic Phallological Museum
Rainbow Street
Hallgrímskirkja
Sundhöllin
Hotel Lotus
National Museum of Iceland
Reykjavik Domestic Airport (RKV)
Reykjavik
Google

Nauthólsvík Geothermal Beach, Reykjavík.

A parting gift: Sýlingarfell erupting south of Keflavík. The smoke and steam are being blown by a 40 knot west wind. Photo courtesy of Adam Prost.

REFERENCES

Aabech, J.S. Gullfoss. Accessed 22 October 2020. http://www.vulkaner.no/n/gullfoss/egullfoss.html

Abdelghafoor, M. 2007. *Geological and geothermal mapping in Sveifluháls area, SW-Iceland*. Geothermal Training Programme, United Nations University Reports 2007 no. 3, Reykjavik, 23 p.

Allison, M.S., J.R. Hall, M.T. Papadopoulos, J. Eiríksson, and S.M. Jones. 2025. Updated stratigraphy and structure of the Barmur Group (Tjörnes beds, Iceland) and implications for the opening of the Bering Strait, *In* Hernández-Molina, F. J., G. Davoli, E.J. Stirling, D. Chiarella, and A.R. Viana, (eds) *Oceanic Gateways: Modern and Ancient Analogues and their Conceptual and Economic Implications.* Geological Society, London, Special Publication 553, https://doi.org/10.1144/SP553-2023-152.

Andi Snæfellsness, Snæfellsnes Regional Park. Accessed 21 October 2020. https://www.snaefellsnes.is/english

Another Iceland, Eldborg Crater. Accessed 21 October 2020. https://www.anothericeland.com/guides/eldborg-crater

Bischoff, T. 2014. Geysir and Gulfoss, in *Iceland Field Guide*, Caltech Enrichment Trip, Pasadena. 59 p.

Buz, J. 2014. Glaciers and Jökulhaups, in *Iceland Field Guide*, Caltech Enrichment Trip, Pasadena. 59 p.

Cars Iceland, Fjaðrárgljúfur. Accessed 22 October 2020. https://www.carsiceland.com/post/fjadrargljufur-ca nyon-iceland

Claus, N. 2020. Vestmannaeyjar Walking Tracks – The Most Beautiful Walks on Heimaey. Accessed 22 October 2020. https://wildernesscoffee-naturalhigh.com/vestmannaeyjar-walking-tracks-best-walks-o n-heimaey/

Cohen, B. 2018. Falls to Fall For: Öxarárfoss in Þingvellir National Park in Iceland. Accessed 22 October 2020. https://thegate.boardingarea.com/falls-to-fall-for-oxararfoss-in-thingvellir-national-park-in-iceland/

Denk, T., R. Zetter, F. Grímsson, and L.A. Símonarson. 2011. Introduction to the Nature and Geology of Iceland. Accessed 22 October 2020. https://www.researchgate.net/publication/251136570

Eiriksson, J., and B.J. Bjornsson. 1974. Geological investigations in Grindavik SW-Iceland. The Icelandic Harbour Authority, Reykjevik. 11 p. plus map 1:25,000.

Evans, D.J.A., M. Ewertowski, and C. Orton. 2017. Skaftafellsjökull, Iceland: glacial geomorphology record-ing glacier recession since the Little Ice Age. *Journal of Maps* v. 13 no. 2, p. 358–368.

Everest, J., T. Bradwell, L. Jones, and L. Hughes. 2017. The geomorphology of Svínafellsjökull and Virkisjökull-Falljökull glacier forelands, southeast Iceland. *Journal of Maps* v. 13 no. 2, p. 936–945.

Extreme Iceland. 2016. Hellisheidi Geothermal Power Station – South Iceland. Accessed 22 October 2020. https://www.extremeiceland.is/en/attractions/hellisheidi-geothermal-power-station

Fun Iceland. Driving in Iceland. Accessed 22 October 2020. https://www.funiceland.is/facts/travel-guide/ driving-in-iceland/

Fun Iceland. Lóndrangar. Accessed 22 October 2020. https://www.funiceland.is/nature/rock-pinnacles/lond rangar/

Guide to Iceland, Hellnar. Accessed 22 October 2020. https://guidetoiceland.is/travel-iceland/drive/hellnar

Guide to Iceland, Lóndrangar. Accessed 22 October 2020. https://guidetoiceland.is/travel-iceland/drive/lond rangar

Guide to Iceland, Vatnshellir Cave. Accessed 22 October 2020. https://guidetoiceland.is/travel-iceland/drive /vatnshellir-cave

Heimaslóð, Stafsnes. Accessed 22 October 2020. http://www.heimaslod.is/index.php/Stafsnes

Hit Iceland, Gullfoss. https://hiticeland.com/places_and_photos_from_iceland/gullfoss-waterfall-in-iceland

Hit Iceland, Öxaráfoss. Accessed 22 October 2020. https://hiticeland.com/places_and_photos_from_iceland/ %C3%B6xar%C3%A1rfoss-waterfall

Jóhannesson, H. 2014. *Bedrock geological map of Iceland*. Icelandic Institute of Natural History, Reykjavik, 1:600,000. Accessed 21 October 2020. https://en.ni.is/resources/publications/maps/geological-maps

Katla Geopark, Dyrhólaey. Accessed 22 October 2020. http://www.katlageopark.com/geosites/dyrholaey/

Katla Geopark, Eldgjá. Accessed 22 October 2020. http://www.katlageopark.com/geosites/eldgja-ofaerufoss/

Katla Geopark, Thorsmork. Accessed 22 October 2020. http://www.katlageopark.com/geosites/%C3%BEor smork/

Katla Geopark, Fjaðrárgljúfur. Accessed 22 October 2020. http://www.katlageopark.com/geosites/fjadra rgljufur/

Katla Geopark Project, Geological Report. Accessed 22 October 2020. http://www.katlageopark.com/media /39154/Geological-report.pdf

Khubaeva, O. 2007. *Geothermal mapping in the Krysuvik Geothermal Field*. Geothermal Training Program, United Nations University Reports 2007 no. 8, Reykjavik, p. 145–156.

Klescheva, M. 2014. Recent Volcanism, in *Iceland Field Guide*, Caltech Enrichment Trip, Pasadena. 59 p.

Kristjansson, L., I.B. Fridleifsson, and N.D. Watkins. 1980. Stratigraphy and paleomagnetism of the Esja, Eyrarfjall and Akrafjall mountains, SW Iceland. *Journal of Geophysics* v. 47, p. 31–42.

Lonely Planet, Katla Geopark. Accessed 22 October 2020. https://www.lonelyplanet.com/iceland/the-south/ attractions/katla-geopark/a/poi-sig/1500535/1318065

Lonely Planet, Stórhöfði. Accessed 22 October 2020. https://www.lonelyplanet.com/iceland/heimaey/activiti es/storhofdi/a/poi-act/1027236/1003580

Martin, E., and O. Sigmarsson. 2007. Crustal thermal state and origin of silicic magma in Iceland: the case of Torfajökull, Ljósufjöll and Snæfellsjökull volcanoes. *Contributions to Mineralogy and Petrology* v. 153, p. 593–605.

Mawejje, P. 2007. *Geothermal exploration and geological mapping at Seltún in Krýsuvík Geothermal Field, Reykjanes Peninsula, SW-Iceland*. Geothermal Training Programme, United Nations University Reports 2007 no. 12, Reykjavik, p. 257–276.

Maruszczak, A. 2018. *Geochemical variations in off-axis alkaline glasses – Snaefellsnes Peninsula, Iceland*. University of Iowa Geoscience, MSc thesis, 116 p.

McPhee, J. 1989. *The control of nature.* Noonday Press, Farrar, Straus, Giroux, New York. 272 p.

Nakajima, M. 2014. Reykjavik, in *Iceland Field Guide*, Caltech Enrichment Trip, Pasadena. 59 p.

National Energy Authority of Iceland, Iceland Deep Drilling Project. Accessed 22 October 2020. https://nea .is/geothermal/the-iceland-deep-drilling-project/

National Energy Authority Iceland, Geothermal Energy Exhibition. Accessed 22 October 2020. https://www .on.is/en/geothermal-exhibition/

Nature Iceland, Geology of Iceland. Accessed 22 October 2020. https://www.nat.is/geology-of-iceland/

Oregon State University, Eldgja Fissure System. Accessed 22 October 2020. http://volcano.oregonstate.edu/ oldroot/volcanoes/volc_images/europe_west_asia/eldgja.html

Prestvik, T. 1979. *Geology of the Öræfi District, southeastern Iceland.* Nordic Volcanological Institute, University of Iceland, report 79-01, Reykjavik. 28 p.

Prost, G.L., and B.P. Prost. 2018. *The geology companion: Essentials for understanding the earth.* CRC Press, Boca Raton, FL, 486 p.

Renyard, L. 2019. A Complete Guide to Kirkjufell Mountain. Accessed 23 October 2020. https://www.extrem eiceland.is/en/attractions/kirkjufell-mountain

Saga Geopark. 2015. Proposed SAGA Geopark. Accessed 17 June 2024. https://www.west.is/static/files/sagag eo/saga_geopark_annex3_application_dossier_revised_250116.pdf

Sæmundsson, K. 1992. Geology of the Thingvallavatn Area. *Oikos* v. 64 no. 1/2, Ecology of Oligotrophic, Subarctic Thingvallavatn (May–June, 1992), p. 40–68.

Share, J. 2018. The Geologic Evolution of Iceland: Part I. Accessed 23 October 2020. https://written-in-stone -seen-through-my-lens.blogspot.com/2018/06/the-geologic-evolution-of-iceland-land.html

Sinton, J., K. Grönvold, and K. Sæmundsson. 2005. Postglacial eruptive history of the Western Volcanic Zone, Iceland. *Geochemistry Geophysics Geosystems* v. 6 no. 12, 34 p. https://doi.org/10.1029/2005GC001021

Slotznick, S. 2014. Þingvellir: Geologic Setting, in *Iceland Field Guide*, Caltech Enrichment Trip, Pasadena. 59 p.

Smithsonian Institution, Global Volcanism Program. Accessed 23 October 2020. https://volcano.si.edu/volca no.cfm?vn=370030

Stefánsson, Á.B. 2010. *The Vatnshellir Project - a first for Iceland.* Proceedings 14th International Symposium on Vulcanospeleology, Undara Volcanic National Park, Queensland, Australia. p. 115–122.

Sooflauschig. 2018. Geology of Iceland - Part 2: The Iceland Hotspot. Accessed 23 October 2020. https:// steemit.com/geology/@sooflauschig/geology-of-iceland-part-2-the-iceland-hotspot

Summit Adventure Guides, 2020. Vatnshellir Cave Tour. Accessed 23 October 2020. https://www.summitgu ides.is/vatnshellir-cave-op1r6

Sykora, S. 2016. How to Explore Iceland Through the Eyes of a Geologist. Accessed 23 October 2020. http:// exploringtheearth.com/2016/07/05/geological-journey-iceland-land-fire-ice-part-1/ and http://exploring theearth.com/2016/09/26/iceland-glaciers-volcanoes-south/

Thompson, A. 2010. How Did Iceland Form? Accessed 23 October 2020. https://www.livescience.com/8129-i celand-form.html#:~:text=The%20pocket%20of%20magma%20that,to%20San%20Francisco's%20Exp loratorium%20museum.

Thordarson, T., and Á Höskuldsson. 2008. Postglacial Volcanism in Iceland. *JÖKULL* no. 58, p. 197–228.

Trønnes, R.G. 2003. Geology and Geodynamics of Iceland. Accessed 23 October 2020. https://www.seman ticscholar.org/paper/Geology-and-geodynamics-of-Iceland-Tr%C3%B8nnes/70f7585576ef67abbb34ff9 550058552595417ff

Tsokonombwe, G. 2012. *Geological and structural mapping of the Midfell- Flúðir low-temperature geother-mal field, S-Iceland.* Geothermal Training Programme, United Nations University Report 2012 no. 34, Reykjavik, p. 861–879.

UNESCO, 2017. Katla UNESCO Global Geopark (Iceland). Accessed 23 October 2020. http://www.unesco. org/new/en/natural-sciences/environment/earth-sciences/unesco-global-geoparks/list-of-unesco-global -geoparks/iceland/katla/

UNESCO. 2019. Accessed 23 October 2020. https://whc.unesco.org/en/list/1604/

USGS. 2014. Understanding Plate Motions. Accessed 23 October 2020. http://pubs.usgs.gov/gip/dynamic/ understanding.html

USGS, Glaciers and Icecaps. Accessed 23 October 2020 https://www.usgs.gov/special-topic/water-science -school/science/glaciers-and-icecaps?qt-science_center_objects=0#qt-science_center_objects

Vargas Morales, J.R. 1992. *Geology and geothermal considerations of Krýsuvík Valley, Reykjanes Peninsula, Iceland.* Geothermal Training Programme, United Nations University Reports 1992 no. 13, Reykjavik, 33 p.

VisitSouthIceland, Katla Geopark. Accessed 23 October 2020. https://www.south.is/en/what-to-see-do/nature/katla-geopark

Ward, P.L. 1971. New interpretation of the geology of Iceland. *Geological Society of America Bulletin* v. 82 no. 11. p. 2991–3012.

Wikipedia Axlar-Björn. Accessed 23 October 2020. https://en.wikipedia.org/wiki/Axlar-Bj%C3%B6rn

Wikipedia, Blue Lagoon. Accessed 23 October 2020. https://en.wikipedia.org/wiki/Blue_Lagoon_(geothermal_spa)

Wikipedia, Djúpalónssandur. Accessed 23 October 2020. https://en.wikipedia.org/wiki/Dj%C3%BApal%C3%B3nssandur

Wikipedia, Eldgjá. Accessed 23 October 2020. https://en.wikipedia.org/wiki/Eldgj%C3%A1

Wikipedia, Eyjafjallajökull. Accessed 23 October 2020. https://en.wikipedia.org/wiki/Eyjafjallaj%C3%B6kull

Wikipedia, Geology of Iceland. Accessed 23 October 2020. https://en.wikipedia.org/wiki/Geology_of_Iceland

Wikipedia, Geysir. Accessed 23 October 2020. https://en.wikipedia.org/wiki/Geysir

Wikipedia, Geothermal Power in Iceland. Accessed 23 October 2020. https://en.wikipedia.org/wiki/Geothermal_power_in_Iceland

Wikipedia, Gullfoss. Accessed 23 October 2020. https://en.wikipedia.org/wiki/Gullfoss

Wikipedia, Heimaey. Accessed 23 October 2020. https://en.wikipedia.org/wiki/Heimaey

Wikipedia History of Iceland. Accessed 23 October 2020. https://en.wikipedia.org/wiki/History_of_Iceland

Wikipedia, Jökulsárlón. Accessed 23 October 2020. https://en.wikipedia.org/wiki/J%C3%B6kuls%C3%A1rl%C3%B3n

Wikipedia, Katla Geopark. Accessed 23 October 2020. https://en.wikipedia.org/wiki/Katla_Geopark

Wikipedia, Kirkjufell. Accessed 23 October 2020. https://en.wikipedia.org/wiki/Kirkjufell

Wikipedia, Kleifarvatn. Accessed 23 October 2020. https://en.wikipedia.org/wiki/Kleifarvatn

Wikipedia, Lóndrangar. Accessed 23 October 2020. https://en.wikipedia.org/wiki/L%C3%B3ndrangar

Wikipedia, Ocean-birth. Accessed 23 October 2020. https://commons.wikimedia.org/wiki/File:Ocean-birth_hg.png

Wikipedia, Reykjavík. Accessed 23 October 2020. https://en.wikipedia.org/wiki/Reykjav%C3%ADk

Wikipedia, Reynisdrangar. Accessed 23 October 2020. https://en.wikipedia.org/wiki/Reynisdrangar

Wikipedia, Skaftafell. Accessed 23 October 2020. https://en.wikipedia.org/wiki/Skaftafell

Wikipedia, Sólheimajökull. Accessed 23 October 2020. https://en.wikipedia.org/wiki/S%C3%B3lheimaj%C3%B6kull

Wikipedia, Þingvellir. Accessed 23 October 2020. https://en.wikipedia.org/wiki/%C3%9Eingvellir

Wikipedia, Tholeiitic Magma Series. Accessed 23 October 2020. https://en.wikipedia.org/wiki/Tholeiitic_magma_series

Wikipedia, Vatnajökull National Park. Accessed 23 October 2020. https://en.wikipedia.org/wiki/Vatnaj%C3%B6kull_National_Park

Witze, A., and J. Kanipe. 2014. Island on Fire – The Extrordinary Story of a Forgotten Volcano that Changed the World. Pegasus Books, New York, London. 224 p.

Worland, J. 2017. A solution to our clean energy problem may lie right beneath our feet. *Time Magazine*, July 25, 2017. Accessed 23 October 2020. https://time.com/4844086/geothermal-energy-iceland-deep-drilling-project/

Zijlstra, A. 2016. The Eldgja Eruption: Iceland's Baptism by Fire. Accessed 23 October 2020. https://www.volcanocafe.org/the-eldgja-eruption-icelands-baptism-by-fire/

2 Pyrenees of Spain, France, and Andorra

Sorrosal Falls, near Broto, Spain. Photo courtesy of U.S. Geological Survey, https://commons.wikimedia.org/wiki/File:Waterfall_taken_in_Broto,_located_in_the_Spanish_Pyrenees_(the_area_of_the_Pyrenees_mountain_range_that%27s_located_within_Spain).jpg

OVERVIEW

The majestic Pyrenees were formed by the slow but inexorable collision of the African and European tectonic plates. Part of the Alpine Orogeny, these mountains have been uplifted over the past 50 million years or so. It is a story of continental collisions, of fold-thrust belts, of salt in motion, and of evolving sedimentary basins. It is an area of stunning landscapes rich in history, both geologic and human, and well worth exploring.

Starting and ending in Barcelona, this geo-tour guides us on a loop trip through the tectonic zones of the eastern and central Pyrenees. We traverse the southern foreland basin, cross the central range from south to north, then explore the northern foreland before crossing the east end of the mountains. We examine thrust faults, deformation related to both thrusting and salt tectonics, and we examine the interplay between tectonics and sedimentation during early rifting, later compression, and finally basin inversion. We visit peridotites brought to the surface by crustal extension, thrust-related growth strata, evaporite detachment horizons, and large fan deltas derived from uplifted blocks. This geological tour, roughly 1,250 km (780 mi), takes us through two UNESCO Global Geoparks (Sobrarbe Pirineos and Orígens), national parks (Montserrat, Parc National des Pyrénées), natural parks (Parc Natural de la Muntanya de Montserrat, Muntanya de Sal de Cardona,

DOI: 10.1201/9781003435594-2

and El Migmón Cadí Moixeró, Collserola), and World Heritage Sites (Fumanya, Monte Perdido/ Pyrénées – Mont Perdu). We see hot springs, waterfalls, glacial valleys, and jagged peaks. The scenic Pyrenees of Spain and France, caught between colliding continents, has a geologically complex and intriguing story to tell.

Major stops on the Pyrenees geo-tour.

ITINERARY

    Begin – Barcelona
    Stop 1 Montserrat Area
            1.1 Satin Slates, Catalan Coastal Ranges
            1.2 Montserrat Fan Delta Conglomerate
            1.3 Monastery of Montserrat
            1.4 Montserrat View
    Stop 2 Súria Anticline and El Migmón
    Stop 3 Cardona Salt Diapir
    Side Trip 1 Sant Llorenç de Morunys Growth Strata and Progressive Unconformities
            ST1.1 Progressive Unconformities, Les Cases Altes de Posada
            ST1.2 Growth Strata, Santuari de Lord

GEOLOGY OF THE PYRENEES

The Pyrenees Mountains extend east-west along the border of Spain and France for 450 km (280 mi), from the Bay of Biscay in the west to the Golfe du Lion in the Mediterranean. In fact, related structures extend eastward into southern Provence and are finally cut off by the arc of the Western Alps (Wikipedia, Geology of Pyrenees). The Pyrenees are 80 to 125 km (48 to 75 mi) wide and reach their highest point of 3,404 m (11,168 ft) at Aneto Peak.

The range is a result of uplift and thrusting associated with the collision of the microcontinent of Iberia with the southwestern corner of the European tectonic plate. This Pyrenean Orogeny, part of the larger Alpine Orogeny, lasted from Cretaceous to Eocene, roughly 100 to 25 Ma. (I use the terms Pyrenean Orogeny and Alpine Orogeny interchangeably.) Following collision and uplift the range underwent intense erosion and localized extension. We will focus on the central and eastern Pyrenees.

Tectonic Zones

The overall architecture of the Pyrenees is of an asymmetric uplift with shallow south-directed thrusts on the south and steep, north-directed thrusts on the north flank of the range. The mountain belt has been subdivided into five tectonic zones. From south to north these are:

Pre-Alpine (Late Cretaceous) configuration of the Iberian and European plates. Modified after Teixell et al., 2016.

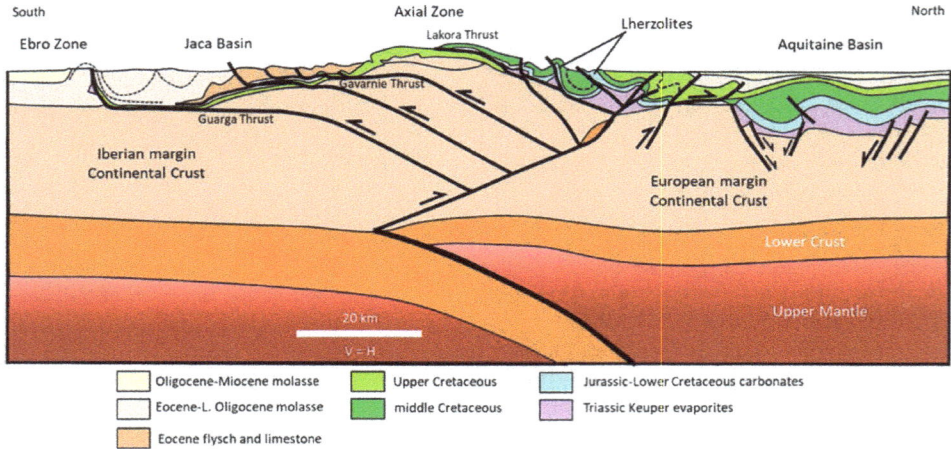

Post-Alpine collision configuration of the Pyrenees. SPF is the South Pyrenean Fault; NPF is the North Pyrenean Fault. Modified after Teixell et al., 2016.

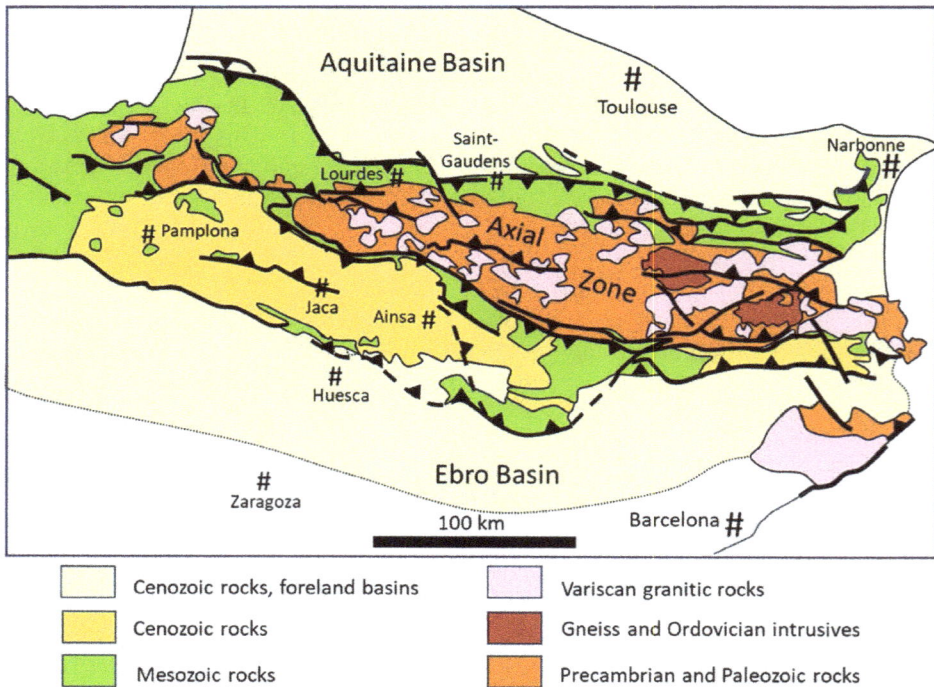

Major geological zones of the Pyrenees. Modified after Clariana et al., 2021 and Teixell, 2000.

(1) The Ebro Basin

This is the southern foreland basin of the Pyrenean Orogen (the area affected by mountain building). The basin formed by flexing of the Iberian lithosphere under the weight of the South Pyrenean thrusts and sediments. The Ebro Basin is a molasse basin. A molasse basin, also called a synorogenic foreland basin, is one that formed at the same time as the mountain range and is filled with sediment eroded off the mountains. The Ebro Basin contains sediments of Cretaceous and Tertiary age that rest directly on Variscan (in some places still called Hercynian) basement (Schellart, 2002). Variscan basement is the old, metamorphosed rock that underlies more recent sediments. These rocks were metamorphosed during the Devonian-Carbonifereous Variscan Orogeny (380–280 Ma).

The Ebro foreland basin developed as an underfilled marine basin from 55 to 37 Ma, and then as an overfilled continental basin from deposition of the Cardona evaporite (~37 Ma) to the end of shortening in the Oligocene. Around ~37 Ma (Eocene) an intermontane basin formed, bounded by the Pyrenees, the Catalan Coastal Ranges, and the Iberian Range. A network of rivers brought sediments to the Ebro Basin, which contained a large central lake. The end of deformation at the Pyrenean fold-thrust front occurred during upper Oligocene (~24.7 Ma; Vergés et al., 2002).

The southern margin of the Ebro Basin is the reverse-fault-bounded Catalan Coastal Ranges. The northeast-southwest trending Catalan Coastal Ranges are considered to be the result of Eocene–Oligocene inversion of the Pyrenean Rift system (Vergés and Garcia-Senz, 2001; Grool et al., 2018). Basin inversion refers to the process whereby a region that was previously a basin, or low-lying area of sediment accumulation, is uplifted (and eroded) due to compressional forces.

Late Jurassic–middle Cretaceous Pyrenean rifting formed several basins between the Bay of Biscay in the Atlantic and the Tethys Ocean to the east.

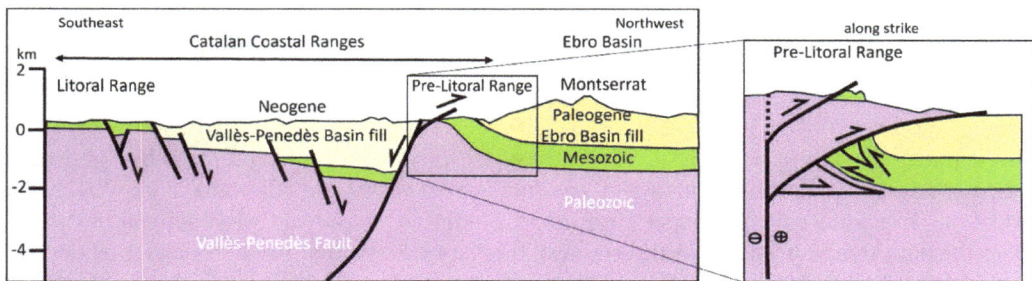

Two cross--sections along the north flank of the Catalan Coastal Ranges showing the change from Paleogene north-directed thrusting to Neogene normal faulting. Modified after López-Blanco, 2006.

The Ebro Basin has been divided into a Southern Folded Foreland section in the northeast (Catalan sector) and an undeformed section in the south and west. In the deformed sector fold intensity decreases away from the thrust fronts. The folds are more or less parallel to the east-west thrust fronts, but turn northeast near Oliana.

Above the Paleozoic basement are Upper Cretaceous to Paleocene redbeds (red sandstones and shales) and Eocene limestones, marine marls (a clayey limestone or limey claystone), and upper Eocene evaporites (salt and gypsum of the Cardona Formation). Evaporites form when minerals (usually salt and gypsum) dissolved in a water body precipitate out due to evaporation. The lower Oligocene contains conglomerates that merge southward into lake and evaporite deposits. The folded Paleogene series of the Southern Folded Foreland are topped by an unconformity, or erosional break in the rock sequence, which is, in turn, overlain by flat-lying nonmarine Miocene and Pliocene strata. The Ebro Basin deepens northward toward the South Pyrenean Frontal Thrust, where it contains between 3,000 and 5,000 m (9,800 to 16,400 ft) of sediments (Wikipedia, Geology of Pyrenees). The South Pyrenean Frontal Thrust (or South Pyrenean Front) separates the Ebro Basin from the South Pyrenean Zone.

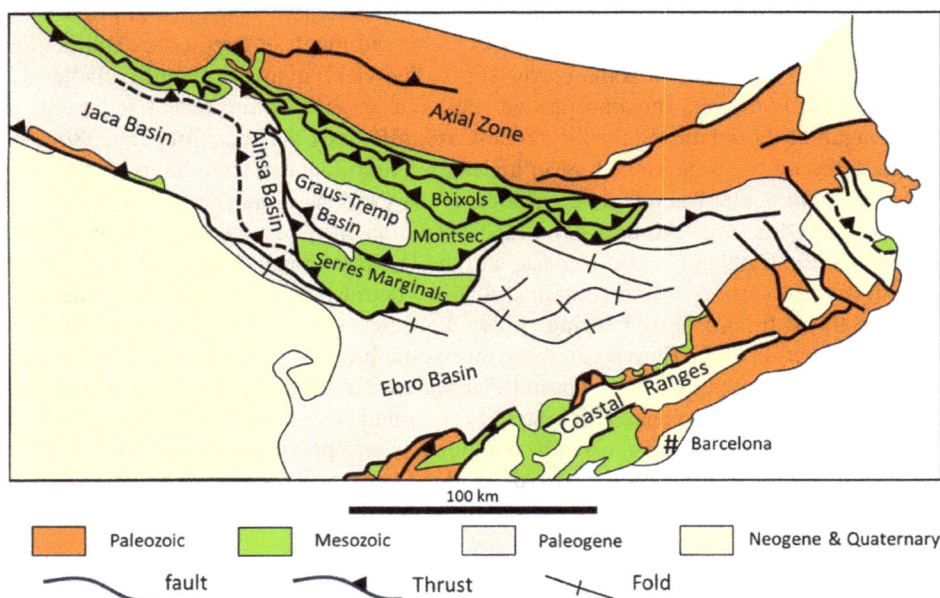

Disposition of the Ebro Basin and South Pyrenean Zone. Modified after Lopez-Blanco et al., 2003.

(2) The South Pyrenean Zone

This segment consists of a Mesozoic–Eocene sedimentary sequence that is detached on Middle or Upper Triassic evaporite horizons and was transported southward (Clariana et al., 2021). This part of the Triassic is called "Keuper" in Europe. Southward movement was confined in the west by north-south-trending folds and thrusts near the Cinca River (the Mediano and Boltaña anti-clines, large uplifted folds) and in the east by northeast-southwest strike-slip faults near the Segre River. This constriction forced the sedimentary cover to pile up into several stacked thrust sheets such as the Monte Perdido and Cotiella nappes. (A nappe is a thrust sheet, the section of rock car-ried on a given thrust fault.) The less constrained, centrally located Bóixols Thrust sheet contains both foreland (southward) and hinterland (northward) directed thrusting. The Montsec Thrust sheet, south of the Bóixols Thrust, is laterally equivalent to the Lower Pedraforca Thrust just to the east (Wikipedia, Geology of Pyrenees).

In the Eastern Pyrenees three stacked thrust sheets comprise the South Pyrenean Zone. From top (oldest) to bottom (youngest), they are the Upper Pedraforca, the Lower Pedraforca, and the Cadí thrusts. The Upper and Lower Pedraforca sheets contain mostly Jurassic to Paleocene units detached above Triassic evaporites. The Upper Pedraforca thrust contains ~1,500 m (4,900 ft) of sediments originally deposited in a Lower Cretaceous basin. The Lower Pedraforca thrust sheet contains ~1,700 m (5,600 ft) of post-rift and synorogenic Upper Cretaceous conglomerates, marls, and limestones, capped by ~700 m (2,300 ft) of latest Cretaceous to Paleocene redbeds of the Tremp Formation (Grool et al., 2018). Synorogenic sediments are those that accumulate during active mountain building and associated erosion.

In the Central Pyrenees, a large imbricate thrust system contains three main thrusts: from north to south, from top to base, and in order of emplacement these are the Bóixols, Montsec, and Serres Marginals thrust sheets. They were emplaced from the Late Cretaceous to the Oligocene (Clariana et al., 2021). These thrusts and the piggyback basins they carried are collectively known as the South Central Unit (Vergés, 1993).

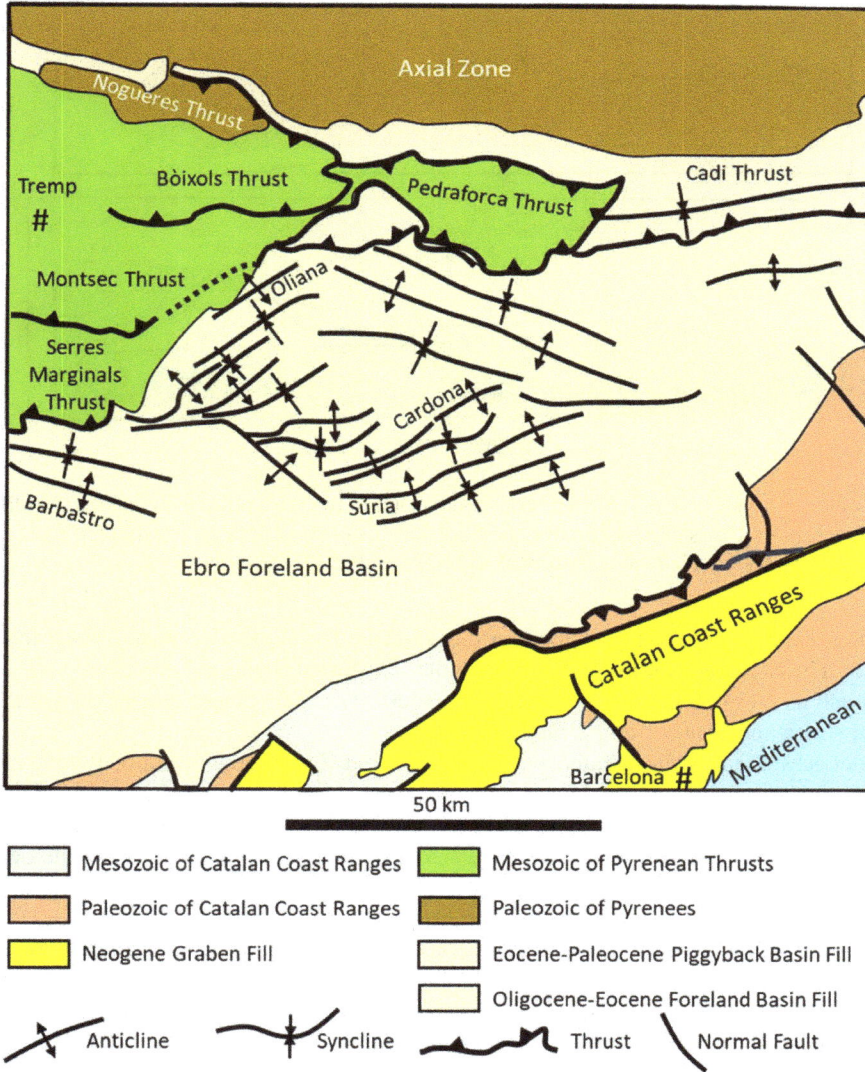

General geologic map of the Catalan Coastal Ranges, Ebro Foreland Basin, and South Pyrenean Zone. Modified after Verges, 1993.

The Eocene seas advanced and retreated four times in the South Pyrenean Zone. Each cycle contains nonmarine sands, silts, and shales deposited ahead of the advancing South Pyrenean Thrust front. These units grade southward into moderate to deep marine carbonates and mudstones that, in turn, grade into shallow marine platform carbonates along the southern margin of the basin. During the last cycle the basin was marls of the Eocene Igualada Formation bounded on the south by coarse fan delta deposits, the Montserrat Formation, sourced from the uplifting Catalan Coastal Ranges. At the end of cycles 2 and 4, thick evaporites of the Beuda and Cardona formations, respectively, were deposited in the central parts of the basin. The Ebro foreland basin became fully continental during deposition of the uppermost Cardona Formation at ~36 Ma (Grool et al., 2018).

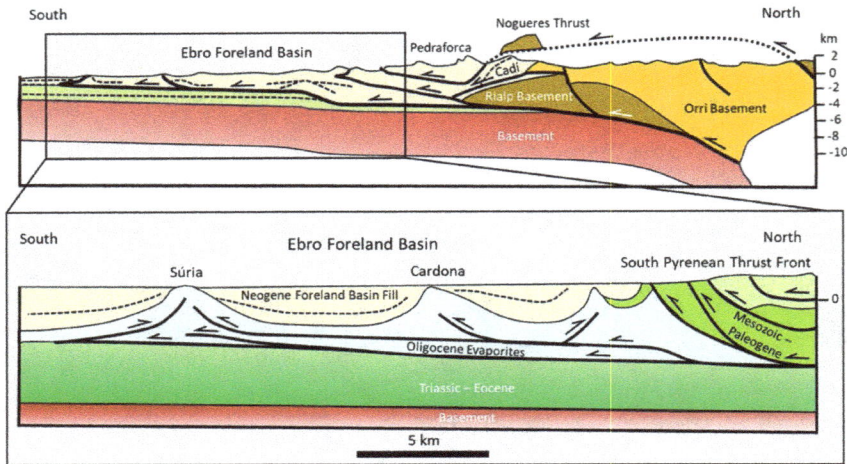

Top: Cross-section through the South Pyrenean Zone showing the major thrust complexes. Modified after Verges, 1993. Bottom: Cross-section through the Ebro Basin showing types of thrust-cored folds. Modified after Sans, 1996, Bulletin of Canadian Petroleum Geology, CEGA © 1996.

The South Pyrenean Zone is bounded on the south by the South Pyrenean Thrust that carries the Montsec Thrust Sheet over the Serres Marginals. The thrust sheets are estimated to have moved south as much as 30 to 50 km (18 to 30 mi). The South Pyrenean Thrust carries piggyback basins such as the Ainsa and Graus-Tremp basins.

The Paleogene Jaca Basin, located between the Axial Zone and the South Pyrenean Frontal thrust, contains thin pre-orogenic Mesozoic rock layers; in contrast, the Paleogene sedimentary basin fill is up to 9 km (29,500 ft) thick and conforms to a typical flysch-to-molasse (turbidite to syntectonic sediment) foreland basin sequence (Bosch et al., 2016). Turbidites are the deposits of sediment-laden currents that flow down marine slopes into deep water. These turbidity currents are triggered by events such as earthquakes or submarine landslides.

The geometry of the South Pyrenean front varies from east to west. West of Barbastro, the South Pyrenean Thrust is emergent and superimposes Mesozoic rocks on undeformed foreland of the Ebro Basin. In the central and eastern Pyrenees, the Ebro foreland basin is deformed into fault-related folds and thrusts over a buried thrust front, in some places thought to be a triangle zone. Triangle zones are defined as the strata between a basal blind thrust (a thrust does not come to the surface) that terminates toward the foreland, the hinterland-verging thrust that initiates at the termination point, and the most foreland-verging thrust that intersects the surface. Vergence refers to the direction a fold in inclined, which indicates where the compressive push is directed.

Erosion and differential unloading alone may trigger diapirism (upward movement) in areas where salt has accumulated in anticlinal cores. Diapirs triggered by such unroofing tend to be elongated parallel to the fold axes (Sans, 2003).

The South Pyrenean thrust zone is detached mainly in Triassic evaporites over undeformed Variscan basement. The evaporite décollement (detachment surface) comes to the surface at the South Pyrenean Frontal Thrust of the Serres Marginals. Deformation on this thrust system began in the early Eocene and continued until the early Miocene (Schellart, 2002).

The Serres Marginals (Border Ranges) of the southern South Pyrenean Zone consist of a Mid-Late Triassic to lower Eocene sedimentary section. Units of the Serres Marginals were thrust south into the Ebro Basin and were unconformably covered by Oligocene and Miocene units of the Ebro Basin. In the east, the Serres Marginals are represented by the tectonically related Port del Comte and Cadí Thrust sheets. The Serres Marginals are overthrust in the north by the Montsec Thrust Sheet of the South Pyrenean Zone (Wikipedia, Geology of Pyrenees).

(3) The Axial Zone

As the name implies, the Axial Zone is in the highest, central part of the Pyrenees. The Axial Zone is a Precambrian and Paleozoic-cored basement uplift that was folded and metamorphosed during the Devonian-Carbonifereous Variscan Orogeny. It is an antiformal stack (a descriptive term for a fold whose limbs are inclined downward from the fold axis) consisting of several south-directed basement-involved thrust sheets. The stack was actively uplifted from late Eocene to early Oligocene time (Bosch et al., 2016; Grool et al., 2018; Clariana et al., 2021; Wikipedia, Geology of Pyrenees). Silurian black shales form the lower detachments; Triassic evaporites form the upper décollements. The zone contains three major south-directed thrust sheets, which are, from bottom to top, the Nogueres, Orri, and Rialp sheets in the central and eastern Pyrenees. In the western part of the Central Pyrenees these are the Lakhoura, Gavarnie, and Guarga Thrust sheets (Schellart, 2002).

The basement is cut by major east–west-oriented, late Variscan fracture zones that were reactivated as thrust faults during the Alpine Orogeny. Schistosity, a metamorphic alignment of minerals that is a result of deep burial and compression, was imposed on both the basement as well as sedimentary cover, implying an Alpine origin. These fractures account for an overall 10 to 20 km (6 to 12 mi) of crustal shortening. To the east, in northern Catalonia, the Axial Zone was downfaulted in Neogene and Quaternary time, forming grabens. The zone finally ends somewhere beneath the Mediterranean. The central and eastern sections of the Axial Zone are bounded on the north by the North Pyrenean Fault, a system of roughly east-west, steeply dipping reverse-faults. The trace of the North Pyrenean Fault becomes diffuse west of Lourdes. The southern margin of the Axial Zone is a reverse-fault that overthrusts the South Pyrenean Zone (Wikipedia, Geology of Pyrenees).

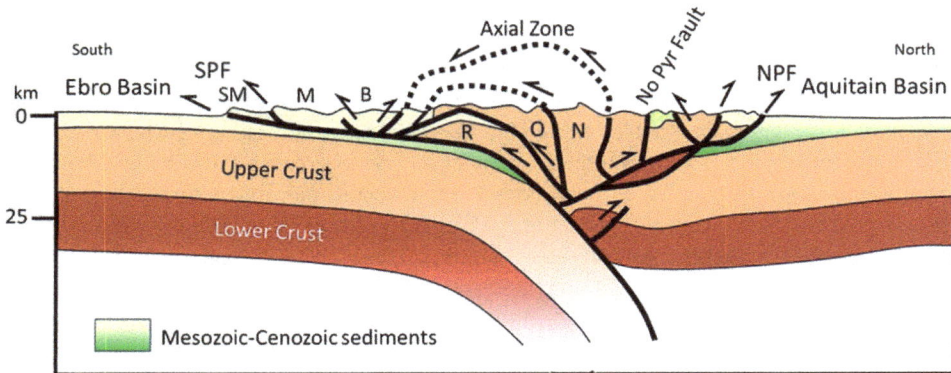

North-south cross-section through the Pyrenees showing the position of the Axial Zone and foreland basins. SPF = South Pyrenean Front; SM = Serras Marginals; M = Montsec Thrust; B = Boixols Thrust; R = Rialp Thrust; O = Orri Thrust; N = Nogueres Thrust; NPF = North Pyrenean Front. Modified after Clariana et al., 2021; Muñoz, 1992; Berástegui et al., 1993.

The Axial Zone was intruded by late Variscan granites and granodiorites around 310–290 Ma. These Paleozoic rocks are unconformably overlain by Triassic redbeds and evaporites, Jurassic carbonates and shales, and a highly variable Cretaceous sequence dominated by carbonates and mass flow deposits. These are overlain by Cenozoic terrestrial and marine sediments.

The boundary between the Axial Zone and the North Pyrenean Thrust System is the North Pyrenean Fault, a high-angle reverse fault. This major boundary fault is interpreted to be where the Iberian Plate plunges beneath the European Plate. It is a suture zone where two continental plates were welded together (Diaz et al., 2018; van Hinsbergen et al., 2020).

(4) The North Pyrenean Zone

The North Pyrenean Zone is only about 10 km (6 mi) wide in most areas. It is characterized by strong folding and steeply dipping thrusts. The zone is bounded on the south by the North Pyrenean Fault, a south-directed reverse fault/thrust, and is bounded on the north by the North Pyrenean Front, a north-directed thrust over the Aquitaine Basin, part of the European tectonic plate (Wikipedia, Geology of Pyrenees). The North Pyrenean Frontal thrust was initiated in Late Cretaceous time and was active into the Paleogene. Molasse deposition in the Aquitaine Basin, derived from Pyrenean uplifts, extends from the late–middle Eocene to the Miocene (Bosch et al., 2016).

The 6,000+ m (19,700+ ft) sedimentary sequence of the North Pyrenean Zone consists mainly of Jurassic and Cretaceous rocks detached above Upper Triassic evaporites and shales (Clariana et al., 2021). Interestingly, an older basement penetrates the sedimentary cover in several faulted uplifts. Basement rock in the uplifts consists of Precambrian gneisses and Paleozoic igneous and metamorphic rocks (Grool et al., 2018). "Ribbons of lithospheric mantle" were uplifted and eroded along this suture zone during middle to Late Cretaceous time (Diaz et al., 2018). There are several small *"massifs nord-pyrénéens"* (north Pyrenean basement uplifts) between Lourdes and Perpignan.

A narrow, roughly 5 km (3 mi) wide strip just north of the North Pyrenean Fault underwent high temperature/low pressure metamorphism during middle Cretaceous time (~110 Ma). This metamorphism affected the Mesozoic sedimentary cover, transforming limestones into marble and shales to hornfels. Scattered within the metamorphic zone are several lherzolites (including the type locality at Lers). This coarse-grained ultramafic (low silica) igneous rock originated in the upper mantle and moved upward along deep-penetrating faults (Wikipedia, Geology of Pyrenees).

Pre-Alpine peridotites (olivine-rich upper mantle rock) are located east of the Lers body. Most of the peridotites form blocks and lenses embedded in debris flows consisting of Mesozoic metasediments. The blocks were emplaced as a result of uplift related to extreme thinning of the continental basement. The peridotites indicate a hyper-extended Cretaceous passive margin.

Metamorphic rocks in the southern part of the North Pyrenean Zone include carbonate breccias enclosing peridotite fragments in anastomosed tectonic lenses. The Pyrenean peridotites are variously altered, with some consisting of fresh peridotite, whereas others are completely serpentinized (de Saint Blanquat et al., 2016).

A shallow-water carbonate platform (think of the present-day shallow water limestone deposits around the Bahamas) developed during a period of tectonic quiet that lasted from Early Jurassic to the end of Lower Cretaceous. Middle Cretaceous deep marine sediments mark the beginning of the North Pyrenean Basin, a 400 km (240 mi) long east–west-oriented pull-apart basin that filled with upwards of 3,000 m (9,800 ft) of turbidites/flysch between Triassic and Late Cretaceous. The North Pyrenean Zone was a rapidly subsiding Cretaceous basin between the European and Iberian margins. The rifting has been related to the opening North Atlantic Ocean based on magnetic seafloor anomalies of the Bay of Biscay and Atlantic. This basin was completely inverted during the Alpine/Pyrenean Orogeny (Bosch et al., 2016; Wikipedia, Geology of Pyrenees).

The onset of Alpine convergence between Europe to the north and Iberia and Africa to the south, beginning at ~84 Ma (Late Cretaceous), resulted in a north-south shortening direction with a minor strike-slip component that was mainly focused along the North Pyrenean Fault. This was followed by subsidence slowing and ending in the northern foreland during the Paleocene (66–59 Ma). Tectonic activity returned in the late Paleocene with distributed basement-involved shortening across the North Pyrenean tectonic wedge. Widespread continental sedimentation (Carcassonne Group) commenced at the end of the early Eocene and continued until at least the early Oligocene. The youngest evidence for deformation in the North Pyrenean foreland is a slight folding of the upper Carcassonne Group during the late Eocene (Grool et al., 2018).

(5) The Aquitaine Basin

North of the Pyrenees, the Aquitaine Basin is the second largest sedimentary basin in France after the Paris Basin. The basin sits on Variscan basement that was eroded to a flat surface during the Permian and began subsiding in the Early Triassic. The basin contains as much as 11,000 m (36,000 ft) of Alpine uplift-derived sediments (Wikipedia, Geology of Pyrenees).

The Aquitaine foreland basin is located on the European plate, where it was formed by downward flexing under the weight of North Pyrenean thrusts (Schellart, 2002).

The southernmost part of the Aquitaine Basin contains the Sub-Pyrenean Zone. Below 6,000 m (~20,000 ft) of mostly Mesozoic basin fill is Paleozoic metamorphic basement. Near the Cretaceous–Tertiary boundary the mainly marine Mesozoic units change to continental redbed deposits. In Paleocene/Eocene times, the sea advanced again from the Atlantic into the SubPyrenean Basin, a downwarp north of the slowly rising Pyrenees. Sedimentation stopped in late Eocene due to major compression and uplift, the Pyrenean Main Phase. Upper Cretaceous and Paleogene sediments were folded during the Eocene and thrust northward. The near-surface was deformed into simple west-northwest–east-southeast folds. The subsurface, however, contains Triassic salt diapirs and north-directed thrusts (Wikipedia, Geology of Pyrenees).

Geologic History

The geologic history of the Pyrenees goes back long before the Alpine Orogeny. Rare Precambrian outcrops are mainly gneisses and sediments metamorphosed to amphibolite and granulite (high temperature–low pressure) facies. Cambro-Ordovician metamorphic rocks comprise amphibolite facies migmatites (partially melted banded rocks), mica schists, and low temperature–low pressure phyllites (foliated mica-rich rock). The Late Precambrian and Lower Paleozoic mostly mudstone-sandstone successions had been overprinted by metamorphism applied during the Variscan Orogeny.

During the Variscan Orogeny, the present-day Axial Zone and South Pyrenean Zone were part of Iberia. The Serres Marginals were part of the Ebro Block, a northeastern section of Iberia. The North Pyrenean and Aquitainian zones were part of the microcontinent Aquitania. Iberia and Aquitania were both part of the northern margin of continental Gondwana (the Southern Continent), which became the southern foreland of the Variscan Orogen. At the close of the Variscan Orogeny, Iberia was still connected to northwest France (the Armorican Massif) and most likely was attached to Aquitania.

The first indications of Variscan-related uplift and erosion appear in the western Pyrenees in Early Carboniferous (Mississippian) time. Pre-Variscan passive sedimentation evolved into thick, synorogenic sedimentation, primarily turbidites derived from Variscan uplift (Wikipedia, Geology of Pyrenees). The main Variscan Orogeny is expressed as a major unconformity dated about 310 Ma. Sediments deposited above the unconformity are not metamorphic or only weakly metamorphosed, whereas the sediments below the unconformity experienced high temperature–low pressure metamorphism. Compressional stresses imposed Schistosity and folded the Paleozoic sediments. Late Variscan magmatism emplaced granodiorites and biotite granites between 310 and 270 Ma (Late Carboniferous to Early Permian).

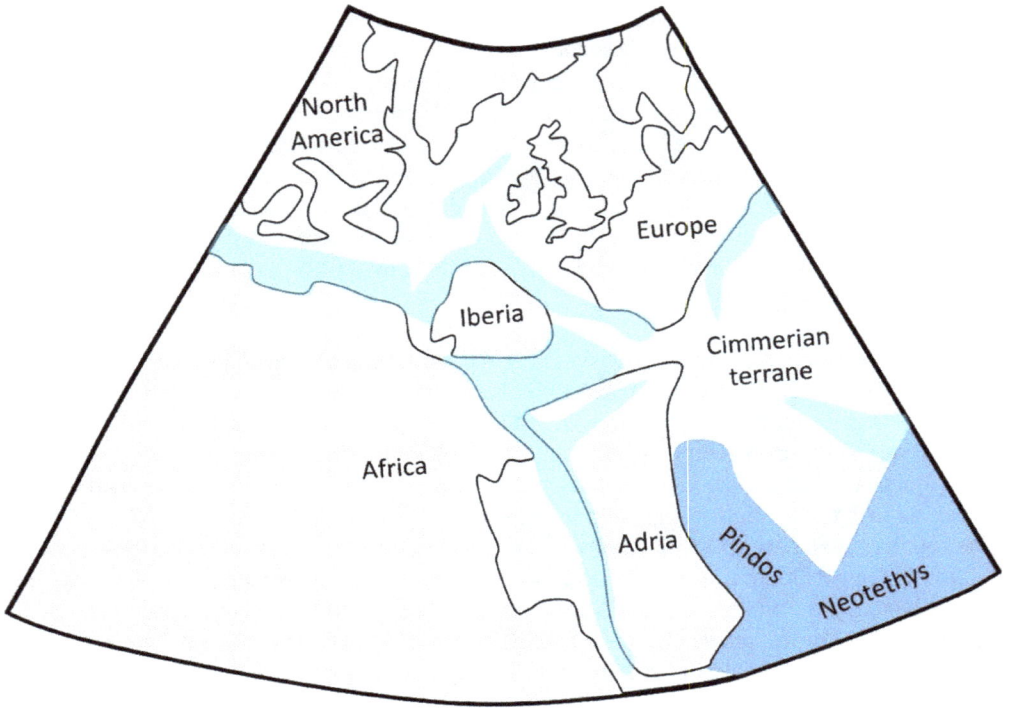

Early Jurassic (200 Ma) plate tectonic reconstruction showing deep basins to extensions of the ocean to shallow continental shelves as dark to light shades of blue. Modified after Angrand et al., 2020.

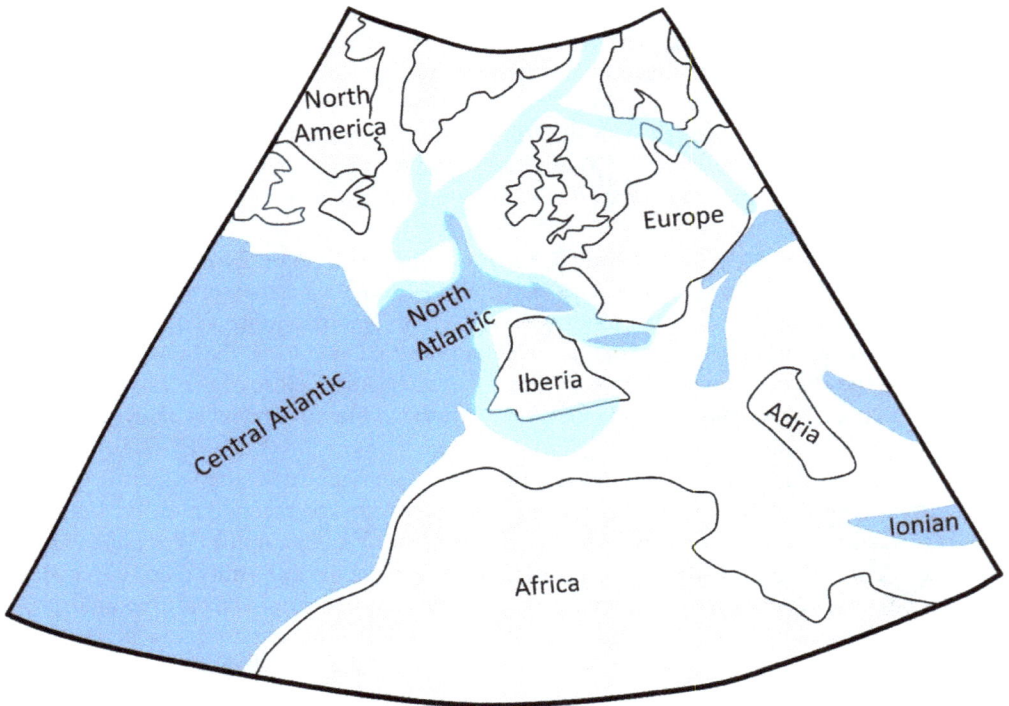

Early Cretaceous (100 Ma) plate tectonic reconstruction. Modified after Angrand et al., 2020

Middle Triassic to Upper Jurassic sediments are similar on both sides of the Pyrenees because most of the area was continuous and below sea level. By the Upper Triassic, the sedimentation spread over the entire Pyrenees area. About 220 million years ago evaporites (gypsum and salt) were deposited in lagoons and shallow basins where evaporation exceeded seawater inflow. The structurally weak and ductile evaporites would later become major detachment and thrust horizons. Upper Triassic basalts erupted along fissures in the Pyrenees area and southern Aquitaine Basin to form pillow lavas. In earliest Jurassic time sea levels rose and fossil-bearing limestones were deposited. Carbonate platforms indicate shallow marine conditions far from land. The basin was undergoing roughly north-south extension during this period, and west–northwest-elongated fault-bounded uplifts and basins developed, more or less following the old Variscan fracture trends. The northern margin was rimmed by the stable Aquitanian continental shelf.

Falling sea levels in Early Jurassic time deposited yet more evaporites; later these dissolved, creating dolomite breccias, layers of broken rock. Later in the Early Jurassic sea levels rose again and fossil-bearing limestones were deposited in open marine settings while oolitic limestones (limestone composed of small spherical grains of calcium carbonate) accumulated in shallow waters. Sea levels fell during the Middle Jurassic. The basin developed a deeper western area open to the Atlantic and a shallow, enclosed eastern domain (Wikipedia, Geology of Pyrenees).

Minor Late Jurassic sedimentation is represented by shallow platform sediments in the west and by dolomite deposition in the east. Latest Jurassic saw a complete withdrawal of the sea.

During Late Jurassic a rift propagated towards Aquitainia (the Aquitaine Basin region of southern France) from the spreading Mid-Atlantic Ridge that was located along the continental margin of present-day northwestern France. The rift wedged Iberia southward, separating it from the Armorican Massif and opening the Bay of Biscay by Cretaceous time. Paleomagnetic studies indicate a 35° counterclockwise rotation of Iberia during Lower Cretaceous. Rotation of Iberia caused rift basins along the North Pyrenean Zone in middle Cretaceous time (105–100 Ma; Vergés and Garcia-Senz, 2001; van Hinsbergen et al., 2020; Wikipedia, Geology of Pyrenees). Pyrenean rifting created an arm of the sea that connected the Bay of Biscay in the Atlantic with the Tethys Ocean (sometimes called the Alpine Ocean) to the east (Vergés et al., 2002).

During times of falling sea levels evaporitic, dolomitic, lagoonal, and lake sediments were deposited. Most of Cretaceous time is characterized by the interplay of carbonates, shales, and evaporites due to sea level fluctuation between shallow marine shelf and near-shore settings. The source region of sediments at this time was the Aragon/Pyrenees area that was undergoing early, broad regional uplift. River delta sediments were carried north, and up to 1,000 m (3,300 ft) of conglomerates are interpreted as delta front deposits (Wikipedia, Geology of Pyrenees).

Just before Late Cretaceous time the Pyrenees region had shelf carbonates deposited in the shallow South Pyrenean and Axial zones, and flysch/turbidites deposited in the deeper North Pyrenean Zone. A strongly subsiding North Pyrenean Basin developed along east–west-trending Variscan fracture zones. The basin deepened toward the Atlantic and shallowed to the east. The North Pyrenean Basin was bounded on the north by the relatively stable and shallow marine Aquitaine Shelf.

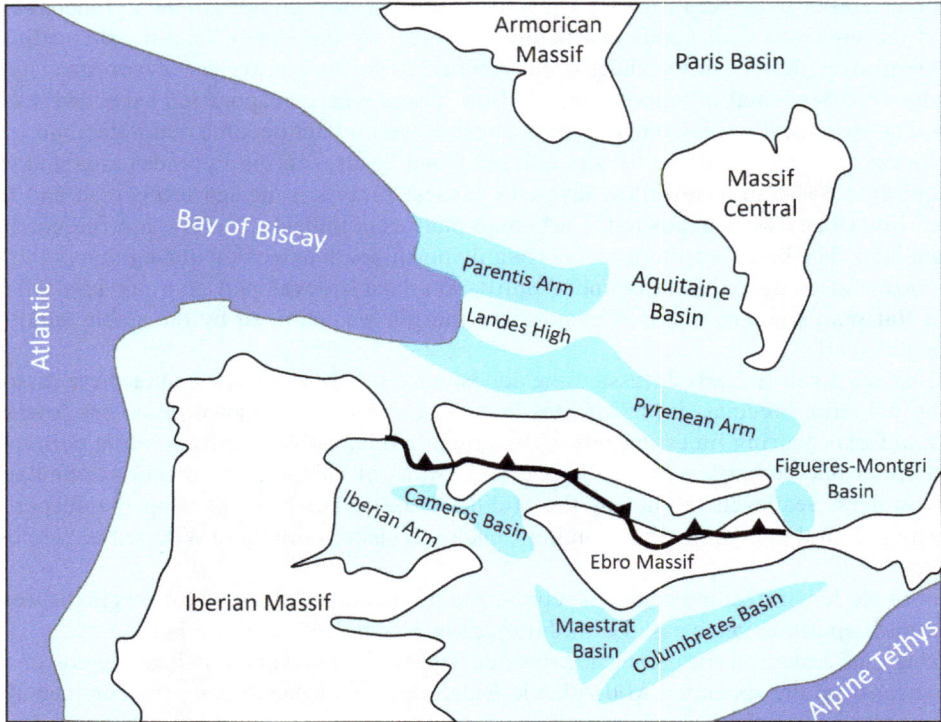

Late Cretaceous Pyrenean rift system. Modified after Vergés et al., 2002.

About 90 million years ago (Late Cretaceous) the extension was replaced by compression. North-moving Iberia is thought to have been partially subducted beneath (forced under) Aquitania. When the subduction process ended, parts of the northern Axial Zone and the North Pyrenean Zone had been pushed northward over the Sub-Pyrenean Zone of the Aquitaine Basin. Basin inversion (uplift) caused originally extensional faults to be reactivated as reverse faults and thrusts. Along the northern margin of Iberia, a thick wedge of deformed sediments, an orogenic wedge, formed and moved slowly to the north. Sediments eroded from the orogenic wedge were deposited in a deep foreland basin to the north, with both wedge and foreland basin migrating northward with continued Pyrenean uplift. Middle Cretaceous submarine lavas erupted in the western part of the North Pyrenean Zone, while farther east magmas intruded Upper Cretaceous strata. On the Spanish side of the orogen, the first thrust sheets (Upper Pedraforca, Bóixols, and Turbón thrusts) were emplaced. Around 55 Ma the Pyrenees suffered strong compression, the Pyrenean Main Phase that lasted until about 47 Ma, when the Lower Pedraforca and Montsec thrust sheets were emplaced in the Southern Pyrenean Zone. Other compressional deformational phases followed during the Oligocene and the Pliocene. Estimates of the overall shortening across the Pyrenean orogen range from 100 to 165 km (60 to 100 mi; Schellart, 2002; Wikipedia, Geology of Pyrenees). In the western Central Pyrenees, total shortening is estimated on the order of 75 to 80 km (45 to 48 mi). This difference is the result of greater convergence between Iberia and Europe in the east than in the west (Schellart, 2002; van Hinsbergen et al., 2020).

The first Alpine shortening and uplift affected mainly the eastern Pyrenees. Marine sedimentation continued in the western Pyrenees during the Eocene. Limestones, marls, and sandstones were deposited both north and south of the rising mountains. During the early Eocene the first conglomerates were deposited, a key indicator of the Pyrenean Main Phase strong deformations and uplifts ~50 to 40 Ma. On the south side of the Pyrenees in Catalonia, folded conglomerates have been dated as middle Eocene (44 to 37 Ma). Outstanding examples of upper Eocene–lower Oligocene growth strata in the Sant Llorenç de Morunys area are related to early Pyrenees deformation (Vergés et al., 2002). The Pyrenean Main Phase manifested itself on both sides of the Axial Zone as steep reverse faults and low-angle, large displacement thrusts. North of the Axial Zone thrusting was to the north; south of the Axial Zone thrusting was south-directed. Less intense deformation followed the Pyrenean Main Phase. For example, along the northern margin of the Ebro Basin folded Oligocene is overlain unconformably by flat-lying Miocene continental units, indicating Oligocene deformation (~25 Ma; Wikipedia, Geology of Pyrenees).

Since the Neogene, the Pyrenees have undergone post-orogenic extensional collapse as indicated by graben structures in the east and volcanism associated with the extension in the Golfe du Lion and opening of the offshore Valencia Trough. Since the middle Oligocene opening of the Valencia Trough, an extensional fault system has developed parallel to the coast of northeast Spain. Crustal-mantle extension and thinning produced uplift of the Catalan Coastal Ranges and the southeast margin of the Ebro Basin. Over 1.5 km (4,900 ft) of uplift is responsible for significant erosion of this margin creating the resistant Eocene conglomerate fan deltas and alluvial fans of Montserrat and Sant Llorenç del Munt (Vergés et al., 2002).

Post-Alpine extension in the eastern Pyrenees and slight compression in the western Pyrenees is thought to be the cause of medium-sized earthquakes such as the magnitude ≥ 6.0 quake near Arette in 1967, the magnitude 5.1 quake near Arudy in 1980, and the magnitude 5.0 earthquake near Lourdes in 2006 (Vergés et al., 2002; Wikipedia, Geology of Pyrenees).

From the Miocene onward the Pyrenees were subjected to severe erosion, expressed as thick molasse deposits (shale, sandstone, and conglomerate deposited in front of rising mountains) in both the northern and the southern foreland basins. Renewed uplift began in Pliocene time, leading to the formation of huge alluvial fans and conglomerates along the mountain front.

During the Quaternary, the Pyrenees experienced several glaciations. These were by and large less intense than in the Alps. Today about 20 small glaciers remain. All these glaciers have been retreating since about 1850 (Wikipedia, Geology of Pyrenees).

The Pyrenees and adjacent basins have been the target of petroleum exploration, and they are a classic locality for studying the interaction of thrust belts and sedimentation.

Hydrocarbons

Oil seeps have been known since Roman times, but the first gas discovery wasn't made in this area until 1939. Whereas several potential oil and gas source-rocks occur in Paleozoic rocks, the best are in Mesozoic syn- and post-rift strata, as well as in the Paleogene fill of the foreland basins.

A small petroleum system in the southern Pyrenees in Spain contains Eocene source rocks deposited during the early Pyrenean Orogeny in the marine foreland basin. These source rocks generated a small gas field. Early Eocene turbidites and calcareous breccias of the Hecho Group are the reservoir rock for the Serrablo gas field on the northern flank of the Jaca Basin. This field produced 930 million m³ (32.8 bcf) gas. The hydrocarbon potential of the South Pyrenean basins is low because of the lack of an extensive source rock, intense deformation, and deep erosion (Vergés and Garcia-Senz, 2001; Kendall et al., 2020).

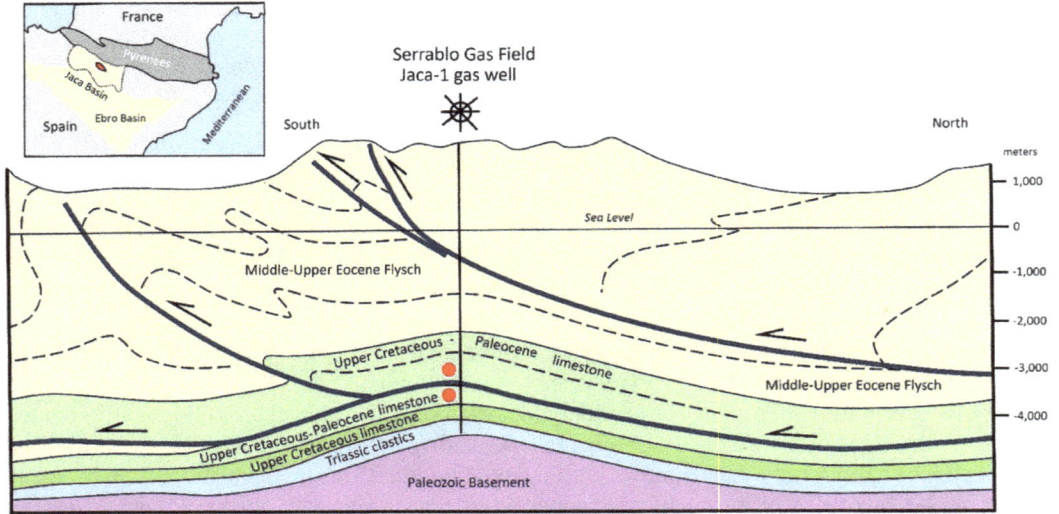

Cross-section through the Serrablo Gas Field, Jaca Basin. Producing interval shown as red dots. Modified after Meléndez-Hevia and de Buergo, 1996.

The Aquitaine Basin in France has a significant petroleum system in and adjacent to the Pyrenees. It is the largest oil and gas province in France. Most of the oil and gas fields in the Aquitaine Basin are located in Late Jurassic (e.g., the Deep Lacq Gas Field) and Late Cretaceous carbonate reservoirs (e.g., Upper Lacq Oil Field). Reservoir porosities range from 2% to 20%, with lower porosities enhanced by fracturing. The main source-rocks are Late Jurassic (e.g., Lons Formation) and Early Cretaceous marine shaly limestones and shales. Interbedded shales and anhydrites provide effective seals. Traps are Early Cretaceous rift-bounding faults, partly modified by salt diapirism, that were reactivated during Pyrenean compression. Hydrocarbon generation, expulsion and migration began

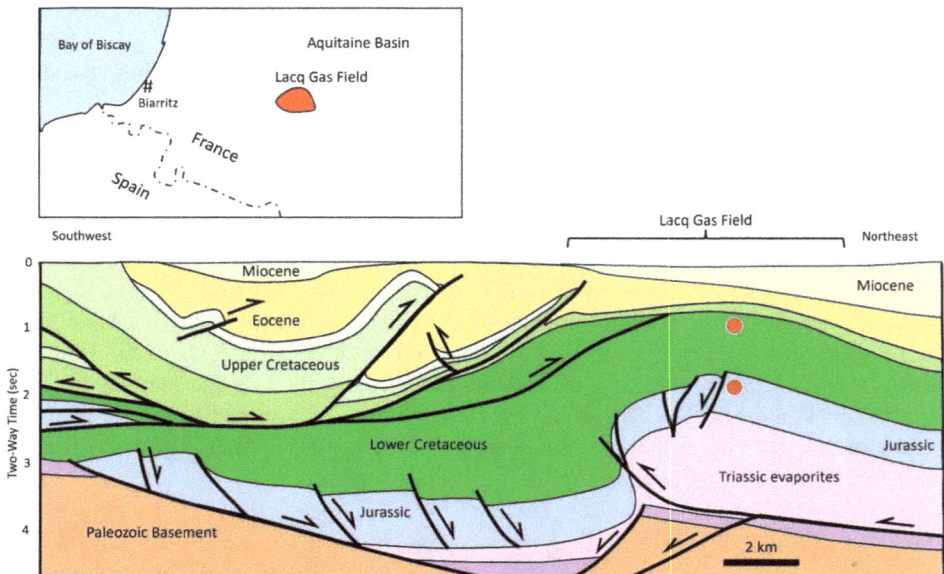

Location and cross-section through the Deep Lacq Gas Field, Aquitaine Basin, France. Producing intervals shown as red dots. Modified after Bahnan et al., 2021.

in Early Cretaceous time beneath 5–6 km (16,000–20,000 ft) of overburden, and continued until the Late Tertiary. Preexisting structural traps were modified during basin inversion and uplift, causing re-migration of the oil and gas to new traps. North Pyrenean fields contain ultimate recoverable reserves of roughly 22.5×10^6 tonnes (161 million barrels) of oil and condensate and about 287×10^9 m^3 (10.135 trillion cubic feet) of gas (Vergés and Garcia-Senz, 2001; Bourrouilh et al., 2004; Kendall et al., 2020).

The St. Marcet Gas Field was discovered in 1939 (8×10^9 m^3 or 290 billion cubic feet of gas). The discovery well was drilled on a surface anticline in the North Pyrenean foothills. (An anticline is a fold with the oldest rocks exposed in the center and usually, but not always, with the center uplifted and sides inclined downward.) This was followed by discovery of the Upper Lacq Oil Field in 1949, the giant Deep Lacq Gas Field in 1951 (260×10^9 m^3 or 9.2 tcf gas), and the Meillon Gas Field in 1965 (65×10^9 m^3 gas – 2.3 tcf). Several smaller fields (Ucha/Lacommande, Rousse, Cassourat) were also discovered during the same period. In the 1970s five fields (Pecorade, Vic Bilh, Lagrave, Castera Lou, Bonrepos-Montastruc) were discovered near the basin margin.

Mining

Classical sources mention important gold resources in the Pyrenees (Silva-Sánchez and Armada, 2023), although it has never been a notable producer of gold. Ken Degorgue, who works in a shop that sells prospecting equipment, in 2020 found a 2.25 gram gold nugget while panning in a (not named) river in the Pyrenees foothills. Most of the gold has been found as fine flakes, mainly by prospectors panning in rivers along the northern foothills of the Pyrenees (Connexion, 2020).

There is evidence of Roman silver coins being minted using Pyrenees silver (Mining Heritage of Sobrarbe). The Aulus Mountain area, on the French side of the divide near Aulus-les-Bains, has ore deposits hosted in Devonian dolomite and Late Cretaceous limestone. This area contains one of the more important silver mines, Castel-Minier. The ore minerals were remobilized into fractures in the dolomite during the Variscan Orogeny and intrusion of a Carboniferous (Late Pennsylvanian) granite massif. The Alpine/Pyrenean orogeny may have contributed fracturing and remobilization of metals. About 30 mines are reported by the French Geological Survey in the Aulus Mountain area. The main mineralization consists of cerussite (lead carbonate), massive and dispersed galena (lead sulfide), and gray copper (copper in galena containing silver and antimony). At the L'Argentière lead-silver mine the vein ore consists of galena and sphalerite (zinc sulfide) with minor chalcopyrite (copper-iron sulfide) and tetrahedrite (antimony sulfide with silver) in Late Cretaceous limestone (MacroStrat, 2015; Hansson et al., 2019; Py-Saragaglia et al., 2017).

Discovery of a silver-rich lead vein in 1830 brought wealth to the Biros Valley in the western corner of Ariège Department. For more than a century (1850–1950) zinc and silver-bearing lead were mined from sites located between 1650 m and 2070 m above sea level at Bulard and Bentaillou. Bulard, also known as Macchu Picchu, was the highest mine in Europe. Located in Precambrian to Ordovician age amphibolite-gneiss or Cambro-Ordovician metamorphosed limestone, the first galleries were excavated using pickax and black powder. Tunneling became easier after the invention of dynamite in 1867, and in 1910 the pneumatic drill was a tremendous improvement. Mining was dangerous: in addition to roof falls, miners were subject to "lead colic" and silicosis. Women sorted the ore at the bottom of the valley of Bocard d'Eylie. In 1907, at the height of mining, the Bentaillou Mine alone employed close to 500 miners working 12-hour days. In 1926 the zinc market collapsed, setting off a huge strike and closing the mine (Ariège Pyrenees).

Lead (galena) and zinc (sphalerite) deposits in the Spanish Pyrenees near Bielsa on Liena Mountain occur in Late Cretaceous dolomite and limestone near the contact with Permo-Triassic conglomerate and sandstone. Lead-zinc deposits near Serveto and Ana and in the Chistau Valley west of Gistain are also found in Late Cretaceous dolomite.

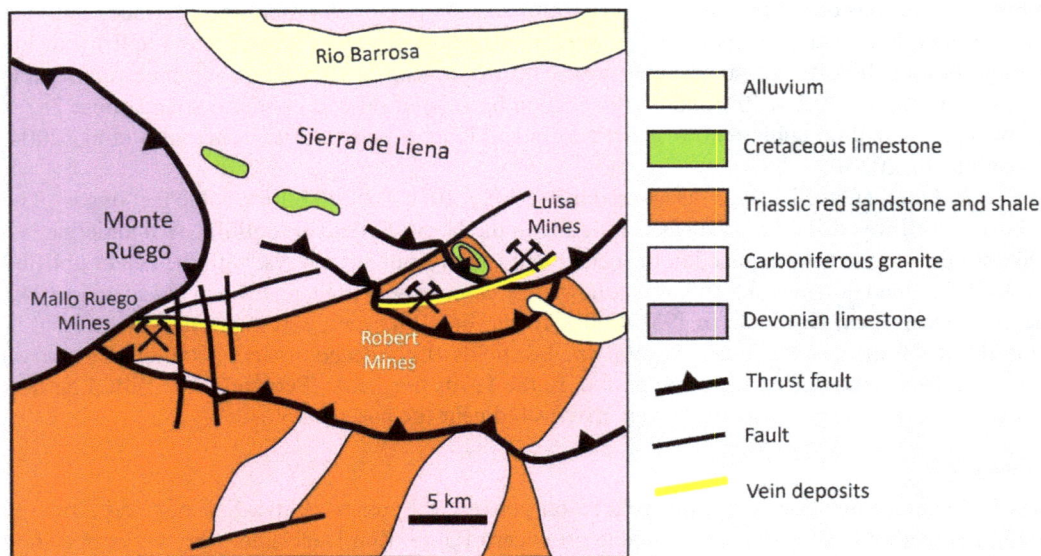

Geologic map of the mines in the Liena Mountain district. Modified after Mining Heritage of Sobrarbe.

Cobalt and nickel ore occurs in Late Cretaceous dolomite in the Chistau Valley near San Juan de Plan (Mining Heritage of Sobrarbe).

Copper mining in the Pyrenees goes back to the Bronze Age. Research has identified prehistoric copper mines in several areas (O'Brien, 2014). The Mine de Cuivre Gallo-Romaine du Goutil (43.001880, 1.386804) appears to be localized along an east-west fault in Middle-Upper Triassic dolomite in the northern Pyrenean foothills. Catalan-branded copper became common around 2,500 BCE, with iron following about a thousand years later (P-O Life, 2024).

Iron mines were located from the Vicdessos Valley to the Canigou Mountains (Py-Saragaglia et al., 2017). Iron had been mined in the Auzat and Sem valleys since the Middle Ages. Localized near the faulted contact between Precambrian migmatite and Jurassic to Upper Cretaceous impure carbonates, the iron was also exploited in Lercoul, Miglos, Larnat, and Château Verdun in the French Pyrenees. The most productive iron mining was on Mener Mountain near Parzan in the Bielsa area. Iron ores include Hematite and goethite in Late Cretaceous dolomite and limestone. The mines closed during the 1930s as the business became unprofitable in the face of increased competition (Ariège Pyrenees; Mining Heritage of Sobrarbe).

The iron mines of Rancié, located in the Nasser Rebaï, Ariège (central French Pyrenees) and Puymorens (in Cerdagne in the Pyrénées-Orientales) have been exploited since pre-Roman times. The first written documents mentioning the iron mine at Rancié are charters granted by the County of Foix in 1272. During the Napoleonic wars, desperate for iron for cannons, an engineer was assigned to manage the mine while up to 300 miners worked in teams, each one exploiting its own vein as quickly as possible. Iron was in high demand, and the number of forges tripled. Rancié increased its production sixfold between 1811 and 1841. Ultimately, the mine was abandoned in 1929 following a collapse that killed two people.

The Puymorens iron mine (Pimorent in Catalan) is in the commune of Porté-Puymorens. Mining activity dates to at least the 13th century. The mine easily supplied the Catalan forges, which were then in full development. The Catalan Forge, a medieval Spanish innovation, was designed to increase the efficiency and yield of iron smelting. The appearance of the English blast furnace in the department at Tarascon in 1868 led to abandonment of the classic Catalan Forge. After a long history of staff reductions, mining formally stopped in 1960 due to competition with iron from Lorraine and Mauritania (Saqalli et al., 2022).

In and around the Canigou Massif, there are two forms of iron ore hosted in Cambrian limestone formations. Originally deposited as carbonate iron ores (ferric carbonate, $FeCO_3$), some of these have been altered into brown hematite (hydrated ferric oxide Fe_2O_3). Canigou ore is of a particularly high quality. The earliest iron surface outcrops were known and worked by the Romans, and archaeological evidence has dated sites in the village of Taurinya, on the northern slopes of the Canigou, as the oldest Roman workings. Monastic archives in the region reveal that the monks were aware of iron resources and were involved in iron mining as early as 855. Jump forward 1,100 years and, in the early 1960s, the French government decided to invest in tourism in the coastal areas of the region rather than put more money in a declining mining industry. Mines were already closing: Escoums in 1961, Aytua in 1962, Taurinya in 1963. In spite of protests, the Canigou iron mining ended when the Batère Mine closed in the 1980s (Carter and Claughton, 2018).

Marble quarries in the Couserans date back to Roman times, and, since then, Aubert à Moulis marble has been highly sought after. With contrasting black limestone and white calcite, this stunning Upper Triassic–Lower Jurassic marble breccia adorns some of the great monuments in the world, from the Hagia Sophia in Istanbul to Les Invalides in Paris.

The largest open talc quarry in the world is located in Lower Cretaceous schist at Luzenac, 7.3 km west of Ax-les-Thermes. At an altitude of 1,800 m, it produces more than 400 tons of talc per year (Ariège Pyrenees).

Perhaps as important as any metal, salt has been mined in the Pyrenees since pre-Roman times. Salt was mined from the Triassic Keuper and Eocene Cardona evaporites. One of the largest mines was near the village of Salinas de Trillo. Kaolinite and gypsum were also mined from the same units. We visit an Eocene salt diapir that is being mined at Cardona, an Eocene salt-cored anticline at Salent, Triassic Keuper salt springs at Gerri de la Sal, and Triassic Keuper salt springs at Les Salines de Vilanova de la Sal.

A BRIEF BACKGROUND HISTORY OF THE PYRENEES REGION

The Pyrenees region includes parts of Spain, France, and Andorra. The local people speak Spanish and French, Basque, Aragonese, Catalan, and Occitan. Basque, found on the Atlantic coast and adjacent mountains, is considered the oldest language in Europe and does not belong to the Indo-European language family. Aragonese is spoken locally south of the mountain divide in Aragon; Occitan is spoken north of the divide; both are local idioms. Catalan is the official language in Catalonia, Valencia, the Balearic Islands, and Andorra. Catalan developed out of common Latin, just as did French and Spanish. Thus, Catalan often sounds like a mixture of French and Spanish words. Neither Castilian (Spanish) nor French are indigenous in the Pyrenees. They are spoken as a result of the language policies of governments in Madrid and Paris (Edelmayer, 2018).

The region's creation myth centers on Pyrene, a mythical princess and daughter of Bebryx, a king in Mediterranean Gaul (France). Hercules, during his quest to steal the cattle of Geryon (one of his Labors), stays with Bebryx, gets drunk, and rapes his host's daughter. Pyrene gives birth to a serpent and escapes to the forest, fearing her father's anger. Distraught, she tells her story to the trees, but attracts the attention of wild beasts who tear her to pieces. After his victory over Geryon, Hercules returns to the kingdom of Bebryx and finds the girl's remains. The now sober Hercules is struck with remorse and demands that the nearby mountains join in mourning and preserve her name: "struck by the Herculean voice, the mountaintops shudder at the ridges; he kept crying out with a sorrowful noise 'Pyrene!' and all the rock-cliffs and wild-beast haunts echo back 'Pyrene!'" (Mountain Bug, History of the Pyrenees).

The first documented human presence in the Pyrenees, near La Margineda (Andorra), is from the Epipaleolithic period (20,000 to 10,000 years ago). Around the fifth millennium BCE the Neolithic Revolution marked the transition from a hunter-gatherer economy to an economy based on agriculture and livestock.

From approximately 1100 BCE to the middle of the 3rd century BCE, contact with Mediterranean civilizations was mainly with Phoenicians along the coast. Their territories extended from the peninsula's South Atlantic coast to Iberia's Mediterranean coast. Greek settlements stretched from the mouth of the Ebro River to the Gulf of Roses in northeasternmost Spain. The interior was inhabited by various tribes, mostly herders who migrated from pastures in the northern highlands in the summer to the southern central plateau in the winter. The earliest written records about the peninsula date from this period (La Moncloa, Social History of the Pyrenees). Celtic Iberian tribes mentioned in ancient sources include Bebrics in the Central Pyrenees, Iacetans in the Aragonese Pyrenees, Cerretani in Cerdanya, Andosins in Andorra, Airenosis in the Aran Valley, Castellanis in Ripollès, and Basques (called Vascones by the Romans) in the southern Pyrenees (Mostaslavski, 2019).

Between 300 and 200 BCE the Phoenicians and Greeks were replaced by Romans and Carthaginians.

Romans came to Iberia to take control of the western Mediterranean from Carthage during the 2nd century BCE. The rich agricultural and mineral resources in the south were a strategic objective. The Roman conquest of the peninsula took place between 218 and 19 BCE While Hannibal was making his legendary journey across the Alps (218 BCE), Roman legions were attacking his Spanish base at Carthago Nova (Cartagena). Hannibal's defeat by Publius Scipio in 202 BCE was the beginning of the end for Carthage (La Moncloa, Social History of the Pyrenees). Gradually the Romans extended their influence with military campaigns, by building new Roman cities and by doling out land to veterans. This mainly affected the Mediterranean coast and valleys: the introduction of Roman culture in the Pyrenees was slower. By 44 BCE the Romans had conquered Gaul (France) and the Pyrenees were integrated into the Roman Empire (Mostaslavski, 2019).

With the fall of Rome in 476 CE, various Germanic tribes invaded the peninsula and settled there. Successive invasions of the Alemanni, the Vandals, and the Visigoths swept the region. By the mid 500s the Suebi were in the west (Galicia), the Cantabrian-Pyrenean shepherds were in the north, and the Byzantines (Eastern Roman Empire) were in the south (Baetica). The Visigoths had their capital at Toledo. Integration of the Visigoths and local Romans was facilitated by the conversion of King Reccared to Christianity at the Third Council of Toledo in 589. Both cultures had a landed aristocracy that believed in the autonomy of the nobility, which led to conflicts. Eventually one aggrieved clan, the Witiza family, asked for help from the Arab/Berber armies across the Strait of Gibraltar. Discord in the Visigoth kingdom allowed the Muslims to make deals with various nobles. In 711 CE, Ṭāriq ibn Ziyād landed on the Iberian Peninsula with an army primarily composed of Berbers, initiating the Umayyad conquest.In 711 CE, Ṭāriq ibn Ziyād landed on the Iberian Peninsula with an army primarily composed of Berbers, initiating the Umayyad conquest. In 711 CE, Ṭāriq ibn Ziyād landed on the Iberian Peninsula with an army primarily composed of Berbers, initiating the Umayyad conquest. In 711 CE, Ṭāriq ibn Ziyād landed on the Iberian Peninsula with an army primarily composed of Berbers, initiating the Umayyad conquest.

The invaders were Muslims from the the northwest coast of Africa and created the state of Al-Andalus. It is often forgotten that, in one form or another the Muslims occupied the Iberian Peninsula for 781 years, contributing much to later Spanish culture and language. By the mid-8th century, the Muslims had occupied much of Iberia and the Umayyad prince Abd al-Rahman became emir based in Cordoba. The Frankish king Charles Martel beat back an Umayyad invasion of Aquitaine at the Battle of Tours (732). The Moors were again repulsed by the Frankish emperor Charlemagne, son of Charles Martel, in 795. A buffer zone, the Marca Hispanica (the Spanish March), was created in and south of the Pyrenees to be the frontier between the Franks and Moors. The County of Barcelona (801) and the kingdoms of Aragon and Pamplona (825) were carved out of this zone. The Principality of Catalonia extended north and south of the main ridge of the mountains until the 17th century. To the west the Kingdom of Navarre controlled major roads through the Pyrenees until the 16th century. Pilgrimage routes to Santiago de Compostela, the reputed burial site of Saint James, were also of great importance in the Pyrenees region. The alleged grave was discovered in Galicia in the 9th century. Most Christian pilgrims had to cross the Pyrenees, and many used the route through Saint-Jean Pied-de-Port (Edelmayer, 2018).

In the 10th century, the Umayyad leader Abd al-Rahman III expanded Al-Andalus and became the first Spanish caliph (Mostaslavski, 2019; La Moncloa, Social History of the Pyrenees). The Christian enclaves along the Pyrenees became fiefdoms of the caliph. Muslim Spain's strength was based on economic power derived from trade, a highly developed crafts industry, and farming methods that were superior to those then used in Christian Europe. The Caliphate of Cordoba was the first urban and commercial economy to flourish in Europe since the fall of the Western Roman Empire. Hispano-Muslim scholars were noted for their contributions to medicine, mathematics, and astronomy. The Caliphate of Cordoba was weakened, however, during the 11th century as a result of intense military campaigning against the Christian north combined with fiscal pressures (La Moncloa, Social History of the Pyrenees).

The first Christian uprising had occurred in the 8th century in the mountains of Asturias, north-western Spain. By the 10th century Christian expansion had created the Kingdom of Castile. In 1143 the Atlantic margin of the peninsula became the Kingdom of Portugal. In 1230 the Kingdom of Castile united with the Kingdom of Leon under Ferdinand III. During the 12th and 13th centuries four main Christian kingdoms existed in Iberia: Portugal, Castile-Leon, Navarre, and Aragon-Catalonia.

The Muslim presence was eventually reduced to the Kingdom of Granada when, in January 1492, they were finally conquered by the armies of Ferdinand and Isabella of Castile–Aragon, completing "the Reconquest." The conquest of Granada allowed the newly united Spain to concentrate on maritime exploration. Economic interests, a seafaring tradition, and competition with neighboring Portugal had encouraged Castile to search for new commercial trade routes to the Indies. In the Treaty of Alcaçovas (1479) Afonso V of Portugal gave up his claim on the Crown of Castile and recognized Castilian possession of the Canary Islands. In return, Castile recognized Portuguese possession of the Azores and the Atlantic west of Portugal, the Cape Verde Islands, and Madeira. The Canary Islands provided Castile with a base for alternative routes to the Far East. Columbus took advantage of the Canaries to sail to the Indies and inadvertently discover America (La Moncloa, Social History of the Pyrenees).

South of the Pyrenees during the 1400s Aragon controlled both the Kingdom of Aragon, based in Zaragoza, and the Principality of Catalonia with its capital Barcelona. Both kingdoms had been ruled by the Castilian Trastámaras since 1412. After Ferdinand II of Aragon married Isabella I of Castile in 1479, they merged their realms.

The Kingdom of Navarre, with its capital at Pamplona, lay west of Aragon. Navarre extended north beyond the dividing ridge of the Pyrenees. In 1425, Blanca I of Evreux–Navarre, daughter of King Charles III of Navarre and Leonor Trastámara, took the throne of Navarre. She then married John II of Aragon. In 1484, the Kings of Navarre gained control of the Pays de Foix (near Foix) and Bigorre territories (southeast of Lourdes) in the northern Pyrenees. Blanca's granddaughter Catherine of Navarre wanted her son Henry II to marry a daughter of Louis XII of France, but Ferdinand II of Aragon did not and used this as an excuse to occupy Navarre in 1512. In 1515, the joint kingdoms of Castile and Aragon absorbed Navarre into their kingdom. The Pays de Foix and Bigorre territories were not returned to France until the 17th century (French Investment Property, Midi-Pyrénées History).

Disputes over Navarre provided an ongoing conflict between the House of Habsburg (Austria–Spain) and the House of Valois (France). Negotiations to return Navarre to the House of Albret–Foix failed, as did French attempts to reconquer the area. In 1530 Charles V, King of Spain and Archduke of Austria, withdrew from Lower Navarre because he lacked the funds to defend the area. The resulting border is still recognized today.

When the Valois line ended in France in 1589, Henry III of Navarre became heir to the French throne as Henry IV of France. From then until 1789 (the French Revolution), his successors held the title "King of France and Navarre," while his rivals south of the Pyrenees called themselves "King of Castile, Leon, Aragon, Navarre, etc."

After the outbreak of the Thirty Years' War (1618–1648), the Spanish kings Philip III and Philip IV supported their relative, Ferdinand II of the Holy Roman Empire (in central Europe). This increased tensions with France, which entered the war in 1635. The war, previously mainly in central Europe, now included the Pyrenees region. As part of this conflict Castilian troops marched into Barcelona. This led the Catalans to sign a mutual assistance pact with France and, in 1641, they declared Catalonia a republic under French protection. When the Treaty of the Pyrenees was finally signed in 1659, the Spanish had to relinquish Roussillon (along the southeast French coast) and the northern part of Cerdagne to France (Edelmayer, 2018). The main result of this treaty was that northern Catalonia became part of France, whereas Andorra continued as a vassal of the bishop of La Seu d'Urgell and the Comte de Foix.

After the French Revolution, the French Pyrenees were divided into the departments of the Pyrénées-Atlantiques, Hautes-Pyrénées, Haute-Garonne, Ariège, and Pyrénées-Orientales, also known as Northern Catalonia (Mostaslavski, 2019). The first thermal springs in the Pyrenees (e.g., Ax-les-Thermes) were transformed into full-fledged spa resorts under the Second Empire and the Third Republic (French Investment Property, Midi-Pyrénées History).

Throughout history the Pyrenees were a refuge to escape persecution. In the 13th century the Cathars, who dared to establish a Christian sect outside of Catholicism, used the Pyrenees as a safe haven. Catholics fled to Spain from France during the religious wars of the 16th century, as did many French nobles escaping the French Revolution. In the 20th century, left-wing and Republican Spaniards crossed the Pyrenees to France to escape Franco's troops (Edelmayer, 2018). During World War II, airmen shot down in France crossed the mountains to reach the relative safety of neutral Spain (Mountain Bug, History of the Pyrenees).

In 1978 the Constitution of Spain created the autonomous communities of Euskadi, Aragon, and Catalonia. Navarre continued as a Chartered Community. Likewise, in 1979 the French constitution created the regions of Aquitaine, Midi-Pyrénées, and Languedoc–Roussillon.

The Principality of Andorra, one of the smallest states in the world, has preserved its independence since 1278 despite the efforts of the French and Spanish monarchs. It was governed, in part, by the bishop of La Seu d'Urgell in Catalonia and by the Counts of Foix. Since 1589 Andorra has been ruled by the French king (now the French President) and the bishop of La Seu d'Urgell (in Catalonia). This arrangement allowed the country to retain its independence (Edelmayer, 2018). In 1991 Andorra became a parliamentary democracy (Mostaslavski, 2019).

BEGIN – BARCELONA

Ahh, Barcelona: great food, white sand beaches, and history. Hans Christian Andersen famously called Barcelona the "Paris of Spain" (Rodriguez, 2022). Barcelona is the capital and largest city of Catalonia and the second largest city in Spain. With a population of 1.6 million (2016), the city is located on the Mediterranean coastal plain between the Llobregat and Besòs rivers, and it is bounded on the north and west by the Serra de Collserola mountains, part of the Catalan Coastal Ranges. The mountains protect the city from the cold winds that blow out of the north and west, providing the city with a warm climate (average temperature is 16°C, or 61°F) and about 600 mm (23 in) of rain per year. Spanish is the most spoken language, followed by Catalan (Wikipedia, Barcelona; Rodriguez, 2022).

Barcelona was founded by either the Phoenicians or the Carthaginians, who had trading posts along the Catalonian coast. The name Barcelona comes from the ancient Iberian Baŕkeno, which evolved to became Barcino and then Barcilonum in Latin. Ruins of an early settlement here include tombs and dwellings that date to before 5,000 BCE. Under the Romans, the community was a colony called Colonia Faventia Julia Augusta Pia Barcino. Roman relics can be seen in the Plaça del Rei, part of the Barcelona City History Museum. The city was laid out in a typically Roman grid that is still visible in the historical center, the Barri Gòtic (Gothic Quarter). Some of the original Roman walls were incorporated into the Catedral Basílica Metropolitana de Barcelona, originally built in 343.

After being conquered by the Moors (North African Muslims) in 717, Moorish Barjelūnah became a prime target of the Franks. It was retaken in 801 by Charlemagne's son Louis, who made Barcelona the seat of the "Spanish March" (Marca Hispanica), a buffer zone ruled by the Counts of Barcelona. The Ebro River became the northern boundary of Moorish Spain. Barcelona eventually expanded its territory to include much of modern Catalonia. In 985 Barcelona was sacked but not occupied by the Moorish army of Almanzor. Most of the population was killed or enslaved and the city left in ruins. In 1137, the Kingdom of Aragon and the County of Barcelona merged after the marriage of Ramon Berenguer IV and Petronilla of Aragon. Barcelona became the economic and administrative center in the Kingdom of Aragon until its union with Castile in 1492 (Wikipedia, Barcelona).

Barcelona began an economic decline when Naples became the capital of the Catalan-Aragonese kingdom in 1442. The advent of the Habsburg monarchy, the rise of Turkish power in the Mediterranean, and the discovery of America all furthered this decline (Rodriguez, 2022).

The Catalan Revolt (1640–52) against Philip IV of Spain was an early attempt at creating an independent Catalonia. The revolt led to a brief period of independence under French protection that ended with Spanish reoccupation of Catalonia and a royal amnesty promising to respect Catalan customs. As if revolts and wars weren't enough, a plague from 1650 to 1654 killed half the city's population (Wikipedia, Barcelona).

After 1705, when the Catalans permitted the archduke Charles III of Austria to establish his court in Barcelona during the War of the Spanish Succession, Philip V of Spain laid siege to Barcelona and finally took it in 1714. This led to a period of prosperity spurred by the development of the cotton industry (Rodriguez, 2022). Barcelona found itself occupied again between 1808 and 1813, this time by Napoleon's troops. The war with the French left the province in ruins, but the postwar period saw the start of industrialization. By 1833 Catalonia's textile industry was using steam power, and Barcelona soon became the most important industrial city in the Mediterranean basin.

During the Spanish Civil War, Catalonia, in general, and Barcelona, in particular, were fiercely Republican. A Catalan republic was declared in Barcelona in 1931. It was the main center of Republican strength when the Civil War broke out in 1936. The fall of the city to the Nationalists in 1939 caused an exodus of civilians fleeing to the French border. The resistance of Barcelona to Franco's fascist government was not without consequences: autonomous institutions were abolished, and the use of the Catalan language was suppressed (Wikipedia, Barcelona). Only in 1977 was the Generalitat, an autonomous Catalan government, restored. Agreements with the Spanish national government, signed in 1979, outlined new areas for self-government (Rodriguez, 2022).

In 1992, Barcelona hosted the Summer Olympics, increasing the city's prestige. The Olympics are credited with driving major changes in what had until then been a largely industrial city. In preparation for the games, many industrial buildings along the seafront were demolished and 3 km (2 mi) of public beach were created. Road capacity was increased, sewage plants were built, and the amount of green space and parks increased. By 2012 Barcelona was the 12th most popular city destination in the world, and the 5th most popular in Europe. It is the most popular city to visit in Spain. It is renowned for Catalan cuisine (think jamon ibirico, paella, pastas, stews, sausages), some of the best beaches in the world, a warm climate, and eight monuments that are UNESCO World Heritage Sites, among them architectural works of Antoni Gaudí and Lluís Domènech i Montaner. Barcelona was voted the 24th most "livable city" in the world in 2015, according to the magazine *Monocle*

(Wikipedia, Barcelona). In 2017, 32 million tourists visited the city. In fact, overtourism has become a source of resentment for many inhabitants, who are being priced out of the housing market.

Since industrialization of the textile industry, manufacturing has played a large role in Barcelona history. Barcelona is now Spain's major Mediterranean port and commercial center: the Barcelona area provides about one-fifth of Spain's industrial output. Historically, the textile industry dominated Barcelona's economy, but its relative importance has declined. Chemicals, pharmaceuticals, automobiles, electronics, appliances, publishing, telecommunications, energy, and metallurgy are among Barcelona's leading industries (Rodriguez, 2022). SEAT, the largest Spanish automobile manufacturer, is headquartered in its suburbs, and a Nissan factory is located in the industrial area. Derbi, a manufacturer of motorcycles, scooters, and mopeds that is now part of Italy's Piaggio Group, is located there. According to the Global Language Monitor's annual ranking, in 2015 Barcelona was the seventh most important fashion capital in the world, after Milan and before Berlin (Wikipedia, Barcelona). Between industry and tourism, the city and Catalonia in general are thriving.

Stop 1 Montserrat Area

Montserrat is part of the Catalan Pre-Coastal Range (or Pre-Littoral Range). The highest point is Sant Jeroni, at 1,236 m (4,055 ft). It is the site of the Benedictine abbey, Santa Maria de Montserrat. "Montserrat" literally means "serrated mountain" in Catalan. Montserrat became a National Park in 1987 (Wikipedia, Montserrat).

The abbey of Santa Maria de Montserrat was founded in the 11th century and still functions, with over 70 monks in residence. It is Catalonia's most important religious retreat, and groups of young people often hike overnight to watch the sunrise. The Virgin of Montserrat is Catalonia's most popular saint, and her shrine is located in the Mare de Déu de Montserrat, next to the monastery. A basilica houses a museum with works of art by prominent painters. The Publicacions de l'Abadia de Montserrat is one of the oldest publishing houses in the world, having published its first book in 1499.

Legend has it that a statue of the Virgin of Montserrat was found here around 880. Around 1011 a monk from the monastery of Santa Maria de Ripoll came to the mountain and built the monastery of Santa Maria de Montserrat at the site of an old hermitage containing the statue of the Virgin.

The basilica of Montserrat was built in the 16th century. During Napoleon's invasion of Spain, the abbey was burned and sacked by Napoleon's troops, and many of its treasures were lost. Reconstruction began in 1811. On 11 September 1881, to coincide with the Catalan national day, Pope Leo XIII proclaimed the Virgin of Montserrat to be the patron of Catalonia. After the Spanish Civil War, a new façade of the church was built between 1942 and 1968. From the 1940s onward, Santa Maria de Montserrat Abbey became a symbol of Catalan nationalism. In fact, during Franco's rule Santa Maria de Montserrat was a sanctuary for scholars, artists, politicians, and students sought by the fascist regime.

The monastery has a museum divided into three different sections:

- Modern painting, with works by artists from Catalonia such as Salvador Dalí and Joan Miró, and non-Catalans like Pablo Picasso; as well as French impressionists such as Renoir, Monet, and Degas.
- Archeology of the biblical East, showing objects of Egypt, Cyprus, Mesopotamia, and the Holy Land.
- Painting by the likes of El Greco, Caravaggio, and Giordano.

Other collections include Montserrat iconography and religious goldsmithing (Wikipedia, Montserrat Abbey).

Geology

The Catalan Coastal Ranges, characterized by north-northwest verging folds and thrusts, were uplifted during the middle Eocene. As a result, two conglomeratic fan deltas, the Montserrat and Sant Llorenç del Munt systems, were shed northwest into the Ebro Basin (Vergés and Burbank, 1996; López-Blanco, 2006).

Montserrat and Sant Llorenç del Munt are both erosion-resistant conglomerate-dominated mountains on the southeast margin of the Ebro Basin. The two delta systems cover about 5,000 km^2 (1,930 mi^2) and are up to 1,000 m (3,280 ft) thick. The laterally continuous outcrops allow us to visualize the main depositional systems (alluvial fan, near-shore, offshore, carbonate platform) in 3D and to observe lateral changes.

The Catalan Coastal Ranges, parallel to the Mediterranean coast, formed during the Alpine Orogeny. Late Cretaceous to early Miocene north-south compression uplifted the ranges. The Ebro foreland basin is primarily a flexural response to loading by Pyrenean thrusts, but some flexure was also induced by tectonic loading of the Catalan Coastal Ranges on the southeast margin of the basin. After convergence of the Iberian and Eurasian plates was completed in Oligocene time, both compression and extension affected eastern Iberia (e.g., the Valencia Trough) during late Oligocene.

Northwest-directed basement-involved thrusting in the Coastal Ranges has been related to left-lateral northeast-southwest-oriented strike-slip faults. The thrusts come to the surface along a narrow northeast-southwest zone, the Prelitoral Range, or Cordillera Prelitoral. The folded and thrusted Prelitoral Range is the northwesternmost part of the Paleogene Catalan Coastal Ranges (López-Blanco, 2006).

The main features of the present Catalan Coastal Ranges developed during the late Oligocene–Miocene opening of the Valencia Trough in the western Mediterranean and exhibit a faulted horst-and-graben structure (uplifted and downdroped crustal blocks). The Catalan Coastal Ranges and the southeast margin of the Ebro Basin display uplift and regional north and northwest tilt of the Paleogene strata (Instituto Geológico y Minéro de España (IGME), Sabadell Sheet).

The two main conglomerate units, at Montserrat and Sant Llorenç del Munt, were deposited during the middle-late Eocene. They overlie the La Salut Sandstone Formation, a red sandstone with interbedded mudstones, siltstones, and conglomerates of Eocene age interpreted to be distal (far from the source) alluvial and river deposits (because of good sorting and rounding of the cobbles). The Montserrat Conglomerate at Montserrat is composed of dominantly Triassic limestone cobbles cemented by calcite (Wikipedia, Montserrat). The unit is interpreted as a proximal (near-source) alluvial fan because of the poor sorting and angular rock fragments. The Sant Llorenç del Munt Conglomerate is found at the Sant Llorenç del Munt and Obac massifs. This conglomerate contains cobbles mostly derived from the Paleozoic basement. It is also interpreted as a proximal alluvial fan deposit.

Laterally between the two conglomeratic units are the Vacarisses facies, consisting of red mudstones and siltstones interbedded with conglomerates. They are interpreted as distal alluvial fan deposits.

The Santa Maria Group is the marine extension of the Montserrat and Sant Llorenç del Munt Conglomerates. That is, the group represent the middle Eocene coastal and submarine parts of the fan delta complexes.

The onset of the Montserrat fan coincided with increasing tectonic subsidence in the Ebro Basin. Increasing sedimentation and subsidence recorded by the fan delta deposits point to rapid uplift in the Coastal Ranges, which resulted in increased sediment supply that eventually exceeded the available space created by both tectonic subsidence and sediment loading (López-Blanco, 2006).

Geologic map of the Montserrat area and stops. From Instituto Geológico y Minéro de España (IGME), Sabadell Sheet. Source: © NC Geological Survey of Spain (IGME).

Start – Barcelona El Prat Airport to Satin Slates: *From airport car rental (Alquiler De Coches) exit onto B-22 toward Barcelona; merge right onto C-31 toward Barcelona; take Exit 196AB toward Girona/Lleida and merge onto B-10 northbound; continue straight onto the A-2; take Exit 582A toward Olesa de Montserrat and merge onto C-55 toward Manresa; just past km 7 turn right (northeast) at the sign for La Puda and park in the parking area on the left. This is* **Stop 1.1, Satin Slates** *(41.566789, 1.880454) for a total of 45.2 km (28.1 mi; 33 min).*

Stop 1.1 Satin Slates, Catalan Coastal Ranges

At this stop we see the pre-Carboniferous (possibly Ordovician) gray to black slate that was metamorphosed during the Variscan Orogeny and carried north on the Alpine Frontal Thrust of the Catalan Coastal Ranges. These ancient rocks are similar to those found in the axial zone of the Pyrenees. Locally, Variscan (Late Carboniferous–Early Permian) granites/gneisses intruded these Paleozoic units (Instituto Geográfico Nacional, Catalan Coastal Ranges).

Satin Slates to Montserrat Fan Delta: *Continue driving north on C-55 for 2.5 km and turn left onto B-113; pull over onto a dirt track on the left and walk back 100 m (320 ft) north. This is* **Stop 1.2, Montserrat Fan Delta Conglomerate** *(41.574904, 1.858857) for a total of 2.6 km (1.6 mi; 3 min).*

Satin Slates.

Stop 1.2 Montserrat Fan Delta Conglomerate

The roadcuts at this stop provide excellent exposures of the red sandstone and conglomerate sequence near the base of the Eocene Montserrat conglomerate.

Montserrat Fan Delta to Monestir de Montserrat: *Return to C-55 and turn left; drive north on C-5 to BP-1121 and signs for Centre Ciutat Montserrat; turn left onto BP-1121 and drive to the parking area on the left. This is* **Stop 1.3, Monestir de Montserrat** *(41.596501, 1.838506) for a total of 12.6 km (7.8 mi; 17 min). Take in the view, look at the rocks, and tour the monastery.*

Stop 1.3 Monastery of Montserrat

Take a moment to soak in the awe-inspiring backdrop to this religious complex. You are standing in the heart of the Montserrat conglomerate fan delta.

Montserrat is composed of a prominent Eocene red-to-pink conglomerate. The resistance of the Montserrat conglomerate to erosion explains its topographic elevation (Wikipedia, Montserrat).

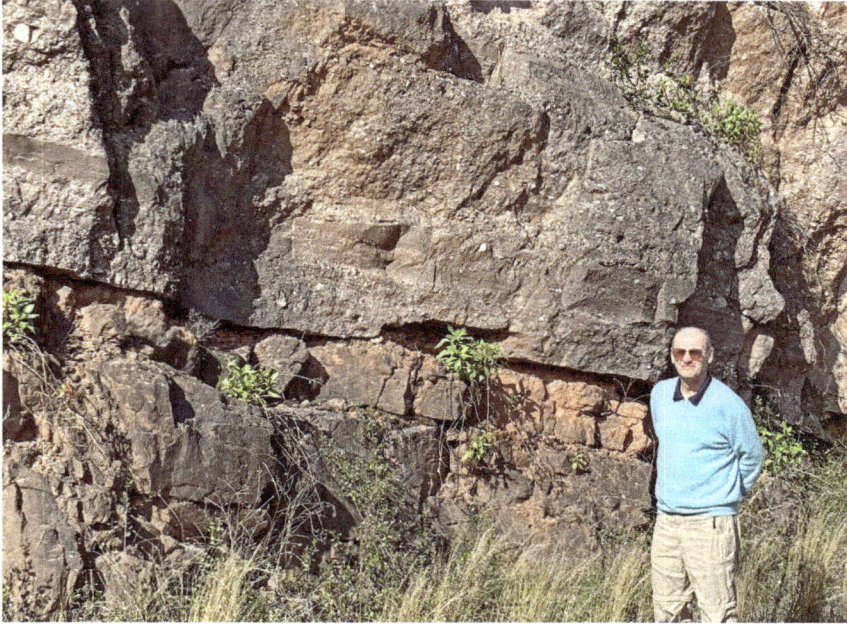

Roadcut in red Montserrat conglomerate near the intersection of C-55 and B-113.

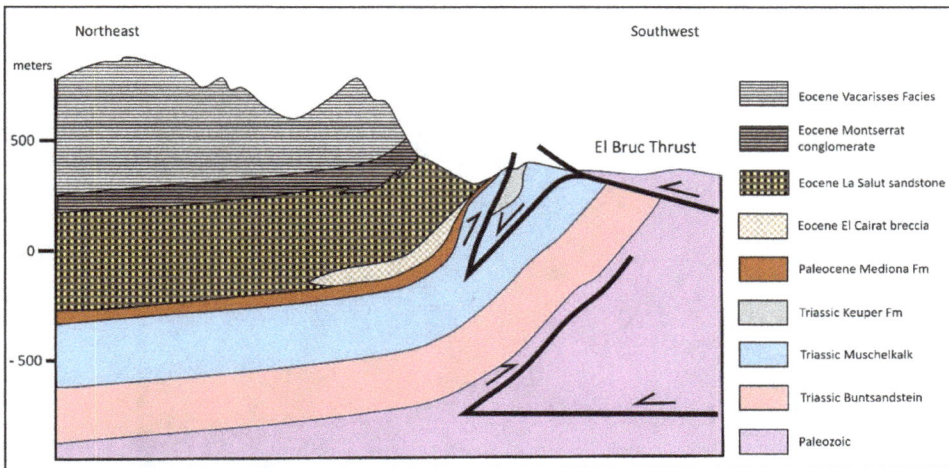

Cross-section showing units that make up Montserrat. Modified after López-Blanco, 2006.

Monestir de Montserat to Montserrat View: *Return on BP-1121 to C-55 and turn left (north); almost immediately turn right (east) onto BP-1121 and drive to Castellbell iI El Vilar; bear right to stay on BP-1121 to Estació Cementiri El Vilar; turn left (north) onto BV-1273 to Rellinars Terrassa; bear right to stay on BP-1273 to Cementiri Municipal; turn right (northeast) onto B-122 and pull over on the left. Walk back southwest to the curve and view. This is* **Stop 1.4, Montserrat View** *(41.637925, 1.874252) for a total of 15.1 km (9.4 mi; 22 min).*

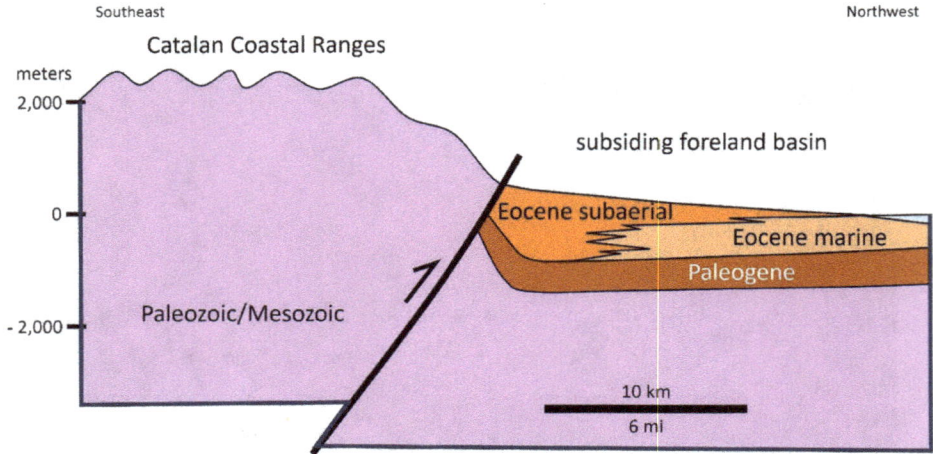

Diagram showing relationship of tectonics and sedimentation at Montserrat. Modified after López-Blanco, 2006.

Monestir de Montserrat and conglomerate spires.

STOP 1.4 MONTSERRAT VIEW

From this stop we look west to the syntectonic conglomerate pile that rises 1,300 m (4,000 ft) above the Catalan coastal plain. Notice how many of the massive resistant units at the south end become thinner and that some die out to the north. This reflects the basinward migration of the fan and the increasing distance from the source of sediments.

Looking west to the Montserrat fan delta system from Montserrat View. Notice the northward tilt caused by uplift of the Coast Ranges and subsidence of the Ebro Basin.

Montserrat View to El Migmón: Return south on BV-1273 to BP-1121; turn right (west) onto BP-1121 and drive to C-58; turn left onto the ramp to Terrassa and merge onto C-58; in 1 km merge onto C-55 north to Manresa; in Manresa use the left lane to merge onto C-25 west to Leida; in 1 km take Exit 132 to C-55 to Solsona/Andorra; take the exit to Súria Nord; at the roundabout take the 2nd exit to Carrer Masia Can Sevila; at the roundabout take the 2nd exit to C-1410z; drive to the parking area on the right. This is **Stop 2, El Migmón** *(41.837570, 1.743893) for a total of 41.48 km (25.8 mi; 38 min). Carefully walk south up the road a few meters for the view.*

STOP 2 SÚRIA ANTICLINE AND EL MIGMÓN

The town of Súria straddles the northern Súria Anticline. A thrust cutting the core of this anticline can be seen at El Migmón; it is one of the main symbols of Súria and is on the town shield. Local tradition has it that the two halves of the world separated here, forming these particular folds. The North and South Súria anticlines are mined for potash, a potassium salt used mainly for fertilizer, at depths of 450–850 m (1,480 to 2,800 ft) below the surface (All Pyrenees, 2018).

Putting this area into its geological context, we are now in the eastern Ebro foreland basin. This part of the basin has three fold trends: northwest-trending folds in the north, northeast folds in the center, and west-northwest-trending folds in the southwest. All of the folds are detached over Eocene–Oligocene evaporites, that is, the layers above the salts are folded like a rug pushed against a wall. In the Súria area the southernmost Pyrenean thrust front is marked by the northeast-trending Santa Maria d'Oló, El Guix, and Súria anticlines (Sans and Vergés, 1995). Thrust trends are related to the orientation of evaporite pinchouts: where the salt tapers out, the overlying layers can't slip and fold any more. The Santa Maria d'Oló, Súria, and Sanaija anticlines all contain deformed lower Oligocene deposits, indicating that deformation occurred after lower Oligocene time. Indications are that thrust deformation migrated southwest over time (Sans et al., 1996).

These southernmost frontal folds of the Pyrenean fold-thrust belt display an *en echelon*, staggered, or parallel and overlapping geometry. From east to west, the north-vergent (north-leaning) Santa Maria d'Oló anticline merges with the south-vergent El Guix anticline. Farther west, the Súria anticline is doubly vergent (see section on Geology of the Pyrenees, Tectonic Zones).

The Súria Anticline is a frontal structure of the fold-thrust belt transported south above the Cardona Salt and localized above the Cardona Salt pinchout. This salt goes from a typical thickness around 300 m (1,000 ft) to zero just south of Súria.

The Súria structure is characterized at the surface by two northeast-trending folds of opposite vergence: a south-verging salt-cored anticline in the north and a fold carried on a north-directed backthrust (a thrust that moves rocks opposite to the main direction of thrusting) in the south. The southern Súria anticline is a fault-bend fold in the hanging wall (thrust sheet) of a north-directed thrust. Cleavage perpendicular to bedding is developed near the thrust zones (Sans and Vergés,

1995). The southern Súria structure contains a south-directed bedding-parallel thrust within the Cardona Formation and a north-directed roof backthrust forming a "triangle zone." Salt is highly ductile and becomes mobile under pressure. Folding is characterized by an early stage when salt moves into the fold core. With increased shortening, the structure evolved to a detachment anticline and a north-directed thrust that broke through the northern anticline limb (Carola et al., 2017).

The axis of the northern Súria Anticline can be mapped for at least 35 km (21 mi). The east end of the northern anticline is symmetric and contains small, north-directed thrusts; the central section of the northern fold is inclined to the south and cored by an array of thrusts (Sans and Vergés, 1995). The northern Súria Anticline has two thrusts: a lower thrust that cuts upsection from the Cardona Formation to an intermediate detachment level in marls and limestones, and an upper thrust that cuts upward from the intermediate detachment to a higher detachment surface in shales (Sans et al., 1996). The present-day double fold geometry developed during a late stage of folding, when the northern fold formed, and the earlier southern structure tightened (Carola et al., 2017).

Shortening related to the Súria anticline increases from east to west, as does displacement of the major backthrust. The minimum displacement is estimated at 350 m (1,150 ft) in Súria, near the center of the thrust. The salt pinch-out in this area increased friction on the base detachment, generating north-directed backthrusts. Salt migrated into the cores of the anticlines during thrusting. Late-stage compression tightened the anticlinal crest and steepened the thrust. Seismic data suggests that maximum detachment and shortening occurred where salt is thickest (Sans and Vergés, 1995).

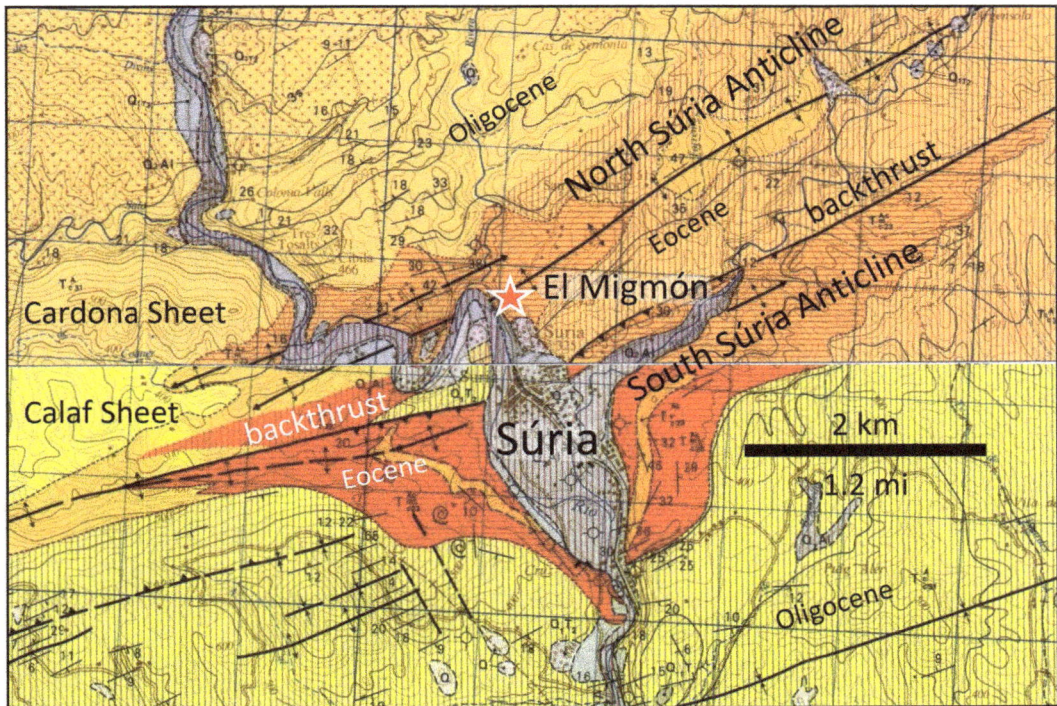

Geologic map of the North and South Súria anticlines. From Instituto Geológico y Minéro de España (IGME), Cardona and Calaf Sheets. Source: © NC Geological Survey of Spain (IGME).

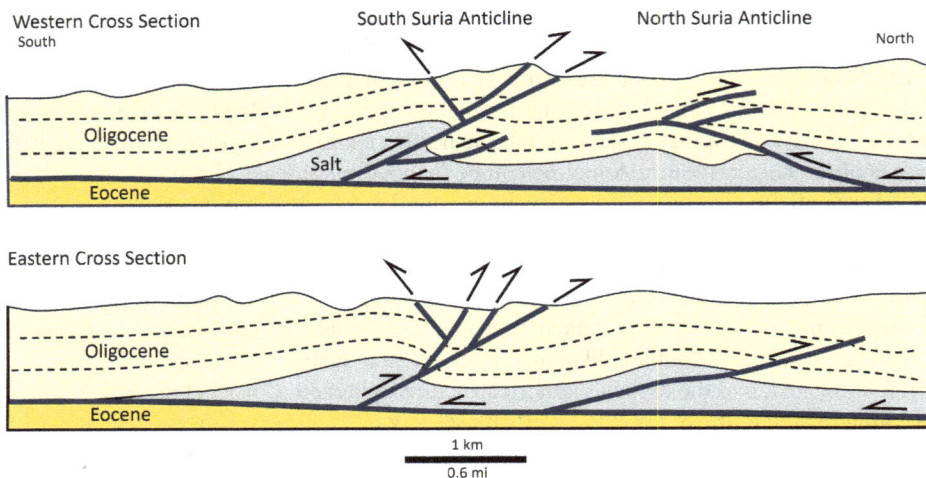

Cross-section through the Súria salt-cored anticlines. Modified after Sans et al., 1996, Bulletin of Canadian Petroleum Geology, CEGA © 1996.

Age	Sequence	Formation (environment)
Oligocene	Solsona	Berga Fm. (alluvial)
		Solsona Fm. (alluvial)
Eocene	Cardona	Barbastro Fm. (cont. evaporites)
		Cardona Fm. (marine evaporites)
	Milany	Tossa Fm. (reef limestone)
		Milany Fm. (delta front)
		Igualada Marl
	Bellmunt	Bellmunt Fm. (alluvial-fluvial)
		Banyoles Fm. (prodelta marls)
	Beuda	Beuda Fm. (marine evaporites)
	Campdevànol	Campdevànol Fm. (turbidites)
	Armàncies	Armàncies Fm. (slope marls)
		Penya Fm. (carbonate platform)
	Corones	Corones Fm. (outer platform)
	Cadí	Cadí Fm. (carbonate platform)
		Sagnari Fm. (marine marl)
		Orpí Fm. (carbonate platform)
Paleocene		Tremp Fm. (redbeds)
		Mediona Fm. (redbeds)

Paleogene stratigraphy of the southeast Ebro Basin. Stratigraphic information derived from Vergés et al., 1992.

View east across the El Cardener River valley at the North Suria Anticline. Dashed lines follow bedding. Headframes are part of the potash mining operation.

The El Migmón roadcut exposes the upper Eocene Cardona Formation gray and red marls inter-bedded with lacustrine (lake) limestones and sandstones. From north to south there are gentle north-inclined layers, then a fault zone that is inclined to the north, then steeply south-dipping layers. The fault is a zone of thrusts that cut the core of the North Súria Anticline.

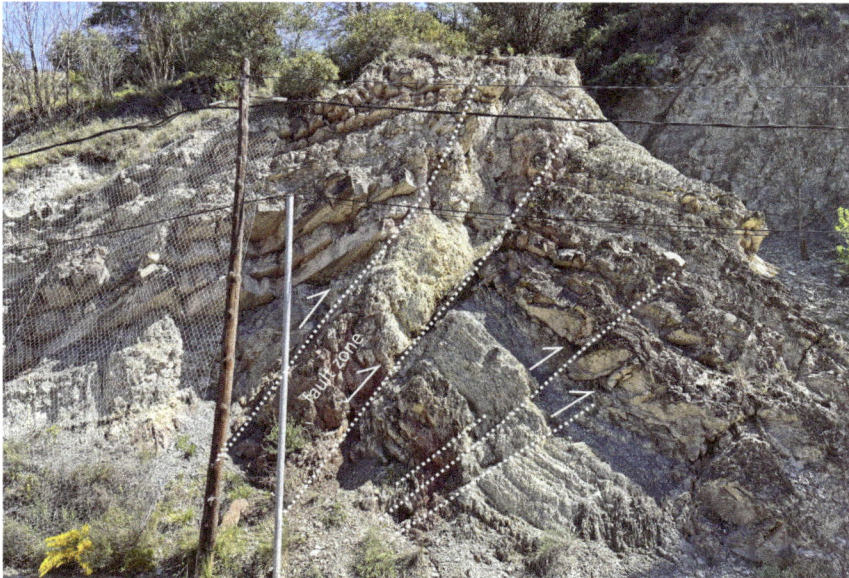

El Migmón is a break between opposite-dipping layers in the North Suria Anticline that is exposed in the C-1410z roadcut. Dotted lines are fault slivers in the Eocene Cardona Formation.

El Migmón to Cardona Salt Diapir: *Return on C-1410z to C-55; merge onto C-55 north to Cardona and take the exit to Cardona/BV-3001/SU; drive southwest on BV-3001 through Cardona; follow the signs to the Parc Cultural de la Muntanya de Sal; continue west on Carretera de la Mina to the park entrance and parking area. This is **Stop 3, Cardona Salt Diapir** (41.905971, 1.674382) for a total of 19.1 km (11.9 mi; 25 min).*

STOP 3 CARDONA SALT DIAPIR

The salt outcrops at Cardona have been worked since Neolithic times. Before refrigeration, salt was used to preserve food. It was so valuable that in places people were paid in salt, which is the origin of the word "salary." It has been the main business and source of income in Cardona since at least Roman times (Bond, 2021).

The Pinòs-Cardona Anticline extends northeast-southwest through the town. The south-inclined anticline has a gently dipping north-west flank and an almost vertical southeast limb (Mindat, Salt Mountain, Cardona). The Cardona diapir, an upwelling and thickening of salt in the core of the fold, broke through the Oligocene caprock (the rock layer that lies over the salt) and actually outcrops at the surface. Salt, highly soluble in water, rarely exists at the surface. A combination of the arid climate and continued rising of the salt allows the outcrop to remain at the surface.

A "diapir" is formed when a thick bed of evaporite minerals, usually salt but also gypsum, works its way vertically upward along fractures or other zones of weakness in the overlying rocks. This usually creates a dome or anticline. The salt, originally precipitated from evaporating seawater as relatively thin horizontal layers, eventually became buried beneath younger sediments. Since salt is less dense than other rocks, and is highly ductile when squeezed, over time and with burial pressure it tends to move up toward the surface, doming the rocks above it. Occasionally it will break through, as at Cardona.

This has been called Europe's only example of an outcropping salt diapir (Carreras, 2022). The salt is derived from seawater that filled the Ebro Basin about 40 to 36 Ma, during the Eocene. The Cardona Anticline began forming during early Oligocene time (Vergés and Burbank, 1996). The salt mountain, extending up to 100 m (328 ft) above the surrounding area, is thought to have broken through only in the last 2 Ma (Carreras, 2022). The salt diapir is roughly 1,800 m long by 600 m wide (5,900 ft by 2,000 ft) and covers an area of approximately 100 ha (250 ac; Second Wiki, Cardona Salt Mountain).

The Mine

The salt at Cardona has been exploited for thousands of years. By the Middle Ages, Cardona had become the most important salt producer in Iberia, with salt being sent across Europe from the port at Barcelona (Bond, 2021).

At the beginning of the 20th century potash, used to make fertilizer and explosives, was discovered. This led to the first underground mine, opened between 1902 and 1905. In 1986 the mines employed 560 workers. By the time the mine finally closed in 1990, a total of 37,900,000 tonnes (42,000,000 tons) of salt and potash had been mined (Second Wiki, Cardona Salt Mountain; Bond, 2021).

The mine workings extend to 1,308 m (4,292 ft) depth, with over 42 km (25 mi) of tunnels (Mindat, Salt Mountain, Cardona). Temperatures in the lowest galleries reach over 50°C (122°F; (Second Wiki, Cardona Salt Mountain). Today the mine is a tourist attraction.

Cardona Castle

Guarding the salt mines is Cardona Castle, home to the Dukes of Cardona from the 11th to the 15th centuries. In the 17th century it was transformed into a major fortification that was partially destroyed in 1714. It was rebuilt in the 18th and 19th centuries such that even Napoleon's army was unable to breach the fortress. The castle has become a symbol of Catalan nationalism and strength. Within the walls is the Church of Sant Vincenç of Cardona, built between 1019 and 1040 in the Lombard Romanesque style. Guided tours are available; see https://www.cardonaturisme.cat/en/activitats/general-interpretative-guided-visit-to-cardona-castle/ (Bond, 2021; Carreras, 2022).

Environmental Effects

Salt Mountain is part of the Sierra de Garrigues. The area is drained by the Salat, a tributary to the Cardener River. The Cardener flows into the Llobregat, which supplies 35% of Barcelona's drinking water. Salt mining in the Llobregat Basin was causing progressive salinization of the drinking water, but since 1989 a public works operated by Aigües de Barcelona (AGBAR) has removed salt and there has been a significant improvement in the quality of the river water (Carreras, 2022).

Visit

Starting in 1997 the former Cardona Salt Mines were converted into the Parc Cultural de la Muntanya de Sal. The site has the Art-Sal exhibition, a showcase of intricate salt sculptures (Bond, 2021).

There is also a museum where visitors can check out the equipment and machinery used in the mines, but the real attraction is the tunnel into the salt mountain. Guests are given hard hats and allowed to walk through 500 m (1,640 ft) of the old mine tunnels to depths of 86 m (282 ft) (Harman, 2013; Carreras, 2022; Second Wiki, Cardona Salt Mountain). Tickets for hour-long guided tours of the salt mountain are available online (Bond, 2021)

Address: Parc Cultural de la Muntanya de Sal, Carretera de la mina s/n, 08261 Cardona
Phone: 93 869 24 75
Email: informacio@cardonaturisme.cat

You can also contact the Oficina de Turisme, Avinguda del Rastrillo, s/n, 08261 Cardona

Phone: 93 869 27 98
Email: oficinaturisme@cardona.cat
Website: https://entrades.cardonaturisme.cat/en/producte/visita-pcms

The castle is open from 10:00 to 13:00 and 15:00 to 20:30.
Check the website for mine hours.

Entrance fees:
Cardona Ticket is a pass that covers three different guided tours for a single price. The tours include the Salt Mountain, Cardona Castle, and the Historic Center.
Tickets (2025) – Adults 22.00 €; Retired 18.00 €; 12–16 years 20.00 €; 7–11 years 11.00 €; 5–6 years 6.00 €; 2–4 years 3.00 €; under 2 years free.
    If you are staying in Cardona, ask at your hotel for the promotional code (applicable only to the Adult rate).

Cardona salt diapir.

Detail of salt deformation, Cardona salt mine.

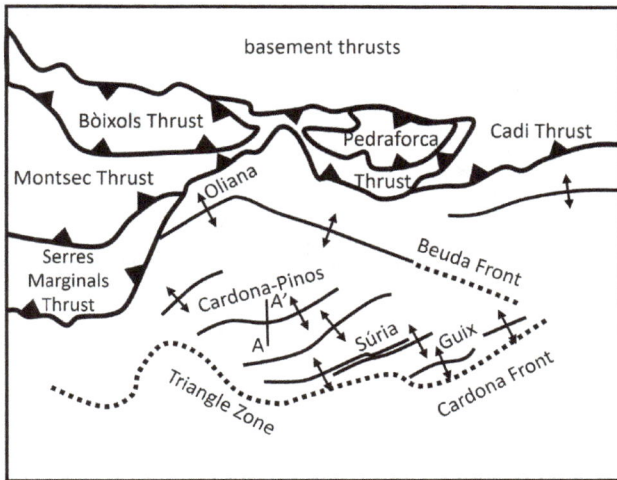

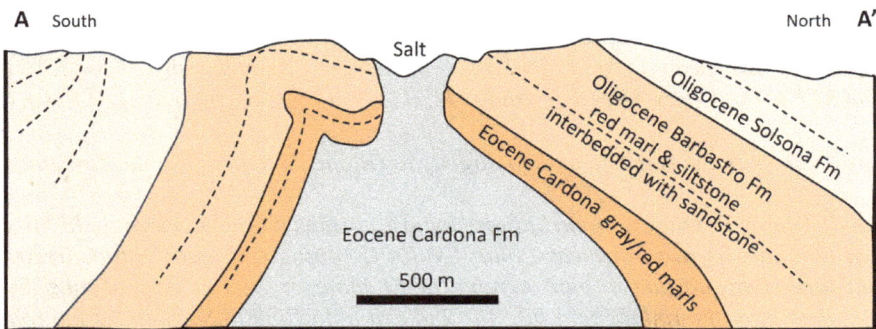

Structure map (above) showing the positions of the Cardona, Suria, and Guix salt-cored folds; (below) detailed cross-section of the Cardona Anticline and diaper. Modified after Sans et al., 1996, Bulletin of Canadian Petroleum Geology, CEGA © 1996.

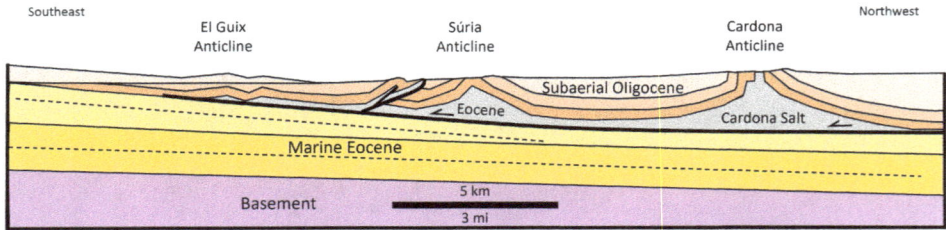

Cross-section showing the relationship of the El Guix, Suria, and Cardona anticlines to the Eocene salt detachment. Modified after Sans, 2003.

Cardona castle as seen from the salt mine.

Side Trip 1, Cardona Salt Diapir to Les Cases Altes: *Return northeast on Carretera de la Mina; at the roundabout take the 1st exit (northeast) onto BV-3001; continue straight onto Carretera Raval de St Joan; turn right (northeast) onto Arrabal de St Joan; at the roundabout take the 1st exit (north) onto B-420; at the roundabout take the 2nd exit onto C-55 to Solsona and Andorra; at the roundabout south of Solsona take the 3rd exit (north) onto C-26/Carretera de Manresa; turn left (north) onto C-462; just past km 17 pull off on the left (west) side and drive 75 m (250 ft) north. This is **Stop ST1.1, Les Cases Altes de Posada** (42.111877, 1.599246) for a total of 40 km (24.8 mi; 42 min).*

If you prefer to skip this side trip and go directly to Oliana Anticline, use the directions below.

Cardona Salt Diapir to Oligocene turbidites: *Return northeast on Carretera de la Mina; at the roundabout take the 1st exit (northeast) onto BV-3001; continue straight onto Carretera Raval de St Joan; turn right (northeast) onto Arrabal de St Joan; at the roundabout take the 1st exit (north) onto B-420; at the roundabout take the 2nd exit onto C-55 to Solsona; at the roundabout in Solsona take the 2nd exit (west) onto C-26; at the roundabout take the 1st exit and stay on C-26; continue northwest on C-26 and at the end of the road turn right (north) on C-14 and drive 1.75 km (1.1 mi) and pull off on a dirt track on the right. This is **Stop 4.1, Oligocene Turbidites, south flank Oliana Anticline** (42.030797, 1.294215) for a total of 45 km (27.9 mi; 43 min).*

Side Trip 1 Sant Llorenç de Morunys Growth Strata and Progressive Unconformities

Outcrops near Sant Llorenç de Morunys are among the best exposed examples of growth structures and progressive unconformities in the world. These occur in the northeastern Ebro Basin near the leading edge of the South Pyrenean frontal thrust.

Growth strata are layers of rock deposited over the top or against the flanks of growing structures. They allow us to decipher the development history of structures. This information is provided by layer thickness and dip changes with respect to fold hinges, and by the position of unconformities (Suppe et al., 1997). In general, layers get thinner toward the growing structure and thicker away from it. The steepness of the layers increases toward the growing structure, and flattens away from it. Progressive unconformities are erosion surfaces that cut across the inclined bedding, with more rock removed by erosion near the structure and decreasing erosion away from the structure.

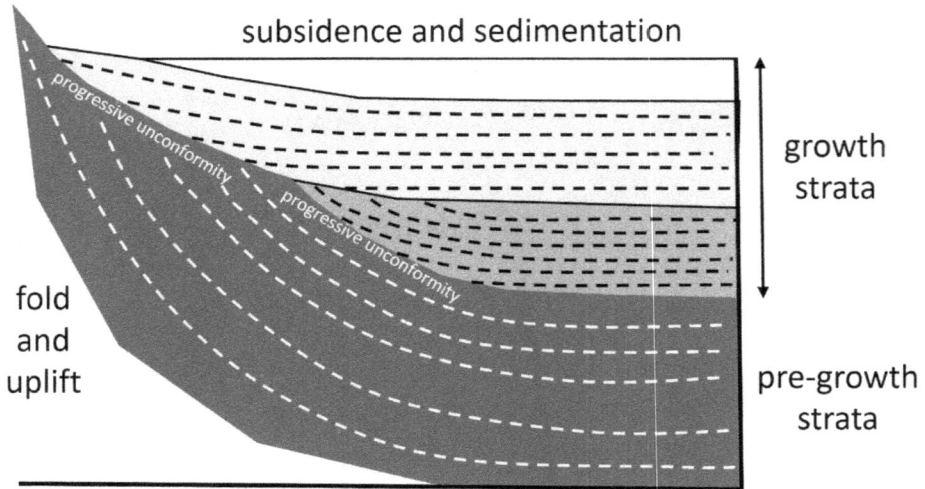

Diagram of a progressive unconformity in growth strata. Modified after Rafini and Mercier, 2002.

As previously mentioned, the Ebro foreland basin subsided at least partly in response to the emplacement of a stack of thrust sheets during south-directed thrusting from the latest Cretaceous to Oligocene time in this area. The basin fill sediments in this area record early to middle Eocene marine conditions followed by late Eocene to Oligocene continental deposition.

The Sant Llorenç de Morunys growth structure lies below the north-dipping Vallfogona blind thrust (a "blind" thrust does not come to the surface; it is mapped only in the subsurface using seismic or well data). The growth structure contains a conglomerate-rich succession, the Berga Conglomerate Formation. The > 2.4 km (7,900 ft) thick conglomerate succession is interpreted to be late Eocene-Oligocene (Ford et al., 1997).

The Sant Llorenç de Morunys structures consist of an east-west trending anticline-syncline pair. The Sant Llorenç de Morunys Anticline is north of the syncline and verges (leans) to the south. It is likely a fault-propagation fold. (Fault-propagation folds are folds that form as thrusts propagate upward through sedimentary layers; the folding is localized at the tips of these faults where displacement goes to zero.) The anticline exposes Eocene Banyoles–Igualada marls and alluvial conglomerates. The southern limb is vertical to overturned and passes into the Busa Syncline containing Eocene–Oligocene growth strata. The growth strata of the Berga Conglomerate Group consist of conglomerates, sandstones, and siltstones of alluvial, fluvial (river), and lacustrine (lake) origins (Ford et al., 1997; Carrigan et al., 2016).

Foreland basin growth strata are ideal recorders of deformation. They provide information on how sediment accumulation and distribution are affected by growing structures. Alluvial fan conglomerates accumulated by repetitive mass flows (landslides) as this structure grew, and this is reflected in (1) thinning of layers as they approach the axis of the anticline, (2) successive erosion surfaces ("progressive unconformities") that affected multiple layers deposited during active uplift of the folds, and (3) thickening of layers into the syncline during periods of tectonic subsidence. These features can be seen today in the near vertical to overturned limb of the anticline (Williams et al., 1998).

Carrigan et al. (2016) were able to date the onset of folding at 33.85 Ma, and the end of deformation as younger than 31.06 Ma. Sedimentation was nearly constant before the first unconformity, but folding rates were variable and appear to be separated into long periods of quiet punctuated by relatively rapid faulting and folding events.

This much everyone can agree on, and it is fairly easy to see the growth layers in the field. There seems to be a debate, however, over the mechanism of the folding.

A group of works that include Suppe et al. (1997) and Novoa et al. (2000) favors the process of kink-band migration to produce the geometry seen in the folds. Migration of the fold hinge (the kink band) across the limb of the fold produces things like segmented fold hinges linked by unconformities, and thickness changes in each growth bed localized in the anticlinal hinge. It is claimed that fold growth above thrust sheets with such peculiar unconformities linked by anticlinal hinges are characteristic of folding by kink-band migration and fault-bend folding (Suppe et al., 1997). Widening of the fold limb through time without significant limb rotation is seen as support for fold growth by kink-band migration (Novoa et al., 2000).

The other camp (Riba, 1973, 1976; Ford et al., 1997) interpreted the progressive changes in dip from north to south as suggestive of progressive limb rotation. Limb rotation related to fault-propagation folding explains the rounded fold hinges, beds that thicken into the syncline, and limb dip variations seen in the common limb of the fold pair. As the fold pair evolved, older strata became progressively overturned. At each stratigraphic level both hinges tightened through time as shown by cleavage developed in the syncline and shallowing of the anticlinal axial plane in older strata.

Regardless of how this debate is resolved, these outcrops are well worth seeing for the excellent exposures of growth strata.

ST1.1 Progressive Unconformities, Les Cases Altes de Posada

From Les Cases Altes de Posada you can look west across the Presa de La Llosa del Cavall (reservoir) to progressive unconformities developed in the Pont de les Cases Formation. You can see the steep common limb of the anticline-syncline fold pair and the nearly flat-lying beds of Les Cases Altes Formation in the syncline. The Busa Syncline axis is beneath the flat-topped mesa. The Santuari de Lord stop is at the northwest end of this mesa.

View west across the Val-Llonga to the southeastern slopes of Tossal de Val-Llonga and the Sant Llorenç de Morunys growth structure. From here you see the Busa Syncline. Dashed lines indicate bedding

*Les Cases Altes to Santuari de Lord: Continue north on C-462; at the stop turn left (west) on C-462/LV-4241 to Sant Llorenç de Morunys; at the roundabout take the 2nd exit (southwest) onto LV-4241/Carr. Solsona; at the curve bear left toward Santuari de Lord; turn left at the sign to Santuari de Lord and park at (42.118077, 1.581744) for a total of 10.2 km (6.3 mi; 17 min). Walk 366 m on the path to **Stop ST1.2, Santuari de Lord** (42.116438, 1.581154).*

A few kilometers north of Les Cases Altes, as you approach and go through several tunnels, the highway goes through roadcuts through the near-vertical to overturned common limb between the Busa Syncline to the south and the Sant Llorenç de Morunys Anticline to the north. The outcrops go from near-horizontal to near-vertical over a few tens of meters. These are Eocene–Oligocene syntectonic conglomerates of the Berga Conglomerate Group, rocks deposited as alluvial fans in front of the rising Pyrenees.

Steeply south-dipping El Castell and Camp de Val-Llonga formations along C-462 south of Sant Llorenç de Morunys. This is the common limb of the anticline-syncline pair. Google Street View northwest.

ST1.2 GROWTH STRATA, SANTUARI DE LORD

This stop has the best views of the growth strata and progressive unconformities that this area is famous for. Growth strata provide insights to the interaction of fold growth with active sedimentation, and they can provide clues to the folding mechanism.

These are some of the largest and best-preserved growth strata anywhere, covering several square kilometers and with over 500 m (1,640 ft) of relief. The size and quality of these outcrops are directly comparable to similar structures seen in the subsurface on seismic data (Vergés, 2022). If you take the short (but steep) trail to the top of the mesa you will see the alluvial fan conglomerates up close.

*Santuari de Lord to Oligocene Turbidites, Oliana Anticline: Return north to Sant Llorenç de Morunys and take C-462 south to Solsona; at the east side of Solsona turn left onto LV-4241/ Carretera de Sant Llorenç de Morunys; at the roundabout take the 1st exit (west) onto C-26/Av Pont; at the roundabout take the 1st exit (north) onto C-26/Carretera de Bassella; at the intersection with C-14 turn right (north) and drive 1.75 km (1.1 mi) and pull off on a dirt track on the right (east) side of the road. This is **Stop 4.1, Oligocene Turbidites, South Flank Oliana Anticline** (42.030797, 1.294215) for a total of 47.5 km (29.5 mi; 56 min). Carefully walk 120 m (400 ft) north along the side of the road to the roadcut on the right.*

Growth strata looking west from Santuari de Lord. From here you see the complex growth geometries in the anticlinal hinge. Dashed lines indicate bedding.

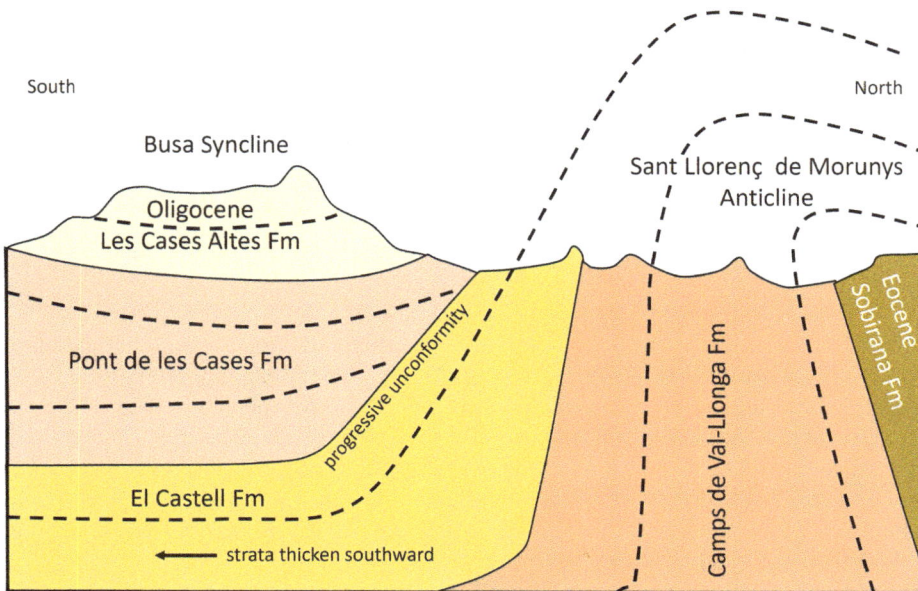

Diagram of the Sant Llorenç de Morunys anticline-syncline pair, growth strata, and progressive unconformity. Modified after Carrigan et al., 2016.

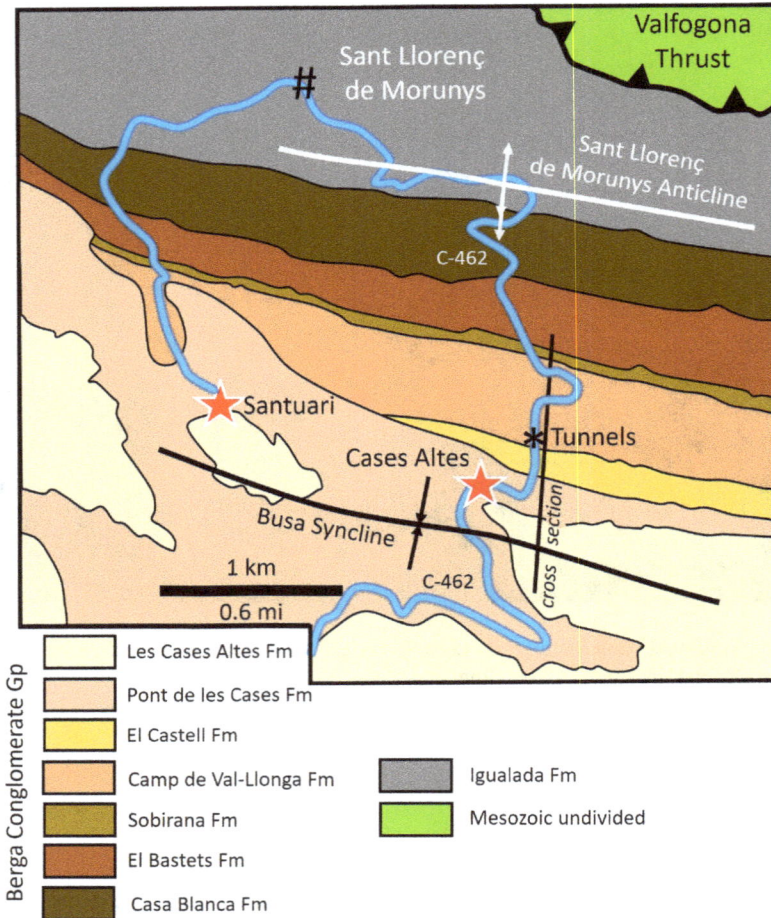

Geologic map of Eocene-Oligocene units in the Sant Llorenç de Morunys area. Modified after Carrigan et al., 2016.

STOP 4 OLIANA ANTICLINE

The Oliana Anticline is a 15-km (9 mi) long, northeast-trending, doubly plunging anticline, that is, both ends of the fold dive downward. It is cored by a duplex structure (consisting of stacked thrust faults) that has repeated the Eocene Igualada Marls (Shackleton et al., 2005). The flanks of the structure comprise four or five upper Paleogene syntectonic units consisting of conglomerates, sandstone, siltstone, and shale. The displacement of the Pyrenean thrust sheets and resulting folding caused angular unconformities in the conglomerates that border the thrusts. Between middle Eocene and early Oligocene, the marine units in this area were followed by continental deposits (Vergés and Burbank, 1996).

Eastern extensions of the Montsec and Serres Marginals thrusts merge with the northeast-south-west-striking Segre Thrust, which is the eastern oblique ramp of the South Central Unit (Burbank et al., 1992). The South Central Unit comprises all of the south-directed thrust sheets and piggyback basins in this area (Vergés, 1993). A ramp structure is where a thrust fault ramps, or cuts upward from one stratigraphic level to a higher level; an oblique ramp is a ramp more-or-less parallel to the thrust transport direction.

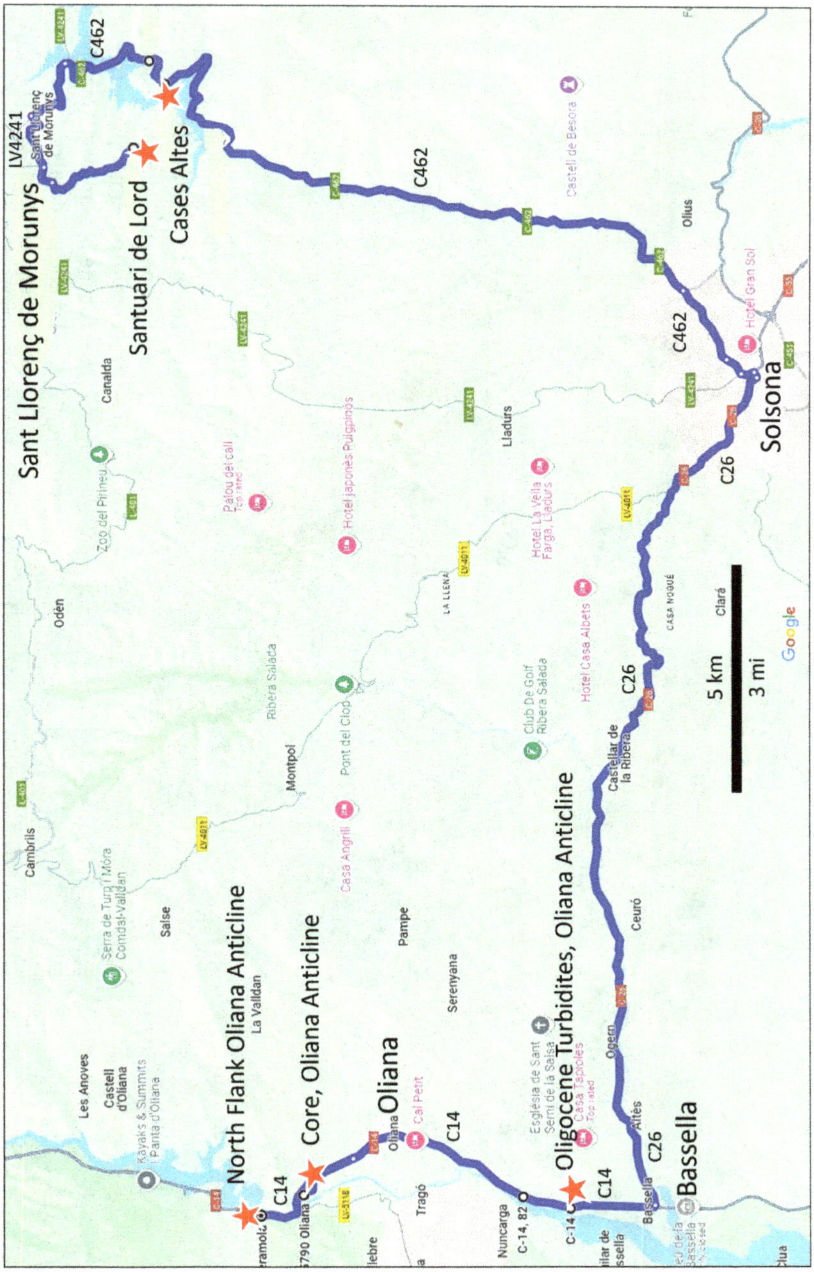
Sant Llorenç de Morunys
LV4241
C462
Santuari de Lord
Cases Altes
C462
C462
Solsona
C26
C26
5 km
3 mi
North Flank Oliana Anticline
Core, Oliana Anticline
Oliana
C14
C14
Oligocene Turbidites, Oliana Anticline
C14
C26
Bassella

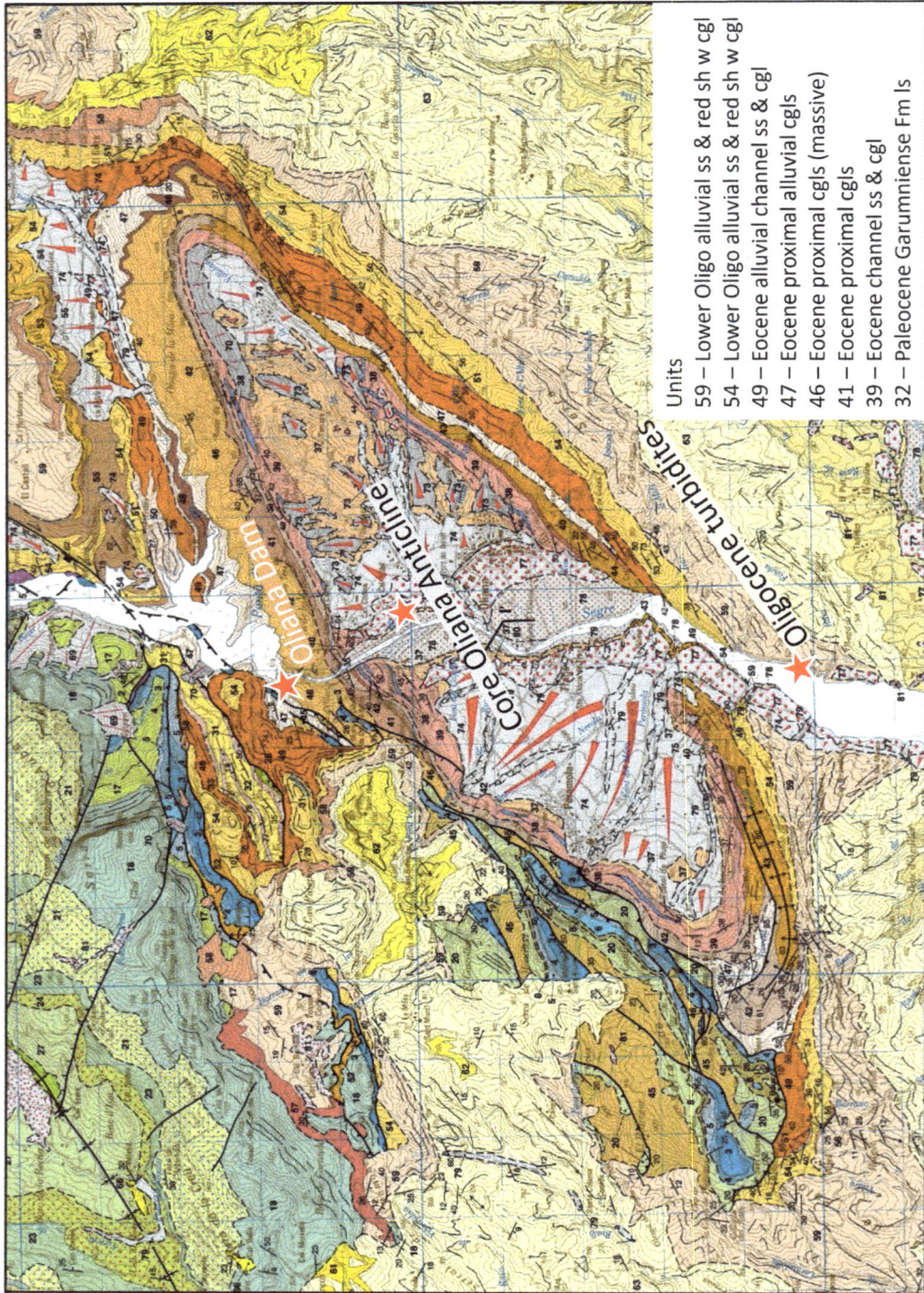

Geologic map of the Oliana Anticline and stops. Modified after Instituto Geológico y Minero de España, Oliana sheet. Source: © NC Geological Survey of Spain (IGME).

Units
59 – Lower Oligo alluvial ss & red sh w cgl
54 – Lower Oligo alluvial ss & red sh w cgl
49 – Eocene alluvial channel ss & cgl
47 – Eocene proximal alluvial cgls
46 – Eocene proximal cgls (massive)
41 – Eocene proximal cgls
39 – Eocene channel ss & cgl
32 – Paleocene Garumniense Fm ls

Oliana Dam

Oliana Anticline

Core Oliana Anticline

Oligocene turbidites

The Oliana Anticline is parallel and adjacent to the Segre Thrust. The anticline has been interpreted as a stack of three thrust sheets (Maestro-Maideu and Serra Roig, 1996). It is the northernmost major structure in the foreland. North of the Oliana Anticline, the Pyrenean thrust sheets are represented from north to south by the Boixols, Montsec, and Serres Marginals thrust sheets. Deformation along the Boixols Thrust occurred in latest Cretaceous time. The Montsec Thrust stopped moving at ~36.2 Ma, and the Serres Marginals thrusting began in this area ~39.8 Ma and continued into the Oligocene (Burbank et al., 1992). So, the thrusting gets younger to the south, toward the foreland. The total amount of shortening in the basin, including all south-directed structures south of the North Pyrenean Fault, is ~88.5 km (~53 mi; Vergés, 1993). Total shortening across the Oliana duplex structure is ~21.4 km (13 mi), about ¼ of the total shortening across the basin (Vergés and Burbank, 1996).

The Oliana Anticline and its eastern continuation, the Puig-reig Anticline, formed at the southern limit of the Eocene Beuda Formation evaporites, which served as the detachment surface for thrusting. At the termination of the Beuda evaporites, the detachment ramps up to shallower Cardona Formation evaporites. Earlier foreland thrusts near Oliana were abandoned as the deformation front progressed south and southwest. A backthrust along the south limb of the Oliana Anticline formed an early Oligocene triangle zone (Sans et al., 1996).

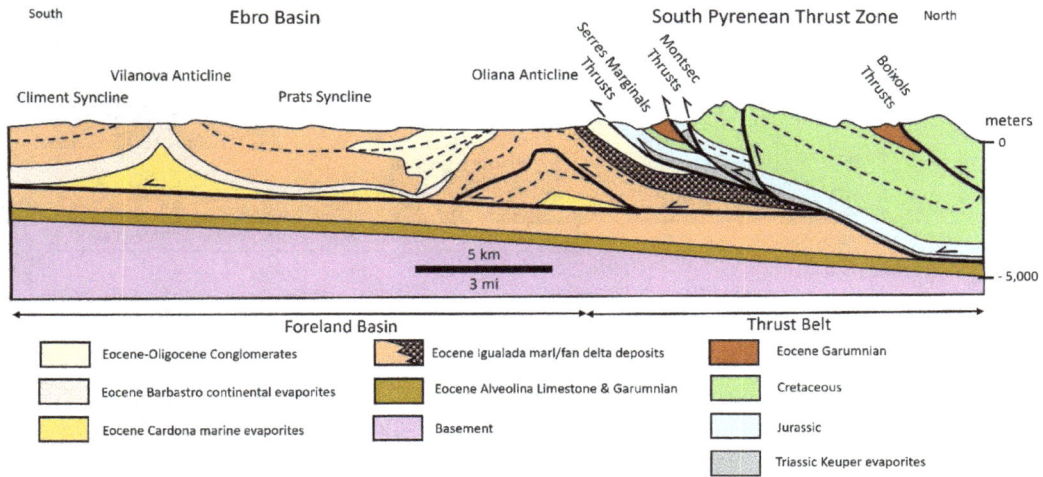

Cross-section through the Oliana Anticline. The backthrust at the south flank of the anticline is not shown. Modified after Burbank et al., 1992.

The Oliana Anticline developed between 36.5 and 34 Ma (earliest Oligocene; Vergés and Burbank, 1996). During the Oligocene, the whole structure was transported ~11 km (7 mi) south on a detachment in the Eocene Cardona evaporites (Shackleton et al., 2005).

By the late Eocene, the Ebro Basin became increasingly restricted by the encroachment of thrust sheets from both the north and the south. Marine deposition (mainly marls) was replaced by terrestrial deposition (conglomerates, sandstones, shale) after 37.2 Ma (Vergés and Burbank, 1996). Growth strata are exposed along both flanks of the Oliana Anticline as evidenced by progressively rotated joint sets and angular unconformities in the conglomeratic units (Shackleton et al., 2005).

Igualada Marl crops out in the core of the Oliana Anticline. The upper part of the marly sequence contains reef limestone in the southern limb of the anticline and alluvial fan delta conglomerates along the northern limb. They are succeeded by Cardona marine evaporites (gypsum and salt), indicating a transition from marine to terrestrial conditions. These evaporites are overlain by >1,000 m (>3,300 ft) of upper Eocene–lower Oligocene continental conglomerate, which prograded to the south (the shoreline migrated seaward due to sedimentation). The continental conglomerates interfinger with evaporite and lacustrine strata to the south and southeast (Burbank et al., 1992).

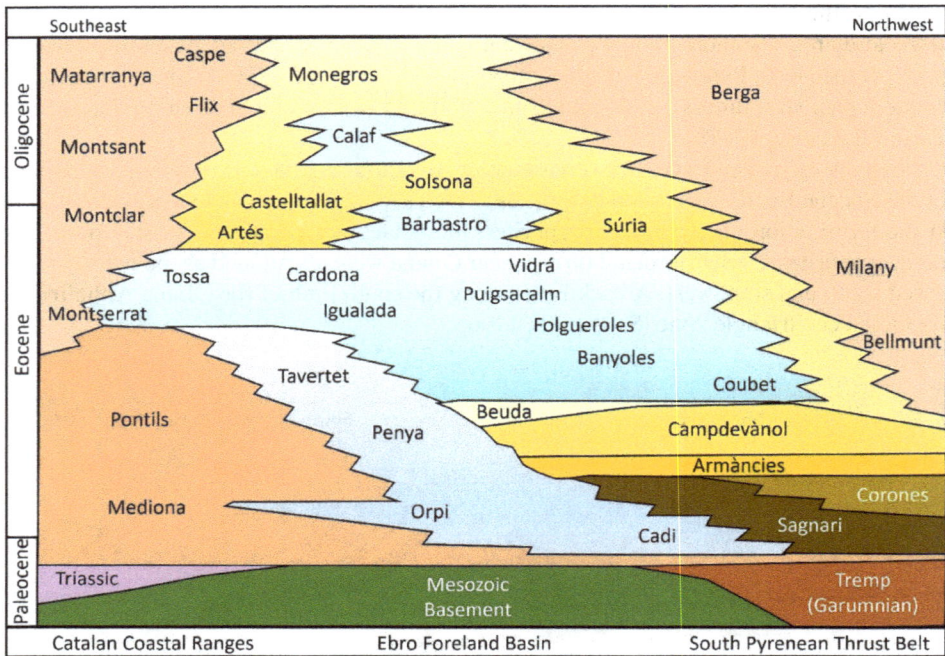

Stratigraphy of the eastern Ebro Basin, Catalan Coastal Ranges, and Cadi Thrust sheet. Modified after Vergés, 1993.

The Oliana well drilled on the anticline penetrated over 2,300 m (7,550 ft) of Igualada Marl without encountering evaporites. This is at least three times the normal stratigraphic thickness encountered in the Basella well that was drilled off-structure. The anomalous thickness and the absence of Cardona evaporites in the well, despite being present both north and south of here, suggests that the anticline is a large duplex structure consisting of stacked thrusts (Burbank et al., 1992). It has been proposed that stacked thrusts develop where less ductile gypsum and/or shaley marls form the detachment surface (that is, a "sticky" detachment surface; Sans et al., 1996).

Along the southeast margin of the duplex, a back thrust separates the lowest conglomerate from underlying marls. This thrust and its hanging-wall anticline formed during movement of the duplex across a thrust ramp, that is, where the detachment ramps up from Beuda Formation evaporites to the Cardona evaporites (Burbank et al., 1992).

Several conglomeratic units overlie the Igualada Marl and have been mapped on both flanks of the Oliana structure. On the north limb they are associated with thrusts and consist of coarse-grained breccia and conglomerate. The contacts between units are generally angular unconformities. On the south flank sedimentation was finer grained sandstone and siltstone, and most stratigraphic contacts appear to be conformable (Burbank et al., 1992; Maestro-Maideu and Serra Roig, 1996). This suggests that the north flank was closer to active thrusting and uplift.

Stop 4.1 Oligocene Turbidites, South Flank Oliana Anticline

This roadcut exposes distal (far from the source) turbidites of the Berga Formation. The units are mostly interbedded sandstone, siltstone, and shale. Slight angular unconformities can be seen in the section.

Interbedded sandstone, siltstone, and shale in the Eocene-Oligocene Berga Formation, south flank of the Oliana Anticline. Units are dipping ~20° southeast here. A slight angular unconformity can be seen. An angular unconformity is the erosion surface separating inclined layers below from flat layers above.

Oligocene Turbidites, Oliana Anticline to Core Oliana Anticline: *Continue north on C-14 for 5.8 km (3.6 mi; 6 min) and pull into the restaurant parking lot on the left. This is **Stop 4.2, Core Oliana Anticline** (42.077135, 1.305763).*

Stop 4.2 Core, Oliana Anticline

The core of the Oliana Anticline is eroded out and does not contain good outcrops. Well and seismic data, however, tell us that beneath the valley is an unusually thick section of Eocene Igualada Marl. This has been interpreted as a stack of thrusts, or a thrust duplex, that doubled or tripled the normal thickness of the marl. Conglomerates of the upper Igualada Formation are exposed in a narrow

View northeast up the core of the Oliana Anticline. Anticline is verging to the southeast. Dashed lines indicate bedding.

band along the northwest flank of the anticline and can be found above the Hotel Can Boix and in the Barranc de Peramola.

Steeper dips on the southeast flank indicate that this asymmetric anticline is verging to the southeast.

Core, Oliana Anticline to Oliana Dam: Continue north on C-14 for 3.3 km (2.0 mi; 4 min), turn right onto the dam road and park by the rest area. This is **Stop 4.3, Oliana Dam, North flank Oliana Anticline** *(42.094006, 1.292632).*

STOP 4.3 PROGRESSIVE UNCONFORMITIES, GROWTH STRATA, AND BERGA FORMATION CONGLOMERATES, OLIANA DAM

There are some excellent roadcuts through steep, 40°–60° north-inclined layers on the north limb of the Oliana Anticline as you drive north to the Oliana dam and Presa de Oliana. Pulling over near the dam you can look west and see an example of progressive unconformities and growth strata as the proximal (near-to-source) conglomerates onlap the underlying Igualada marl.

The rest area here happens to be on those massive conglomerates.

Steeply north-dipping upper Eocene massive proximal conglomerates of the Berga Formation. Street View north to Oliana dam.

View west from Oliana dam on the north flank of the Oliana Anticline. This illustrates growth strata and progressive unconformities in the Oligocene conglomerates (cgl) as they were deposited over the Eocene Igualada Formation marls (ls) that were being uplifted in the core of the anticline. Dashed lines outline bedding.

Orígens UNESCO Global Geopark

Established in 2018, the park encompasses 19 villages with 16,400 inhabitants, 20 interpretive centers, 8 geo-tours, and 80 geosites in an area of 2,040 km² (788 mi²). The economy of this rural area is mainly based on tourism, although agriculture still accounts for about 10% of employment. Some unique aspects of agriculture here include the planting of wine grapes of the Costers del Segre Designation of Origin, and the revival of ancient breeds like Xisqueta sheep. Visitors can hike through spectacular canyons such as the Collegats Gorge. The remote setting, far from large cities, makes this a natural dark sky site and allows people to observe some of the starriest skies of Catalonia, earning the park a Starlight Foundation certification.

Orígens Global Geopark is located in the central part of the Southern Pyrenees. This area was designated a geopark at least in part because it contains the north-south ECORS Pyrenees deep crustal seismic profile that runs along the Noguera Pallaresa River valley. In the north it includes a bit of the Axial Pyrenean Zone with the oldest rocks in the range. In the Southern Pyrenean Zone there are sedimentary basins (Tremp Basin, Àger Basin, Oliana Basin) developed during thrusting. The south end of the park extends slightly into the Ebro Basin. Rocks range in age from early Paleozoic to Tertiary, including an essentially complete Mesozoic section. The Upper Cretaceous section contains bones, tracks, and eggs of some of the last dinosaurs in Europe. The park contains well-preserved middle Eocene–Oligocene synorogenic conglomerates that record uplift of the Pyrenees. Other unique geologic aspects of the park include the Ilerdian type section, a part of the Paleogene, located just east of Tremp; exceptional exposures of delta deposits in the Tremp and Àger basins; and the salt spring in Gerri de la Sal that is related to the groundwater solution of underground salt beds. In fact, geologists travel here from around the world to learn about thrust belts and the relationship of thrusting to sedimentation.

The central Southern Pyrenees is characterized by south-directed thrust sheets: from oldest to youngest and from north to south they are the Orri, Rialp, Nogueres, Bóixols, Montsec, and Serres Marginals thrusts. The pre-Carboniferous rocks in the park were deformed during the Variscan Orogeny; all of the rocks have been affected by the Alpine/Pyrenean Orogeny.

Quaternary Ice Ages, characterized by several glaciations, incised the drainage network and generated the spectacular gorges at Mont-Rebei, Terradets, and Collegats.

Visit

Address: Orígens UNESCO Global Geopark, Plaça de la Creu,1 – 25620 Tremp, Lleida
Phone: +34 973 651 088 / +34 636 868 713
Email: info@geoparcorigens.cat
Website: (https://www.geoparcorigens.cat/en/)

Visitor Center:
Address: Passeig del Vall, 13, 25620 Tremp, Lleida.
Phone: +34 973 653 470
Website: epicentre@pallarsjussa.cat

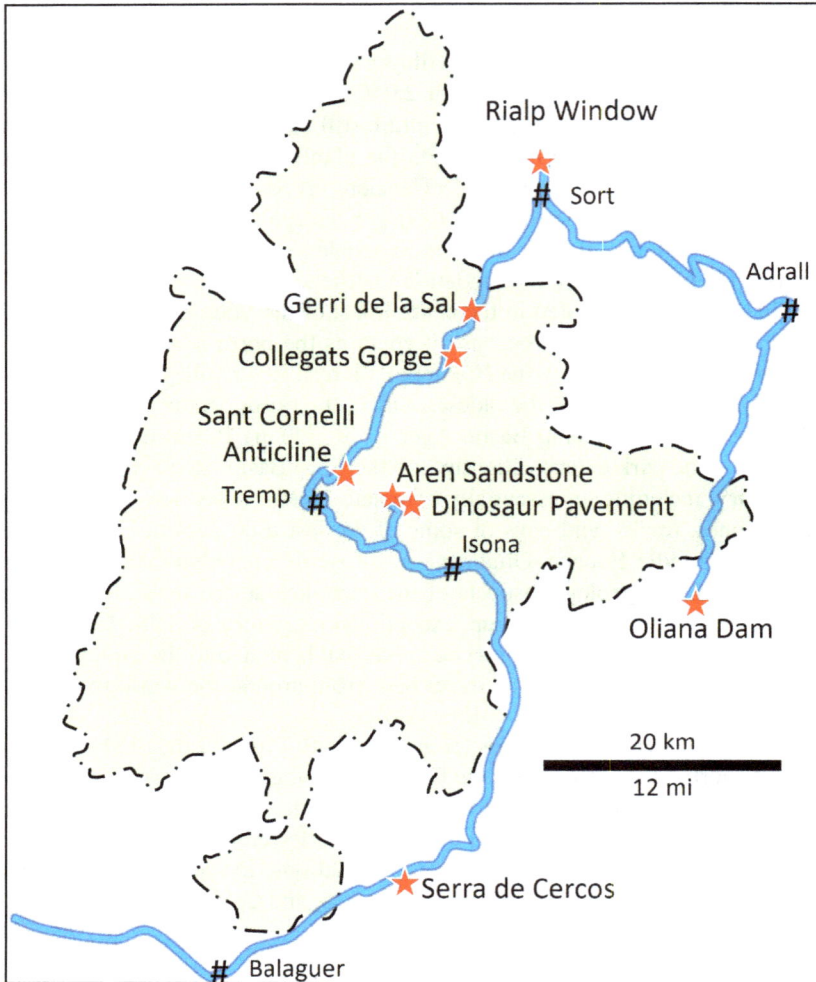

Map of Orígens UNESCO Global Geopark.

Oliana Dam to Rialp Tectonic Window: *Return to C-14 and continue driving north; at the round-about before Adrall take the 2nd exit (west) onto N-260/Carretera de la parróquila/Eix Pirinenc; in Sort turn right (north) onto Av Verge de Montserrat; turn right (north) onto C-13; at the roundabout take the 2nd exit (north) onto LV-5223; from the roundabout drive 2.3 km (1.4 mi; 2 min) north and pull over onto a track on the left before the roadcut. This is **Stop 6, Rialp Tectonic Window** (42.435729, 1.127109) for a total of 77.9 km (48.4 mi; 1 hr 13 min). Carefully walk about 28 m (90 ft) north along the road.*

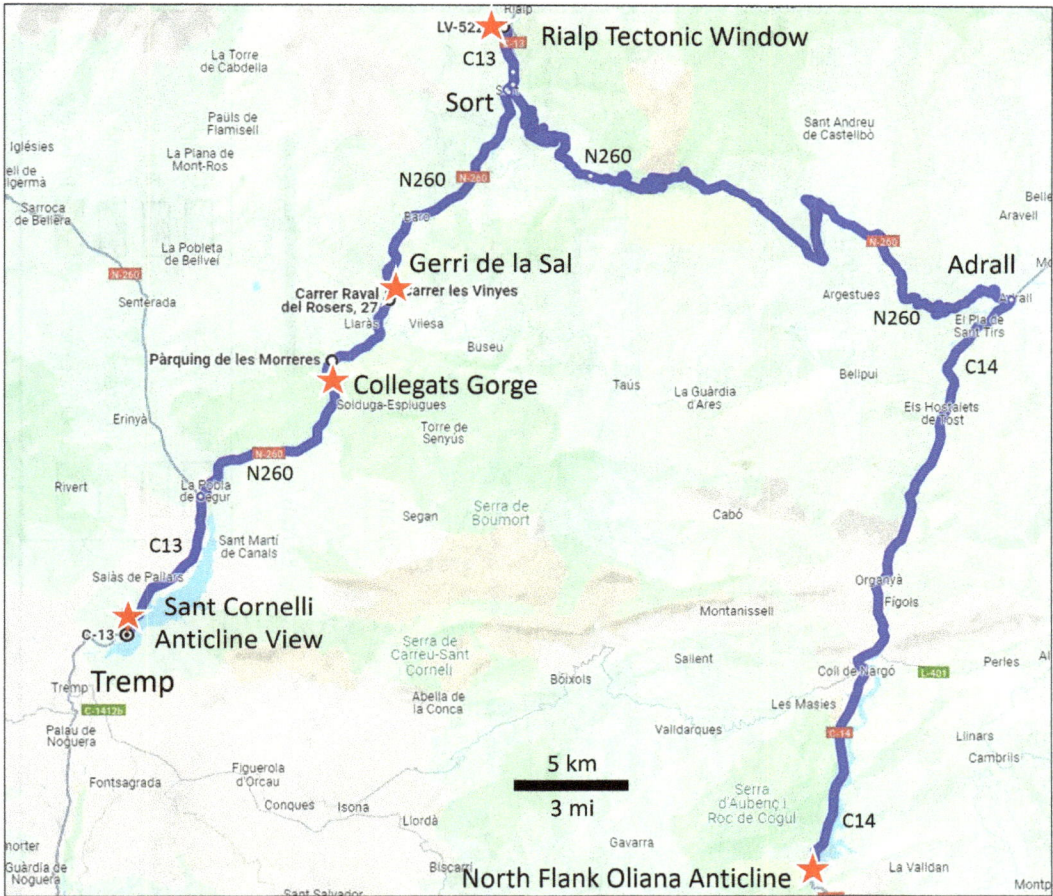

Stop 5 Rialp Tectonic Window

As we drive toward Rialp, we leave the Ebro Basin and enter the South Pyrenean Zone and Orri and Nogueres south-directed thrusts.

A "tectonic window" (also called a "fenster," the German word for window) is an eroded hole, or window in a thrust sheet that reveals the underlying rocks. The Rialp tectonic window exposes Triassic rocks in the footwall (subthrust) of the Orri Thrust. The thrust carries Cambro-Ordovician, Silurian-Devonian, and Carboniferous-Permian units over Middle Triassic (Muschelkalk) rocks (Roberti, 2008).

At this stop a segment of the Orri Thrust puts older Permian undifferentiated rocks over younger Middle Triassic marls, dolomites, and shales.

*Rialp to Gerri de la Sal: Return south on LV-5223; at the roundabout take the 1st exit (south) onto C-13/Avinguda de la Generalitat; in Sort continue south on N-260/Avinguda dels Comtes de Pallars; upon entering Gerri de la Sal look for the sign for Salines Santa Maria and turn left into the parking area; walk north along the river to the Salines del Roser (salt ponds). This is **Stop 6, Salt Spring** (42.325039, 1.065896) **and Museum** (42.323372, 1.064178) for a total of 15.7 km (9.8 mi; 16 min). 5 € parking.*

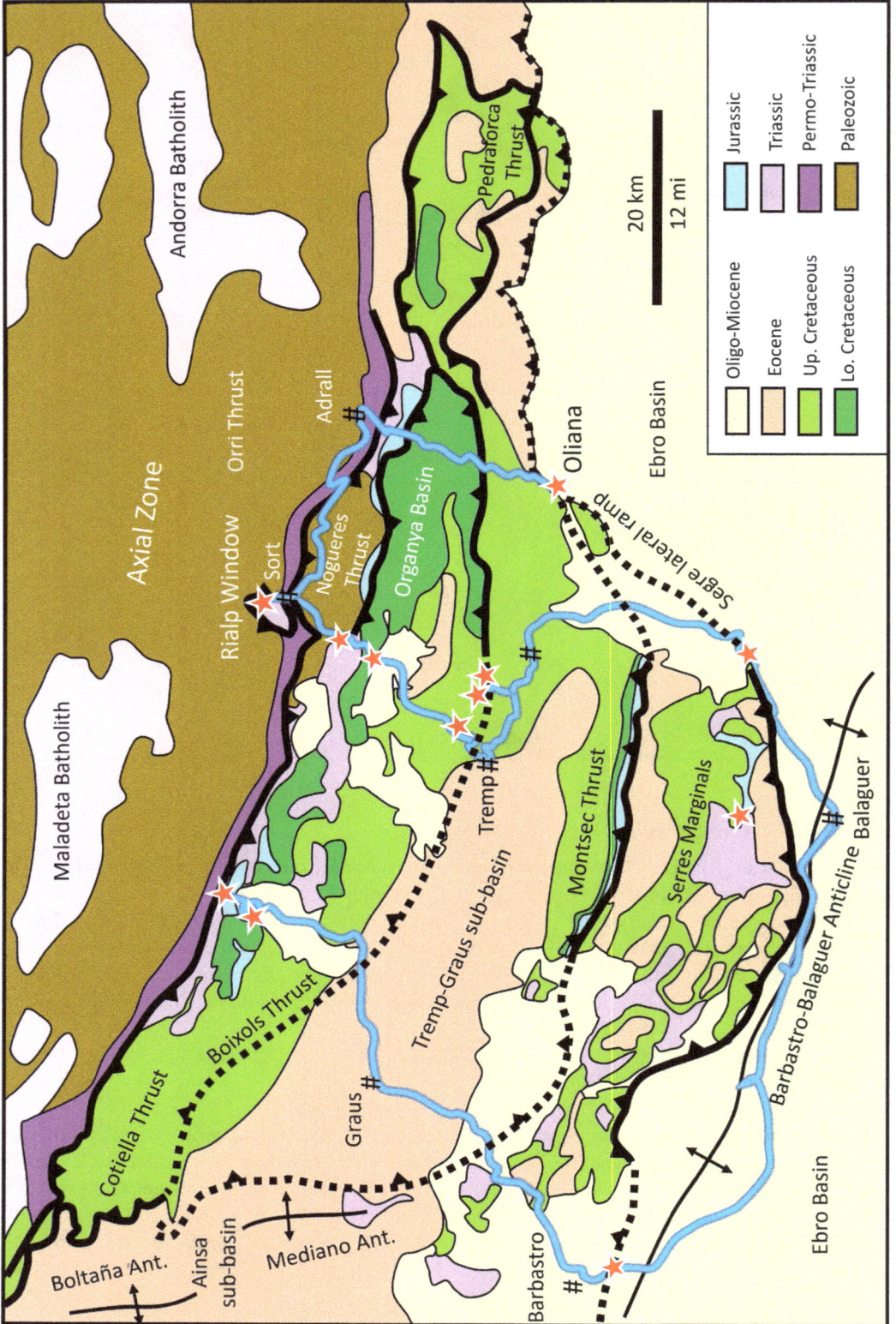

Regional structure of the Southern Pyrenean Zone between Sort and Barbastro. Modified after Burrel et al., 2021.

This segment of the Orri Thrust puts Permian rocks over Middle Triassic rocks. This is the northern edge of the Rialp tectonic window. Dotted line is the Orri Thrust.

STOP 6 GERRI DE LA SAL, MUSEUM OF SALT, AND NOGUERES THRUST

Salt has been produced from a salt spring at Gerri de la Sal since at least the 9th century. Mapping indicates that the source of the salt is a Keuper (Upper Triassic) evaporite up to 3,000 m (10,000 ft) thick in the Senterada-Gerri de la Sal area. The salty springs in Gerri de la Sal confirm that halite (rock salt) is an important component of the Upper Triassic in the subsurface (Burrel et al., 2021).

This spring, along the west bank of the Noguera Pallaresa River, emerges at 21°C (70°F) with a concentration of 120 to 160 parts per thousand sodium chloride (for comparison, the Mediterranean averages 38 ppt). Salt production depends on flow, so the salt extraction season usually lasted from June to August.

The salt was stored in the "Magatzem de la Sal o Real Alfolí," built in the 18th century. It was then carried to the Casa de la Sal where it was weighed, ground, and packed for shipping.

VISIT

The salinas (saltworks, or evaporation ponds) are now in ruins and mostly grass-covered. They were not accessible when I visited in the spring of 2025.

The Museum of Salt is on the west side of the highway up the hill. It is a little tricky to find: you have to look carefully for the signs. You may need reservations to enter the museum.

For museum hours and entry fees, please consult their website: https://sistema.mnactec.cat/en/museums/museu-de-gerri-de-la-sal/

Address: Museu de Gerri de la Sal, Plaça de Angel Esteva, 10, 25590 Gerri de la Sal, Lleida, Spain
Phone: +34 630 056 138 or +34 679 735 262
Email: info@eradortega.org and reservesmuseugerri@gmail.com

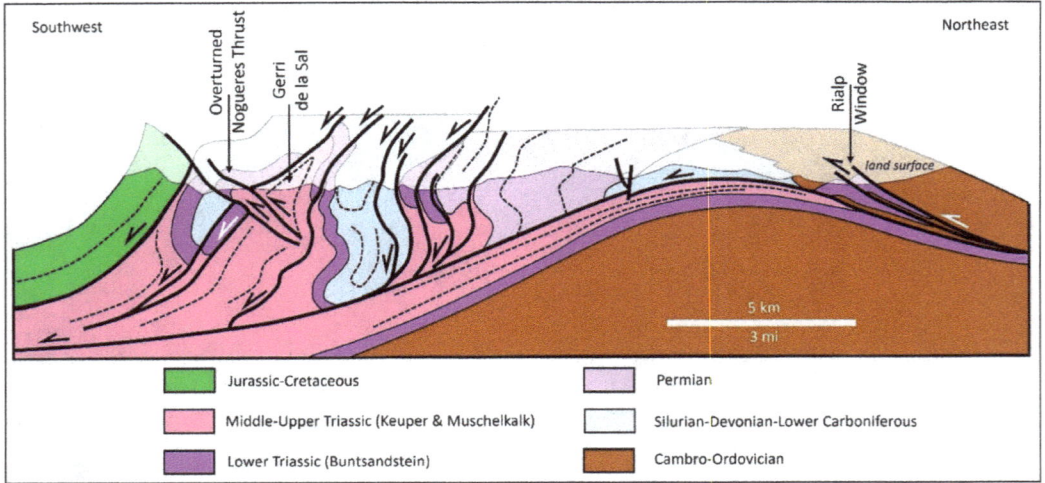

Cross-section through the Rialp Tectonic Window to Nogueres. Modified after Roberti, 2008.

From the old bridge in the center of town you get a great view into the internal structure of the Nogueres Thrust sheet. The frontal part of this thrust sheet is steepened to the point of being overturned. The recumbent (lying on its side) anticline appears to be a syncline, and yet the oldest units are in the core of the fold, indicating it truly is an anticline. Silurian to Lower Carboniferous units have been affected by Variscan folding and thrusting; Permian and Triassic redbeds are separated from the pre-Variscan units by an erosion surface and are unaffected by this older deformation (Saura, 2004; Roberti, 2008).

Frontal Nogueres Thrust. Overturned Triassic redbeds overlie folded and thrusted Paleozoic units. Dashed lines indicate bedding; dotted line is an unconformity (erosion surface).

The valley sits in Upper Triassic units; the redbeds are Lower Triassic; the high topography to the north is undifferentiated Devonian-Carboniferous units.

Nogueres Thrust to Collegats Gorge: *Return south and merge onto N-260 going south; at the sign for "l'Argenteria photographic stop" turn right and park. Walk 300 m (1,000 ft) south to the gorge. This is* **Stop 7, Collegats Gorge/Argentieri** *(42.300330, 1.035138) for a total of 4.8 km (3 mi; 5 min).*

STOP 7 CHEVRON FOLDS AND TRAVERTINE BUILDUPS, COLLEGATS GORGE

In addition to being a stunning cleft in the mountains, this 5 km (3 mi) long, 610 m (2,000 ft) deep gorge also has travertine springs, tightly folded limestones, and massive conglomerates above the Pyrenean unconformity.

We are in the Bóixols Thrust of the Southern Pyrenean Zone. As you walk south into the gorge the northern boundary corresponds to the Morreres Backthrust.

Triassic basalts can be seen in the roadcut. Farther north are Triassic marls. In Collegats Gorge to the south are Lower Cretaceous limestones and breccias (Roberti, 2008). The travertine, a type of limestone deposited by hot springs, is derived from the Lower Cretaceous marine limestone. Chevron folding, characterized by a distinct V-shaped geometry, is caused by compression of alternating strong and weak rock layers. An outstanding example of such folding is visible next to the travertine deposit.

The La Pobla de Segur conglomerate (previously called Collegats Formation) is a syntectonic alluvial fan system that migrated south onto floodplains and into shallow lakes. This conglomerate is dated as middle Eocene to middle Oligocene. The component clasts are sourced primarily from the Pyrenean Axial Zone to the north (Burrel et al., 2021).

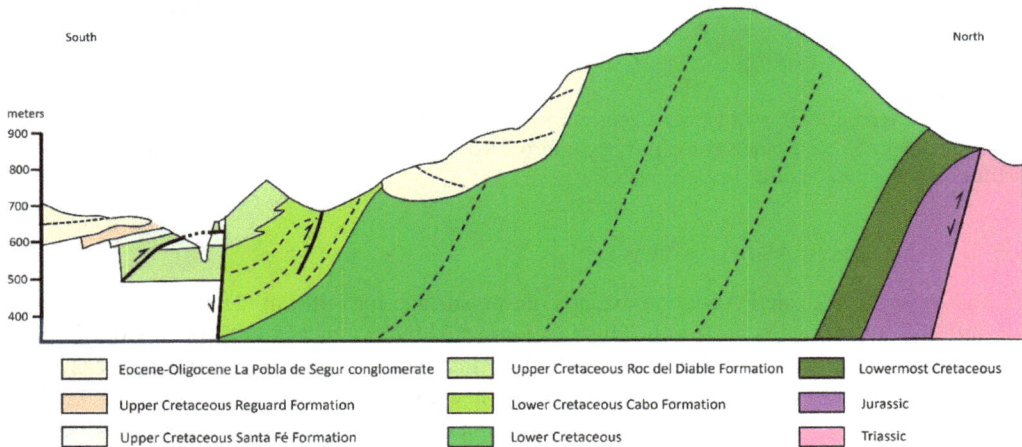

Cross-section through Collegats Gorge. Modified after Roberti, 2008.

Collegats Gorge to Sant Corneli Anticline: *Continue south on N-260; in La Pobla de Segur bear left onto C-13 south; drive to sign for Pantá de St Antoni and a parking area on the left. This is* **Stop 8, Sant Corneli Anticline View** *(42.195237, 0.928496) for a total of 17.5 km (10.9 mi; 16 min).*

Magnificent chevron folding in Lower Cretaceous Roc del Diablo limestone (Cabo Formation) adjacent to a travertine mound, left. Some say the travertine formation inspired Gaudí's design of certain aspects of the Basílica de la Sagrada Familia in Barcelona.

Stop 8 Sant Corneli Anticline View

The east-west Sant Corneli Anticline extends on the surface for ~40 km (25 mi) before it plunges west into the subsurface near Tremp. This is a large fold by any measure. Seismic work indicates that it continues farther west in the subsurface. The anticline, up to 5 km (3 mi) wide, is asymmetric, with a gentle north flank and a steep to overturned south flank (Shackleton and Cooke, 2005; Mencos et al., 2010; Muñoz, 2017). Like a wave rolling into the shore, this tells us that the compression that formed the fold was a push from the north.

The anticline is a fault-propagation fold developed over a thrust ramp at the leading edge of the Boixols Thrust. The thrust does not breach the surface in this area. The fold has been called an "oblique inversion anticline" because the thrust ramps up along an old, reactivated normal fault (the Organyà Fault). The old fault formed during Early Cretaceous extension, and later shortening was at an oblique angle to the ramp. During the Late Cretaceous Alpine/Pyrenean Orogeny the shortening direction was oriented north-northwest-south-southeast, oblique to the east-west extensional fault (Mencos et al., 2010; Tavani et al., 2011). The Boixols Thrust detachment surface is in Upper Triassic evaporites. The anticline probably became larger and steeper as it was carried south by later movement on the Montsec and Serres Marginals thrusts. This is indicated by middle Eocene–Oligocene growth strata (conglomerates) seen on the back limb of the anticline (Muñoz, 2017). The abrupt fold plunge appears to coincide with a deepening of the thrust in the subsurface. The western, subsurface fold is entirely covered by Late Cretaceous syntectonic sediments (Mencos et al., 2010).

The Sant Corneli anticline, at the western end of the Boixols Thrust sheet, contains a thick sequence of Mesozoic carbonates overlain by thin Paleocene and Eocene marine to continental sedimentary rocks and, locally, by late tectonic upper Eocene–Oligocene alluvial fan deposits.

Early Cretaceous extension in the Organya Basin led to deposition of 5,000 m (16,400 ft) of mostly carbonates. The Sant Corneli Anticline is cored at the surface by Early Cretaceous rift-basin sediments. Exposures of Upper Cretaceous carbonate platforms are found on the northern flank of the anticline. Two types of Upper Cretaceous carbonate units are found: (1) Rudist buildups consisting of a rudist/coral belt along the platform margin, and (2) wedges of limestone composed of sand-sized grains (Pomar et al., 2005).

Sant Corneli Anticline to Arén Sandstone: *Continue south on C-13 to Tremp; at the roundabout take the 3rd exit (east) onto C1412b/Av. Alcalde Altisent; turn right (north) onto LV-5112 and drive to the parking area at Orcau. The road doesn't connect, so you have to walk the remaining 1.7 km (1.0 mi) to the Arén Sandstone viewpoint. This is* **Stop 9, Arén Sandstone at Orcau Castle** *(42.162997, 0.983756) for a total of 22.7 km (14.0 mi; 22 min drive plus 30 min walk).*

STOP 9 ARÉN SANDSTONE, ORCAU CASTLE

Geologists come from around the world to study the Arén Sandstone (also known as the Areny Sandstone) as an example of a beautifully exposed sandstone reservoir with the potential to hold oil and gas in the subsurface.

Late Cretaceous Pyrenean thrusting resulted in the foreland basin being segmented into uplift-bounded minibasins. The Bóixols Thrust front splayed into three distinct thrust slivers that get younger toward the foreland (south and west): the Bóixols proper, the Riu, and the Turbón thrusts. The thrust-bounded minibasins filled with syntectonic sandstones, that is, sandstones that were deposited during thrusting (Ardèvol et al., 2000). The Arén Sandstone is one of these syntectonic sandstones.

The Arén Sandstone is part of an upward-shallowing sequence of deep marine (Vallcarga Formation turbidites), delta (Arén Sandstone), and river/floodplain environments (Tremp Formation).

During phases of thrusting and uplift, thick turbidite deposits overlain by prograding delta sediments accumulated in minibasins in front of thrusts. The Arén delta moved basinward, overlapping and covering deep-water sediments, eventually filling the minibasins. During times of relative tectonic quiet, sedimentation overlapped the growing structures. When the syncline in front of a thrust was filled with sediment, the depocenter (main area of sediment deposition) shifted to the minibasin in front of the next thrusted fold to the southwest. Uplift related to the Bóixols Thrust ended around 78–76 Ma; movement on the Riu Thrust ended ~75 Ma; uplift associated with the Turbón Thrust ended ~72 Ma. Foreland basin deformation in this area ended in the latest Cretaceous.

The Arén Sandstone has been subdivided into four depositional sequences. The lowermost Arén 1 Sequence consists of delta deposits up to 1,500 m (4,900 ft) thick. The sandstone thickness decreases as it onlaps the Corneli Anticline, indicating that the thrust was active and the structure was growing already in the Late Cretaceous. The Arén 2 Sequence is mainly fluvio-deltaic (river and delta) deposits that filled the remaining space in the basin in front of the Bóixols Thrust. It is up to 600 m (1,970 ft) thick. Delta deposits of the Arén 3 sequence accumulated in a depocenter in front of the Riu Thrust. It is up to 750 m (2,460 ft) thick. Delta deposits of the Arén 4 sequence accumulated in front of the Turbón Thrust and are up to 500 m (1,640 ft) thick. The Campanué Thrust was

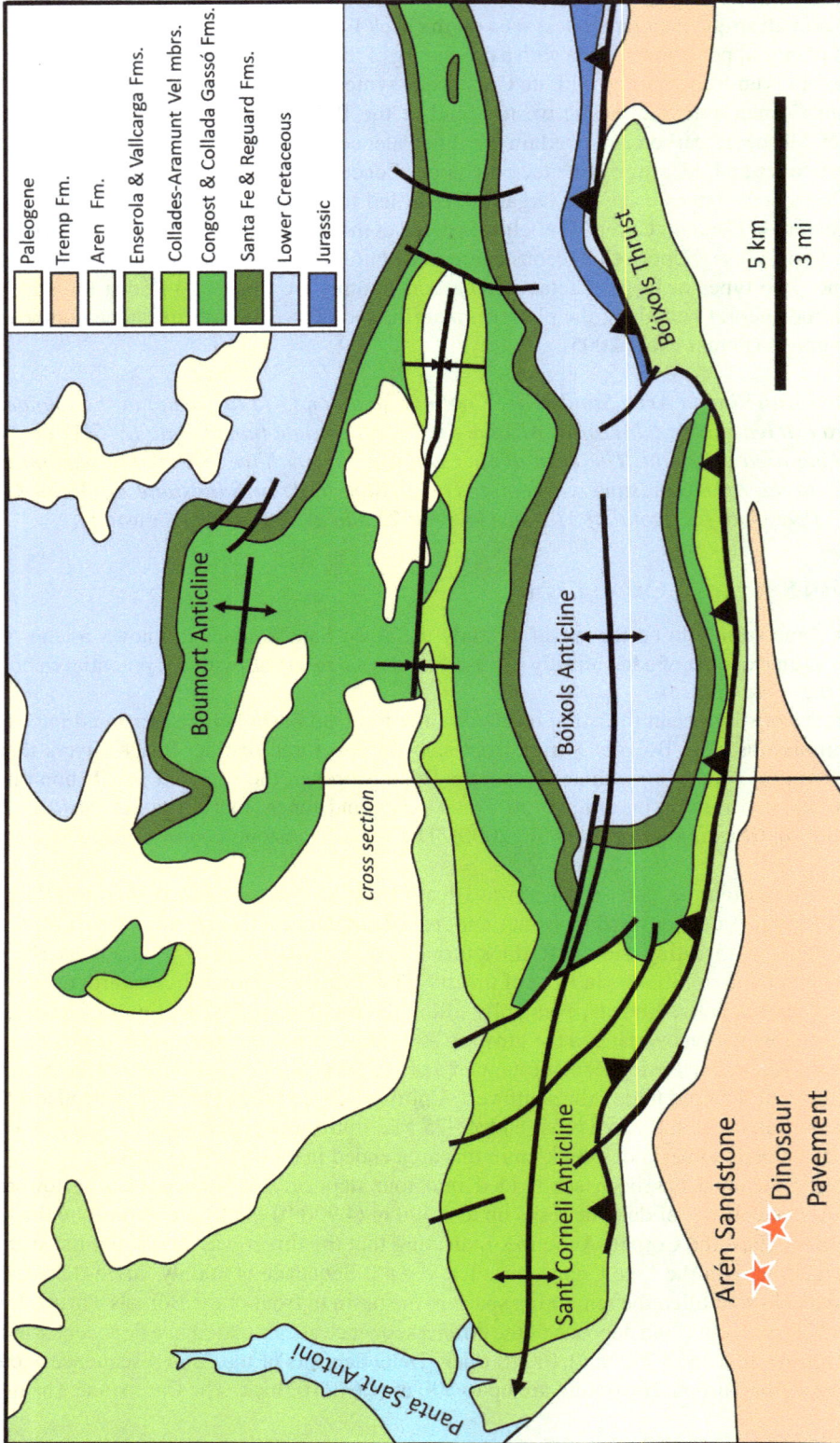

Geologic map of the Sant Corneli Anticline area. Modified after Gutmanis et al., 2017.

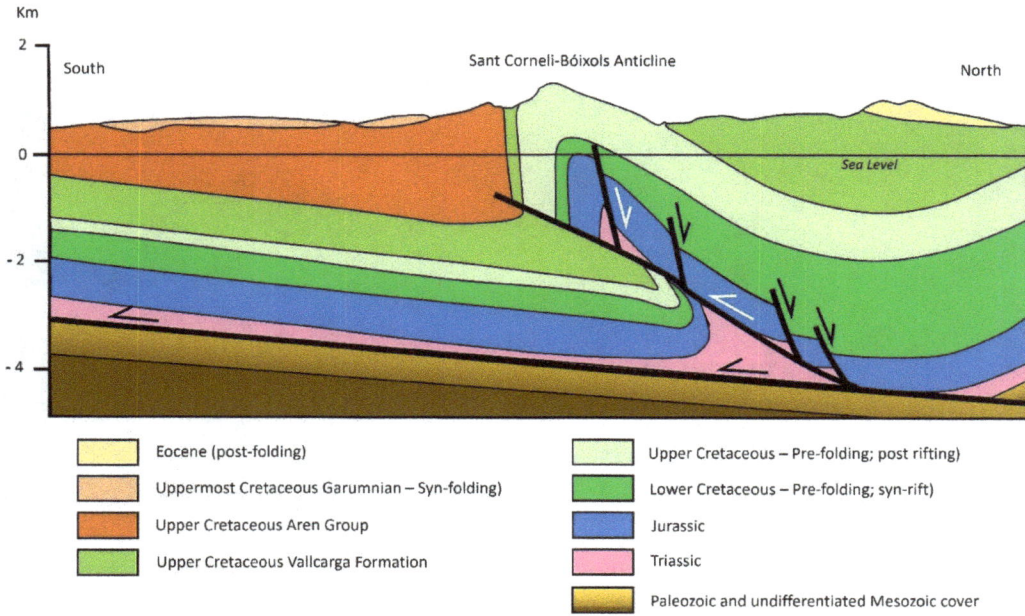

Cross section of the Sant Corneli–Boixols Anticline. The push that formed this fold came from the north. The location is shown on previous map. Modified after Muñoz, 2017.

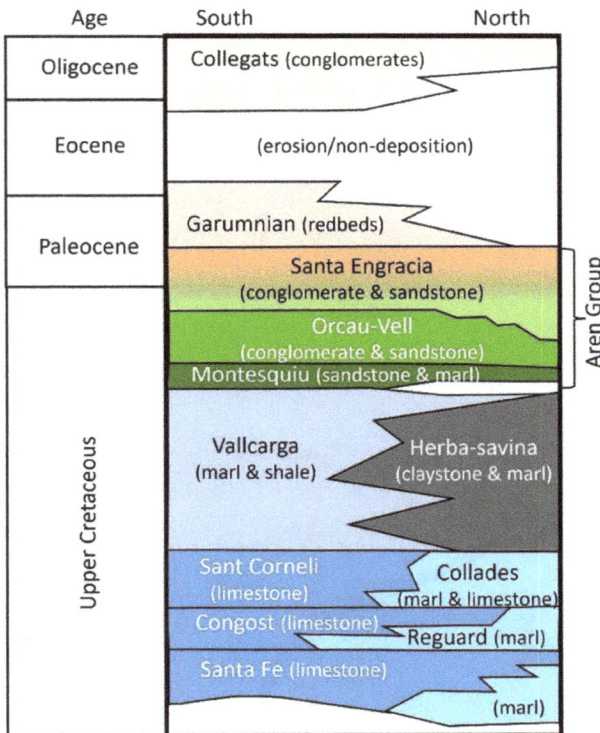

Stratigraphy of the Sant Corneli Anticline area. Stratigraphic information derived from Sanders et al., 2001; DeLisle, 2016; Gutmanis et al., 2017.

View across Pantá de Sant Antoni at the south-verging Sant Corneli Anticline, outlined by the Arén Sandstone (dashed lines). View to the southeast.

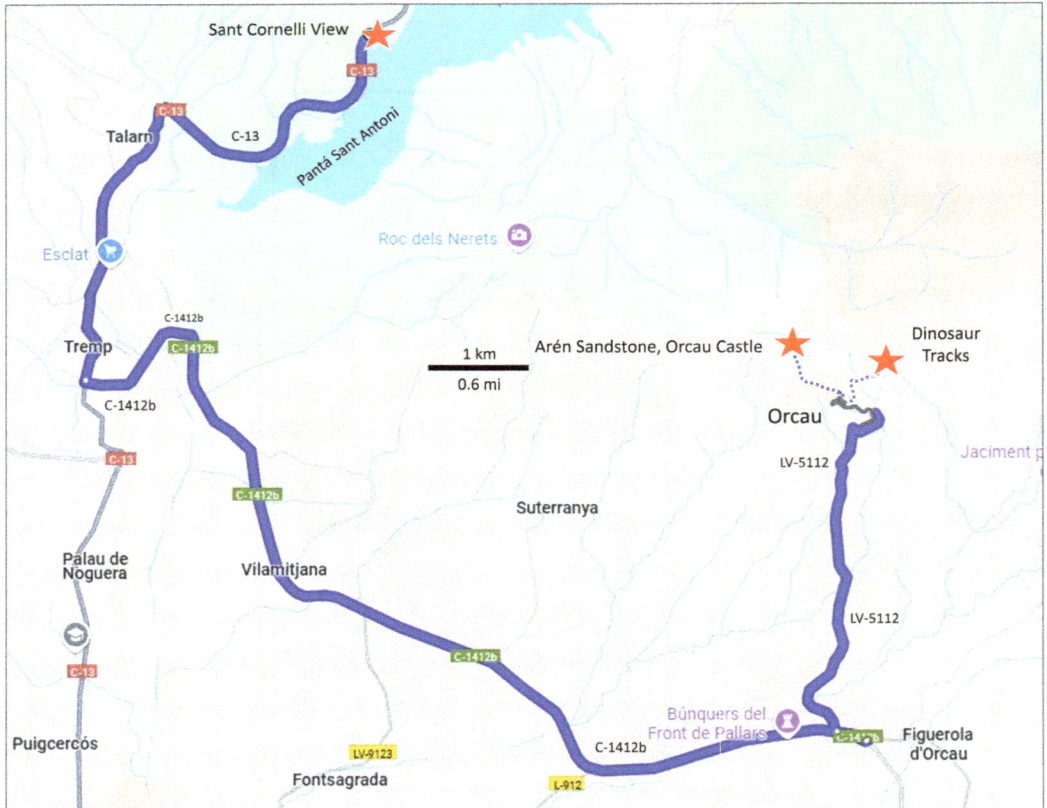

emplaced in late Paleocene–Eocene time and had no effect on the Arén sequences. Deformation progressed from northeast to southwest, as determined by the depositional relationship between dated sequences and growing anticlines (Ardèvol et al., 2000).

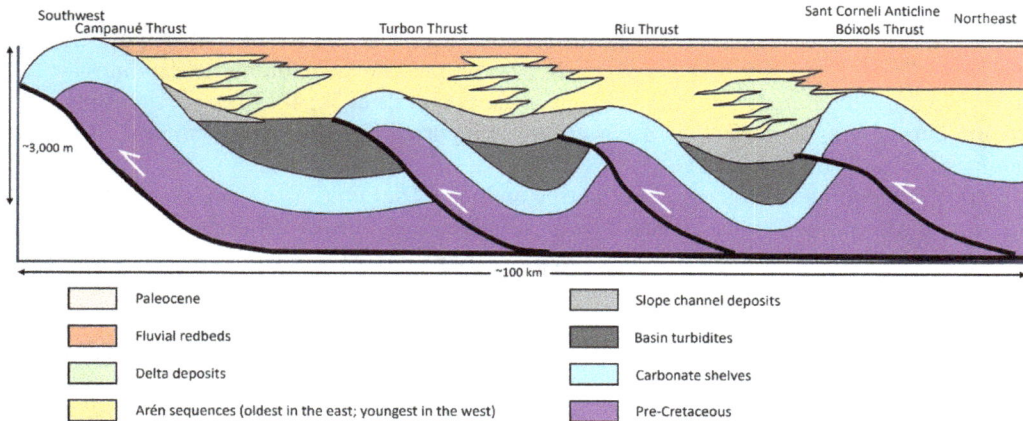

Thrusts and associated mini-basins south of Tremp. The carbonate shelf unit is the last layer before thrust deformation, as indicated by the uniform thickness. The Arén Sandstone at Orcau Castle is part of the delta sequence on the southwest flank of the Sant Corneli Anticline. Modified after Ardèvol et al., 2000.

The Arén Sandstone (Arén 1 Sequence) at Orcau Castle is up to 400 m (1,300 ft) thick. It was formed during a period of high sea level (highstand). The delta and beach deposits get younger to the west, as seen in the foreset beds that move to the west at Orcau Castle. (Foreset beds are the inclined strata at the end of the river channel where sediment is deposited along the basinward-sloping delta front.) The Arén Sandstone here overlies deep marine turbidites and shallow marine marl of the Vallcarga Formation. It is overlain, in turn, by lagoon, estuary, beach deposits, and river and lake deposits of the Tremp Formation (Roberti, 2008).

Salàs Marl and Arén Sandstone north of Orcau. View east to the ruins of Orcau Castle (on the high ridge, center). Photo courtesy of PMRMaeyaert, https://commons.wikimedia.org/wiki/File:Orcau-PM_25584.jpg

Arén Sandstone to Dinosaur Tracks: From the parking area at Orcau take the dinosaur trail ~200 m east. This is Stop 10, Dinosaur Tracks Orcau (42.162997, 0.983756) for a total of 1.5 km (1 mi; 5 min).

STOP 10 ORCAU DINOSAUR TRACKS

This is our last stop in Orígens Geopark, and it contains the tracks of some of the last dinosaurs in Europe. The Tremp Formation, in which they occur, preserves one of the richest dinosaur track records ever identified in the latest Cretaceous of Europe (Vila et al., 2013).

Dinosaur remains in the Tremp Basin were first reported in 1916, during building of the Sant Antoni reservoir (Vila et al., 2011).

The Tremp Formation is a marginal marine and terrestrial unit, about 800 m (2,600 ft) thick, deposited during Late Cretaceous and Paleocene time. The Cretaceous section contains a gray, coastal mudflat/lagoon unit (shallow lake limestone, lagoonal coals, mudflat mudstones, and river sandstones) at the base, and a river section (red mudstones and sandstones) above. Dozens of dinosaur footprints are preserved on the surface of an extensive limestone bed. This layer formed on the bottom of a shallow coastal lake (Catalonia Museums; Vila et al., 2013).

The Orcau-2 site has tracks made by ornithopods (three-toed *Orcauichnites garumniensis* and *Ornithopodichnites magna*) and a third, more abundant type made by sauropods (more rounded prints). The most common prints were left by a group of large, herbivorous, quadruped dinosaurs known as titanosaurs, among the largest dinosaurs ever to walk the earth (Catalonia Museums; Vila et al., 2011).

Dinosaur tracks, Upper Cretaceous Arén Sandstone. Prints are highlighted by red paint.

Dinosaur Tracks to Serra de Cercos: Return to C1412b and turn left (south); drive south on C1412b; continue straight onto L-512; continue straight onto C-14; at Artesa de Segre take the ramp on the right to Balaguer/Leida and merge onto C-14; continue straight onto C26 and pull onto a dirt farm road on the right. Walk back east ~250 m. This is **Stop 11, Serra de Cercos** (41.860556, 0.985000) for a total of 53.6 km (33.3 mi; 44 min).

STOP 11 UNCASTILLO/SARIÑENA FORMATION, SERRA DE CERCOS

South of Orcau we encounter the northern margin of the Ebro Basin near the frontal thrusts of the South Pyrenean Zone. Outcrops in this area are few and far between. A roadcut exposes gray and salmon-colored Oligo-Miocene Uncastillo/Sariñena Formation sandstone interbedded with gray marls and red shales. These are river and floodplain deposits. The northwest-dipping layers in this roadcut are on the southeast flank of the Serra de Cercos Syncline. The syncline is a foreland mini-basin in front of (south of) the south-directed Serra d'Arquells Thrust.

Oligo-Miocene Uncastillo/Sariñena Formation sandstone and shale. View north.

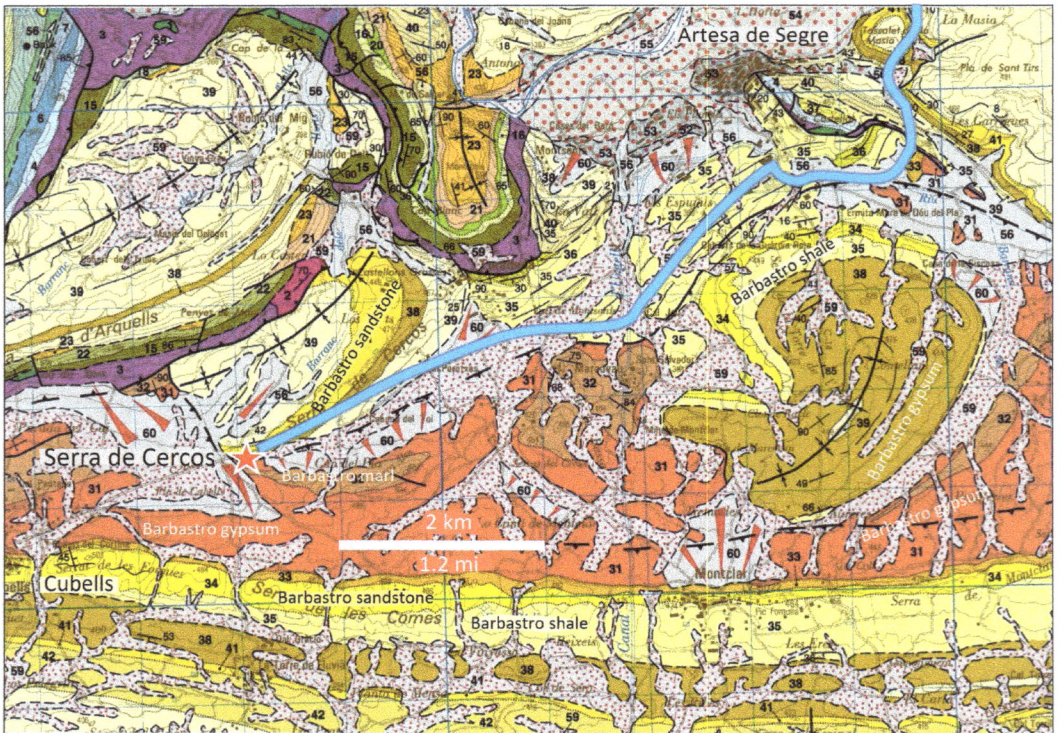

Geologic map, Serra de Cercos area. From Instituto Geológico y Minéro de España (IGME), Artesa de Segre Sheet. Source: © NC Geological Survey of Spain (IGME).

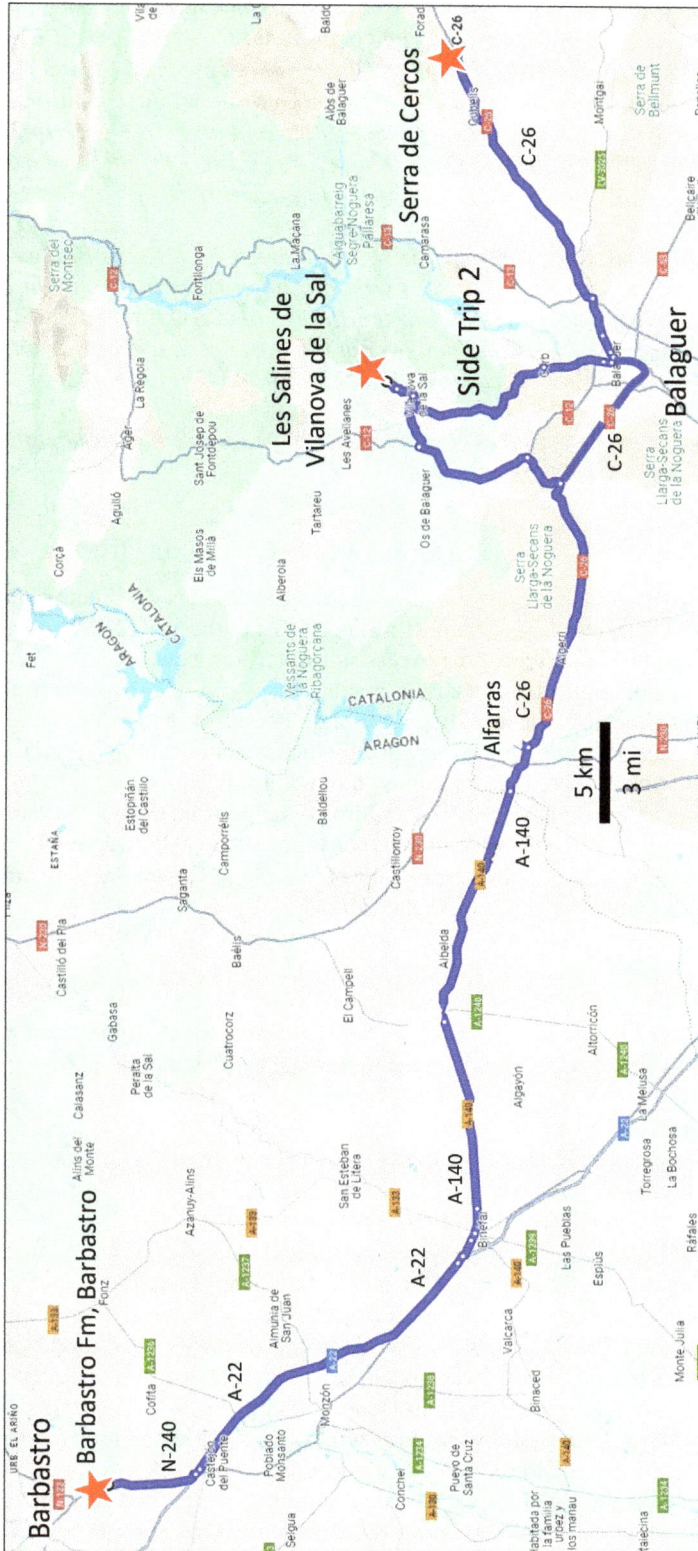

*Serra de Cercos to Side Trip 2, Les Salines: Continue southwest on C-26; at Sant Jordi de Muller bear right onto LV-9227; in Balaguer turn right (west) onto C-148a/Carrer d'Urgell; in Balaguer bear right onto C-12/Ctra de Tremp to Áger; at the roundabout take the 1st exit (north) onto LV-9047/Carr. de Gerb; drive north on LV-9047; in Gerb turn left (north) onto LV-9046/Carr. Diputació; in Vilanova de la Sal bear left onto LV-9045 and drive to **Side Trip 2, Les Salines de Vilanova de la Sal** (41.890262, 0.794737) for a total of 31.6 km (19.7 mi; 32 min).*

If you prefer to skip this side trip and go directly to Barbastro:

*Serra de Cercos to Barbastro: Continue driving southwest on C-26 to Balaguer; stay on C-26 westbound; continue straight on A-140; just west of Alfarras continue straight on A-140; at the roundabout west of Binéfar take the 2nd exit (northwest) onto A-22; take Exit 51 for N-240 north to Barbastro; drive another 4.7 km and pull over at the entrance to a driveway on the right. This is **Stop 12, Barbastro Formation, Barbastro** (42.012340, 0.145428) for a total of 88.7 km (55.1 mi; 1 hr 6 min).*

Carefully cross the highway and walk ~335 m (1,100 ft) south on the bike path on the west side of the road to (42.009256, 0.145355).

Side Trip 2 Les Salines de Vilanova de la Sal and Les Avellanes Diapir

An expression of the Triassic evaporites can be seen at Les Salines de Vilanova de la Sal. The salt pans are on LV-9045 approximately 1.6 km (1 mi) northeast of the town of Vilanova de la Sal. There is a path that crosses the ravine where the excess salt water flows out.

A well delivered salt water to evaporation ponds. A few cm of salt water in the salt pans was allowed to evaporate and the salt was then collected. These salt pans were largely abandoned in the late 1970s. The salt pans are currently owned by the Catalan Training Association which, since 2010, has produced and marketed the output as "mountain salt."

These salt evaporation ponds were probably already being used during the late Roman Empire. The first document with a possible reference to them appears in 1035, where they are referred to as the Monte Salabro saltworks. A 1784 document in the Aragon Crown Archive contains a drawing of the salt pans (Wikipedia, Vilanova de la Sal salt pans).

Visit

Just stopping here and looking at the ruins is not that impressive. If, however, you would like a guided tour or to buy salt, you must call ahead to arrange the day and time:

Phone: +34 607 83 87 80
Website: https://ascalfo.cat/les-salines/
Email: ascalfo@gmail.com
Charge (2025):   1 or 2 people, € 12; 3 or more people, € 4 each.

The Middle to Upper Triassic salt is the major décollement (detachment surface) here. Triassic evaporites crop out in the core of anticlines and at the base of exposed thrusts. Salt diapirism was probably triggered in early Eocene by extension along curved thrust fronts, and by erosional unloading and crestal stretching at anticlines. Oligocene conglomerates overlie the diapir.

Regionally, the Les Avellanes Diapir is located at the southern, leading edge of the Serres Marginals Thrust. Diapirs and salt extrusions are concentrated in the external (southern) edge of the thrust belt as deformation and sediment loading squeezed the salt toward the foreland.

The Les Avellanes diapir outcrops as Triassic evaporites, shales, and carbonates. At the western edge of the diapir, an early Oligocene mixed rock sequence records the lateral extrusion of the diapir and its emplacement as a mobilized salt sheet. The salt neck is exposed at the surface in the

Salt evaporation pans, Les Salines de Vilanova de la Sal. Courtesy of Teresa Llena, https://commons.wikimedia.org/w/index.php?curid=51565108.

northeast part of the salt sheet. Ongoing salt extrusion in the northeast caused uplift and incision of local streams. Like lava from a volcano, the salt layers flowed downhill toward the south and west under the influence of gravity, overriding tectonic breccias of the diapir's caprock. Foliation and other shear-related features are observed in the breccia at the base of the salt sheet.

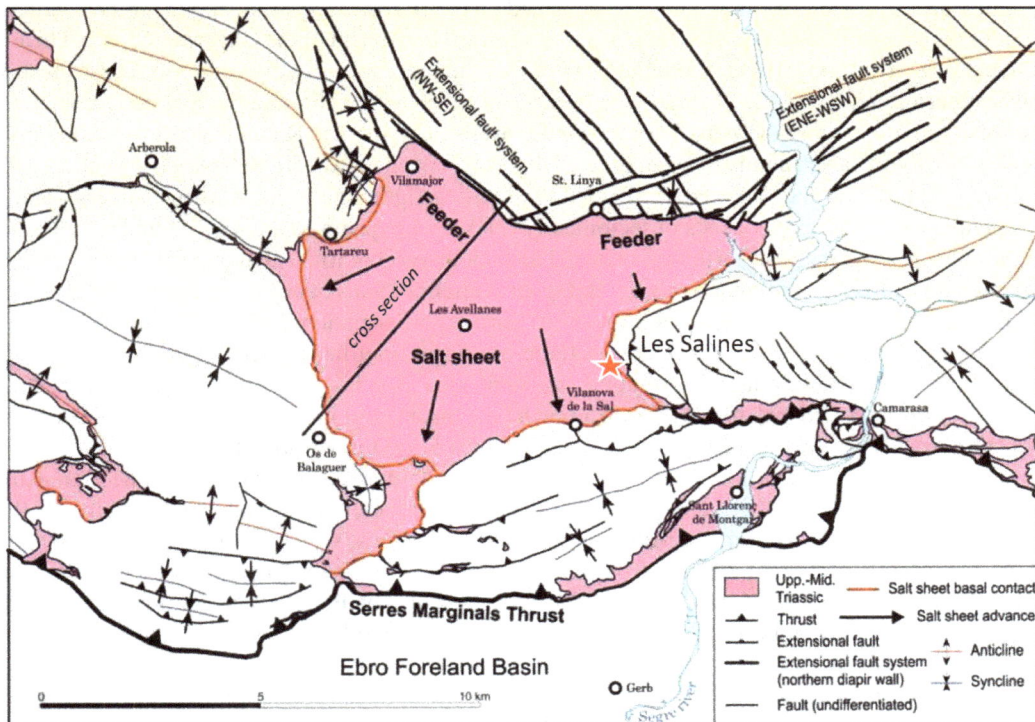

Map of the Les Avellanes Diapir. From Cofrade et al., 2023a.

Although salt is not found at the surface, gravimetry, salt springs, and wells indicate subsurface accumulations of salt in the diapir. Wells penetrating the Triassic in the Ebro Basin encountered salt successions in the Middle Triassic (middle Muschelkalk) and in the Upper Triassic (middle Keuper).

Large fragments of competent, brittle layers, originally interbedded in a layered evaporite sequence, were caught up in and deformed by flow of the salt diapir. These stringers are effective flow markers. The stringers, and thus the diapir itself, has been divided into three structural domains: (1) a feeder zone with mainly vertical flow. This flow rotates to subhorizontal as the salt approaches the surface, (2) a sheet domain (salt glacier) containing subvertical, overturned, and subhorizontal flow, as well as folded stringers, and (3) a sheet front domain characterized by fragments dipping toward the salt front (Cofrade et al., 2023a, 2023b).

ST2, Les Salines to Barbastro Formation Alfarràs: *Return south to Vilanova de la Sal and turn right (west) onto LV-9046/Carr. de la Font; continue west on LV-9046 to Monestir d'Avellanes; turn left (south) onto C-12; turn right (west) onto C-26/Carr. de Balaguer; just west of Alfarras continue straight on A-140; at the roundabout west of Binéfar take the 2nd exit (northwest) onto A-22; take Exit 51 for N-240 north to Barbastro; drive another 4.7 km and pull over at the entrance to a driveway on the right. This is **Stop 12, Barbastro Formation, Barbastro** (42.012340, 0.145428) for a total of 75.9 km (47.2 mi; 1 hr 2 min).*

Carefully cross the highway and walk ~335 m (1,100 ft) south on the bike path on the west side of the road to (42.009256, 0.145355).

Stop 12 Barbastro Formation, Barbastro

We are on the south flank of the Barbastro Anticline looking at a roadcut with highly deformed and partly overturned gypsum in the core of the structure. This large fold lies in the deformed northern margin of the Ebro Basin.

The Barbastro Anticline is an anticline detached on the Barbastro gypsum at or near the southern limit of the evaporite. The detachment allowed southward movement of the Gavarnie-Sierras Thrust, active during the middle Eocene to Oligocene: this is the southernmost and youngest thrust sheet in the Southern Pyrenean Zone. It carries the Serras Marginals in the east and the Sierras Externals in the west. Folding of the Barbastro Anticline and movement on the San Román backthrust, southernmost of the Pyrenean structures, began in late Oligocene (Lucha et al., 2008; Santolaria et al., 2020).

The structure has been defined as a "salt anticline," "an anticlinal stack of thrust slivers," and a "triangle zone." The core consists of tectonically thickened gypsum at the surface and halite (salt) in the subsurface. The west-northwest-oriented and south-verging fold extends for 90 km (54 mi) if you include its lateral continuation as the Balaguer Anticline (Sans et al., 1996; Lucha, 2008).

The late Eocene–early Oligocene evaporitic Barbastro Formation outcrops in the core of the breached Barbastro Anticline. Its thickness there has been amplified by tectonic processes (shortening) and flow. The Barbastro Formation was deposited in a high-salinity lake environment: it consists of gypsum and interbedded halite, anhydrite/gypsum, and marls. The Barbastro Formation lies conformably on the essentially horizontal late Eocene Top Grey Lutite member of the Cardona Formation, which is transitional between marine and continental environments. The Barbastro Formation is overlain by about 2,000 m (6,560 ft) of sandstones, shales, and conglomerates of the early Oligocene Peraltilla and late Miocene Sariñena formations, which can be seen on the flanks of the anticline. These units, derived from the uplifting Pyrenees to the north, represent Pyrenean alluvial systems and the disappearance of the evaporite lake. The Peraltilla Formation on the south limb of the Barbastro Anticline is inclined 30° to 60° south, and locally is near-vertical (Maestro-Maideu and Serra Roig, 1996; Lucha et al., 2008; Santolaria et al., 2020).

Outcrops along the bike path and tunnel provide a view of disharmonic folding developed in the Barbastro gypsum in the deformed core of the Barbastro Anticline.

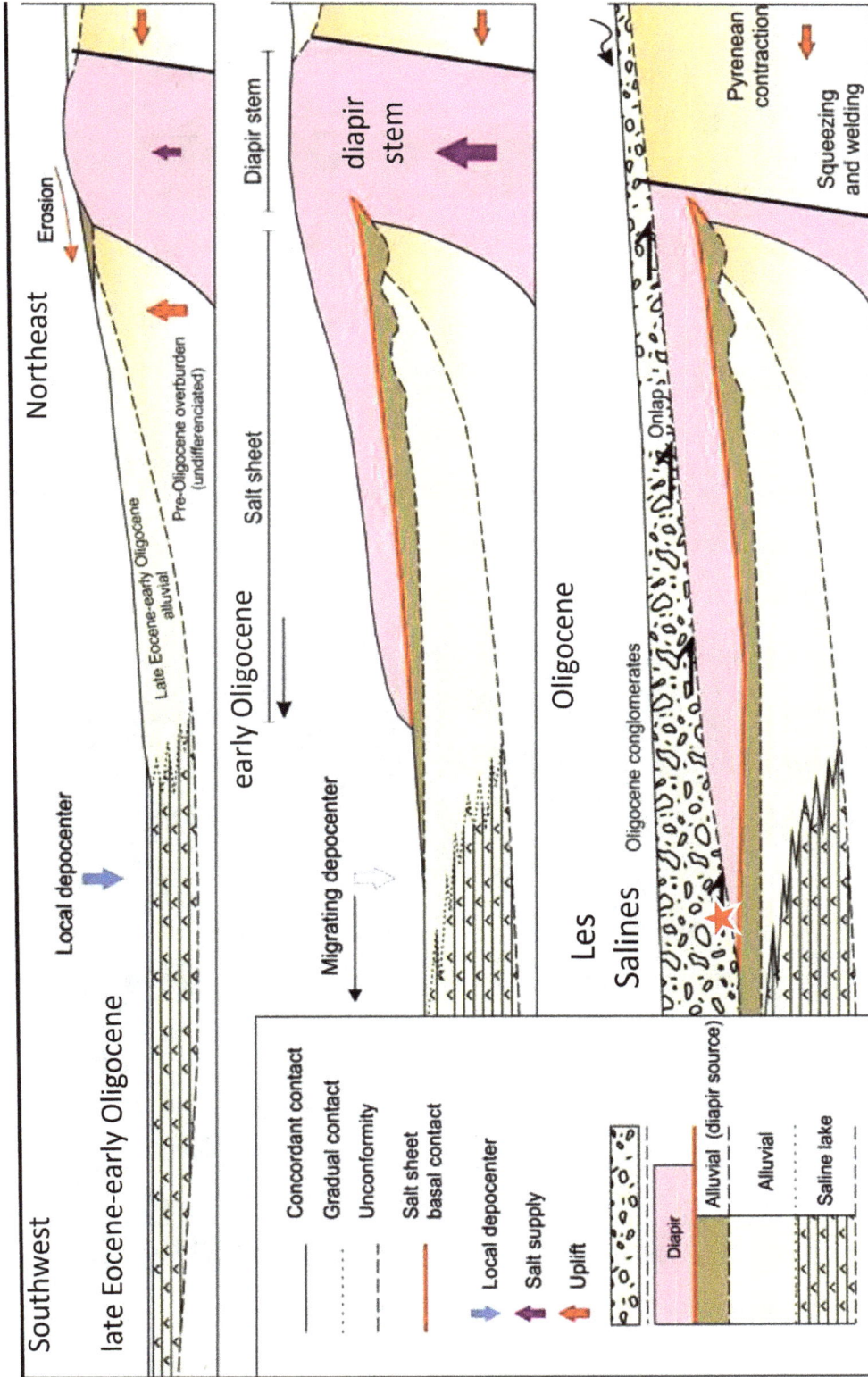

Structural development of the Les Avellanes Diapir. Location of the cross section is shown on the previous figure. From Cofrade et al., 2023a.

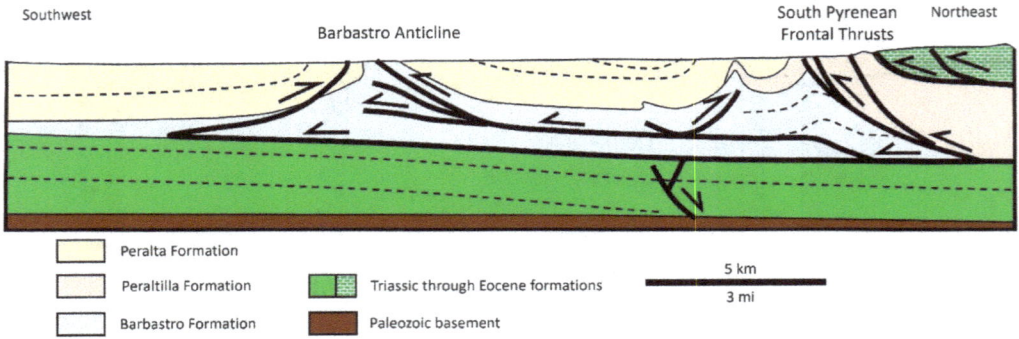

Cross-section showing position of the Barbastro Anticline at the leading edge of foreland thrusting and the result of a triangle zone in the evaporite section. There is an estimated 8 km of shortening along this line. Modified after Sans et al., 1996, Bulletin of Canadian Petroleum Geology, CEGA © 1996.

The roadcut on the east side of the highway ~520 m (1,700 ft) south of this pullout (42.007799, 0.145759) provides an excellent view of disharmonic folding, detachment and fault-propagation folds, and folds with opposing vergence developed in gypsum in the deformed core of the Barbastro Anticline. You get the best view from the west side of the highway, but be very careful as traffic moves fast and there is little shoulder.

Deformed Barbastro gypsum along bike path, south limb Barbastro Anticline.

Deformed Barbastro gypsum along the highway, south limb Barbastro Anticline. Dashed lines follow bedding. Google Street View east.

Barbastro South to Roda Delta Overview: *Continue north on N-240; take the 123 ramp to Graus; at the roundabout take the 1st exit (northeast) onto N-123/Monzón-Fosado Road; at Embalse Barasona take the N-123a exit and continue north on N-123a; at the roundabout in Graus take the 1st exit (east) onto A-1605/C. Valle de Arán; at La Ribagorza turn left (west) to Roda de Isábena and drive to the parking area at the end of the road. This is **Stop 13.1, Roda Delta Overview** (42.291382, 0.529871) for a total of 57.9 km (36.0 mi; 49 min).*

STOP 13 RODA DELTA DEPOSITS

Driving north from Barbastro we enter the Tremp-Graus piggyback basin. This is a basin carried south on the Montsec Thrust in the Southern Pyrenean Zone. The Roda delta section is widely recognized as a classic example of sedimentation affected by growing structures.

Southward movement of the Montsec Thrust during Paleocene–early Eocene time created a piggyback basin, the Graus-Tremp Basin. The stratigraphy of the basin is characterized by a deepening marine unit (the Eocene Alveolina Limestone), followed by marls and a shallowing, coarsening upward succession of shales to siltstones to sandstones and ultimately late Eocene–Oligocene alluvial conglomerates at the top.

The Roda delta brought sediments into this basin. The Eocene Roda delta sandstone is exposed along the north flank of the Graus-Tremp Basin. This delta system grades westward into the San Vicente/Hecho Group of deep marine and delta slope turbidites seen in the Ainsa Basin.

The Roda Sandstone is a predominantly shale-siltstone-sandstone, wave- and tide-influenced, Gilbert-type delta complex. (A Gilbert-type delta is a river delta characterized by steep, wedge or lens-shaped packages of sediments composed of topset (upper proximal), foreset (slope), and bottomset (lower distal) beds. These deltas form when rivers deposit sands in a deep basin with steep slopes.) The Roda delta comprises a series of sediment lobes with steeply dipping clinoforms. Clinoforms are distinct, z-shaped crossbeds characteristic of a delta system. The entire unit comprises six seaward (southwest) stepping delta front packages (Enge et al., 2007). The unit is bracketed by shallow marine carbonates, the La Puebla Limestone Formation at the base and the

Geologic map of the Barbastro area. Star is the location of the stop. From Instituto Geológico y Minéro de España (IGME), Barbastro and Fonz Sheets. Source: © NC Geological Survey of Spain (IGME).

El Villar Limestone at the top. The delta units consist, from bottom to top (deep water to onshore), of prodelta, delta front, and tidal bars, and finally alluvial/delta plain/bay deposits (López-Blanco et al., 2003).

The lower Eocene delta deposits of the Isabena Valley, northern Tremp-Graus Basin, record the start of the main Pyrenean Orogeny in this area. The Roda, St. Esteban, and Campanue fan deltas represent three stacked, upward-coarsening sequences deposited by south-flowing rivers (Bentham and Burbank, 1996). The Roda fan delta in the Isabena valley ranges from ~54.3 to 53.7 Ma; it is interpreted to represent early emplacement of the Cotiella Thrust sheet (Roberti, 2008) or the Bòixols Thrust front (López-Blanco et al., 2003). The Cotiella Thrust comes to the surface just north of the Bòixols Thrust. The St. Esteban delta consists primarily of coarse sandstone and ranges in age from 52.8 to 52.1 Ma. Conglomerates of the Campanue fan delta (~50 Ma) are well developed in the Isabena Valley (Bentham and Burbank, 1996).

The Roda Sandstone was deposited in a shallow marine basin whose depth was modified by a series of gentle, growing folds. This simultaneous sedimentation and folding is suggested by thickening of sedimentary units above synclines and thinning over anticlines; by carbonate platform deposits growing on top of anticlines, and folds tapering off upward in younger units. High sedimentation rates with respect to fold growth meant that the structures were eventually buried by the basinward-moving deltas. There are no noticeable unconformities in the growth strata at outcrop scales.

Folds within the Montsec Thrust sheet inherited their geometry from Early Cretaceous extensional faults below the thrust sheet. These early normal faults were reactivated and inverted (the sense of displacement was reversed) during Late Cretaceous shortening. Folding began in Paleocene time and continued through early Eocene time. Both fault-bend and detachment-style folding occurred during southward displacement of the Montsec Thrust Sheet (López-Blanco et al., 2003).

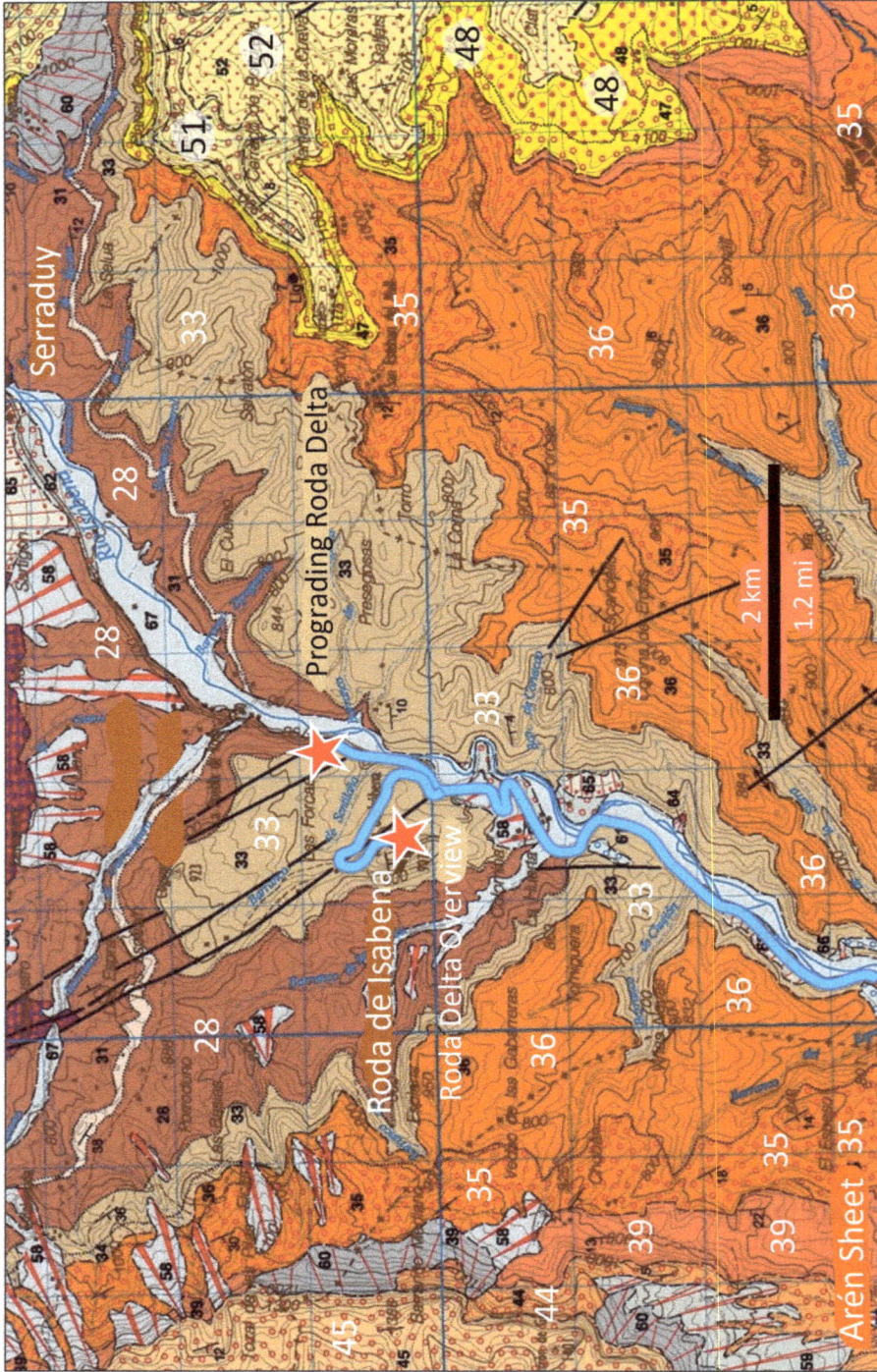

Geologic map of the Roda d'Isabena-Puebla de Roda area. 27 = lower Eocene Alveolina Limestone; 28 = lower Eocene reef limestone; 29 = lower Eocene Riguala Marl; 30 = lower Eocene bioclastic limestone; 33 = lower Eocene Roda sandstone; 35 = lower Eocene San Esteban Formation sandstone; 36 = lower Eocene Castigaleu Formation sandstone-shale; 39 = lower Eocene Castissent Formation sandstone-conglomerate; 44 = lower Eocene Perarrúa limestone; 45 = middle Eocene Campanue conglomerate; 48 = middle Eocene shale-conglomerate; 51 = middle Eocene conglomerate; 52 = middle Eocene conglomerate. From Instituto Geológico y Minéro de España (IGME), Arén Sheet. Source: © NC Geological Survey of Spain (IGME).

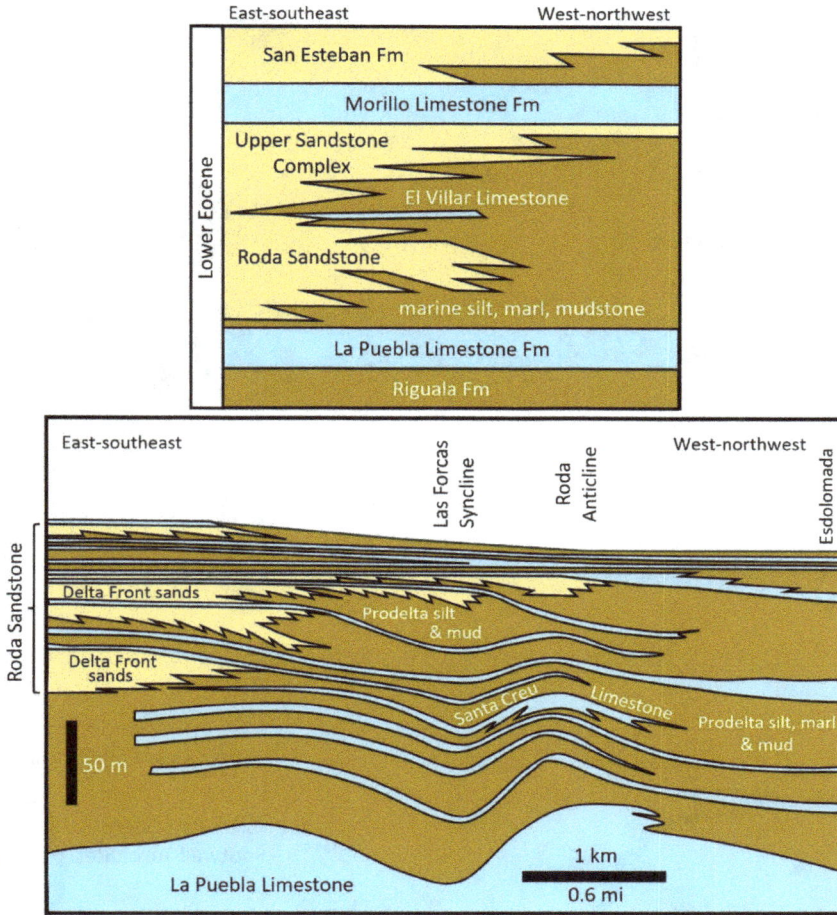

Above: Stratigraphy of the Roda delta deposits. Below: Diagram showing the relationship of Roda delta sedimentation and the seaward migration of sediments as they overrode growing folds. Modified after López-Blanco et al., 2003.

STOP 13.1 RODA DELTA OVERVIEW, RODA DE ISÁBENA

This is a scenic view stop. From this high vantage point you can look northeast and see the south and west prograding sandstones of the lower Eocene Roda delta. Foreset beds (layers deposited on the slope of a delta) are inclined to the south and southwest.

Roda Overview to Prograding Roda Delta: *Return to A-1605 and turn left (north); continue north on A-1605 to Stop 17.2, Prograding Roda delta (42.297973, 0.539282) and use pullout on the right (east) side of highway beyond the guardrail. This is **Stop 13.2, Prograding Roda Delta** (42.297973, 0.539282) for a total of 2.9 km (1.8 mi; 4 min).*

View northeast to the Roda delta (Roda Sandstone) deposits (center and center right).

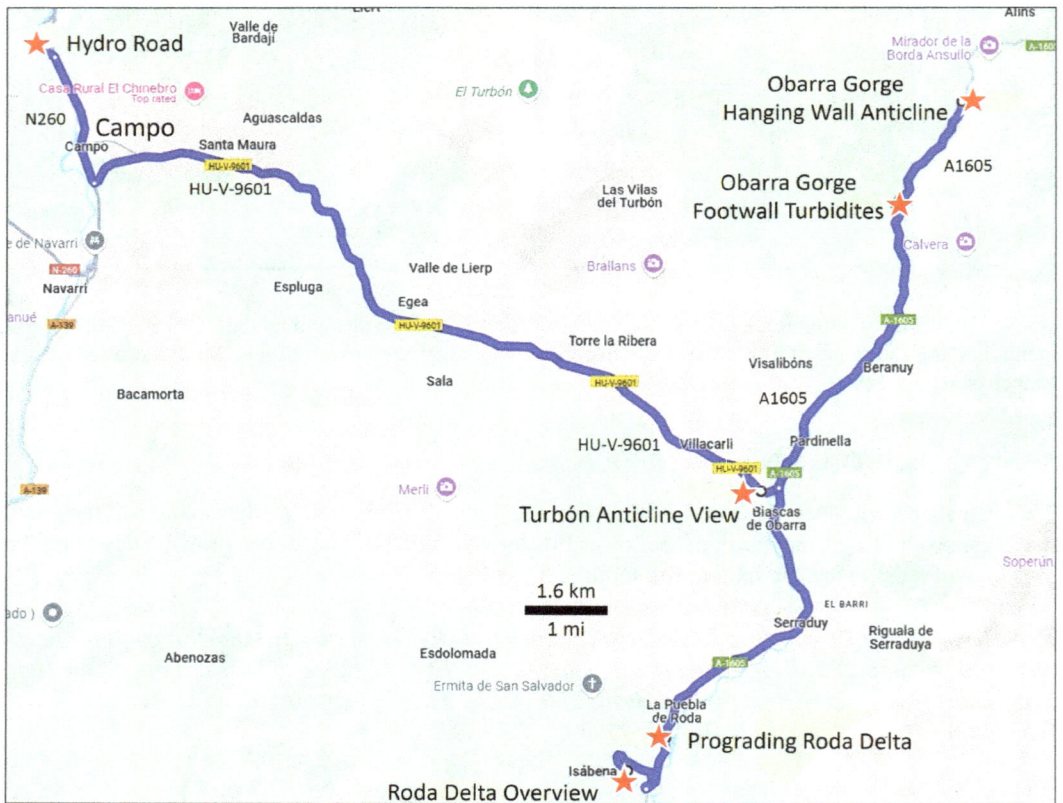

Stop 13.2 Prograding Roda Delta, A-1605 Roadcut

This is the best roadcut section of the delta system. The extensive roadcut reveals Roda delta sands (lower Eocene Roda Sandstone) prograding over prodelta muds. Individual delta lobes are separated by cemented hardgrounds (a seafloor surface that was hardened by precipitation of carbonate cement, forming a rock-like layer). There is both a lateral and a down-dip transition from steeply dipping foreset beds to large gently dipping sub-tidal bars (Enge et al., 2007).

Roda delta south-dipping units in the A-1605 roadcut. Dashed line follows bedding.

Prograding Roda Delta to Obarra Footwall Turbidites: *Continue north on A-1605 for 14.0 km (8.7 mi; 12 min) and pull over on the right. Walk ~100 m (300 ft) south to the roadcut on the west side. This is* ***Stop 14.1, Obarra Gorge Footwall Turbidites*** *(42.397853, 0.596097).*

Stop 14 Obarra Gorge

Scenic Obarra Gorge cuts through the Aras Thrust Sheet, a subset of the larger Bóixols Thrust of the South Pyrenean Zone. This is a good place to examine deformation in the hanging wall (thrust sheet) and the footwall (below the thrust) of the Las Aras Thrust.

Stop 14.1 Footwall Turbidites, Obarra Gorge

This roadcut exposes the Mascarell member of the Vallcarga Formation of Late Cretaceous age. The Mascarell turbidites, up to 1,500 m (4,920 ft) thick, mark the initial sedimentation related to thrusting and foreland basin development in the South Pyrenees Zone, and for that reason are thought to record the change from extension to compression.

The lower section of turbidite sandstones (seen here) includes thin bedded and fine-grained layers with minor thicker and coarser beds, and the strata commonly display well-developed fining upward (Bouma) sequences. The upper part of the section is characterized by channelized, thick-bedded, and coarse-grained sandy turbidites with abundant mudstone fragments (Ardèvol et al., 2000). These turbidites sometimes eroded deep into the underlying units as they were being deposited. The beds are inclined to the south between 33° and 80°. Farther south the beds become overturned and define an overturned syncline. So the beds here are the common flank of a thrusted anticline-syncline pair that verges to the south.

The Mascarell turbiditic unit has been correlated to the Campo Breccia to the west (Saura et al., 2016). The Campo Breccia outcrops along highway N-260 in Esera Canyon north of Campo.

South-dipping Late Cretaceous Mascarell member turbidites just south of the entrance to Obarra Gorge.

Obarra Gorge Footwall Turbidites to Hanging Wall Anticline: *Continue north on A-1605 for 1.5 km (0.9 mi; 2 min) and pull over on the right. This is **Stop 14.2, Obarra Gorge Hanging Wall Anticline** (42.408032, 0.605933).*

Stop 14.2 Hanging Wall Anticline, Obarra Gorge

This stop exposes a small subsidiary thrust and fault-propagation fold in the hanging wall of the Las Aras Thrust. The units are either Upper Triassic or Early Jurassic Keuper Formation limestones and marls.

Small fault-propagation fold on minor thrust in the hanging wall of the Las Aras Thrust Sheet, Obarra Gorge. Dotted lines are thrust faults; dashed lines follow bedding. View southwest.

As can be seen on the cross-section, the large south-verging anticline is cut by the Las Aras Thrust. A thick Jurassic to Lower Cretaceous extensional basin-fill sequence is carried on the thrust sheet. A thin section of syntectonic turbidites lies unconformably above Upper Cretaceous lime-stone on the overturned forelimb of the thrusted anticline.

Obarra Gorge Hanging Wall Anticline to Turbón Anticline: *Return south on A-1605; turn right (west) onto HU-V-9601 and drive another 380 m (1,240 ft) and pull over on the left. This is* **Stop 15, Turbón Anticline View** *(42.343862, 0.562948) for a total of 8.9 km (5.5 mi; 8 min).*

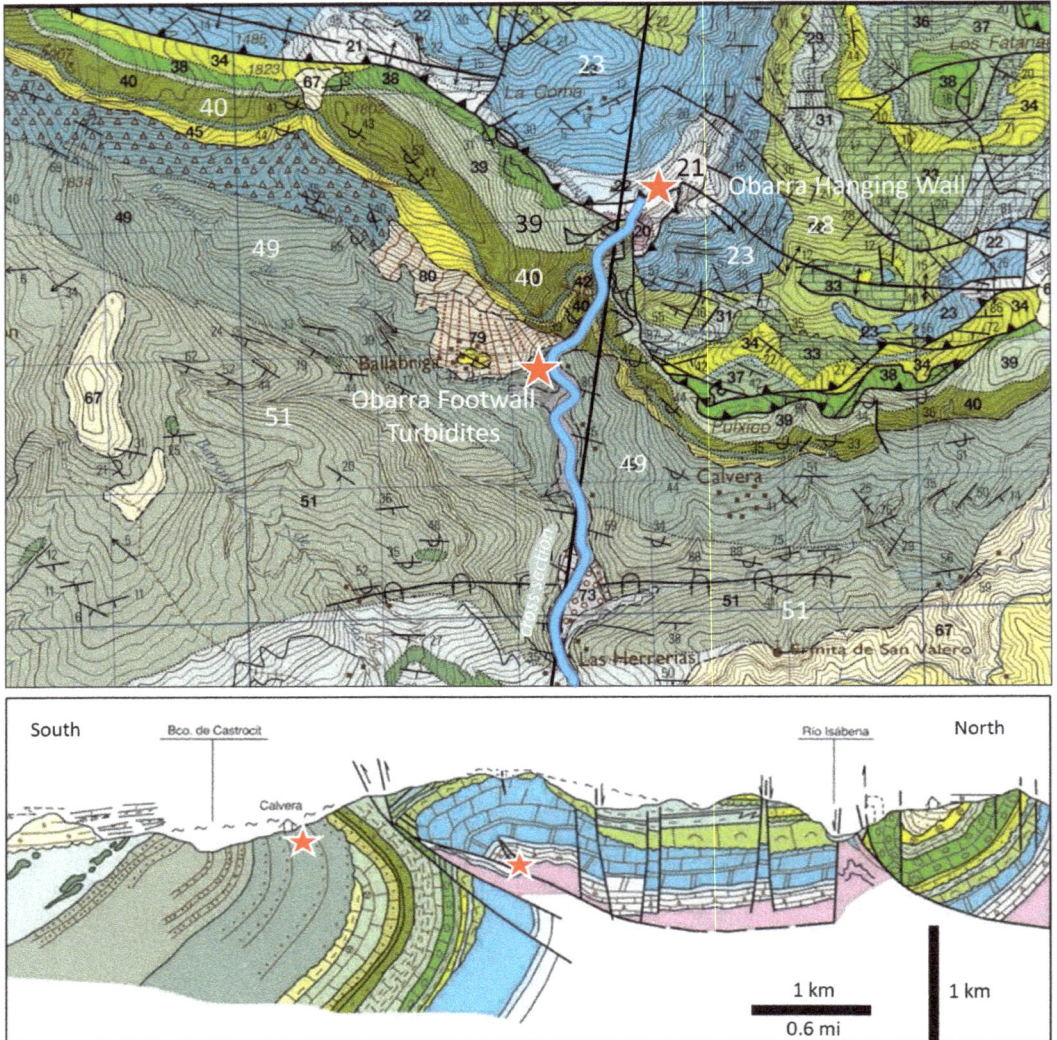

Top: Geologic map of the Obarra Gorge area. Stars show stop locations. Key units include: 21 – Late Triassic–Early Jurassic Keuper limestone-dolomite-breccia; 23 – Upper Jurassic dolomite-limestone; 28 – Lower Cretaceous San Roque Fm limestone; 39 – Upper Cretaceous Reguard Fm limestone; 40 – Upper Cretaceous limestone; 49 – Upper Cretaceous Mascarell member turbidites, Vallcarga Fm; 51 – Upper Cretaceous Mascarell member bioclastic turbidites. Bottom: Cross-section through the Las Aras Thrust, Obarra Gorge. From Instituto Geológico y Minéro de España (IGME), Pont de Suert Sheet. Source: © NC Geological Survey of Spain (IGME).

Stop 15 Turbón Anticline

As we approach the Turbón Anticline we are on the northern margin of the South Pyrenean Zone just south of where the Las Aras/Boixols Thrust comes to the surface. The prominent Turbón Anticline is an asymmetric, southwest-verging fold detached above Upper Triassic (Keuper) evaporites. The fold orientation is controlled by lateral ramps in the underlying Cotiella–Boixols–Montsec Thrust.

A lateral ramp is where the thrust fault cuts up-section along the sides of the thrust sheet. The ramps inherited their orientation from Lower Cretaceous rift-related normal faults. The anticline developed as a result of evaporite flow within the Cotiella–Boixols–Montsec Thrust Sheet, and it is in the footwall of (below) the Las Aras Thrust (López-Mir, 2013; Ramos et al., 2020).

This view of the east limb highlights the 1,000 m (3,280 ft) of structural relief. The anticline stands out because of the resistant Late Cretaceous platform carbonates of the Santa Fe, Pardina, and Congost formations. Jurassic marls and dolomites are exposed in the core (Roberti, 2008).

The strong south plunge of the Turbón Anticline is related to southward tilting during uplift and thrusting in the Axial Zone.

The Turbón and Serrado anticlines developed simultaneously with tectonic inversion of the Cotiella Basin. The Cotiella Basin is a Late Cretaceous, salt-floored, post-rift extensional basin. The Turbón–Serrado fold system consists of contractional evaporite-cored anticlines.

The Cotiella Basin contains several salt withdrawal minibasins bounded by diapirs, salt rollers, and listric faults (salt rollers are small salt bulges characterized by their wavy upper surface; listric faults are faults that flatten at depth). Inversion occurred during the Pyrenean Orogeny. Detachment folding was associated with the deposition of several km of slope sediments, with turbidites and resedimented carbonates onlapping the limbs of the growing anticlines. Tectonic inversion is also revealed by onlap of the synorogenic Campo Breccias member and related turbidites of the Barbaruens Formation (Ramos et al., 2020).

As we drive west toward Campo, watch for the seriously eroded badlands of gray, Upper Cretaceous marls in the Anserola and Barbaruens formations.

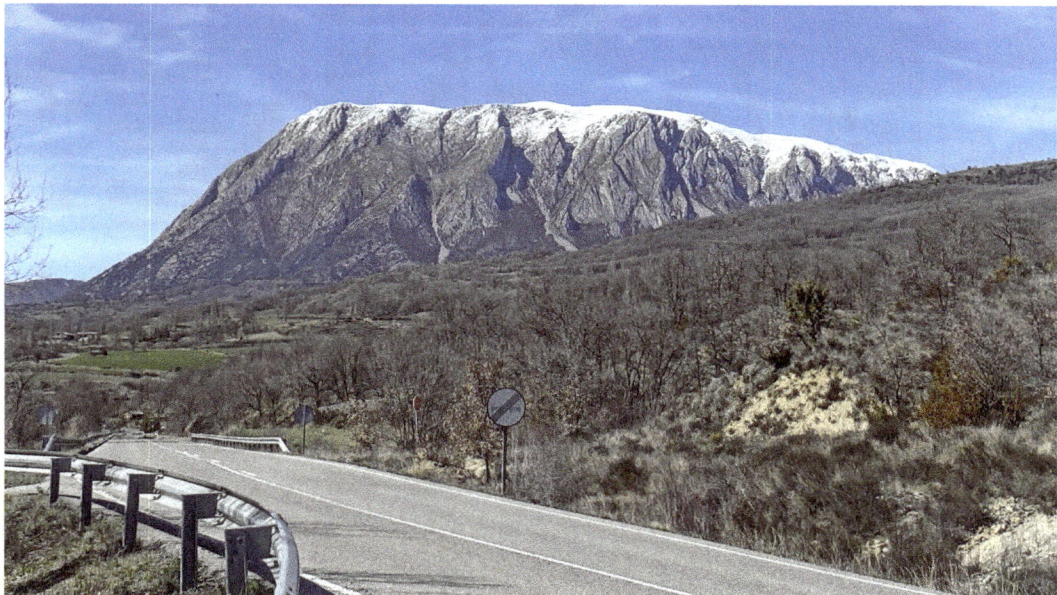

View west to the Turbón Anticline with a dusting of snow.

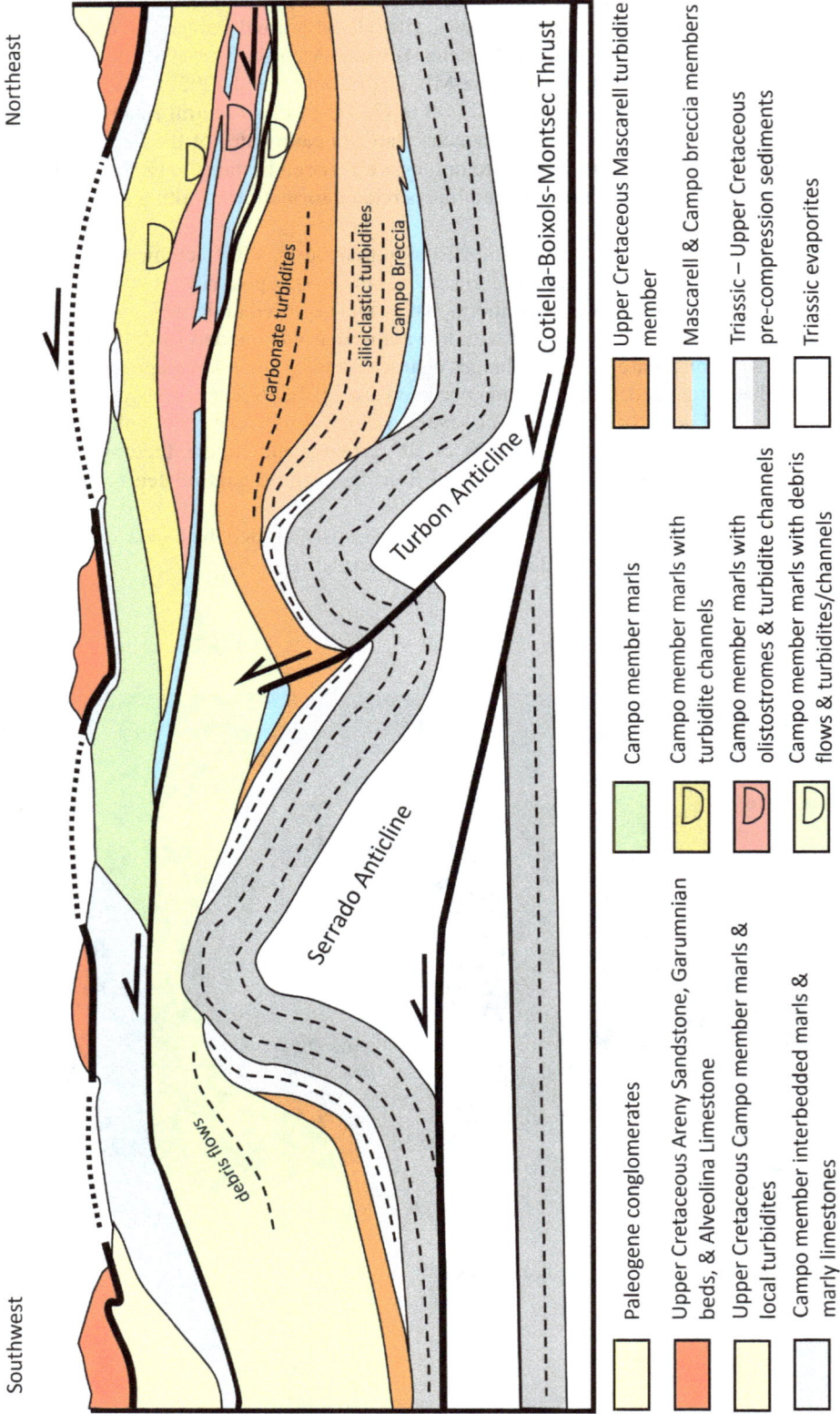

Cross-section of the Turbón Anticline detachment fold above Triassic evaporites. The location is shown in the next figure. Modified after Roberti, 2008.

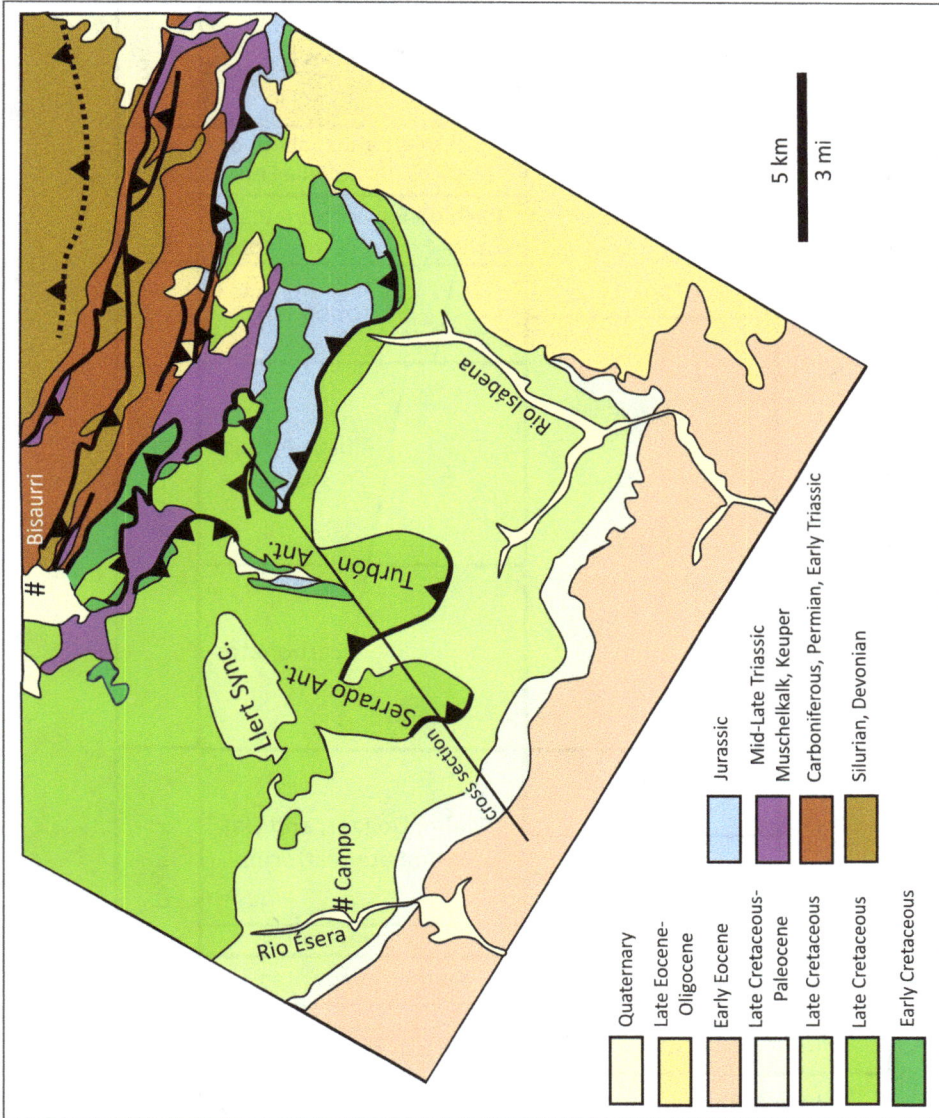

Geologic map of the Turbón Anticline. Geology derived from Roberti, 2008, and Garcia-Senz, 2002.

Age			Unit	Event	
Cenozoic	Oligocene		conglomerates	Syn-orogenic 2	
	Paleogene		sandstones		
			shales		
			carbonates		
Cretaceous	Late		Areny Fm	Syn-orogenic 1	
			Barbaruens Fm		
			Mascarell Turbidites Mb		
			Campo Breccia Mb		
			Anserola, Aguas Salenz, & Maciños del Cotiella Fms	Post-rift 2	
			Pardina Fm carbonates Santa Fe Fm carbonates		
	Early		Turbon Fm	Syn-rift 2	
			Albian-Aptian shales		
			carbonates		
Jurassic	Late		Malm, Dogger, and Lias carbonate platforms	Post-rift 1	
	Middle				
	Early				
			Lias evaporites		
	Triassic		Keuper evaporites	Syn-rift 1	
			Muschelkalk		
	Permian		Buntsandstein		
	Late Carboniferous		basement volcanics		

Stratigraphy and tectonic events chart of the Turbón Anticline, Esera Valley, and Campo area. Stratigraphic information derived from Ramos et al., 2020.

Stop 16 Campo Area

The Esera River valley in the Campo area offers the opportunity to see the deep marine landslide deposits, or turbidites, of the Hecho Group that are related to Pyrenean uplift north of here.

Turbon Anticline to Hydro Road: *Continue driving northwest on HU-V-9601; at the roundabout in Campo take the 1st exit (north) onto N-260 and drive 2.8 km (2.4 mi) and turn left (west) onto the hydro plant road; drive another 140 m (460 ft) and pull over on the right. This is* ***Stop 16.1, Hydro Road*** *(42.426106, 0.386194) for a total of 19.9 km (12.4 mi; 17 min).*

The Campo Breccia at the north end of the Campo Valley represents the first sedimentary response to the Alpine Orogeny in the Pyrenees and subsequent filling of an east-west trough developed in the foreland south of the uplift. Late Cretaceous Pyrenean compression inverted and uplifted the preexisting Cotiella extensional basin, exposing the basin fill to erosion and creating a regional unconformity. The breccia was deposited on this regional angular unconformity (Roberti, 2008;

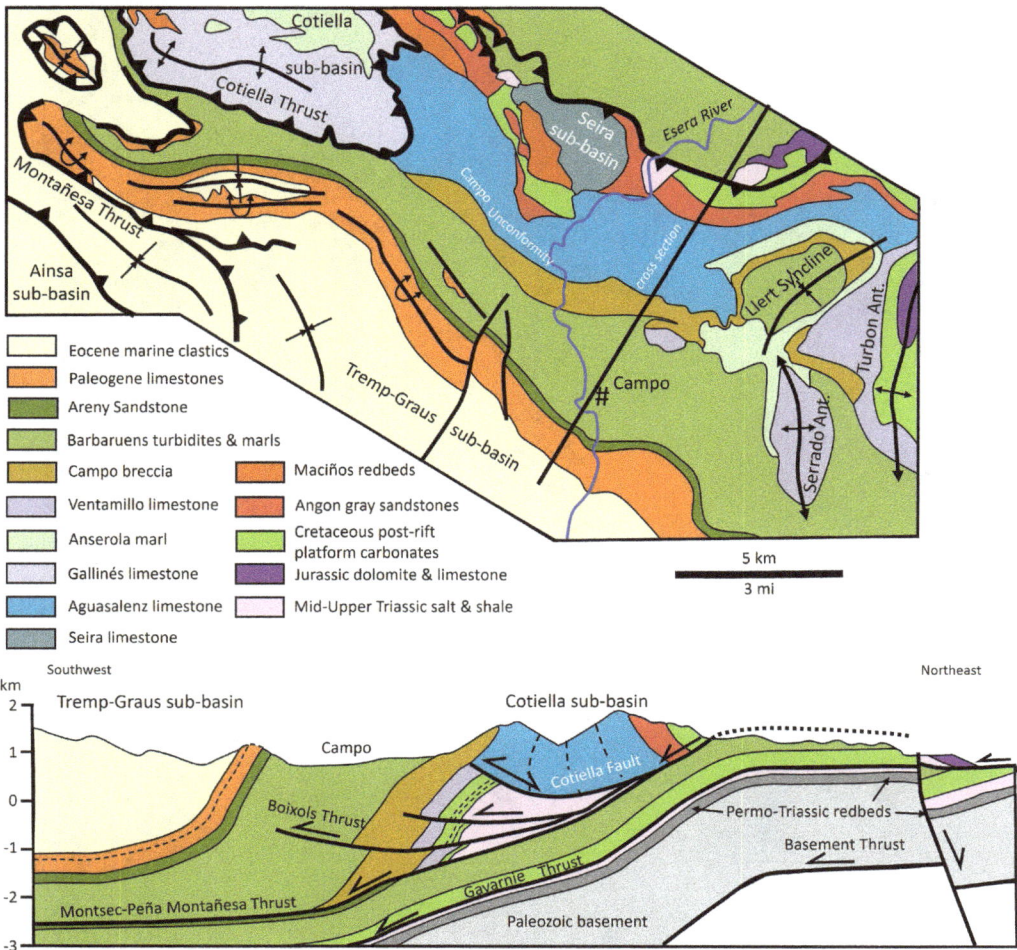

Geologic map and cross-section of the Campo-Jaca section of the Southern Pyrenean Zone. Modified after López-Mir et al., 2016.

Geologic map and stops in the Campo area. From Instituto Geológico y Minéro de España (IGME), Campo Sheet. Source: © NC Geological Survey of Spain (IGME).

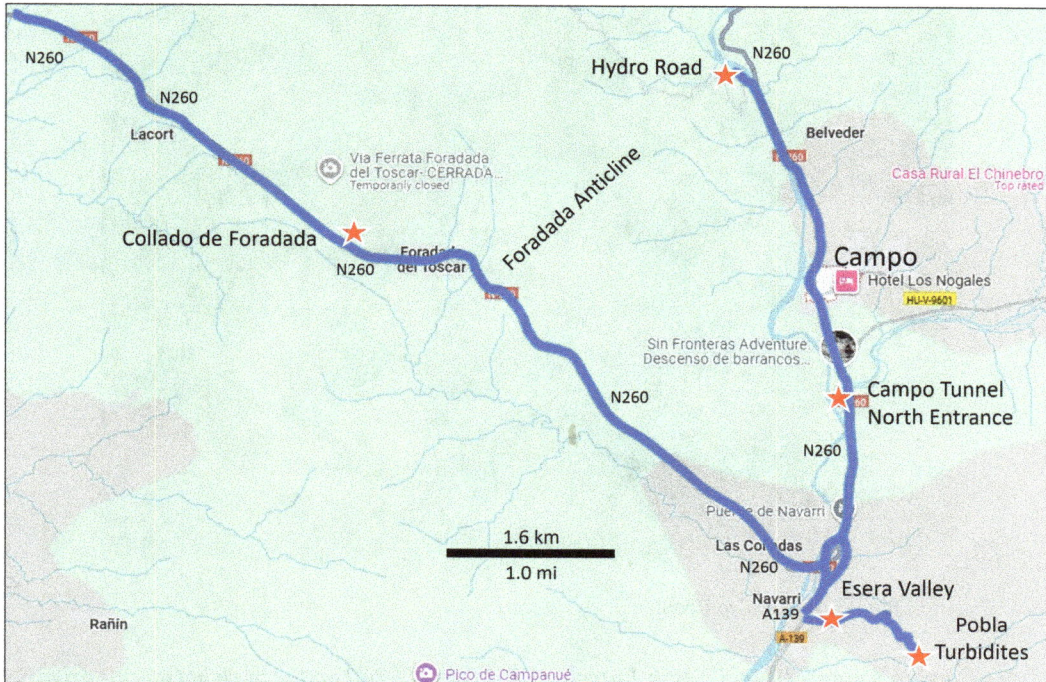

López-Mir, 2013). You can see the breccia in the old road that bypasses the new tunnel (for example, there is a roadcut and pullout on the east side of the road at 42.432844, 0.386445; there is a roadcut and pullout on the west side of the road at 42.429095, 0.383782).

The Late Cretaceous succession, deposited coincident with early uplift of the Pyrenees, has been divided into four main units in this area: the Campo Breccia member, Vallcarga Formation turbidites, Barbaruens Formation turbidites, and Areny (or Arén) Formation sandstone. These units increase in thickness closer to their source to the north (López-Mir, 2013; López-Mir et al., 2016).

The Campo Breccia is made up of angular to sub-rounded fragments derived primarily from Late Cretaceous limestones, with lesser amounts of Lower Cretaceous sedimentary rocks, Upper Triassic diabase, Permo-Triassic redbeds, and Paleozoic metamorphics. The fragments get finer upward and are poorly sorted. In the Ésera River area the Campo Breccia is up to 800 m (2,625 ft) thick. It grades laterally and upward into marls and turbidites of the Vallcarga Formation.

Stop 16.1 Vallcarga Formation Turbidites, Hydro Road, Campo

Upper Cretaceous Vallcarga Formation turbidites are exposed at this stop. The Vallcarga Formation of the Cotiella Basin is also a synorogenic deposit. Together with the Campo Breccia, these units represent the earliest response to the Alpine Orogeny in the Pyrenees.

The Campo Breccia grades upward into marls and turbidites of the Vallcarga Formation. The Mascarell member of the Vallcarga Formation is an alternating series of turbidites and marls whose total thickness can exceed 1,500 m (4,500 ft). As seen here, the Vallcarga Formation consists primarily of thin-bedded, fine-grained turbidites (López-Mir, 2013).

Pyrenean shortening and contraction did not substantially modify the preexisting, extensional Cotiella Basin other than carrying the basin tens of km south on the Montsec–Peña Montañesa Thrust system (López-Mir et al., 2014, 2016).

Vallcarga turbidite outcrop, Hydro Road.

Hydro Road to Campo Tunnel North Entrance: *Return to N-260 and turn right (south); drive 3.2 km (2 mi) and pull over on the right before the guardrail. This is **Stop 16.2, Campo Tunnel North Entrance** (42.398506, 0.398268) for a total of 4.0 km (2.5 mi; 6 min).*

STOP 16.2 ARENY SANDSTONE, CAMPO TUNNEL NORTH ENTRANCE

The Areny (or Arén) Formation sandstone is the resistant ridge-forming sandstone south of Campo. It consists of well-sorted quartz sandstone and limey sandstone that accumulated and advanced southward over the previous turbidite trough, interfingering with the uppermost parts of the underlying turbidites. These nearshore, shallow marine deposits average between 40 and 300 m thick (130 and 1,000 ft thick; López-Mir, 2013). The sandstones thicken to the north and west, and they show growth geometries related to Alpine uplift (López-Mir et al., 2016).

The Upper Cretaceous Areny Sandstone forms the ledge at the top of the ridge. Campo tunnel north entrance, view south.

*Campo Tunnel North Entrance to Esera Valley: Continue south on N-260; take the A-139 exit toward Graus; at the sign for Morillo de Lena/Bacamorta 5 turn left (east) and drive 160 m (535 ft) and pull over on the left at a farm entrance. This is **Stop 16.3, Esera Valley** (42.380326, 0.395411) for a total of 2.3 km (1.4 mi; 3 min).*

STOP 16.3 HECHO GROUP, ESERA VALLEY

At this stop you get a good view of early Eocene turbidites of the Pobla Formation, part of the Hecho Group, off to the west. The mudstone here was deposited "between delta deposits and inner fan sediments" (Gardiner et al., 2003). The inner fan sediments referred to are the coarse-grained, proximal part of a submarine fan.

We are now approaching the western edge of the Tremp-Graus Basin. The boundary between the Jaca and Ainsa basins in the west and the Tremp-Graus Basin to the east appears to be an approximately north-south continental shelf–continental slope break, roughly coincident with the Mediano Anticline. The shelf-slope break separates shallow marine and continental rocks on the east from deep marine rocks to the west. All of these basins are "piggyback basins", areas that subsided and collected sediments while being carried south on various thrusts of the Southern Pyrenean Zone. During early and middle Eocene, there was mainly river and delta sedimentation in the east (Tremp-Graus Basin), and deep marine turbidites of the Hecho Group in the west (Ainsa and Jaca basins). The turbidity currents originated at deltas in the east and flowed west, roughly parallel to the trough axis. Extensive basin floor fan systems were deposited in the trough of the strongly subsiding east-west-oriented Jaca and Ainsa basins. These basins developed more or less parallel to the leading edge of thrust sheets (Labaume et al., 1987; Oms et al., 2003; Martinius, 2012; Roigé et al., 2016).

You also get a magnificent view of the Foradada Anticline from here. The N-260 highway cuts through steeply dipping to overturned Eocene Riguala Formation marls on the flank of the southwest-verging Foradada Anticline. The view to the northwest clearly shows the fold.

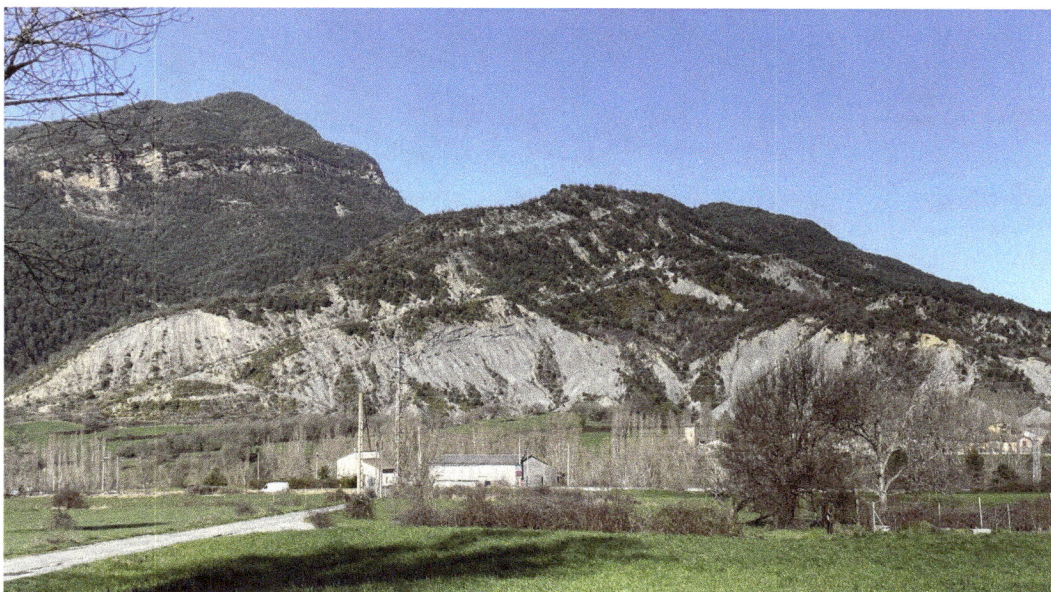

South inclined Eocene turbidites of the Pobla Formation, Hecho Group, Esera Valley. View west.

View northwest to the Foradada Anticline.

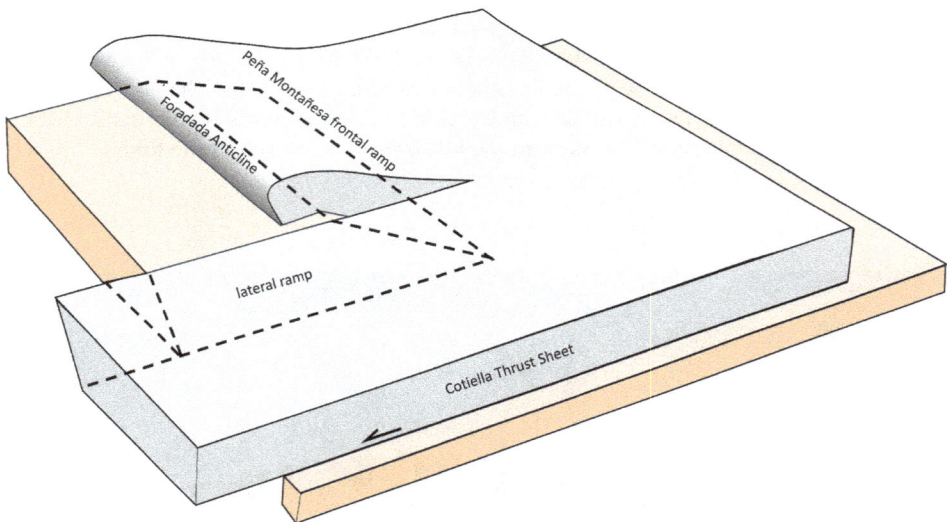

Diagram of the Foradada Anticline over the Peña Montañesa frontal ramp of the Cotiella Nappe and its associated tear fault. Thrusting was post-lower Eocene. Modified after Mutti et al., 1989.

The Foradada Anticline is outlined by the Alveolina Limestone Formation. It verges southwest, has an overturned front limb, and is bounded on the east by a left-lateral tear fault. (A tear fault is one that cuts a thrust sheet and allows changing horizontal displacement of the sheet). It is the frontal ramp anticline of the Cotiella Thrust Sheet. There is a minimum of 610 m (2,000 ft) of relief on this structure. Truly a world-class structure.

Esera Valley to Pobla Turbidites: *Continue driving east on this unnamed road for 1.1 km (0.7 mi; 2 min) and pull over on the right at **Stop 16.4, Pobla Turbidites** (42.377720, 0.404992).*

STOP 16.4 POBLA TURBIDITES, ESERA VALLEY

Here are excellent outcrops of the early Eocene Pobla Formation shale and sandstone turbidites of the Hecho Group. The Hecho Group deep basin sedimentary fill is ~3 km (9,800 ft) thick and is dominated by marls and interbedded turbidite sandstone and shale. The Hecho Group has been subdivided into several sandstone-dominated turbidite sequences separated by thick shale intervals. The thick sandstone intervals are considered to be channel deposits; thin-bedded channel margin deposits are sandstones and interbedded shales; and thick shales with thin turbidites are interpreted to be interchannel deposits (Gardiner et al., 2003).

Above the deepwater Hecho Group are shallow marine delta rocks of the Sobrarbe Formation overlain by river sediments of the Escanilla Formation. These units represent the final, rapid infilling of the basin (Beamud et al., 2003).

Pobla Turbidites to Foradada Anticline: *Return west to A-139 and turn right (north); merge left onto N-260 toward Ainsa; drive west on N-260 for 6.1 km (3.8 mi) and pull over on the right shoulder. This is **Stop 17, Collado de Foradada** (42.412110, 0.341253) for a total of 8.1 km (5.1 mi; 9 min). Or drive another 200 m and pull onto the road on the left.*

Early Eocene Pobla Formation turbidites, Esera Valley. View east.

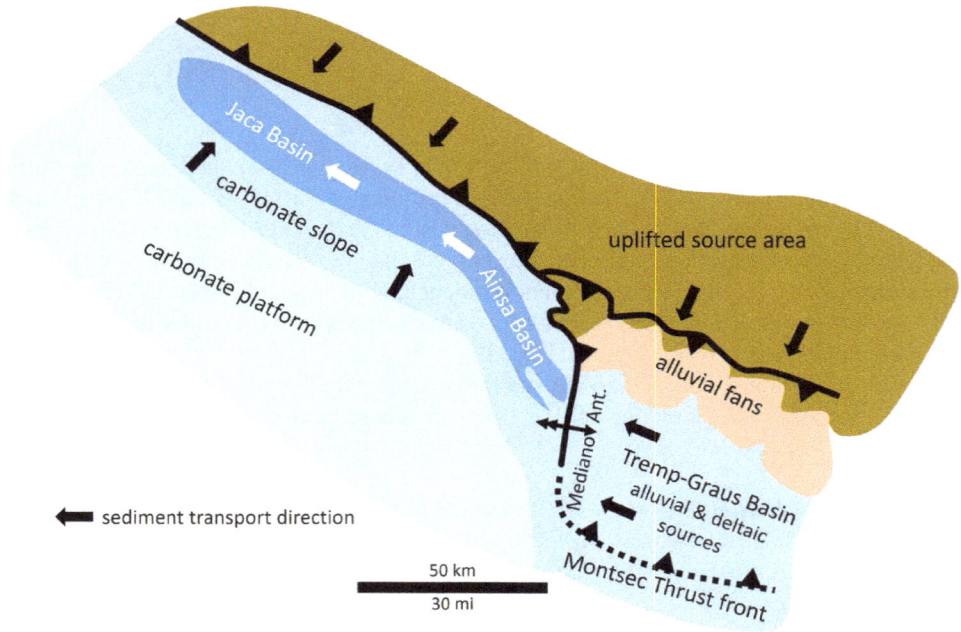

Transition from the deep marine sediments of the Jaca Basin to shallow marine sediments of the Tremp-Graus Basin during the Eocene occurs along a shelf-slope break roughly coincident with the Mediano Anticline. Modified after Dakin, 2016.

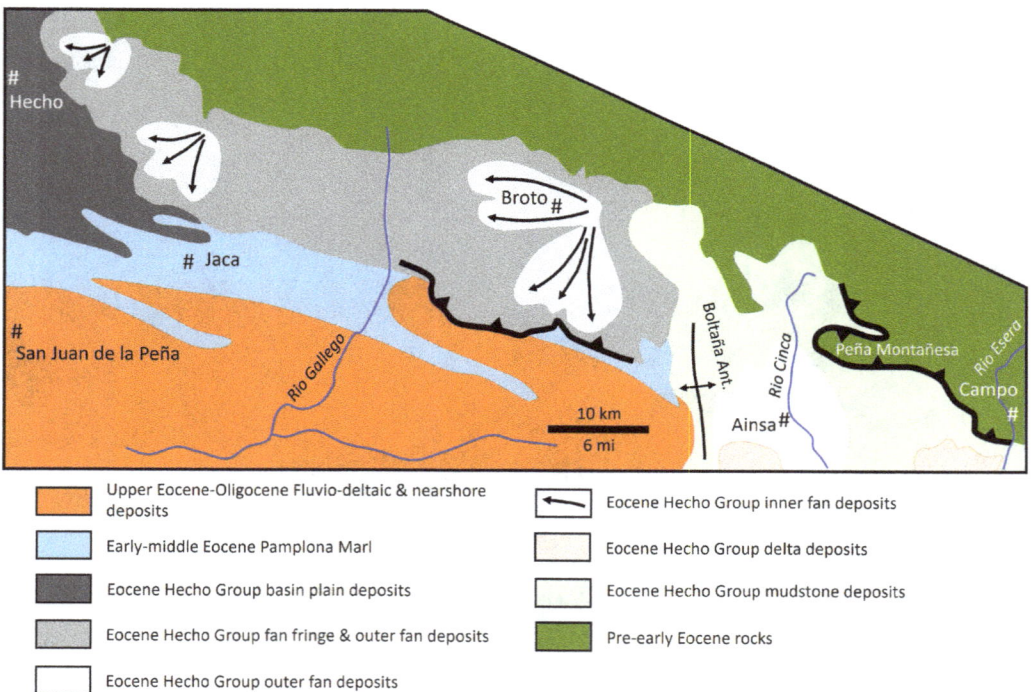

Marine fan deltas of the Eocene Hecho Group, Jaca and Ainsa basins. Modified after Mutti, 1977.

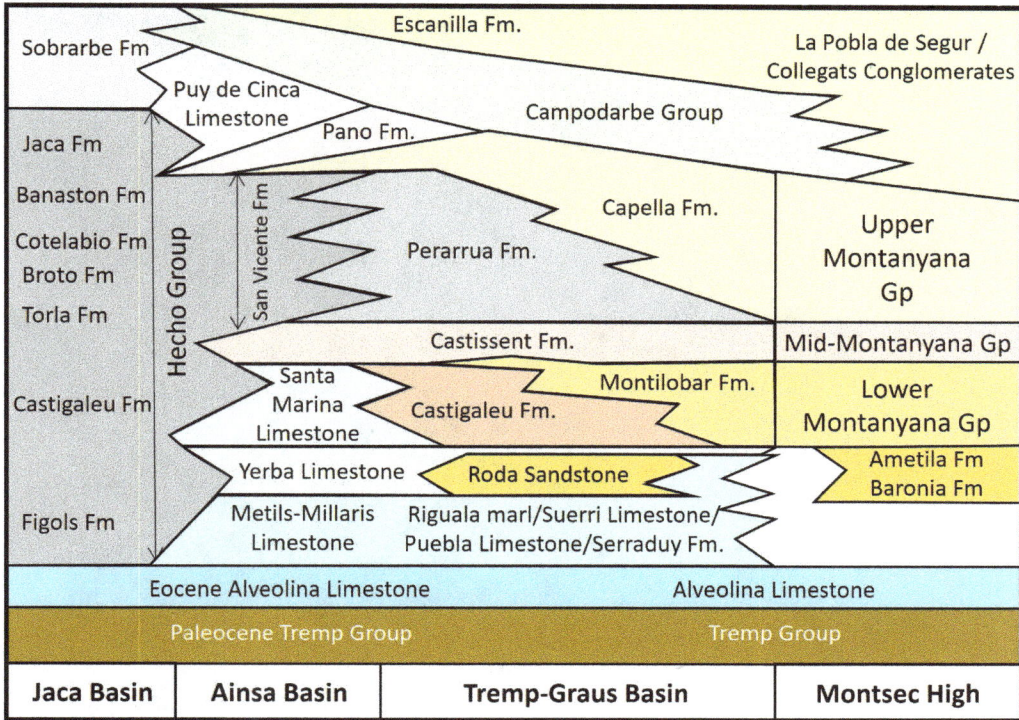

Eocene stratigraphy of the Tremp-Graus and Ainsa-Jaca basins. Stratigraphy derived from Martinius, 2012; Dakin, 2016; Cantalejo et al., 2020.

STOP 17 FORADADA ANTICLINE, COLLADO DE FORADADA

The Collado de Foradada stop is on the overturned west flank of the Foradada Anticline. At road level we see the Eocene Riguala Marl overlain by older Eocene Alveolina Limestone overlain by Paleocene limestone and dolomite at the top of the ridge (Instituto Geológico y Minéro de España, Campo Sheet).

Although not posted, we are very close to the eastern boundary of Sobrarbe Pirineos UNESCO Global Geopark.

SOBRARBE PIRINEOS GEOPARK

Sobrarbe Geopark occupies 2,202 km^2 (850 mi^2) in northern Huesca province. In 2006 the Sobrarbe became a UNESCO European Geopark on account of its spectacular geology, including the Cotiella Massif and Monte Perdido Massif, the highest limestone massif in Europe. Unique aspects of the geology include ancient mine workings (coal, iron, cobalt), a glacial landscape, and limestone karst features such as caves and sinkholes. The lowest point in the Geopark (El Grado reservoir, 475 m or 1,560 ft) and the highest peak (Llardana, 3,375 m, or 11,100 ft) are separated by only a few km, emphasizing the influence of tectonics. The climate ranges from Mediterranean in the lower areas to Alpine in the high mountains.

The age of the geologic formations in Sobrarbe Geopark extend over 500 Ma. Between 500 and 250 million years ago (Paleozoic) most of Sobrarbe was an ocean floor accumulating silt, mud, clay, and sand. These sediments have since become the shale, sandstone, limestone, and quartzite that form the northern mountains and valleys of the Sobrarbe. These rocks were intensely deformed and altered by the Variscan Orogeny, as seen in numerous folds, faults, and granite intrusions.

The west end of the Collado de Foradada roadcut exposes the near vertical to overturned Eocene Riguala Marl on the west flank of the Foradada Anticline.

Between 250 and 50 Ma, the Variscan Mountains were largely eroded and this area was again covered by a shallow tropical sea. Coral reefs and lime mud accumulated on the seafloor.

The marine sedimentation continued during formation of the Pyrenees between 50 and 40 Ma. The sea, which separated Iberia from the rest of Europe, gradually filled with sediments. About 45 Ma this sea became narrower as the uplifting Pyrenees provided a sudden influx of sediments to the basin. Sobrarbe contains exceptional examples of thick turbidites, rocks formed in that sea as a result of submarine landslides (turbidity currents) that dumped sediment-water slurries into deep water.

Convergence of the Iberian and European plates and resulting uplift of the Pyrenees eventually caused the sea to disappear. About 40 Ma a system of deltas moved into the remaining basin. Huge amounts of sandy and muddy sediment accumulated. About 25 Ma, rivers and floods caused large amounts of gravel to accumulate above the delta sediments. Over time these gravels became the conglomerates seen today.

Global cooling over the past 2.5 Ma covered the Pyrenees in ice. The last major Ice Age peaked here about 65,000 years ago. Glaciers covered the valleys and mountains, dramatically carving out the present landscape (UNESCO, Sobrarbe Geopark).

VISIT

The UNESCO Global Geopark Visitor Centre is located in the southeast tower of Ainsa Castle. It contains information on the geology of the district by way of panels, maps, models, rock collections, videos, and interactive displays The Geopark Space takes up four floors of the Castle's tower and describes the geological history of the Sobrarbe District.

In addition, there is a network of 30 self-guided tours that provide access to the best geological sites in the district and explain their origin, meaning, and significance by means of brochures and interpretive displays. All Geo-Routes have been designed for treks and are clearly signposted. Eleven of these geological routes are located in the Ordesa and Monte Perdido National Park portion of Sobrarbe Geopark.

Address: Comarca de Sobrarbe, Av. Ordesa, 79, 22340 Boltaña, Spain
Phone: +34 974518024
Email: geoparque@geoparquepirineos.com
Website: https://www.geoparquepirineos.com/index.php?idi=3

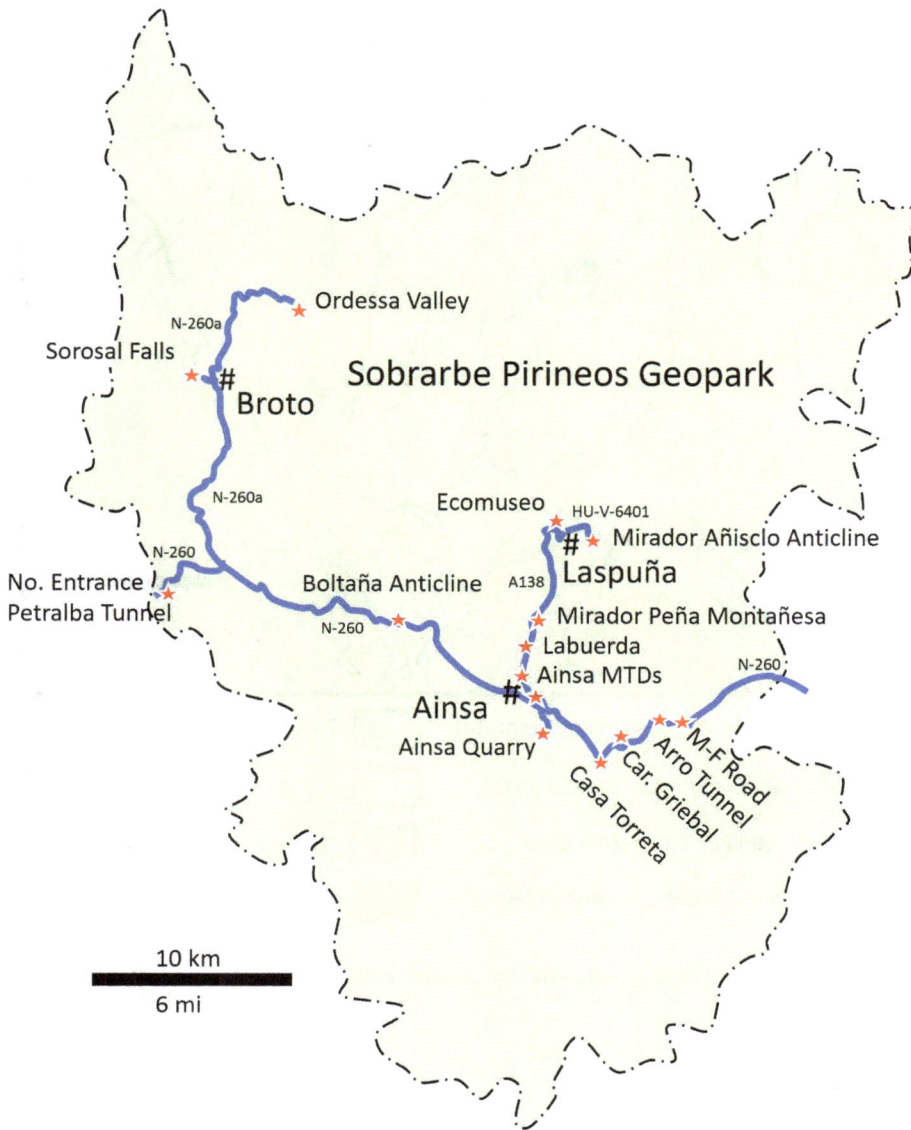

Sobrarbe Pirineos Geopark and stops.

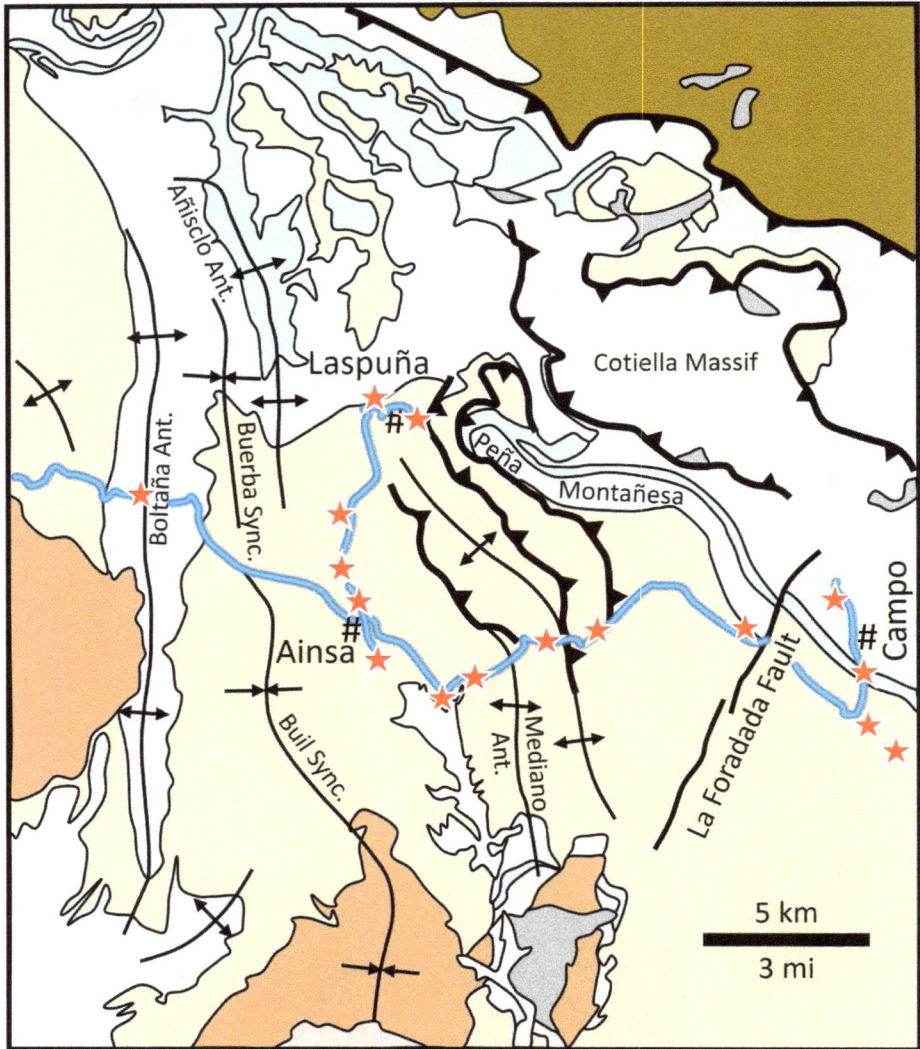

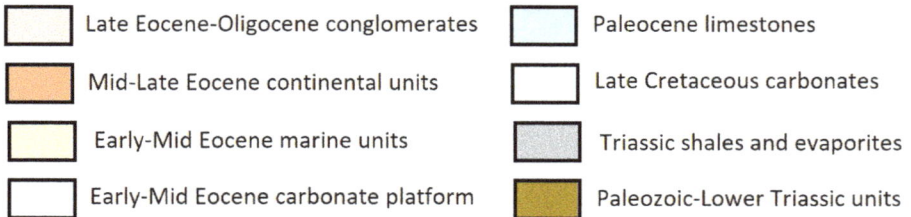

Geologic map of the folds and thrusts in the Ainsa area. Stars indicate stops. Modified after Fernández et al., 2012.

Age	Stratigraphy	Thrusting
Oligocene	conglomerates	
Eocene	continental clastics / San Vicente Fm marine / Boltaña Fm carbonates / Yerba Fm platform carbonate / Metils Fm carbonate / Millaris Fm platform carbonate / Alveolina Limestone	P. Montañesa-Montsec / Gavarnie-S. Exteriores
Late Cretaceous	carbonates	Cotiella-Bóixols
Jurassic	Evaporites/shales	
Triassic		
Paleozoic	Variscan basement	

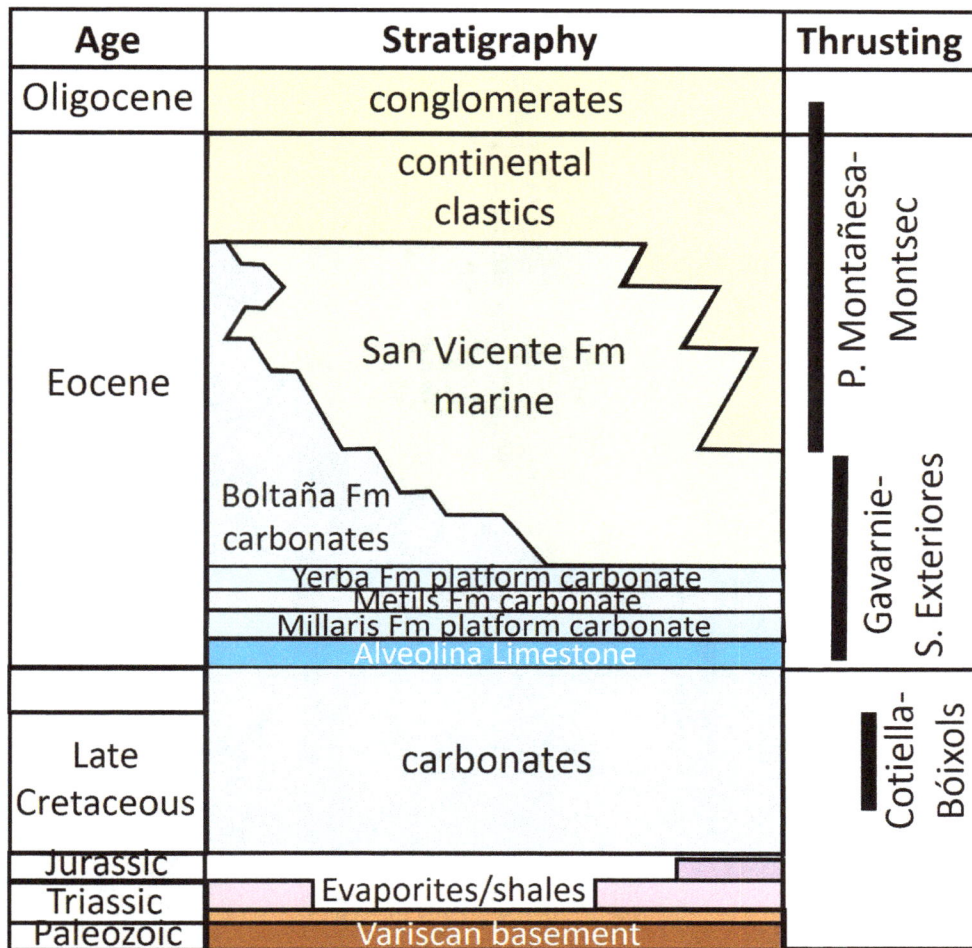

Stratigraphy of the Ainsa Basin. Stratigraphy derived from Fernández et al., 2012

Collado de Foradada to Monzón–Fosado Road: *Continue driving west on N-260; turn right onto the Monzón–Fosado Road and pull over on the right. This is* **Stop 18, Monzón–Fosado Road** *(42.407721, 0.250005) for a total of 9.2 km (5.7 mi; 6 min).*

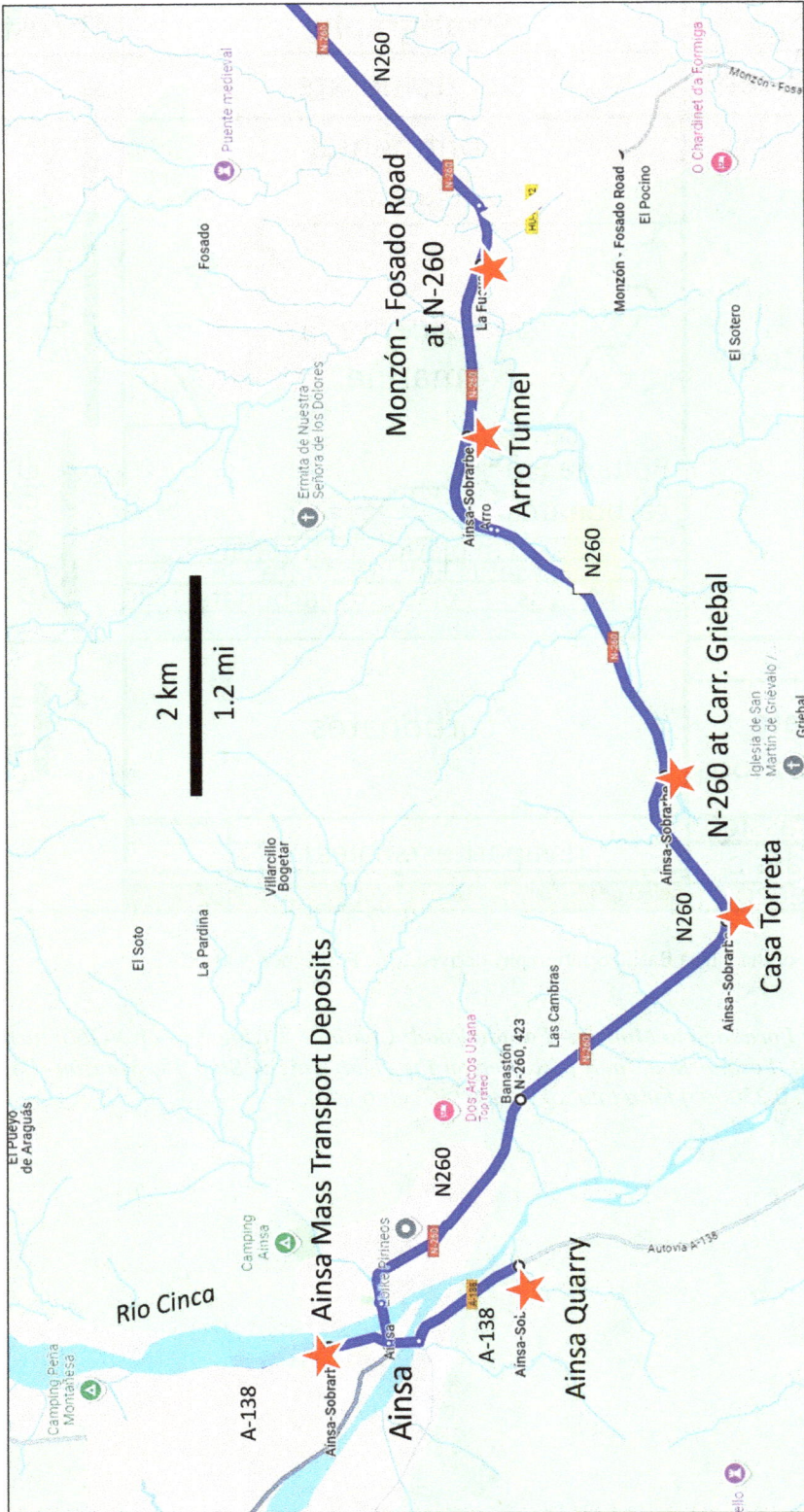

Map annotations:

N260

Monzón - Fosado Road at N-260

Arro Tunnel

N-260 at Carr. Griebal

Casa Torreta

Ainsa Mass Transport Deposits

Ainsa Quarry

Rio Cinca

Ainsa

A-138

2 km
1.2 mi

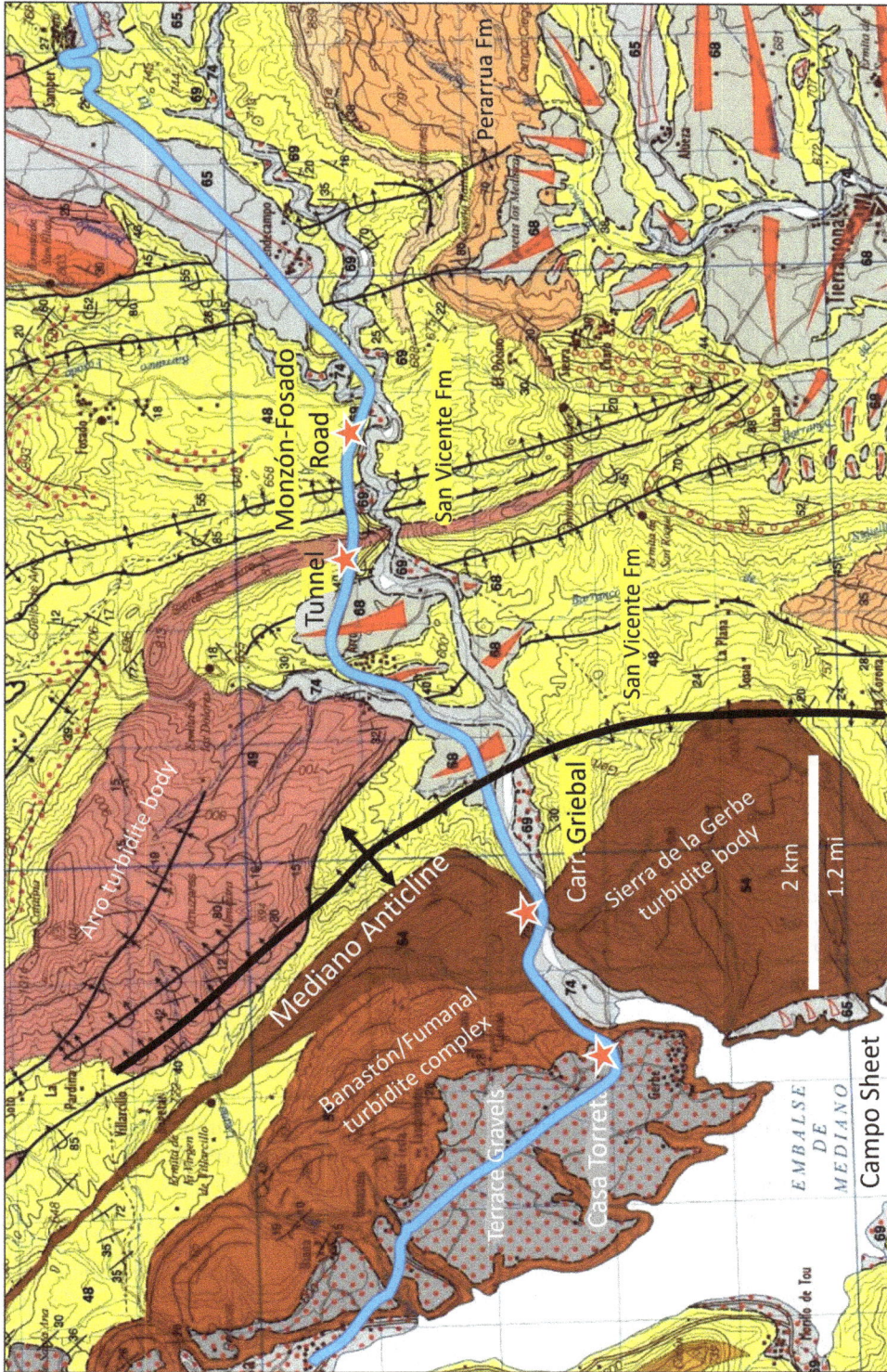

Geologic map of the Rio Lanata Bridge to Barranco de Usana stops. From Instituto Geológico y Minéro de España (IGME), Campo Sheet. Source: © NC Geological Survey of Spain (IGME).

Stop 18 San Vicente Formation Turbidites, Monzón–Fosado Road

This stop has excellent roadcuts exposing Eocene San Vicente Formation (Hecho Group) gray shale with thin, fine-grained turbidite beds (Instituto Geológico y Minéro de España, 1999, Campo Sheet). You can walk up to the outcrop at the intersection and for several hundred meters west along the road. Also look across the valley to the south for a view of the unit.

You are on the east flank of a north-south-trending and west-verging anticline.

N-260 at Monzón–Fosado Road has ~400 m-long roadcut with Eocene San Vicente Formation. View west.

Monzon–Fosado Road to Arro Tunnel: *Continue west on N-260 for 1.6 km (1 mi; 2 min) and pull over on the right at the end of the guardrail. Walk back to the east to the roadcut. This is **Stop 19, Arro Tunnel** (42.408851, 0.231153).*

Stop 19 San Vicente Formation, Arro Tunnel

Another excellent roadcut exposing Eocene San Vicente Formation gray shale with thin, fine-grained turbidite beds.

This stop is in the west flank of the west-verging and north-south-trending Luján Anticline. Walking east toward the tunnel you can see the beds steepen to near-vertical. The tunnel cuts through the overturned section and the southern extension of the Los Molinos (west-directed) Thrust (Instituto Geológico y Minéro de España, 1999, Campo Sheet).

Arro Tunnel to Carreterra Griebal: *Continue driving west on N-260 for 3.6 km (2.2 mi; 3 min) and pull over on the left (south) side of the road by the trash bins. This is **Stop 20, Carreterra Griebal** (42.394433, 0.197198).*

West end of Arro Tunnel. View northeast at Eocene San Vicente Formation roadcut.

STOP 20 SIERRA DE GERBE TURBIDITES AND MEDIANO ANTICLINE, CARRETERRA GRIEBAL

Driving west on N-260 between the Arro Tunnel and Carreterra Greibal, we cross the axis of the north-south-trending Mediano Anticline, one of the major structural features in the region. Although basin boundaries are not well defined, it is generally accepted that the Mediano Anticline marks the western edge of the Tremp-Graus depositional basin and the eastern edge of the Ainsa Basin. The Ainsa Basin is more or less collocated with the regional Buil Syncline. The stratigraphy of the Ainsa Basin is considered transitional between the terrestrial fluvio-deltaic Tremp–Graus Basin to the east and the distal deep marine Jaca Basin to the west (Cantalejo et al., 2020).

The structure of the Ainsa Basin is dominated by three long, north-south-trending anticlines. From west to east they are the Boltaña, Añisclo, and Mediano anticlines. Each of these anticlines is over 10 km (6 mi) long and has structural relief greater than 1,000 m (3,280 ft). The large Buil Syncline lies between the Boltaña and Mediano anticlines. The Mediano Anticline is a detachment fold cored by Triassic evaporites, with a thrust penetrating the steep limb. The Boltaña and Añisclo anticlines are fault-propagation anticlines.

The Mediano Anticline strongly affected the sedimentation in the southern Pyrenean foreland, particularly the Eocene Hecho Group. The anticline has eroded such that we are able to see into the core. Triassic gypsum and clay and Cretaceous limestone and sandstone are exposed there.

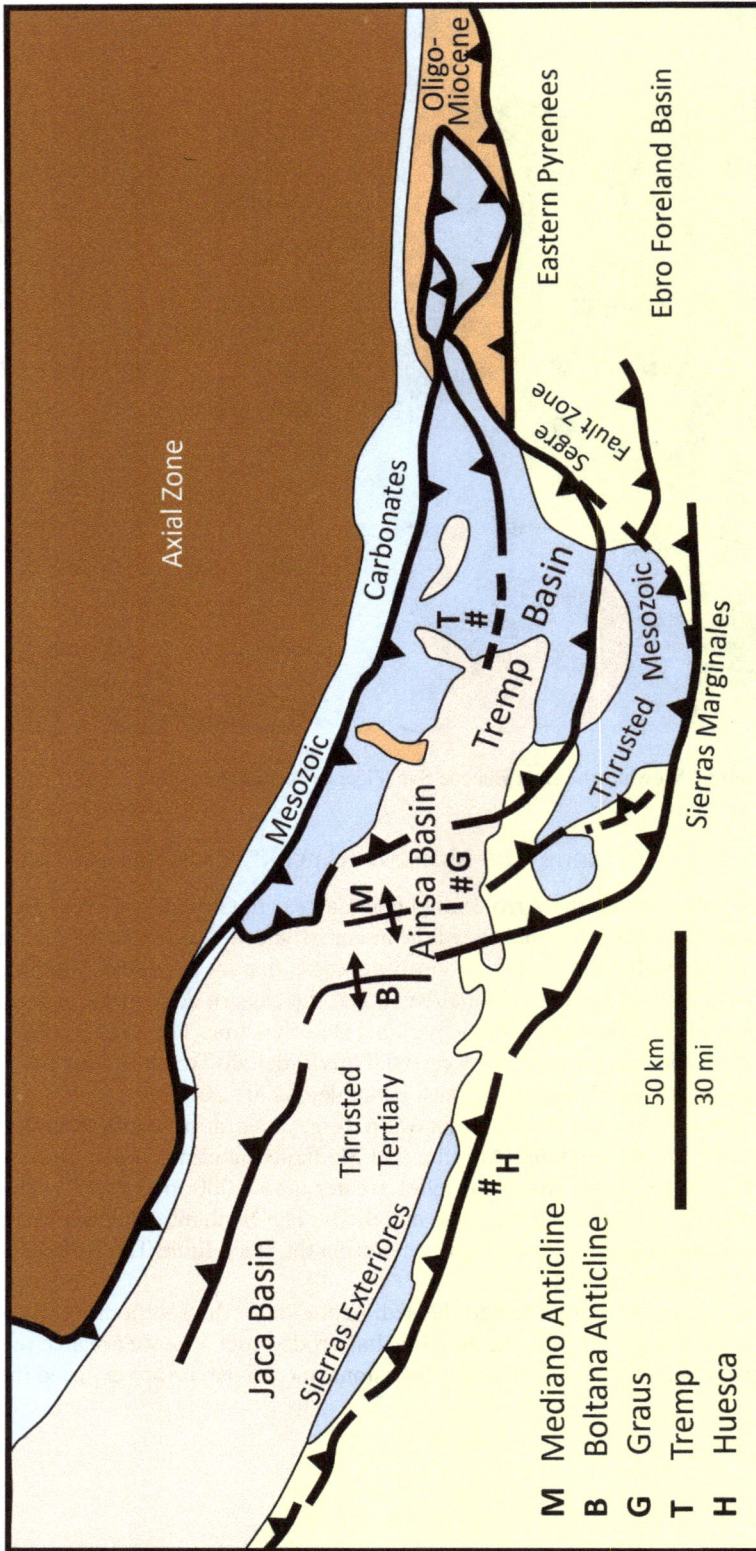

Southern foreland of the Pyrenees showing locations of the Tremp-Graus, Ainsa, and Jaca basins as well as major thrusts and folds. Modified after Bentham et al., 1992.

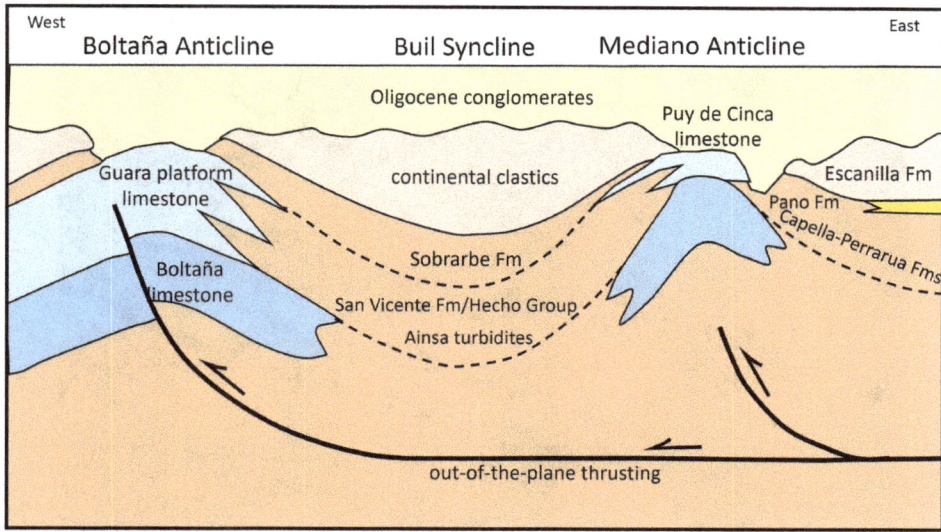

East-west diagrammatic cross-section through the Ainsa Basin. Modified after Bentham et al., 1992. Note in the figure below how different workers have drawn the thrust in the Mediano Anticline.

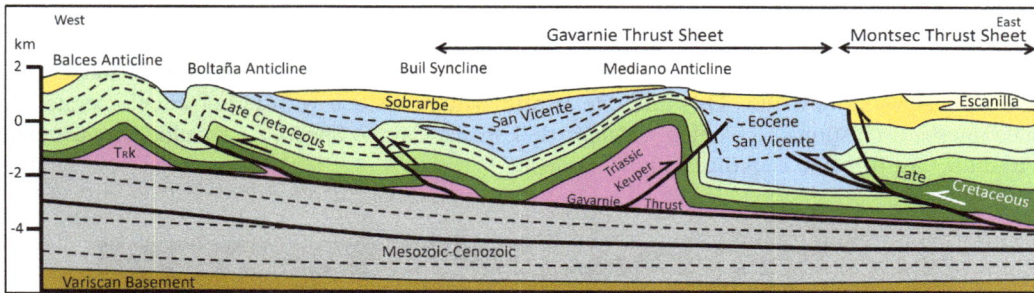

Diagram showing the influence of anticline growth on sedimentation in the Ainsa Basin. Modified after Muñoz et al., 2013.

The Boltaña and Añisclo anticlines have steep western forelimbs and gently dipping back limbs. Both of these anticlines have over 2,000 m (6,500 ft) of structural relief, are west-verging, and are related to blind thrusts at depth. The Boltaña Anticline extends over 25 km (15 mi) with a consistant north-south trend. Then, near the town of Yeba at the north end, the fold axis bends slightly northwest-southeast. This change in trend is also seen in the Añisclo Anticline.

A peculiar feature of these folds is the change in trend of the northern Boltaña and Añisclo anticlines. It has been suggested that the fold system developed as successive, staggered anticlines that grew as the Gavarnie–Sierras Exteriores Thrust Sheet was rotated around a vertical axis. The obliquity of the structures in the Sobrarbe fold system suggests that shortening on the Sierras Exteriores Thrust must be approximately 50 km (30 mi) greater in the eastern Ainsa Basin than in the west (Fernández et al., 2012).

Roadcut in the Eocene Sierra de Gerbe turbidites, Carreterra Griebal.

At this stop there is an excellent roadcut on the north side of N-260 that exposes the Eocene Sierra de la Gerbe turbidite. This turbidite is primarily sandstone with minor conglomerate and shale interbeds (Instituto Geológico y Minéro de España, Campo Sheet). As an added bonus, there is a small, east-directed backthrust in the roadcut.

A small thrust fault (dotted line) at the east end of the Carreterra Griebal roadcut.

Carreterra Griebal to Casa Torreta: Continue driving west on N-260 for 1.5 km (0.9 mi; 1 min) and turn left onto the road to Gerbe. Pull over on the right in front of the sign to Gerbe. This is Stop 21, Casa Torreta (42.389828, 0.180818). Carefully walk back to the roadcut.

STOP 21 RIVER TERRACE CONGLOMERATES, CASA TORRETA

This stop is mainly Pleistocene river terrace consisting of conglomerate, gravel, and sand. West inclined bedding beneath the gravels is mapped as Eocene Sierra de Gerbe turbidites.

Glacial material lies unconformably over a west-dipping Eocene-Oligocene turbidites.

Casa Torreta to Ainsa Quarry: Continue driving northwest on N-260; cross the Rio Cinca in Ainsa and turn left (south) on A-138/Carretera Aragonesa; cross the Rio Ara and at the roundabout take the 2nd exit to stay on A-138/Carretera Aragonesa; drive 1 km (0.6 mi) and pull over on the right at the old quarry. This is Stop 22, Ainsa Quarry (42.405174, 0.147943) for a total of 6.0 km (3.7 mi; 6 min).

STOP 22 AINSA CHANNELIZED SANDSTONE, AINSA QUARRY

The Ainsa Quarry contains Ainsa channelized sandstone, part of the Eocene upper Hecho Group. The quarry wall displays at least two well-defined channels above a Type II mass transport complex (Pickering et al., 2015). Type II (sandy) mass transport deposits occur when the uppermost continental slope and shelf edge suddenly collapse, redepositing unconsolidated sands and gravels in deep water to form deep marine fans (Pickering and Corregidor, 2005).

Ainsa Quarry to Ainsa Mass Transport Deposits: Return north on A-138 to Ainsa; drive a total of 1.8 km (1.1 mi; 3 min) and pull into the parking area on the right at the "leaving Ainsa" sign. Walk 40 m (130 ft) north to the roadcut on the left. This is Stop 23, Ainsa Mass Transport Deposits (42.418976, 0.139872).

Ainsa Quarry channelized sandstone and mass transport deposits.

STOP 23 MASS TRANSPORT DEPOSITS, AINSA

This stop is a roadcut in "Type I mass transport deposits (MTDs)" in the upper Hecho Group (San Vicente/Banastón formations) of the Ainsa Basin (Dakin, 2016). Type I MTDs comprise mudstones, sometimes contorted, and are interpreted to be inter-fan deposits or at the base of sandy submarine fans. At this location we see mudstone deposits, slightly contorted near the top of the roadcut, beneath sandstones.

*Side Trip 3, Ainsa Mass Transport Deposits to Labuerda Eocene Turbidites: Continue north on A-138 for 4.0 km (2.5 mi; 4 min) and pull over on the right shoulder just past the "leaving Labuerda" sign. This is **Stop ST3.1, Labuerda Turbidites** (42.453376, 0.134960).*

Mass transport deposit consisting primarily of mudstone along A-138 at the northern boundary of Ainsa.

If you prefer to skip Sidetrip 3:

Ainsa Mass Transport Deposits to Boltaña Anticline: *Return south on A-138 to Ainsa; turn right (west) onto N-260/Av. Ordesa; drive another 12.9 km (8.0 mi) to a large parking area on the left. This is* **Stop 24, Boltaña Anticline** *(42.467647, 0.018189) for a total of 13.4 km (8.3 mi; 12 min).*

SIDE TRIP 3 EOCENE TURBIDITES, RIO CINCA VALLEY

This digression provides views of the Montsec and Gavarnie thrusts and associated structures, as well as Eocene turbidites of the Ainsa Basin.

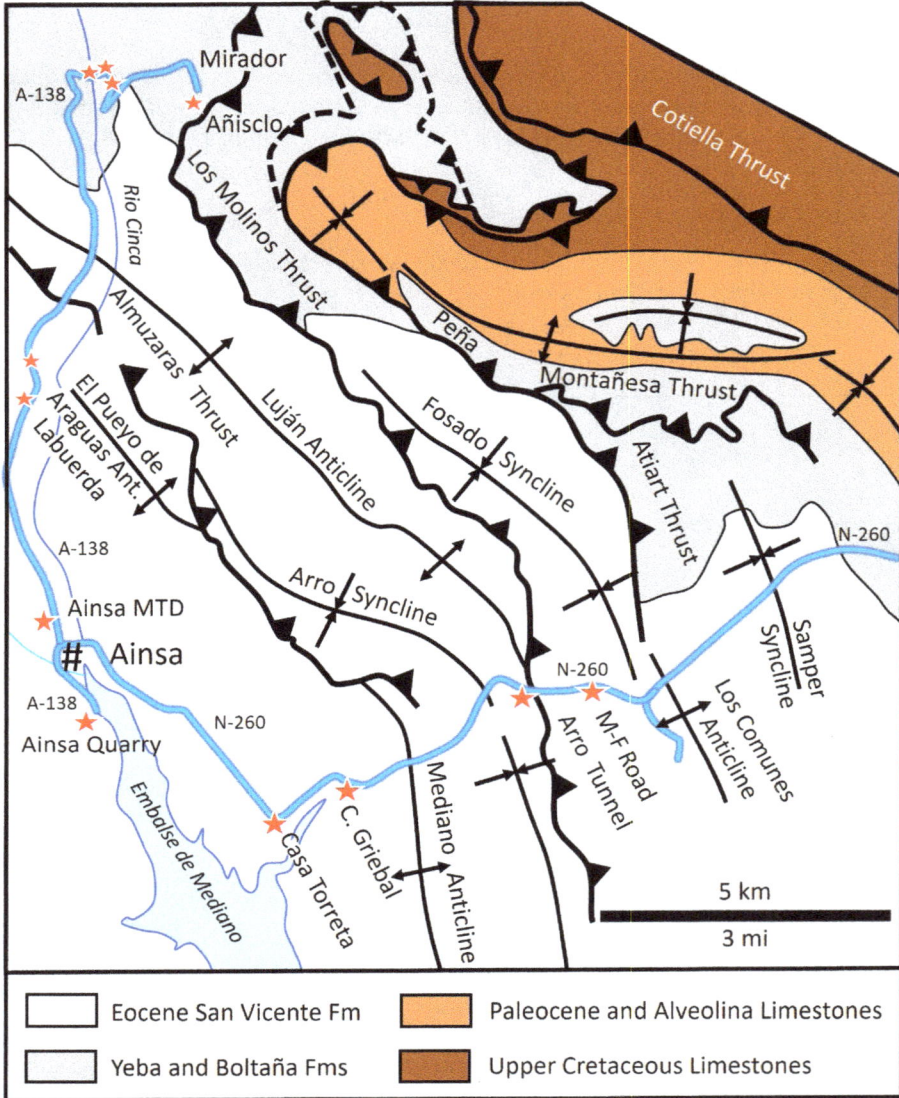

Structure in the Ainsa-Labuerda-Laspuña area. Modified after Fernández et al., 2012.

ST3.1 SAN VICENTE FORMATION TURBIDITES, LABUERDA

This is a remarkably long roadcut exposing Eocene turbidite deposits along A-138 just north of the village of Labuerda.

The Labuerda section starts north of town. The section is a ~1,000 m (3,280 ft) thick interval of syntectonic sediments typical of proximal deep basin submarine fans and slope and basin floor deposits. The lowest part of the system (north end) begins in gravity flow deposits of the Gerbe system and ends in submarine fan deposits at the top (nearest Labuerda). The lower part of the section contains an unconformity related to thrusting. The section above the thrust is fine-grained marl with occasional heterolithic sandstones (interbedded sandstone, siltstone, and shale). Above that are sediment gravity flow sandstones and ultimately overbank or floodplain deposits. Soft sediment slumps are present throughout the section (Läuchli et al., 2021). As mentioned previously, the Ainsa marine basin is stratigraphically between the continental margin Tremp–Graus Basin fluvio-deltaic system and the Jaca Basin distal deep marine system.

If you look carefully, within this roadcut you can see onlap and downlap sequences (bedding terminating against an erosion surface) and small-scale deformation features such as thrusted and slumped beds.

View north at the tilted Labuerda submarine fan deposits.

Small-scale deformation in the Labuerda section. The offset sandstone bed indicates south-directed thrusting prior to rotation of the section.

Side Trip 3, Labuerda Turbidites to Mirador Peña Montañesa: *Continue driving north on A-138 for 1.1 km (0.7 mi; 1 min) and pull over on the right by a driveway. This is* ***Stop ST3.2, Mirador Peña Montañesa*** *(42.462521, 0.136014).*

ST3.2 MONTSEC AND LOS MOLINOS THRUSTS, MIRADOR PEÑA MONTAÑESA

Seen from a distance, the stunning Peña Montañesa is a massive rock face that is the northwestern-most exposure of the Montsec Thrust system. At the base of the massive cliffs, the Peña Montañesa Thrust marks the northeastern boundary of the Ainsa Basin (Hoffman, 2009). The Los Molinos Thrust is lower on the slope, about 1/3 of the way up to the cliff. Sub-Arro mudstones, laterally equivalent to the lower San Vicente Formation, lower Hecho Group, occur below the Los Molinos Thrust.

Peña Montañesa. Dotted lines indicate thrust faults. View northeast.

Side Trip 3, Mirador Peña Montañesa to Ecomuseo: *Continue driving north on A-138 for 5.3 km (3.3 mi) to the turnoff to Laspuna on the right; turn right (east) onto HU-V-6401 and cross the Rio Cinca and pull into the large parking area on the right by the display panels. This is* ***Stop ST3.3, Ecomuseo Displays*** *(42.506170, 0.151745) for a total of 6.1 km (3.8 mi; 6 min).*

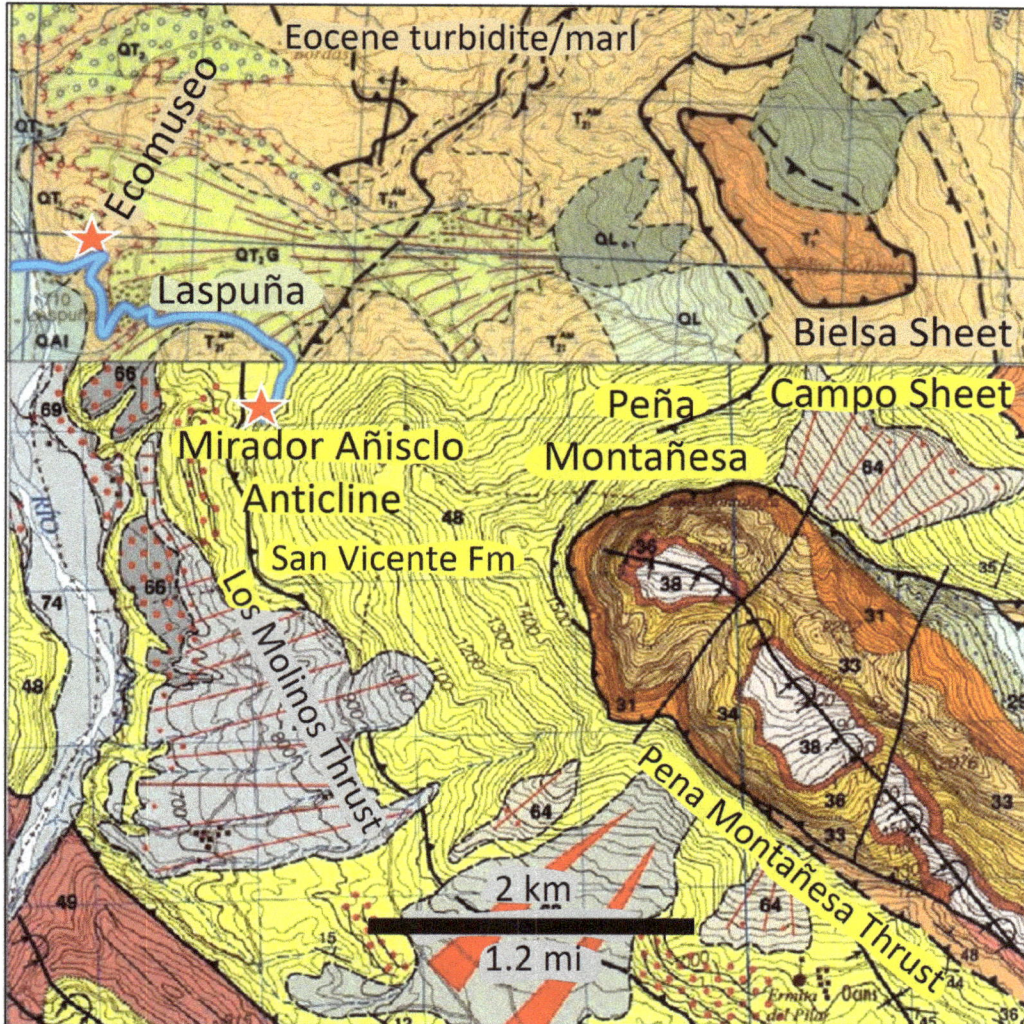

Geologic map showing the traces of the Los Molinos and Peña Montañesa thrusts. From Instituto Geológico y Minéro de España (IGME), Campo and Bielsa Sheets. Source: © NC Geological Survey of Spain (IGME).

ST3.3 SUB-ARRO MUDSTONES, ECOMUSEO DISPLAY

Carefully cross the road and examine the roadcut in the Eocene sub-Arro mudstones, which are considered equivalent to the lower San Vicente Formation of the lower Hecho Group. Here they are dominantly gray shale enclosing fine-grained deep marine turbidites.

*Side Trip 3, Ecomuseo Displays to Mirador Añisclo Anticline: Continue driving east on HU-V-6401 past Laspuña; turn right (south) at the sign for "Los Molinos 8.5" and drive to **Stop ST3.4, Mirador Añisclo Anticline** (42.504048, 0.170954). Pull over on the right for a total of 3.0 km (1.9 mi; 6 min).*

Sub-Arro mudstone, Ecomuseo stop. View north.

ST3.4 Mirador Añisclo Anticline

The Añisclo Anticline is a north-south-oriented fault-propagation fold carried on the Gavarnie Thrust. This type of fold occurs when a propagating thrust fault loses slip and terminates by transferring its shortening to a fold that develops at the fault tip. The anticline extends for about 10 km (6 mi) and has over 2,000 m (6,500 ft) of structural relief. The west-verging anticline and its internal thrust are a result of west-directed Alpine compression and shortening.

The main thrust is detached on Triassic evaporites. Añisclo Anticline is developed in resistant Late Cretaceous and Paleocene limestones.

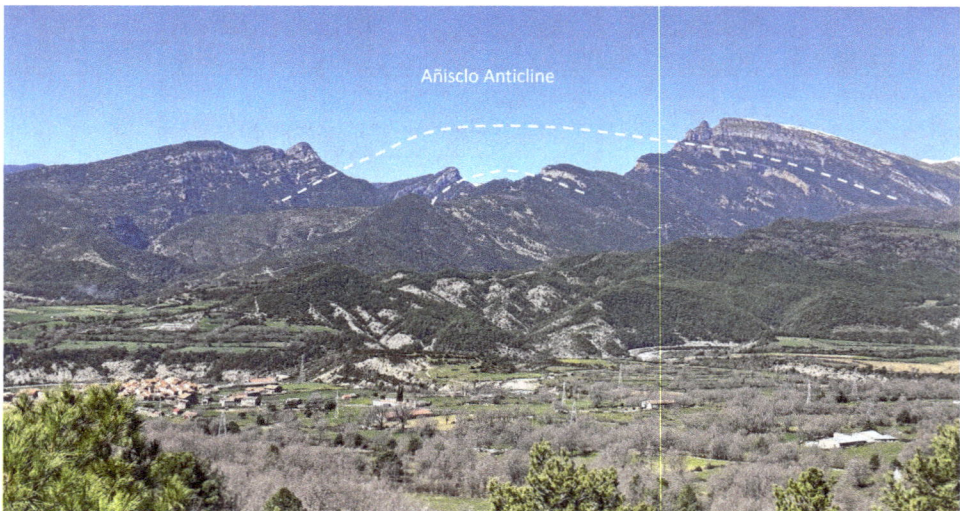

View west across the Rio Cinca Valley to the Añisclo Anticline. The sketched bedding (dashed lines) shows how the anticline is asymmetric, verging to the southwest. The Bellos River flows through the core of the anticline.

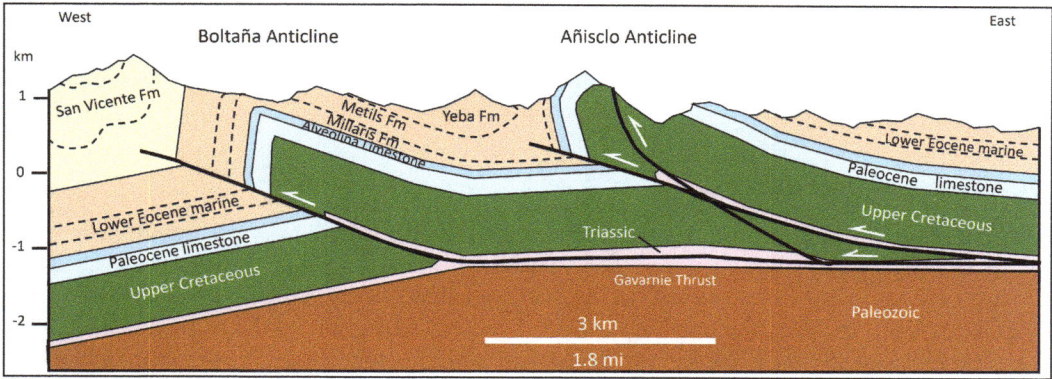

Cross-section through the Añisclo and Boltaña anticlines. Modified after Fernández et al., 2012.

Side Trip 3, Añisclo Anticline to Boltaña Anticline: *Return to A-138 and turn left (south); drive to Ainsa; turn right (west) onto N-260; drive west on N-260 for 12.9 km (8.0 mi) to a large parking area on the left. This is* **Stop 24, Boltaña Anticline** *(42.467647, 0.018189) for a total of 27.7 km (17.2 mi; 29 min).*

STOP 24 BOLTAÑA ANTICLINE

Again, although it is not stratigraphically clear, the Boltaña Anticline is generally taken as the western margin of the Ainsa Basin and the eastern limit of the Jaca Basin (Mochales López, 2011).

As mentioned in the discussion of the Añisclo Anticline, Boltaña Anticline is a 25 km (15 mi) long, north-south-trending asymmetric fault-propagation fold verging to the west. It is part of the Sobrarbe fold system that is carried on the Gavarnie-Sierras Exteriores thrust nappe (locally the Alcanadre-Tozal Thrust), with the main detachment in Triassic (Keuper) evaporites (Mochales López, 2011; Fernández et al., 2012). The fold is cored by a thrust that dies out upward in the San Vicente Formation (Mochales López, 2011). The fold axis is nearly horizontal, with a back limb that dips 20° to 30° east and a forelimb that dips 80° to 90° to the west (Fernández et al., 2012). Look for the change in dip as you drive through the core of the structure. From this stop you can see the full sweep of bedding around the axis as you look south into the anticlinal core. The fold asymmetry is best expressed by beds on the skyline.

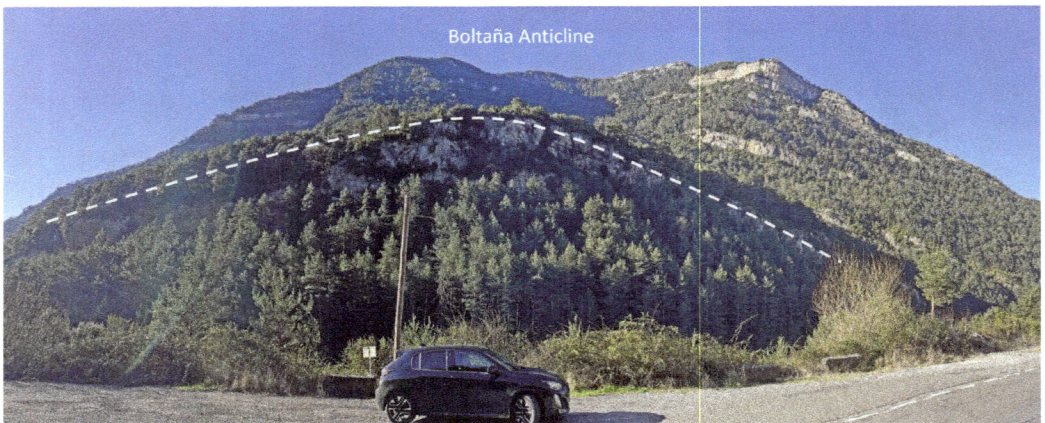

View south into the core of the Boltaña Anticline. Bedding is highlighted by the dashed line. Notice the steeper bedding on the west side indicating west vergence.

Pre-tectonic rocks above the Triassic evaporitic detachment are mainly of Upper Cretaceous and Paleogene age (Mochales et al., 2010). Growth of the Boltaña Anticline influenced deposition of middle and upper Eocene sediments in the Aínsa Basin. Shelf and slope facies derived from the eastern Graus-Tremp Basin interfinger and transition across the Ainsa Basin to deeper marine facies in the Jaca Basin to the west (Mochales López, 2011; Cantalejo et al., 2020). Sediment patterns and thickness variations on the flanks of the anticline indicate that the Boltaña Anticline grew during middle to late Eocene time (43–37 Ma; Muñoz et al., 2013), or just middle Eocene (52–44 Ma; Fernández et al., 2012; Mochales et al., 2012).

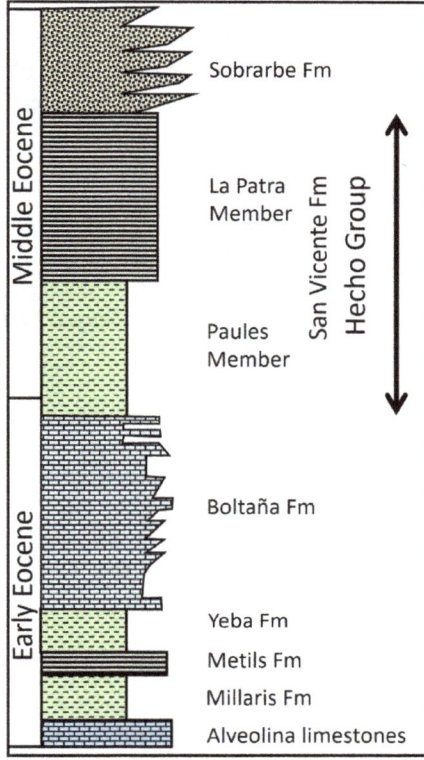

Eocene stratigraphy of the Boltaña Anticline at the boundary between the Ainsa and Jaca basins. Stratigraphic information derived from Mochales López, 2011.

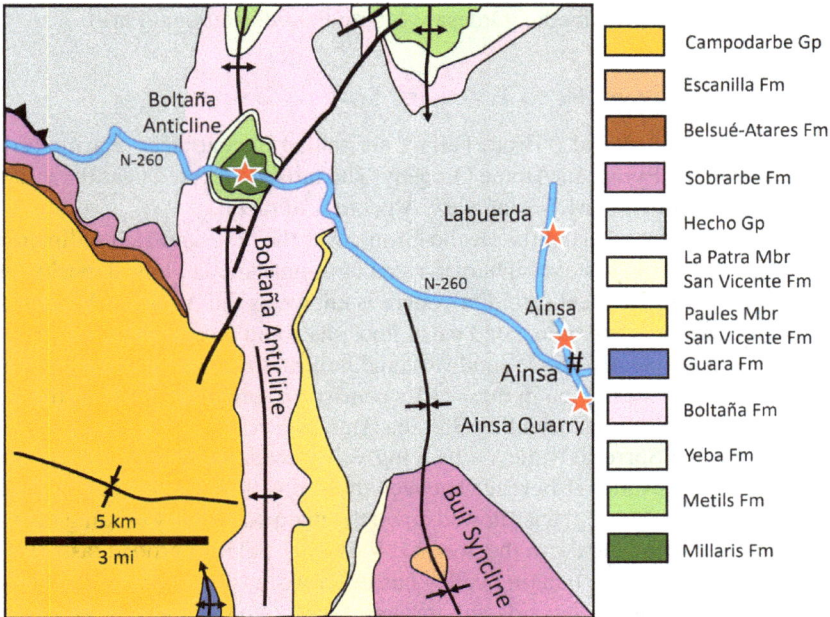

Geologic map of the Boltaña Anticline. Red stars are stops. Modified after Mochales López, 2011.

Whereas most of the anticline is developed in the Boltaña Formation, at this stop the Yeba Formation marl-siltstone-sandstone and Metils Formation marly limestone make up the upper slopes, with Millaris Formation shale-marly limestone and a hint of Alveolina Limestone exposed in the innermost core of the fold (Mochales et al., 2010; Mochales López, 2011; Fernández et al., 2012).

Side Trip 4, Boltaña Anticline to Sorrosal Falls: *Continue west on N-260; at the roundabout on the west side of Fiscal take the 1st exit (north) onto N-260a toward Broto; in Broto cross the Rio Ara and look for the sign on the left to Cascada de Sorrosal; make a hard left at the sign onto Carretera Oto and follow the signs to Cascada de Sorrosal. There is a large parking area on the left; take the path 200 m (650 ft) to the base of the falls. This is* ***Stop ST4.1, Sorrosal Falls*** *(42.605543, -0.128262) for a total of 27.7 km (17.2 mi; 34 min).*

If you prefer to skip Side Trip 4:

Boltaña Anticline to North Entrance Petralba Tunnel: *Continue west on N-260; at the round-about on the west side of Fiscal take the 2nd exit (west) and stay on N-260 toward Jaca and Huesca; drive another 5.4 km (3.3 mi) and pull over on the right before the tunnel entrance. Be mindful of traffic. This is* ***Stop 25, North Entrance Petralba Tunnel*** *(42.487269, -0.164929) for a total of 19.2 km (11.9 mi; 22 min).*

SIDE TRIP 4 SORROSAL FALLS AND ORDESA Y MONTE PERDIDO NATIONAL PARK/WORLD HERITAGE SITE

Ordesa y Monte Perdido National Park (Parque Nacional de Ordesa y Monte Perdido) was established in the Ordesa Valley in 1918. Its area was enlarged in 1982 to cover 156 km² (96 mi²). In 1997 it became part of the UNESCO Biosphere Reserve of Ordesa-Viñamala and was included in the cross-border Pyrénées – Mont Perdu World Heritage Site because of its unique and scenic geology.

The region is dominated by massive cliffs of Cretaceous and Eocene limestone, with karst topography that includes sinkholes and caves. During the Quaternary, repeated glaciations carved cirques and large U-shaped valleys (Wikipedia, Ordesa y Monte Perdido National Park).

ST4.1 PYRENEAN FOLDING AND BROTO TURBIDITES, SORROSAL FALLS

The strata at Cascada de Sorrosal ("Broto Falls") are folded Eocene deep marine turbidites that were deformed during the Pyrenean-Alpine Orogeny. The falls are located near the village of Broto at the edge of Ordesa National Park (Fernández, A geological race).

The Broto turbidite system, part of the Hecho Group, was the first sand-rich sedimentation in the Eocene Jaca Basin. The Broto system is characterized by submarine fan lobe deposits. In the Aínsa Basin just to the east, the equivalent turbidite system is known as the Arró/San Vicente system and consists of turbidite channels and associated basin floor shale and fan deposits.

Folding at this location clearly verges southwest and indicates regional shortening in a northeast-southwest direction. The deformation in these cliffs consists of subsidiary folds on the west flank of the large, regional north-northwest-trending Boltaña Anticline.

The falls connects the Sorrosal Valley, a hanging valley, with the glacial-carved Broto Valley. This waterfall marks an elevation difference between the two valleys formed during the last glaciation, roughly 35,000 years ago. A glacier flowed down the main Ara Valley eroding the valley; however, there was no glacier flowing down the Sorrosal Valley. The glacier eroded faster than the river, thus the river that flows down the hanging valley has to drop over these falls to join the Ara River.

A via ferrata route goes up the left side and through a cave into the gorge at the top.

Cascada de Sorrosal or "Broto Falls." This spectacular double falls has a backdrop of incredible folding in the cliff face. Dashed lines follow the bedding.

Side Trip 4, Sorrosal Falls to Ordesa Valley: *Return to N-260a and turn left (north); at the sign for "Torla, Ordesa" turn right (north) onto Av. de Ordesa; drive through Torla-Ordesa and continue to the large parking area at the end of the road. This is* ***Stop ST4.2, Ordesa Valley*** *(42.649562, -0.059876), for a total of 12.3 km (7.7 mi; 22 min).*

ST4.2 ORDESA VALLEY

Located on either side of the border between Spain and France, Ordesa y Monte Perdido National Park is called Pyrenees National Park on the French side. Summits reach 3,404 m (11,168 ft), with Monte Perdido the third tallest summit in the Pyrenees at 3,355 m (11,000 ft). The Rio Arazas runs through the Ordesa Valley for 11 km (7 mi). Ordesa Valley is known for its views of impressive waterfalls and stunning vistas of Monte Perdido. Mounte Perdido, actually located in France, means "lost mountain" because it is not visible from the French side.

The park is filled with many hiking trails that differ in both length and difficulty. Multiple trails start at the large parking lot at the end of the road. There is a full day hike along the river gorge that provides trekkers with views of massive cliffs and several stunning waterfalls (https://national-parks.org/spain/ordesa-y-monte-perdido). It takes about one hour to reach the first three waterfalls on the Rio Arazas.

As you enter the valley you are on Paleocene limestones. The valley walls are carried on a thrust, likely the Gavarnie Thrust (Labaume and Teixel, 2018) and consist, from base to top, of Upper Cretaceous Marboré sandstone, Upper Cretaceous limestone/dolomite, Paleocene massive

View east up the Ordesa Valley. The massive limestone layers are essentially horizontal.

limestone and marl, and Eocene Hecho Group turbidites. The erosion-resistant limestones and dolomites are the ledge-formers; the slope-formers are sandstones, turbidites (sandstone-shale), and marls (sandy limestone or limy sandstone).

Side Trip 4, Ordesa Valley to Petralba Tunnel: *Return west to Av. de Ordesa and turn left (south); just north of Broto turn left (south) onto N-260a; at the roundabout north of Fiscal take the 1st exit (west) onto N-260; stay on N-260 toward Jaca and Huesca; drive another 5.4 km (3.3 mi) and pull over on the right before the tunnel entrance. Be mindful of traffic. This is* **Stop 25, North Entrance Petralba Tunnel** *(42.487269, -0.164929) for a total of 19.2 km (11.9 mi; 22 min).*

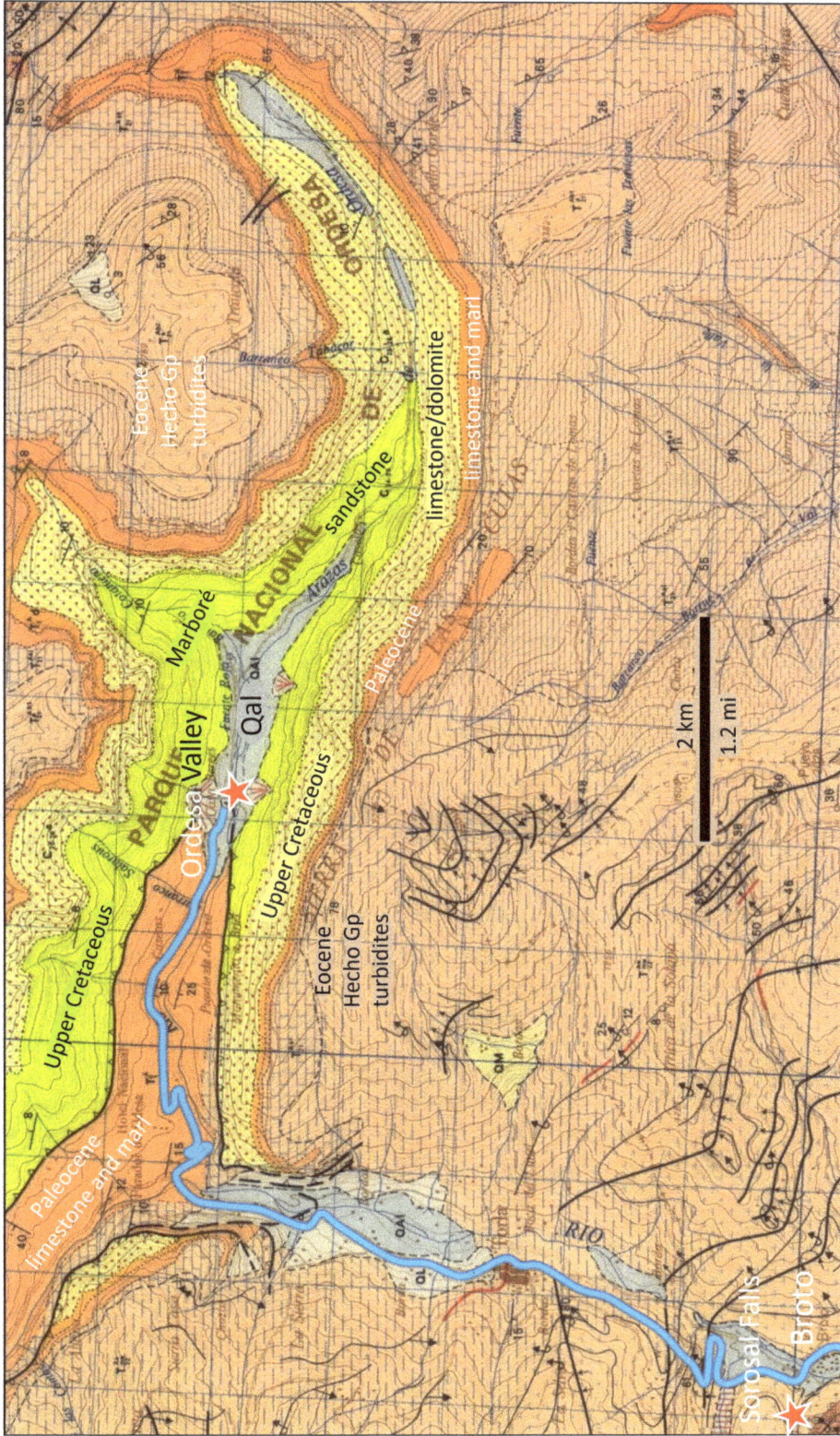

Geologic map of the Broto to Ordesa Valley area. From Instituto Geológico y Minéro de España (IGME), Broto Sheet. Source: © NC Geological Survey of Spain (IGME).

Stop 25 Escanilla Formation and Cancias Syncline, North Entrance Petralba Tunnel

We are now in the Jaca Basin. This stop, at the north entrance to the Petralba Tunnel, is on the north flank of the northwest-southeast Cancias Syncline, so the Escanilla Formation beds in the roadcut are dipping south. The tunnel goes through the axis of the syncline. The syncline has upper Eocene–Oligocene Campodarbe Formation conglomerate in its core, and upper Eocene nonmarine shale, sandstone, and conglomerate of the Escanilla Formation on the flanks (Instituto Geológico y Minéro de España (IGME), 1991, Yebra de Basa Sheet; Labaume and Teixell, 2018; Vinyoles et al., 2020; Coll et al., 2022). This stop is just south of the trace of the Oturia Thrust, which merges with the Broto Thrust at depth.

The Pyrenean Orogeny caused south-directed thrusting that proceeded from the Axial Zone southward. Thrusting carried the Jaca Basin south as a piggyback basin, and also created the External Sierras at the southern, leading edge of deformation (Nichols, 2005; Labaume et al., 2016; Ball, 2020).

The pre-orogenic section lies above largely undeformed Variscan/Paleozoic basement and is relatively thin. The orogenic section, mainly Paleocene-Eocene, is up to 9 km (29,500 ft) thick and consists of a foreland basin succession of deepwater turbidites to shallow marine delta and shelf sediments that reflect erosion of the uplifting Axial Zone of the Pyrenees. The Ainsa and Jaca basins were carried on south-directed thrusts that only come to the surface in the Sierras Exteriores (sometimes called the Sierras Marginals in this area) that separate the Jaca Basin from the Ebro Basin to the south (Bosch et al., 2016; Vinyoles et al., 2020). The main detachment surface for the thrusts is the Triassic Keuper evaporites. From north to south and oldest to youngest these thrusts are the Gavarnie, Broto, Guarga, and Sierras Exteriores, active from Late Cretaceous to Oligocene. The Ebro Basin to the south formed by downward flexing of the crust under the weight of the South Pyrenean thrusts.

Growth of the Boltaña Anticline separated the Ainsa Basin from the Jaca Basin. Sedimentation in the Jaca Basin was ongoing during thrust deformation, as can be seen in growth sequences that thin onto actively growing folds. Deformation began in early Eocene and continued into early Miocene time (Schellart, 2002).

Sediments entered the Jaca Basin from the Ainsa and Tremp-Graus basins to the east and from the uplifting Pyrenees to the north. They formed extensive deep marine basin floor turbidite fans of the Hecho Group (Martinius, 2012; Roigé et al., 2016). Sedimentation evolved from Eocene deep marine to shallow marine and delta environments and ultimately to Oligocene alluvial and fluvial deposits. The transition from marine to terrestrial environments shifted from east to west as the basin filled. Lower Oligocene Bernués Formation conglomerates are the youngest sediments derived from erosion of the Pyrenees and the emerging fold-thrust belt to the north (Roigé et al., 2019).

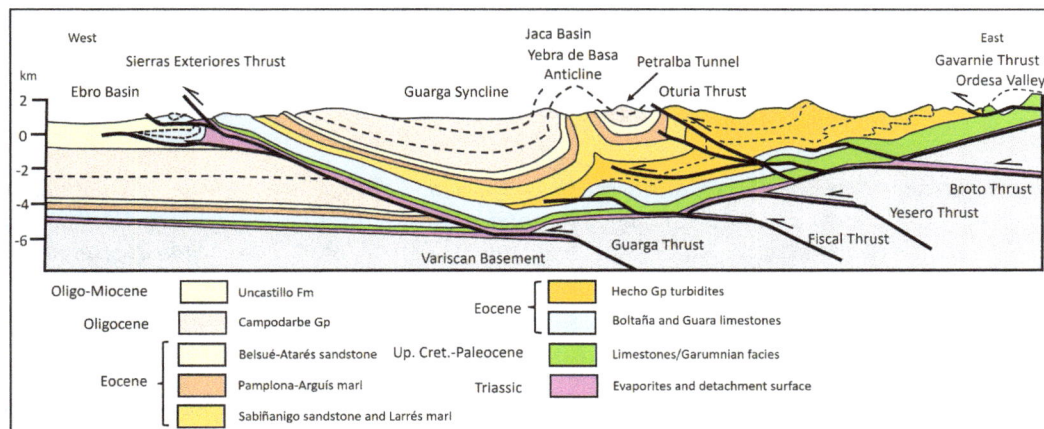

North-south cross-section through the Cancias Syncline–Petralba Tunnel Area. Modified after Labaume and Teixell, 2018.

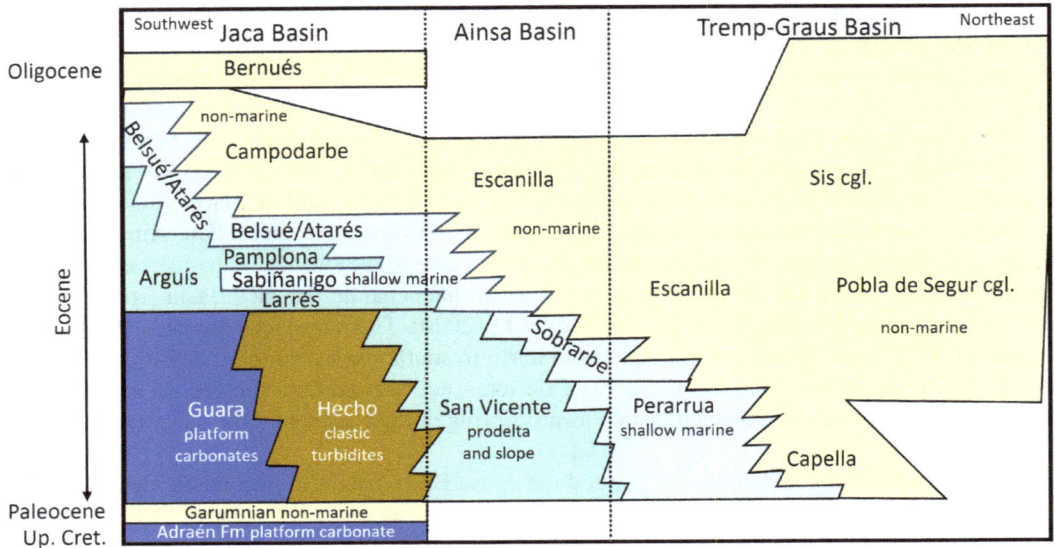

Stratigraphy of the Jaca Basin and relationship to units in the Ainsa and Tremp–Graus basins. Stratigraphy drawn from Vidal-Royo et al., 2013; Roigé et al., 2019; Vinyoles et al., 2020.

Parking area and south-dipping Eocene Escanilla Formation at north entrance to Petralba Tunnel. Panoramic view west.

North Entrance Petralba Tunnel to Barranco Pardinalta: *Continue driving west on N-260; after 7.6 km (4.7 mi) bear right toward Fanlillo in order to turn left; cross the highway onto an old paved road; drive a total of 8.5 km (5.3 mi; 7 min) and pull off the road. This is* **Stop 26, Barranco Pardinalta** *(42.475558, -0.245731).*

STOP 26 SABIÑÁNIGO SANDSTONE AND BASA ANTICLINE, BARRANCO PARDINALTA

We are on the north flank of the west-northwest-trending Basa Anticline. This anticline verges and is overturned to the south. This breached anticline (an anticline whose core has been eroded out) contains marls of the middle Eocene Margas de Larrés Formation in the core and has progressively younger Sabiñánigo Sandstone Formation, Pamplona-Arguís Marl Formation, and sandstone and marl of the Belsúe-Atarés Formation on its flanks.

At this stop we see the ridge-forming glauconitic middle Eocene Sabiñánigo Sandstone Formation (Instituto Geológico y Minéro de España (IGME), 1991, Yebra de Basa Sheet). The middle Eocene Sabiñánigo Sandstone was deposited in outer shelf to delta-front environments with sources to the east and southeast. Despite its name, the Sabiñánigo Sandstone consists of fine sand and silt deposits interbedded with clays and silts. The upper part transitions to the Atarés Delta and the Santa Orosia alluvial fan, with paleocurrent directions indicating a source to the north (Roigé et al., 2016).

North-dipping Sabiñánigo Sandstone at Barranco Pardinalta forms a ridge along the north flank of the Basa Anticline. Here there is an equal amount of sandstone, siltstone, and claystone. View west.

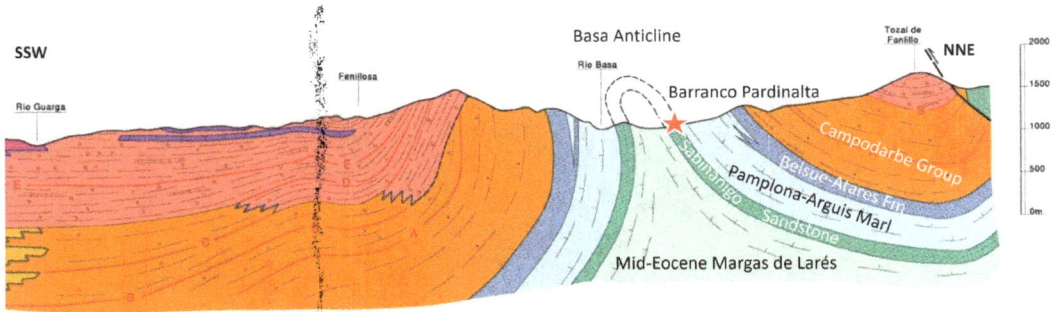

Cross-section through the Basa Anticline near Yebra de Basa. Location shown on next figure. Star shows the position of the Barranco Pardinalta stop. From Instituto Geológico y Minéro de España (IGME), Yebra de Basa Sheet. Source: © NC Geological Survey of Spain (IGME).

Barranco Pardinalta to Rio Basa: *Return to N-260 and turn left (west); drive 9.0 km (5.6 mi) to the Osan exit and make a U-turn; return 1.4 km (0.9 mi) to an unnamed road on the right and turn right (south); drive 150 m (500 ft) and pull over on the left. This is **Stop 27, Rio Basa** (42.489797, -0.312406) for a total of 10.5 km (6.5 mi; 9 min).*

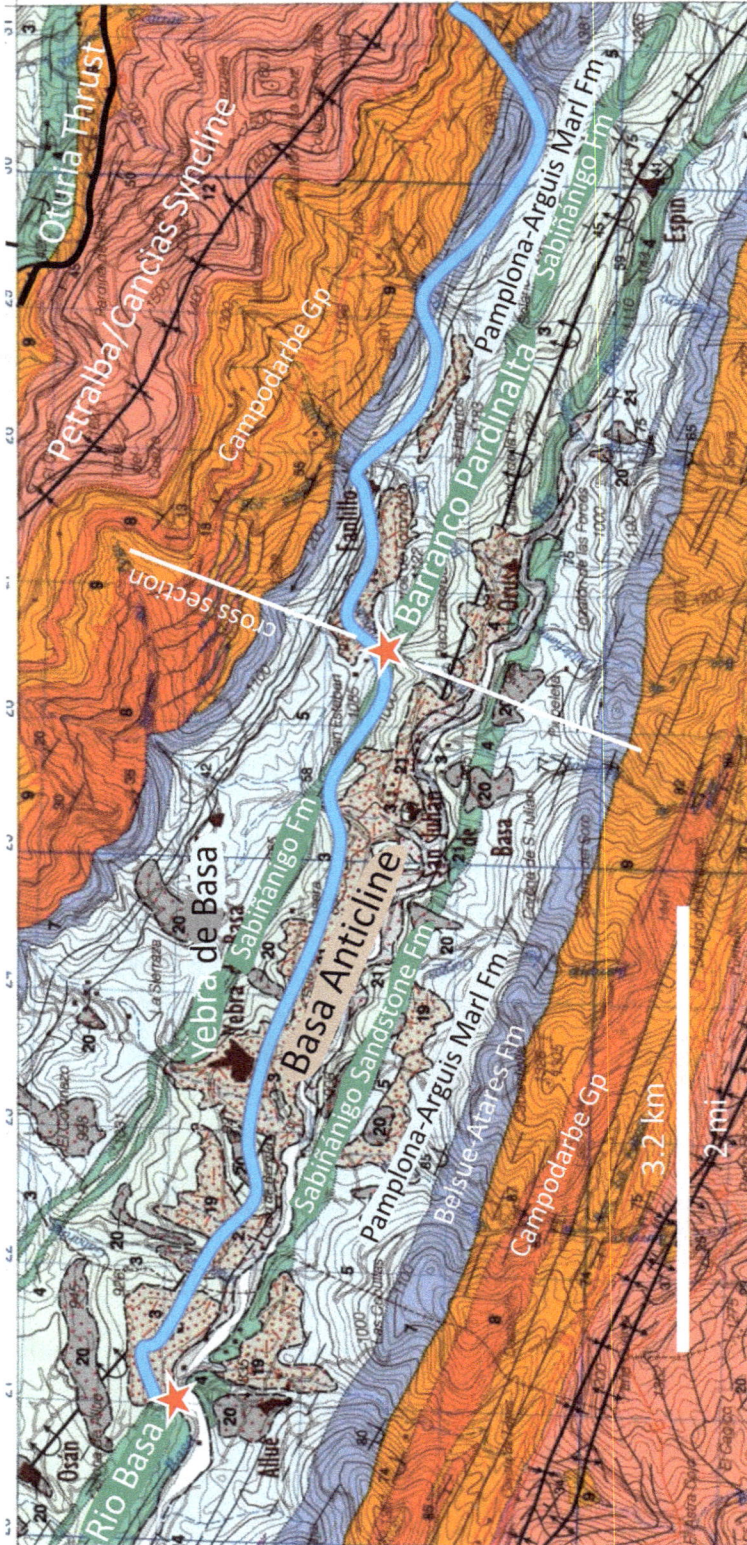

Geologic map of the northeast Jaca Basin and Basa Anticline. From Instituto Geológico y Minéro de España (IGME), Yebra de Basa Sheet. Source: © NC Geological Survey of Spain (IGME).

As you drive west on N-260, look for a classic angular unconformity in the Sabiñánigo Sandstone on the first curve west of the Fanlillo Road (42.475370, -0.247177). The dotted line is the unconformity; dashed lines follow bedding. There is no safe place to stop here.

STOP 27 SABIÑÁNIGO SANDSTONE, RIO BASA

Notice how the units here still dip 70°–80° to the northeast. And yet according to the geologic map we are south of the anticlinal axis. That means that this stop is on the overturned south flank of the Basa Anticline. Once again, the roadcut is in the Sabiñánigo Sandstone.

Overturned Sabiñánigo Sandstone on the south flank of the Basa Anticline.

Rio Basa to Fuente La Valle: *Continue driving west on N-260; at roundabout on the eastern out-skirts of Sabiñánigo, take the 3rd exit (south) onto A-23/E-7 to Huesca; at 15.5 km (9.6 mi) at the roundabout take the 1st exit (west) toward La Guarguera; pull over on the right shoulder just past the sign to Boltana. This is* **Stop 28, Area Recreativa Fuente La Valle** *(42.407981, -0.387630) for a total of 16.3 km (10.1 mi; 13 min).*

STOP 28 BERNUÉS FORMATION, AREA RECREATIVA FUENTE LA VALLE

Oligocene brown shale and channel sands of the alluvial Bernués Formation are exposed at this stop near the axis of the Jaca Basin. By Oligocene time the Jaca Basin has filled and marine deposition has ended. Other than recent alluvium and glacial deposits, this is the youngest unit in the basin.

Beds dip ~13° south into the Jaca Basin (Instituto Geológico y Minéro de España, 1991, Yebra de Basa Sheet; Coll et al., 2022).

Oligocene channel sands, Area Recreativa Fuente La Valle. View north.

Fuente La Valle to Mirador de Monrepós: *Return to A-23/E-7 at roundabout and take the first exit (south); drive south on E-7 to Exit 388 to Caldearenas; continue on the feeder road to the parking area on the right. This is* **Stop 29, Mirador de Monrepós** *(42.361023, -0.380874) for a total of 9.8 km (6.1 mi; 7 min).*

STOP 29 MIRADOR DE MONREPÓS

This viewpoint is in gently north-inclined beds of Oligocene channel sands and red shales of the Bernués Formation and Quaternary gravels (Instituto Geológico y Minéro de España, 1991, Yebra de Basa Sheet).

But the main reason for the stop is the view of the High Pyrenees to the north. The day we were there the view was obscured by rain and clouds. On a clear day the view is panoramic and magnificent.

Mirador de Monrepós to Embalse de Arguis: *Continue south on A-23/E-7; take Exit 381 to Arguis; at the roundabout take the 1st exit (south) to Arguis; at the next roundabout take the 1st exit (south) and drive to parking area on the left. This is* **Stop 30, Embalse de Arguís** *(42.306688, -0.427367) for a total of 8.2 km (5.1 mi; 7 min).*

STOP 30 SIERRAS EXTERIORES, EMBALSE DE ARGUÍS

The stop is on poorly exposed Eocene Arguís Formation gray shale. We are in the Sierras Exteriores (External Sierras) just north of the Ebro Basin.

This stop provides a nice view south across the reservoir to north-dipping beds of the middle Eocene Guara Formation limestone and thin calcareous sandstone. The Guara Limestone Formation in the External Sierras formed on a shallow carbonate ramp environment along the margin of the Jaca Basin in the south Pyrenean foreland.

The Guara Formation consists of foraminiferal (plankton-derived) and algal limestones, with minor shallow marine sandstones and shales. It was deposited around and over a bulge (upwarp) in

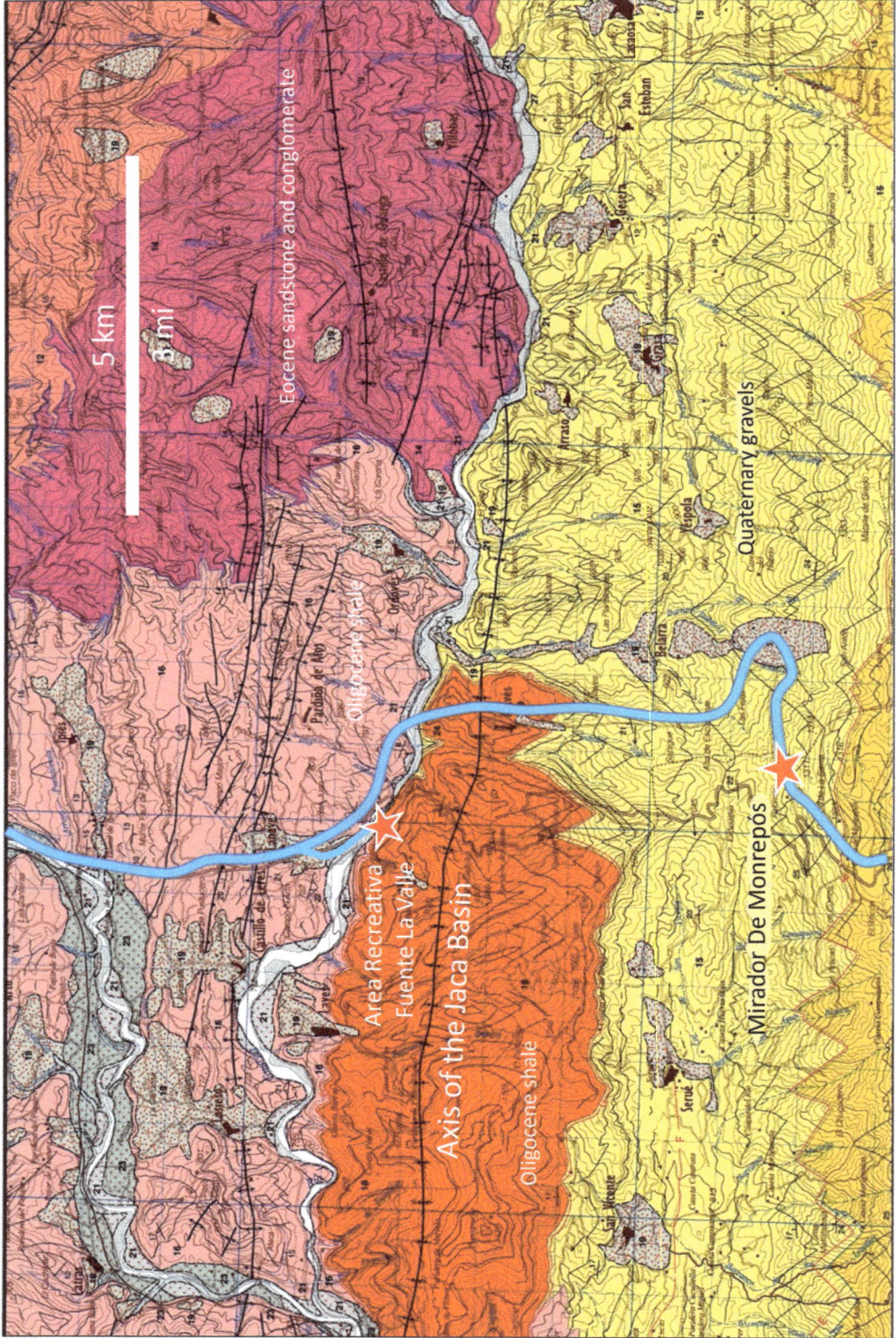

Geologic map of the axial Jaca Basin and stops. From Instituto Geológico y Minéro de España (IGME), Yebra de Basa Sheet. Source: © NC Geological Survey of Spain (IGME).

View from Mirador de Monrepós across the Jaca Basin to the High Pyrenees. Photo courtesy of Esther Blas Ruiz, Google Maps.

the foreland basin in response to loading by Pyrenean thrusting as it progressed southward. Depth indicators show three shallowing and deepening episodes related to tectonics rather than global sea level changes. This formation is the last carbonate platform developed in the Pyrenees (Pedley, 1994; Huyghe et al., 2009).

Northwest-dipping beds of Guara Formation, Embalse de Arguís. This is the west flank of the Sierra de Águila. View south.

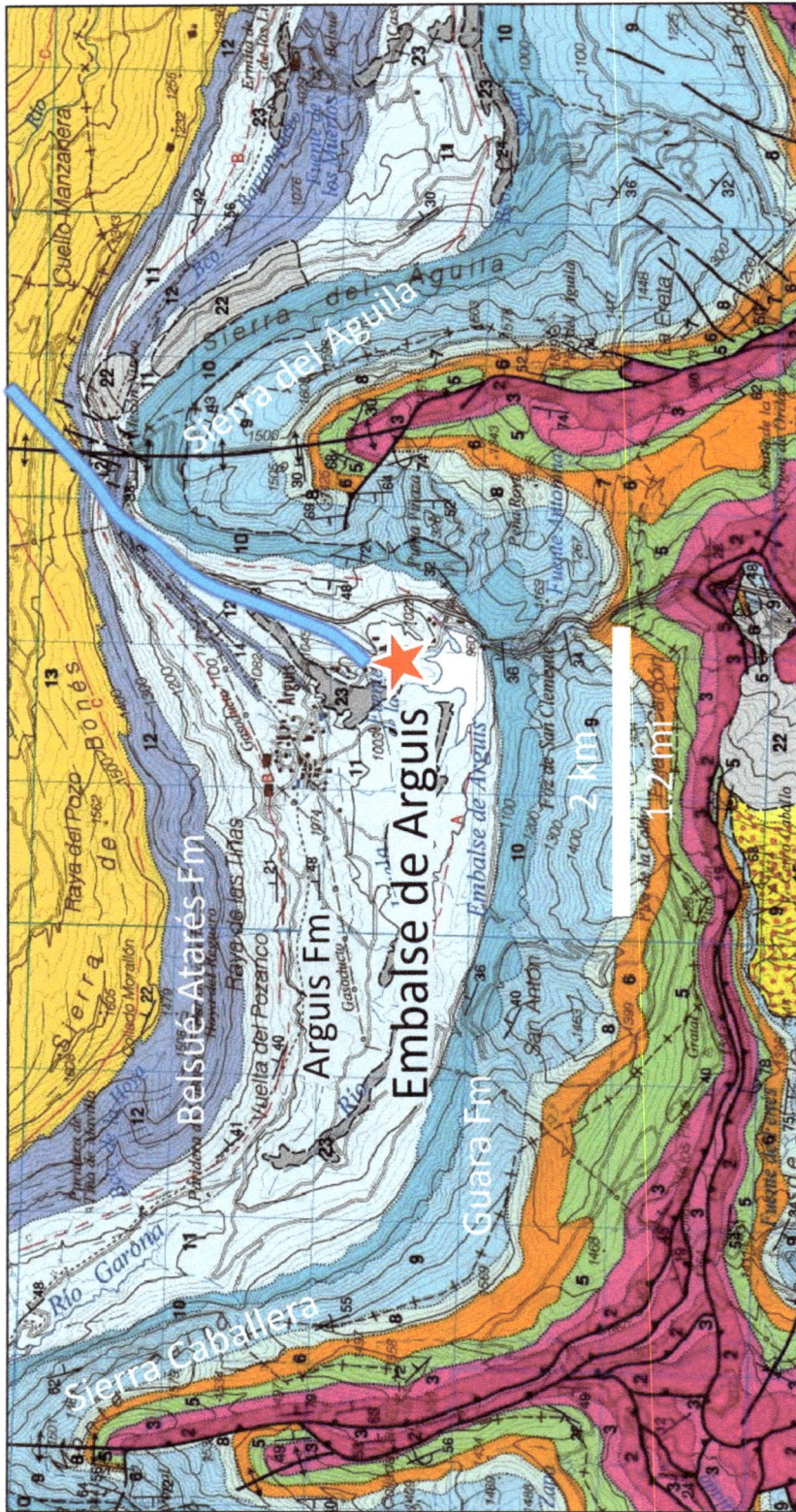

Geologic map of the Sierras Exteriores at the Embalse de Arguís stop. Note the thrusted Sierra del Águila Anticline and how the Arguís and Belsué-Atarés formations thin over this fold. From Instituto Geológico y Minéro de España (IGME), Apies Sheet. Source: © NC Geological Survey of Spain (IGME).

Structurally, this stop is on the west flank of the north-trending and north-plunging Sierra del Águila Anticline. The geologic map pattern is a classic example of a down-plunge projection, that is, a cross-section of the fold in map view. It clearly shows that these are slightly west-verging folds cored by thrust faults that tip out in the Guara Formation. The folds were then tilted to near-vertical by later uplift. The original folding occurred during middle to upper Eocene, at a time when the Arguís and Belsué–Atarés formations were being deposited in shallow marine to delta environments. We know this because these units thin over the anticline and thicken away from it.

Embalse de Arguis to Mallos de Agüero: Return to the A-23 roundabout and take the 1st exit *(south) onto A-23; drive south on A-23/E-7 to Huesca; on the outskirts of Huesca take Exit 360 to the A-132 toward Pamplona; turn right (northwest) onto A-132 toward Ayerbe; at 54.9 km (34.2 mi) turn left (west) onto HU-534/Z-534 and drive 4.3 km (2.6 mi) and pull out on the left. This is* **Stop 31, Mallos de Agüero** *(42.349117, -0.793711) for a total of 59.2 km (36.8 mi; 45 min).*

STOP 31 UNCASTILLO FORMATION ALLUVIAL FAN CONGLOMERATES, MALLOS DE AGÜERO

This is the first of the TransPyrenean GeoDrive stops (or the last, if you are going from north to south). The TransPyrenean GeoDrive is a UNESCO sponsored project that created stops with panels that explain the geology and landscape in terms that are accessible to the general public. We will be seeing many more of these as we cross the Pyrenees, especially on N-330 north of Jaca, Spain, and N-134 south of Oloron-Sainte-Marie in France.

We stand before some of the most impressive scenery in Spain. What *are* these spectacular rock towers and how were they formed? The answer involves simultaneous sedimentation and deformation followed by erosion.

The Mallos de Riglos and Mallos de Agüero are vertical spires of rock, up to 275 m (900 ft) high, that occur at the boundary between the External Sierras to the north and the Ebro Basin to the south. These stunning landscape features are much appreciated for both their scenic beauty and their challenge to climbers. So much so that in 2016 the Council of Aragon declared the Mallos de Riglos, Agüero, and Peña Rueba features as natural monuments (Wikipedia, Mallos de Riglos). *¡Qué maravilloso!*

Mallos de Agüero, view north. These spectacular conglomerate towers are a favorite of climbers.

During early Oligocene to late Oligocene, the Jaca Basin became isolated from the Ebro Basin by the emergence of the External Sierras. The young uplift shed conglomerates to the south. During late Oligocene to early Miocene time, Ebro Basin sedimentation was controlled by two major river systems, the Luna in the west and the Huesca to the east (Friend et al., 1996). The conglomerates were part of alluvial fans associated with the Luna fluvial system (Arenas et al., 2001; Nichols, 2005). The Luna system consisted of proximal (near source) alluvial fan conglomerates and distal (far from the source) river channel sediments deposited into the Ebro Basin (Ball, 2020).

As the External Sierras were uplifted and eroded, large aprons of rock, consisting of sand, pebbles, cobbles, and boulders were shed southward into the Ebro Basin. These alluvial fans were derived from the eroded Sierra, Eocene-Oligocene Jaca Basin, and even the Pyrenean Axial Zone. The northern part of these alluvial fans, along the edge of the Sierras, was caught up in the deformation and rotated by varying amounts. This folding/rotation is thought to have induced fracturing in the cemented conglomerate, and later erosion along the fractures left spires of rock that now form the Mallos (Wikipedia, Mallos de Riglos).

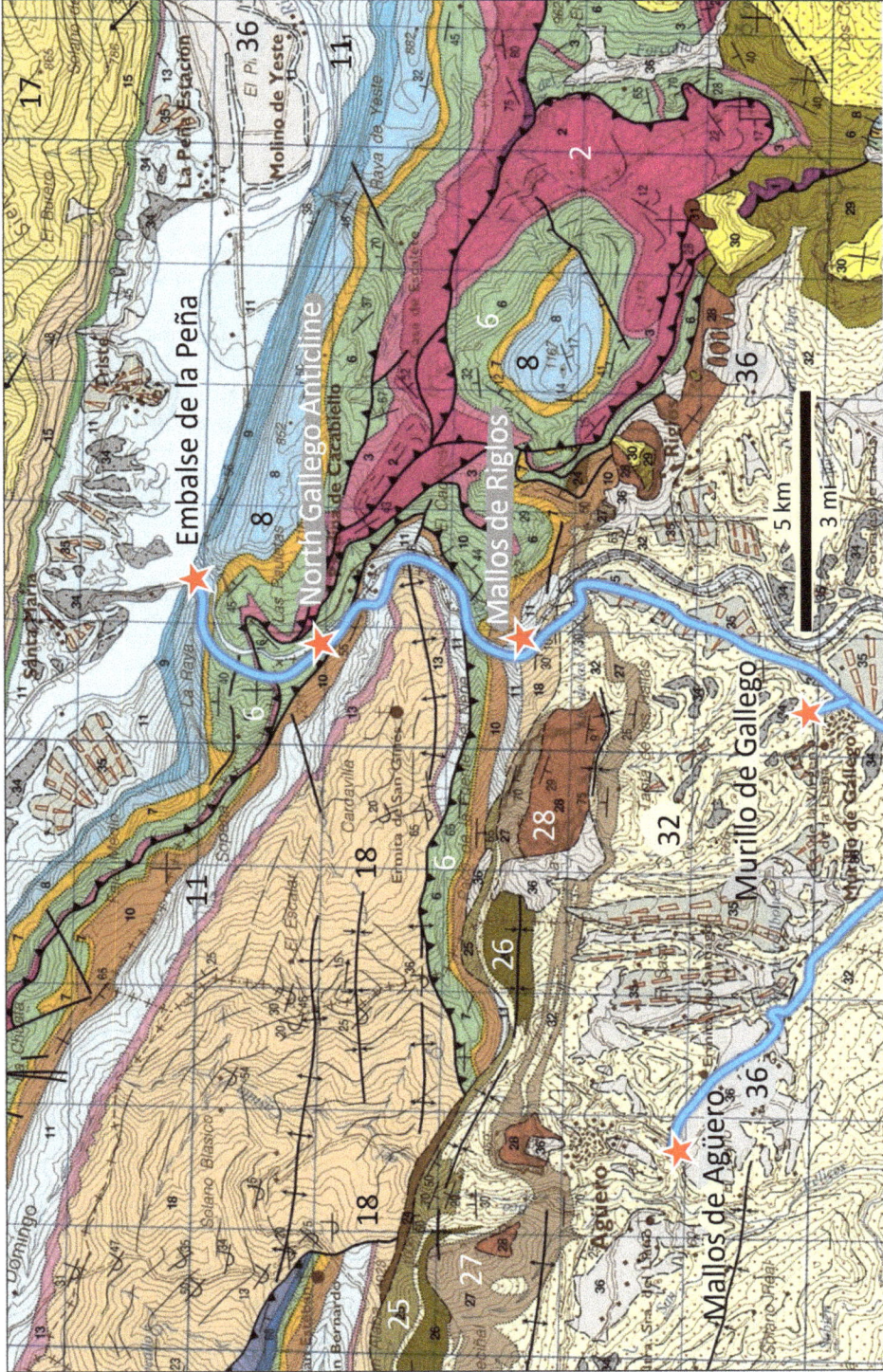

Geologic map of the Mallos de Agüero/Mallos de Riglos area. 36 is Quaternary alluvium; 32 is Oligo-Miocene Uncastillo Fm shale; 25 to 28 are Uncastillo Fm conglomerates; 18 is Eocene-Oligocene red shale and tabular sandstone of the Campodarbe Group; 11 is middle Eocene Arguís Formation marl; 8 is early Eocene Alveolina limestone; 6 is early Eocene Guara Formation limestone-sandstone; 2 is Triassic Muschelkalk limestone and marl. From Instituto Geológico y Minéro de España, 1991, Agüero Sheet. Source: © NC Geological Survey of Spain (IGME).

Near the village of Riglos, the Riglos, Santo Domingo, Gabardiella, and Linas thrusts caused progressive rotation of and unconformitites within the conglomerate layers. The earliest movement was on the Riglos Thrust: conglomerates were deposited on the thrust sheet. Later out-of-sequence thrusting (Santo Domingo, Gabardiella, and Linas thrusts) sequentially cut the thrust sheet behind (north of) the earlier thrust. (A normal thrust sequence proceeds from oldest in the hinterland to youngest in the foreland. Thrusts get younger in the direction of thrust movement. In out-of-sequence thrusting the oldest thrust is the frontal thrust and faults get younger toward the hinterland). Major movement on the Riglos Thrust occurred between 25 and 24 Ma (early Miocene). This was the last major thrusting in the southern Pyrenees (Hogan and Burbank, 1996).

Conglomerates in the Riglos and Agüero area are part of the Uncastillo Formation, a series of coalescing alluvial fans and a braided river system (a network of intertwined, or braided river channels) that formed adjacent to the conglomerate fans. Imbricated pebbles and cobbles in the Uncastillo Formation indicate a flow direction to the south (Ball, 2020).

A typical example of these fans outcrops north of Agüero. The internal features of the fan are nicely exposed by erosion in the present-day Agüero Valley. The fan has sections that coarsen upwards (indicating early development), progressive unconformities (indicating deposition during early deformation), and sections that fine upwards (indicating later, decreasing deformation and deposition farther from the source). Much of the conglomerate deposited on the fan came directly from thrust sheets uplifted within the Sierras immediately to north. The changing composition of the conglomerates indicates several phases of unroofing (erosion) of the thrust sheets (Friend et al., 1996).

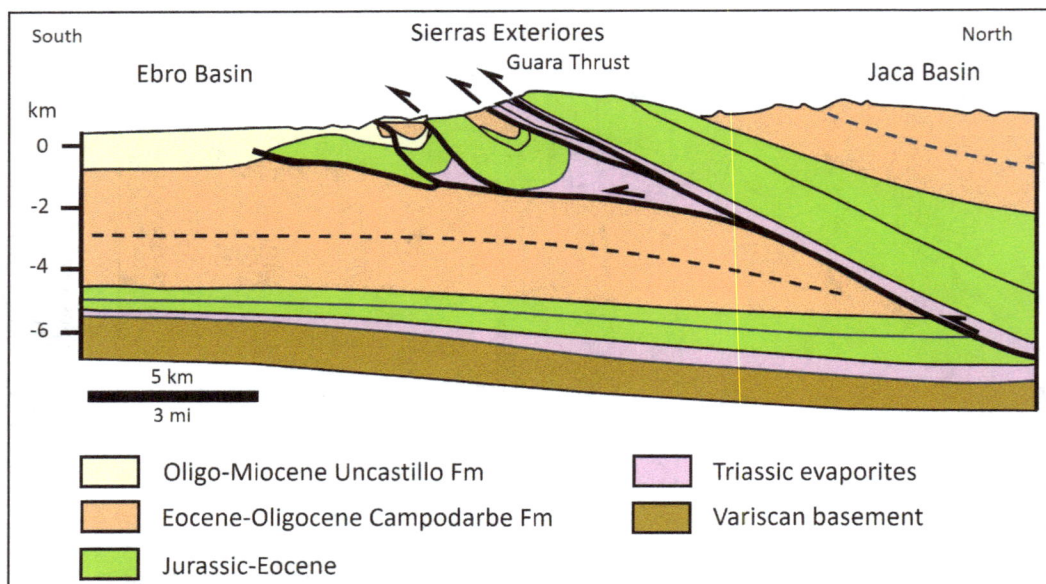

Cross-section from the Jaca Basin across the External Sierras to the Ebro Basin. Modified after Arenas et al., 2001.

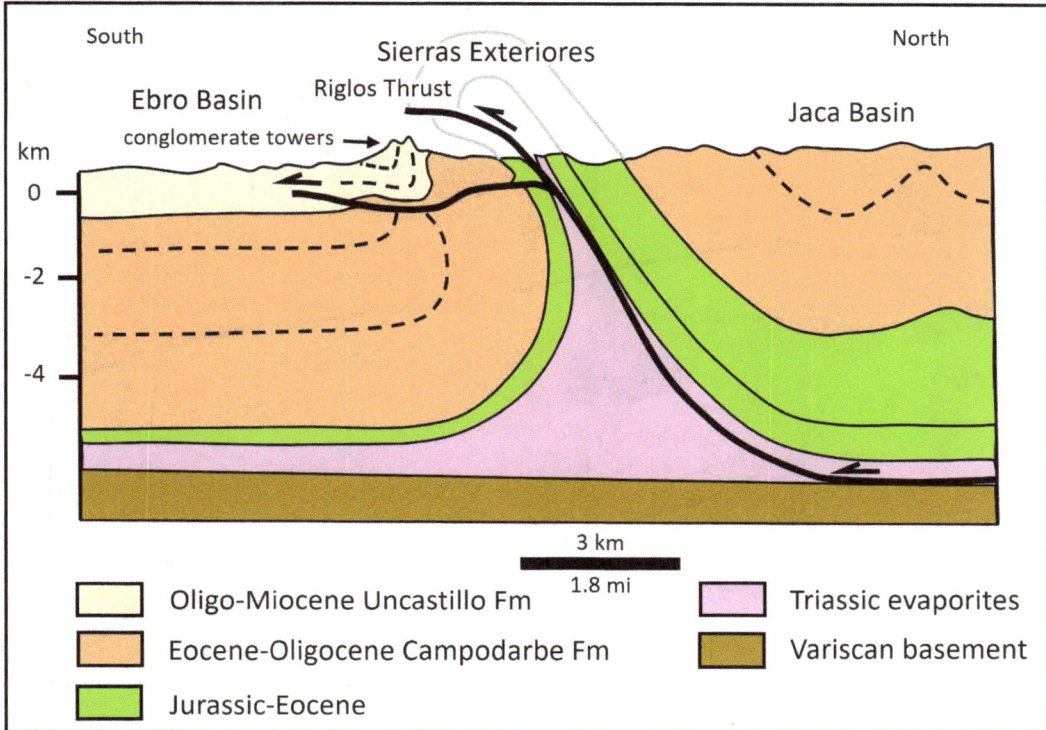

Cross-section through the Mallos de Riglos area. As erosion proceeds in the Sierras Exteriores, the composition of the conglomerates comes from progressively older units. Modified after Arenas et al., 2001; Ball, 2020.

South-southwest-directed shortening during deposition of the fans resulted in syndepositional faulting, folding, and unconformities in the conglomerate fans. Great thicknesses of fan deposits were deposited in the late Oligocene and early Miocene. The last phase of conglomerate deposition occurred between 24 and 23 Ma. Deep downcutting by rivers and streams over the last million years led to the erosion of the fan deposits and their spectacular exposure in the modern landscape (Hogan and Burbank, 1996; Arenas et al., 2001; Nichols, 2005).

Mallos de Agüero to Murillo de Gallego: *Return southeast on HU-534/Z-534 to A-132; turn left (north) onto A-132 and drive to C. Barella; turn left onto C. Barella and drive to Calle Virgen and the geopark display at the top of the hill. This is* ***Stop 32, Murillo de Gallego*** *(42.337222, -0.751639) for a total of 6.2 km (3.9 mi; 15 min).*

Stop 32 Uncastillo Formation Alluvial Fan Conglomerates, Murillo de Gallego

This stop provides a panoramic view from the Mallos de Agüero to the Mallos de Riglos from within the town of Murillo de Gallego. As noted earlier, these are massive sections of coarse Oligo-Miocene conglomerate shed off the rising Pyrenees to the north.

Murillo de Gallego to Mallos de Riglos: *Return south on C. Barella to A-132/Carr. Tarragona San Sebastián; turn left (north) onto A-132 and at 3.8 km (2.4 mi; 6 min) pull into turnout on the right by the TransPyrenean GeoDrive displays. This is* ***Stop 33, Mallos de Riglos*** *(42.359194, -0.744917).*

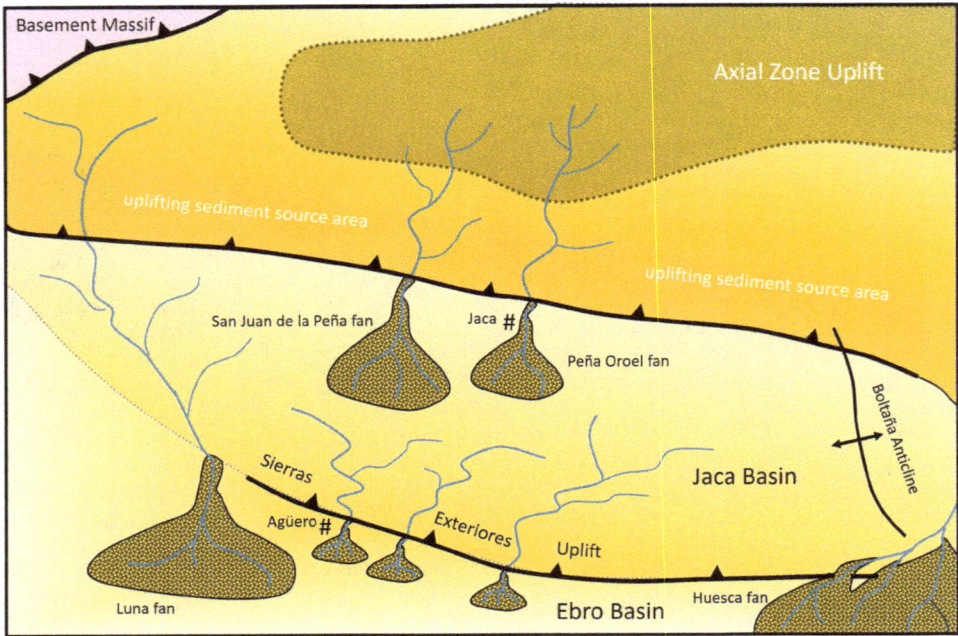

Oligo-Miocene paleogeography with fan deposition. Geology derived from Arenas et al., 2001; Roigé et al., 2019; Ball, 2020.

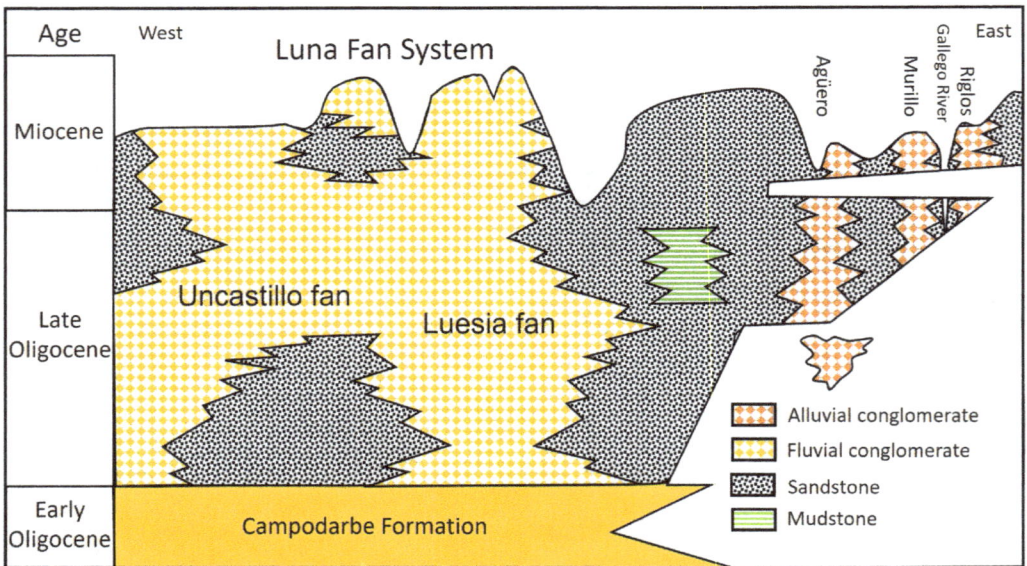

Stratigraphy of the Uncastillo Formation and Luna System. Stratigraphic information drawn from Arenas et al., 2001.

View northeast to the Mallos de Riglos from Murillo de Gallego.

STOP 33 DEFORMED CAMPODARBE AND UNCASTILLO FAN CONGLOMERATES, MALLOS DE RIGLOS

We are now at the entrance to the External Sierras, the southernmost deformation related to the Pyrenean Orogeny. We are at the boundary between the Sierras and the Ebro foreland basin.

According to the geological interpretive panel, Oligocene-Miocene folding that formed the External Sierras led to uplift, erosion, and subsequent deposition of the Campodarbe Formation fluvial sands and later the orange, alluvial Uncastillo Formation conglomerates. These units were caught up in the deformation and folding as they were being deposited.

Mallos de Riglos to North Gallego Anticline: *Continue driving north on A-132 for 2.3 km (1.4 mi; 3 min) and pull over on the gravel track on the right. This is **Stop 34, North Gallego Anticline** (42.3740217, -0.7436931).*

STOP 34 NORTH GALLEGO ANTICLINE, SIERRAS EXTERIORES

As you drive north through the Gallego River Gorge to Embalse de la Peña, you pass through a couple of south-verging overturned folds at the southern edge of the External Sierras. These include the North Gallego Anticline and the Gallego Gorge Antiform. Whereas the North Gallego Anticline is a leading-edge anticline on the North Gallego Thrust, the Gallego Gorge Antiform is actually an overturned syncline: the beds dip away from the fold axis, but the youngest unit (Campodarbe Formation) is in the core of the fold.

From this vantage we get a view to the northwest of the south-verging North Gallego Anticline that is carried on the North Gallego Thrust. This small anticline is interpreted as a fault-propagation fold resulting from late stage out-of-sequence imbricate thrusting (Ball, 2020).

North Gallego Anticline to Embalse de la Peña: *Continue driving north on A-132 for 1.7 km (1.0 mi; 2 min) and pull into the parking area on the right. This is **Stop 35, Embalse de la Peña** (42.383417, -0.738111).*

Top: Mallos de Riglos from geological pullout on the TransPyrenean GeoDrive. View east. Bottom: Interpretation based on the GeoDrive Panel. Dotted line is the unconformity; dashed lines follow bedding; solid lines are thrust faults.

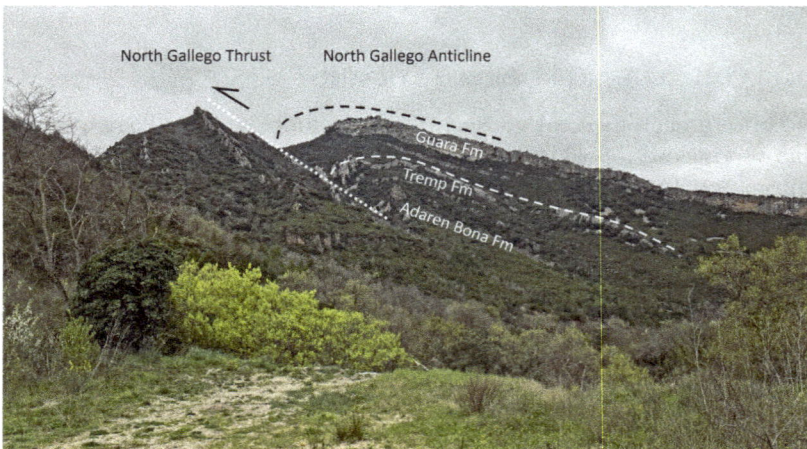

View northwest to the North Gallego Anticline. The Adaren Bona Formation is Upper Cretaceous gray limestone; the Tremp Formation here is Paleocene conglomerate, sandstone, evaporites, and limestone; the Guara Formation is lower Eocene limestone (Ball, 2020).

STOP 35 GUARA FORMATION LIMESTONE, EMBALSE DE LA PEÑA

The rocks seen at this stop are late Eocene Guara Formation limestone on the back (north) limb of the North Gallego Anticline. This 200 m (660 ft) thick gray limestone is a result of the accumulation of millions of tiny shells belonging to calcareous foraminifera, single-celled marine plankton. In particular, *Nummulites* and *Alveolinas* are common. These indicate a shallow, tropical marine environment and a carbonate shelf, similar to the Florida Keys today. This is in contrast to the turbidite basin found farther north and closer to the source of sediments.

The Eocene Guara Formation limestone, Embalse de la Peña. View southwest.

Eocene marine fossils in the Guara Formation, Embalse de la Peña stop. A 2 Euro coin provides scale.

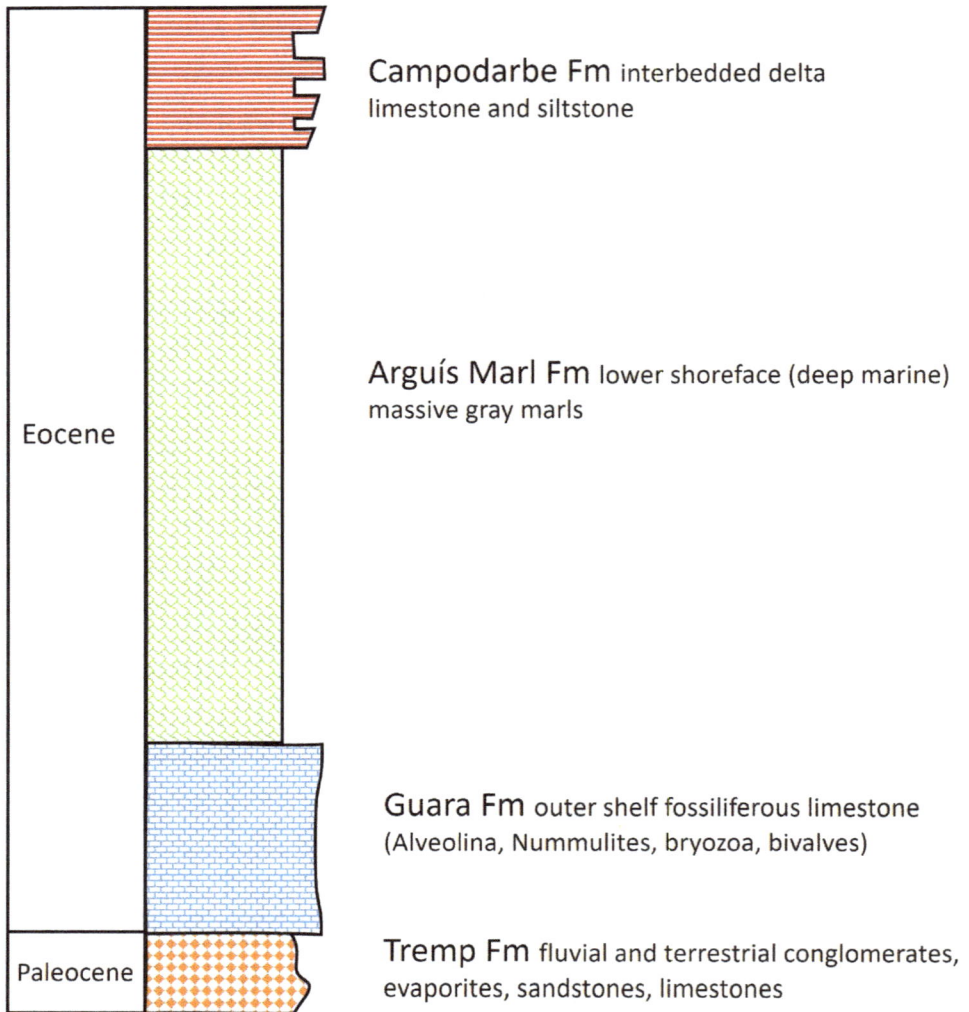

Campodarbe Fm interbedded delta
limestone and siltstone

Arguís Marl Fm lower shoreface (deep marine)
massive gray marls

Eocene

Guara Fm outer shelf fossiliferous limestone
(Alveolina, Nummulites, bryozoa, bivalves)

Tremp Fm fluvial and terrestrial conglomerates,
evaporites, sandstones, limestones

Paleocene

Stratigraphic column for the Embalse de la Peña stop. Stratigraphy based on Ball, 2020.

Stop 36 Bernués Formation Conglomerates, San Juan de la Peña

We are now north of the External Sierras of the Pyrenees and back in the Jaca Basin. The massive conglomerates seen at San Juan de la Peña are age-equivalent to the conglomerates at Mallos de Riglos. Whereas those at Mallos are part of the Oligocene–early Miocene Uncastillo Formation of the Luna alluvial fan delta, those at San Juan de la Peña are in the Oligo-Miocene Bernués Formation, part of the San Juan de la Peña alluvial fan system. The remnants of this fan are preserved in a broad, gentle syncline in the center of the Jaca Basin. The San Juan de la Peña alluvial fan, up to 1,100 m (3,600 ft) thick, is the youngest unit in the Jaca Basin. Pebble composition of the fan indicates an uplifting source just to the north in the eroding Eocene Hecho Group (Roigé et al., 2017; Coll et al., 2022).

The onset of major thrusting and thrust loading led to rapid basin floor subsidence, thereby promoting thick vertical stacking of fan and channel deposits (Turner, 1992).

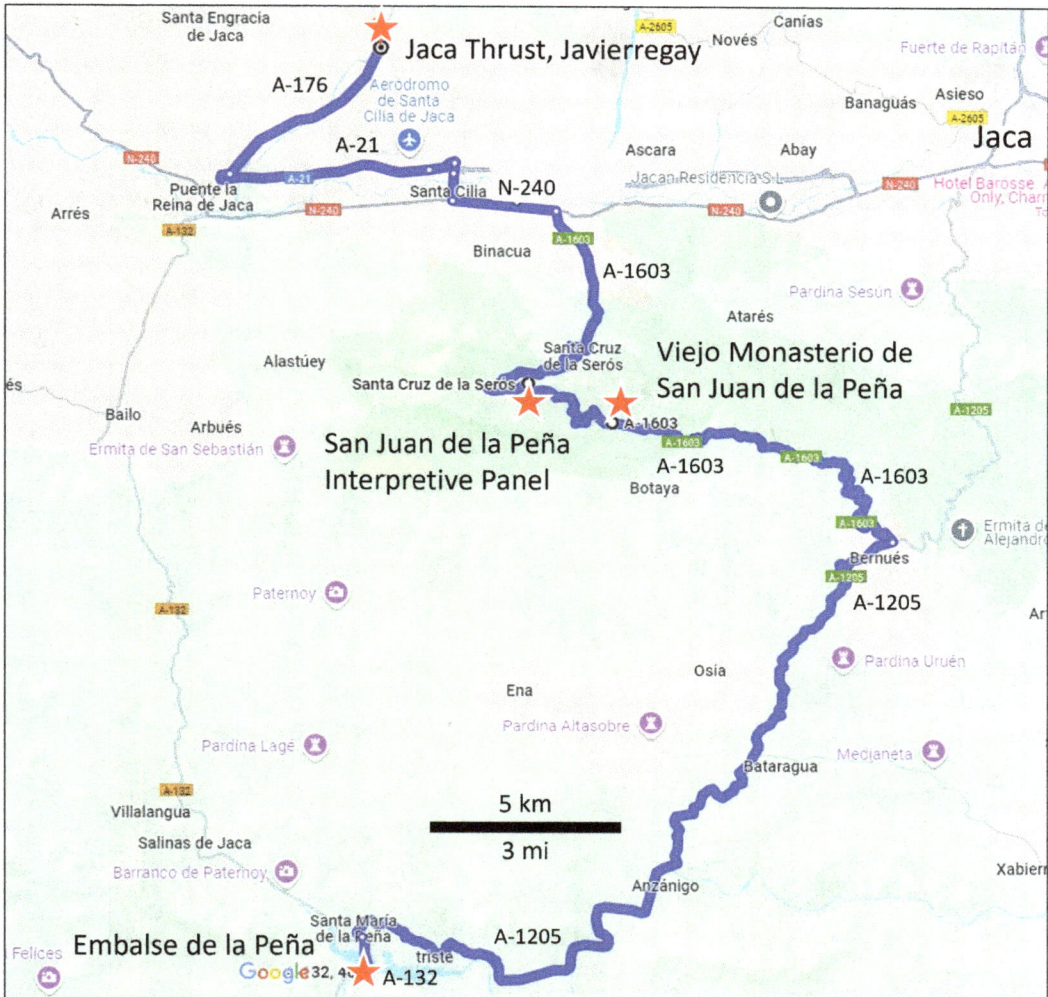

The end of sedimentation in the Jaca Basin is attributed to Oligocene–early Miocene activity on the Guarga Thrust that ultimately resulted in uplift and erosion of the basin. Sediments eroded from the Jaca Basin moved south and were deposited as Uncastillo Formation conglomerates of the Luna alluvial fan in the Ebro Basin (Roigé et al., 2019).

*Embalse de la Peña to San Juan de la Peña: Continue north on A-132 across the reservoir; at the roundabout take the 1st exit (north) onto A-1205/Carr. Jaca-San Juan de la Peña; continue north on A-1205 to the sign for San Juan de la Peña; turn left (west) onto A-1603 and drive to the parking area on the left at the Royal Monastery (42.507790, -0.665429) and purchase tickets for the tour. Continue driving west another 1.3 km (0.8 mi) and pull over on the left. This is **Stop 36.1, Monasterio Viejo de San Juan de la Peña** (42.507669, -0.673659) for a total of 37.2 km (23.1 mi; 45 min).*

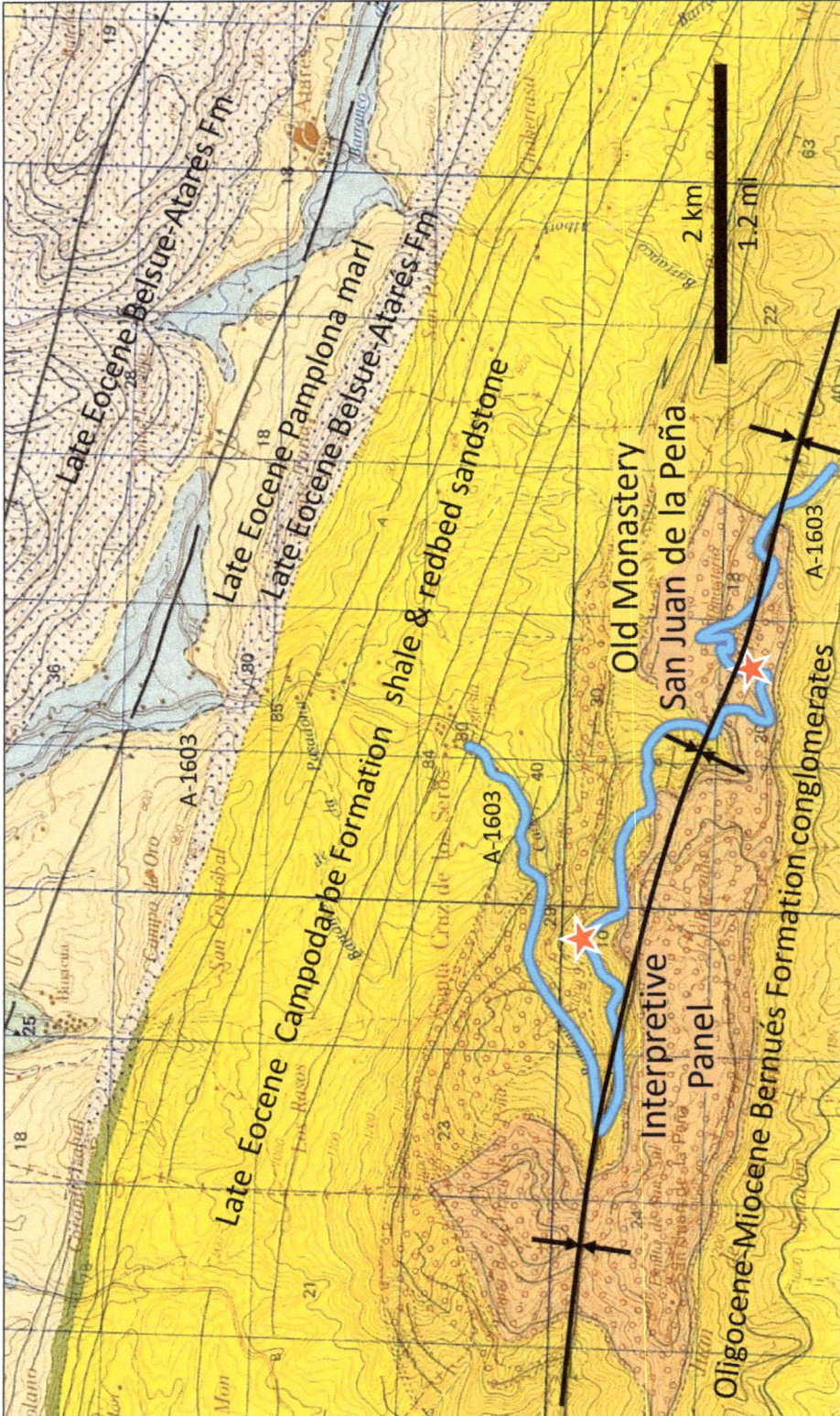

Geologic map of the San Juan de la Peña area. From Instituto Geológico y Minéro de España (IGME), Jaca Sheet. Source: © NC Geological Survey of Spain (IGME).

STOP 36.1 BERNUÉS FORMATION CONGLOMERATES, MONASTERIO VIEJO DE SAN JUAN DE LA PEÑA

San Juan de la Peña means "Saint John of the Cliff." The old monastery (Monasterio Viejo), below a cliff ledge consisting of conglomerates of the same name, was originally built in 920. In the 11th century the monastery became part of the Benedictine Order and was the first monastery in Spain to use the Latin Mass. The cloister, built ca. 1190, contains a series of capitals with biblical scenes. After the fire of 1675, a new monastery (Monasterio Nuevo) was built on the plateau above and a little over a km away. The second floor contains a royal pantheon of the kings of Aragon and Navarre. The present room, with its marble sculptures and stucco medallions recalling historic battles, was built mostly during the reign of Charles III of Spain in 1770. In 1889 the monastery was declared a National Monument. It is also part of the Protected Landscape of San Juan de la Peña and Monte Oroel (Brujulea Place, San Juan de la Peña).

The old monastery's church is partly carved into the Oligo-Miocene Bernués Formation conglomerates of the cliff (Wikipedia, Royal Monastery of San Juan de la Peña). The cloister and pantheon are built of Oligocene sandstone (Gutierrez, 2013).

According to legend, a Roman soldier acquired the Holy Grail, the chalice of the Last Supper, during the 3rd century. He brought the relic back to his home at Huesca. In the early 8th century the Moorish invasion led by Tariq ibn Ziyad conquered most of Christian Hispania, and the cup was sent to San Juan de la Peña to keep it safe from Muslim marauders. It remained there until King Martino V of Aragon took it to his palace in 1399 and placed it in Valencia Cathedral. When the monks asked him to return it, the king sent a replica, which burned in the fire of 1675. A replica of this replica is displayed on premises at San Juan de la Peña to this day (Atlas Obscura – Monastery of San Juan de la Peña).

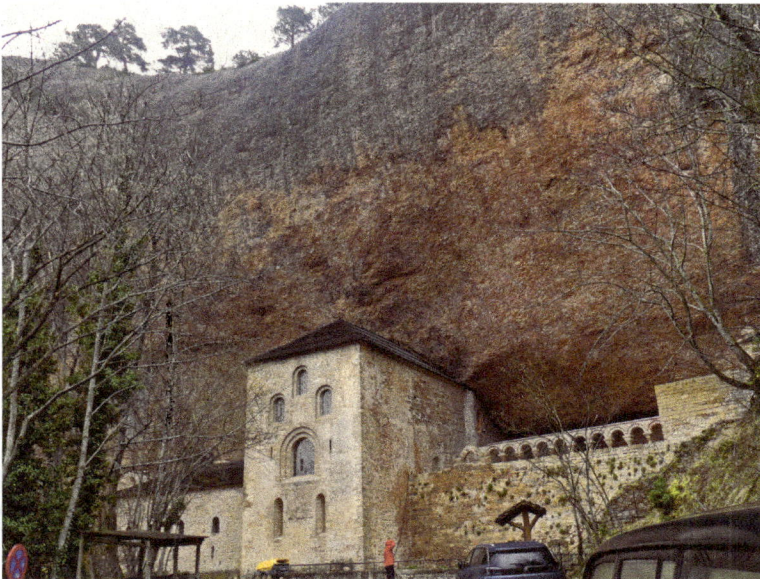

The ancient (Viejo) Monastery of San Juan de la Peña is built into a cliff of Oligo-Miocene Bernués Formation, part of the San Juan de la Peña alluvial fan system.

VISIT

The monastery is open to the public year-round. The car park is free, and there is a restaurant on site. There is a fee for monastery tours.

Entry fee (2025):	12 € (ask about reduced rates for seniors and students)
Hours:	10:00 to 14:00 year-round;
	Afternoon hours:
	16 March – 31 May: 15:30–20:00
	1 June – 31 August: 15:00–21:00
	1 September – 31 October: 15:30–20:00
Address:	A-1603, s/n, 22711 Jaca, Huesca, Spain
Phone:	+34 974 35 51 19
Email:	monasteriosanjuan@aragon.es
Website:	www.monasteriosanjuan.com

San Juan de la Peña to Interpretive Panel: *Continue west on A-1603 for 3.6 km (2.2 mi; 7 min) and pull into the parking area with an interpretive panel on the right. This is **Stop 36.2, San Juan de la Peña Interpretive Panel** (42.516722, -0.690556).*

Stop 36.2 Bernués Formation Conglomerates, San Juan de la Peña Interpretive Panel

This TransPyrenean GeoDrive interpretive panel explains the origin of the conglomerates seen here. Basically, the Oligocene-Miocene uplift of the Pyrenees to the north caused the sea to retreat southward and led to great amounts of very coarse alluvial material being shed off the uplifting mountains. The sandstone and conglomerate here are part of the Oligo-Miocene Bernués Formation.

Conglomerate in roadcut, San Juan de la Peña Interpretive Panel. View southeast.

Interpretive Panel to Jaca Thrust: *Continue north on A-1603 to the N-240/Antigua Carr a Jaca; turn left (west) toward Pamplona and drive to Santa Celia; at the roundabout take the 1st exit (north) onto Camino Somanes; pass under the A-21 and at the roundabout take the 3rd exit (west) onto the A-21 ramp; drive to the A-176; at the roundabout take the 2nd exit onto A-176 north; just north of Javieregay pull onto a small side road and park. Walk along the roadcut **at Stop 37, Jaca Thrust** (42.591923, -0.734363) for a total of 23.5 km (14.6 mi; 23 min).*

Stop 37 Jaca Thrust, Javierregay

Immediately north of Javierregay on A-176 a roadcut exposes the Jaca Thrust. This stop is one of the few opportunities to put your hand on a thrust (the other opportunity on this tour is at Rialp). Not only that, but there is well-exposed deformation in the Eocene Hecho Group turbidites carried by the thrust. Here it is a thrust zone rather than a single surface. The thrust puts older Eocene Hecho Group over younger Eocene Larrés Marl.

Jaca Thrust in roadcut near Javierregay. Heavy solid line is the thrust fault. View southeast.

Hanging wall deformation above the Jaca Thrust near Javierregay. Folds verge south, indicating south-directed transport. Dashed lines follow bedding. View east.

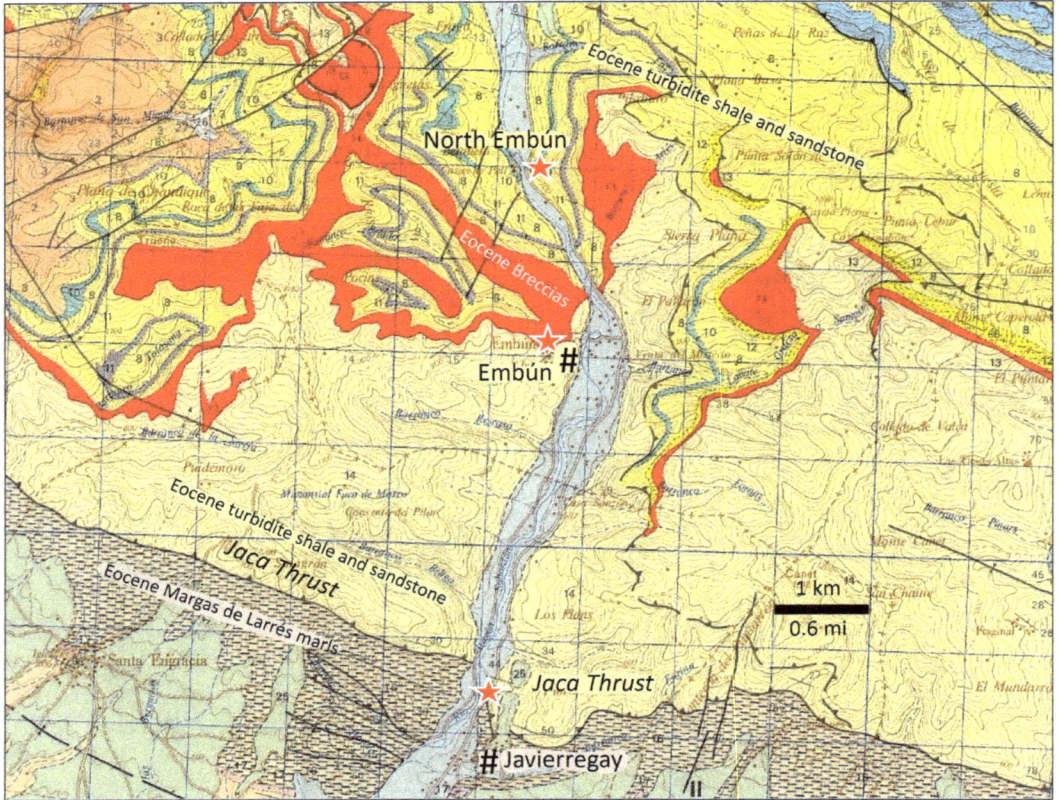

Geologic map of the Javierregay-Embun area. From Instituto Geológico y Minéro de España (IGME), Jaca Sheet. Source: © NC Geological Survey of Spain (IGME).

STOP 38 EMBÚN AREA TURBIDITES AND MEGABRECCIAS

This area contains deformed turbidites of the Eocene Hecho Group. As much as 4,500 m of turbidite deposits have been measured in the Hecho Group. Included in the turbidites are sections of megabreccias, units containing fist-size to km-scale chunks of shallow marine limestone. The megabreccias were derived from a carbonate platform developed along the southern margin of the foreland basin. Occasional instability and submarine landslides, triggered by structural steepening (folding, foreland bulge uplift) caused by thrusting, were accompanied by large-magnitude earthquakes. The breccias were deposited as debris flows that evolved into high-density turbidity currents in their distal portions. Nine of these laterally extensive megabreccias have been mapped in the Hecho Group of the Jaca Basin (Payros et al., 1999; Bauluz et al., 2012).

***Jaca Thrust** to **Embún Turbidites:** Continue north on A-176 to the Embún turnoff; turn left (west) toward Embún and drive to **Stop 38.1, Embún Eocene Turbidites** (42.628067, -0.722951) for a total of 4.9 km (3.1 mi; 5 min).*

STOP 38.1 HECHO GROUP TURBIDITES, EMBÚN

The road to Embún passes by outcrops of Hecho Group turbidites. Shales show pencil cleavage: pencil cleavage is a low-temperature tectonic fabric formed during shortening of clay-rich sediments. (Cleavage is a parting, the tendency of a mineral to break along specific planes of weakness).

Outcrops of shaley turbidite on the road to Embun. Note the pencil cleavage.

Pencil cleavage is a result of the intersection of at least two cleavages at high angles to one another, usually when a diagenetic cleavage, formed during the early stages of sediment hardening, intersects a later tectonic cleavage. The result is pencil-like shards of shale. Cleavage in general indicates shortening in a rock, as might occur during thrusting and bedding plane slip.

Embún Turbidites to North Embún Breccia: *Return to A-176 and turn left (north); drive to a pullout on the left after the curve at 3.3 km (2.0 mi; 4 min). This is **Stop 38.2, North Embún Eocene Turbidites and Breccias** (42.644566, -0.728543). Carefully walk back to the curve to examine the roadcut by the old flume.*

STOP 38.2 TURBIDITES AND MEGABRECCIAS, NORTH EMBÚN

This stop is at an extensive roadcut in the Hecho Group megabreccias. Megabreccias formed by collapse of the carbonate platform margin. Downslope movement of the breccias began perhaps as early as Paleocene, and sediments may have moved up to 30 km north. Breccias contain clasts (angular fragments) up to 2 km wide, although here we see clasts up to only two meters wide.

Top: roadcut in the Hecho Group megabreccias. Box indicates location of clast. Bottom: close up of one of the breccia fragments (arrow).

The episodic instability and mass wasting that caused these breccias are thought to be a result of multiple phases of structural uplift accompanied by large earthquakes.

North Embún Breccia to Aragüés del Puerto Thrust: *Continue north on A-176; turn right (east) toward Aragüés and Jasa onto A-2605; at Jasa continue straight (north) to Aragüés; after 12.6 km (7.8 mi; 14 min) pull over on the right shoulder. This is **Stop 39, Aragüés del Puerto Thrust Deformation** (42.714941, -0.666508).*

STOP 39 GAVARNIE THRUST HANGING WALL DEFORMATION, ARAGÜÉS DEL PUERTO

This country road, Carretera de Lizara, passes excellent outcrops of the middle Eocene Hecho Group turbiditic shales and sandstones. The outcrops are deformed by southwest-directed compression, and they expose subsidiary folds on the south flank of an anticline overturned to the south. The Hecho Group is carried on an unnamed southwest-directed thrust that is probably a splay off the deeper Gavarnie Thrust.

Beds of Garde Cotefablo megabreccia cross the road between this stop and the village of Aragüés del Puerto (Instituto Geológico y Minéro de España, 1989, Anso Sheet).

STOP 40 DEFORMED TURBIDITES, AISA AREA

The outcrops south of Aisa are again deformed middle Eocene Hecho Group turbiditic shales and sandstones that are carried on south-southwest-directed thrust sheets.

Aragüés del Puerto Thrust to Aisa: *Return south to Jasa and A-2605; turn left (east) onto A-2605 and drive to Aisa; at 15 km (9.3 mi; 28 min) on the south side of Aisa pull over into a driveway on the left and walk back north along the roadcut. This is **Stop 40.1, Aisa Deformed Eocene Turbidites** (42.676340, -0.619758). Walk south 50 m (160 ft) along this deformed outcrop.*

STOP 40.1 DEFORMED HECHO GROUP TURBIDITES, AISA

Middle Eocene Hecho Group intricately deformed turbidites are exposed in this roadcut. The structure is consistent with south-southwest-directed thrusting. The mechanical stratigraphy (mechanical properties of different rock layers, including strength, stiffness) of thin-bedded interlayered sandstone and shale makes this a ductile, easily deformed unit even under relatively low overburden pressure (shallow burial).

Aisa to Iglesia San Bartolomé: *Continue driving south on A-2605 for 1.8 km (1.1 mi; 2 min) to **Stop 40.2, Iglesia San Bartolomé Deformed Eocene Turbidites** (42.662479, -0.627814) and pull over on the right.*

STOP 40.2 DEFORMED HECHO GROUP TURBIDITES, IGLESIA SAN BARTOLOMÉ

A small south-verging anticline in middle Eocene Hecho Group turbidites is exposed in this roadcut. As at Aragüés del Puerto and Aisa, south vergence indicates south-directed compression and thrusting.

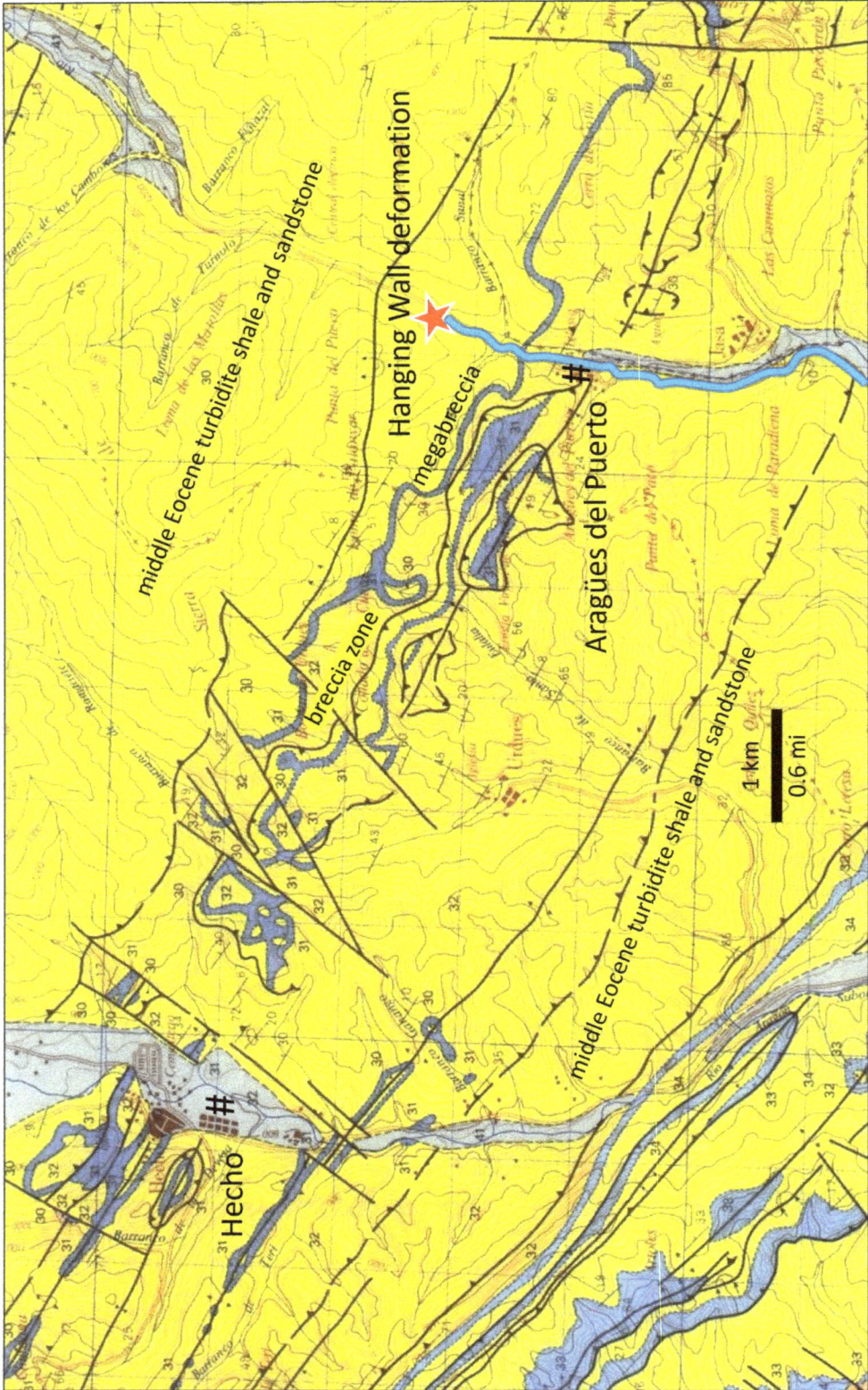

Geologic map of the Aragües del Puerto area. From Instituto Geológico y Minéro de España (IGME), Anso Sheet. Source: © NC Geological Survey of Spain (IGME).

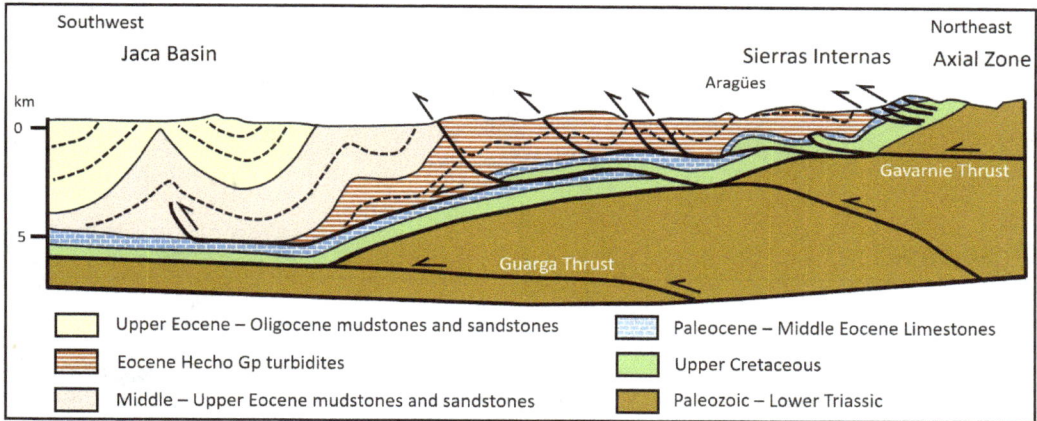

Cross-section through the Jaca Basin showing structure near Aragüés. Modified after Bauluz et al., 2012.

Aragüés del Puerto deformed Hecho Formation. A small-displacement thrust fault (dotted line) and south-west-verging fold is evident here. View southwest.

*Iglesia San Bartolomé to Jaca Interpretive Panel: Continue driving south on A-2605 to N-330a/ Regimiento de Galicia on the northern outskirts of Jaca; turn right (south) on N-330a; at the roundabout take the 1st exit onto N-240/Regimiento de Galicia/Antigue Carret. a Jaca; at the second roundabout turn right toward Prol. Monteano but stay in the roundabout; take the 3rd exit (south) onto Camino dos Callizos; park on the right just after the curve (42.561842, -0.564229) for a total of 23.2 km (14.4 mi; 30 min). Walk back west 280 m (920 ft) to **Stop 41** at the curve, **Jaca Interpretive Panel – Pyrenees Overview** (42.561583, -0.565806).*

Geologic map of the Aisa area and Río Estarrón Valley. From Instituto Geológico y Minéro de España (IGME), Anso and Jaca Sheets. Source: © NC Geological Survey of Spain (IGME).

Deformed middle Eocene Hecho Group turbidites south of Aisa. Dashed lines outline bedding. View southeast.

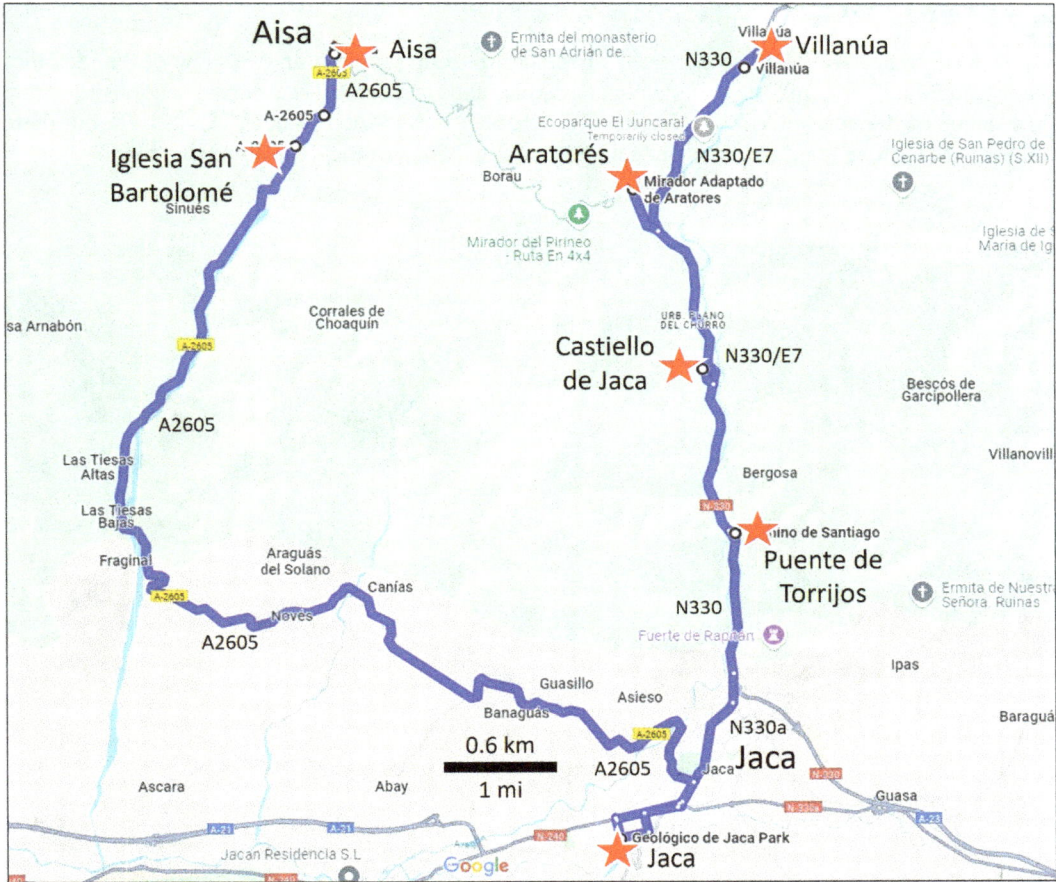

Deformed middle Eocene Hecho Group turbidites deformed into a south-verging fold. View east.

Stop 41 Pyrenees Overview, Jaca Interpretive Panel

The TransPyrenean GeoDrive interpretive panel at this stop provides an overview of the tectonics that caused the Pyrenean Orogeny. Paleogeographic maps and crustal scale cross sections illustrate "the dance of the continents." A number of the displays have faded or been vandalized, but there remains at least one good summary panel and a geologic map of the TransPyrenees route.

The Jaca interpretive panel has a good geologic map showing all the stops on the TransPyrenean GeoDrive and a plate tectonic summary for this part of the Pyrenees.

Jaca Interpretive Panel to Puente de Torrijos: *Drive east on Francisco Viñas; Francisco Viñas turns left (north) and becomes Joaquín Martinez de la Roca; at the roundabout take the 1st exit (east) onto N-240/Av. Nuestra Senora de la Victoria; at the roundabout take the 3rd exit (north) onto N-330a/Regimiento de Galicia; merge onto N-330/E-7 north; at 6.2 km (3.8 mi; 9 min) turn right onto Camino de Santiago and pull over on the right shoulder by the sign for "Punto Geológico." This is* ***Stop 42, Puente de Torrijos, Hecho Group Turbidites*** *(42.606443, -0.543456).*

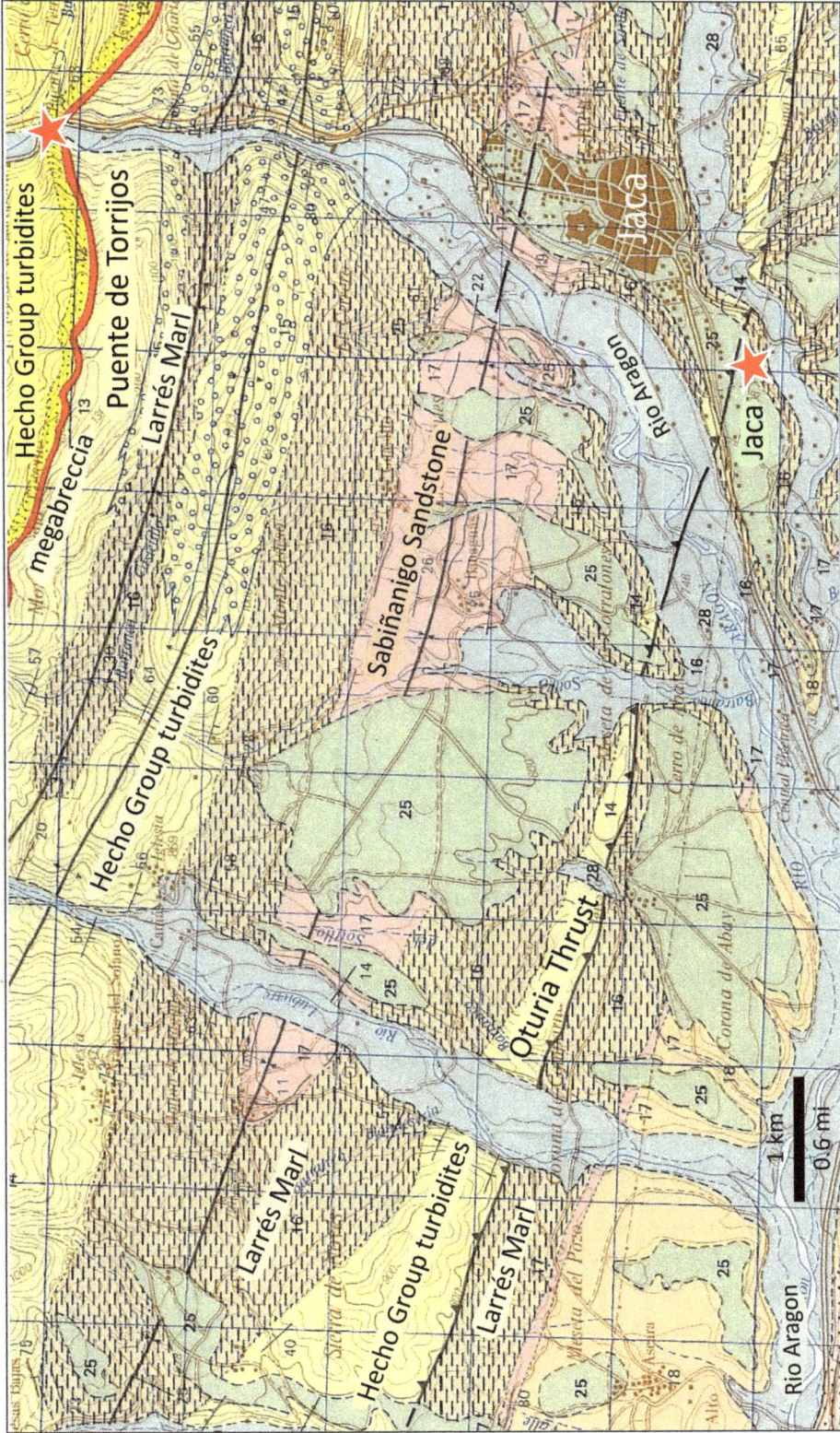

Geologic map of the Jaca–Puente de Torrijos area. From Instituto Geológico y Minéro de España (IGME), Jaca Sheet. Source: © NC Geological Survey of Spain (IGME).

STOP 42 HECHO GROUP TURBIDITES, PUENTE DE TORRIJOS

The panel at this stop on the GeoDrive appears to have been removed or vandalized. It is meant to explain what a turbidite is and how it forms on "the turbulent seabed."

As mentioned previously, a turbidite is a sedimentary rock sequence formed from a turbidity current, a dense, sediment-water slurry that flows down slopes in the ocean or a lake. These currents are triggered by earthquakes, landslides, or storms that disturb sediment that has accumulated on slopes, causing it to rush downwards. As the turbidity current slows, it deposits its sediment load, creating a characteristic sequence of layers with coarser material at the bottom and finer material at the top, or graded bedding.

Puente de Torrijos turbidite outcrop. Turbidites are characterized by multiple fining upward sequences of sandstone to siltstone to shale.

Puente de Torrijos to Castiello de Jaca: *Drive northwest on ramp to N-330/E-7 and turn right (north); at Castiello de Jaca turn left (northwest) onto Calle Santiago (follow signs for Ruta Geológica Transpirenaica); bear left (northwest) just past the city park and park at the interpretive panel. This is* **Stop 43, Castiello de Jaca Flysch Deposits** *(42.630167, -0.549778) for a total of 3.0 km (1.8 mi; 4 min).*

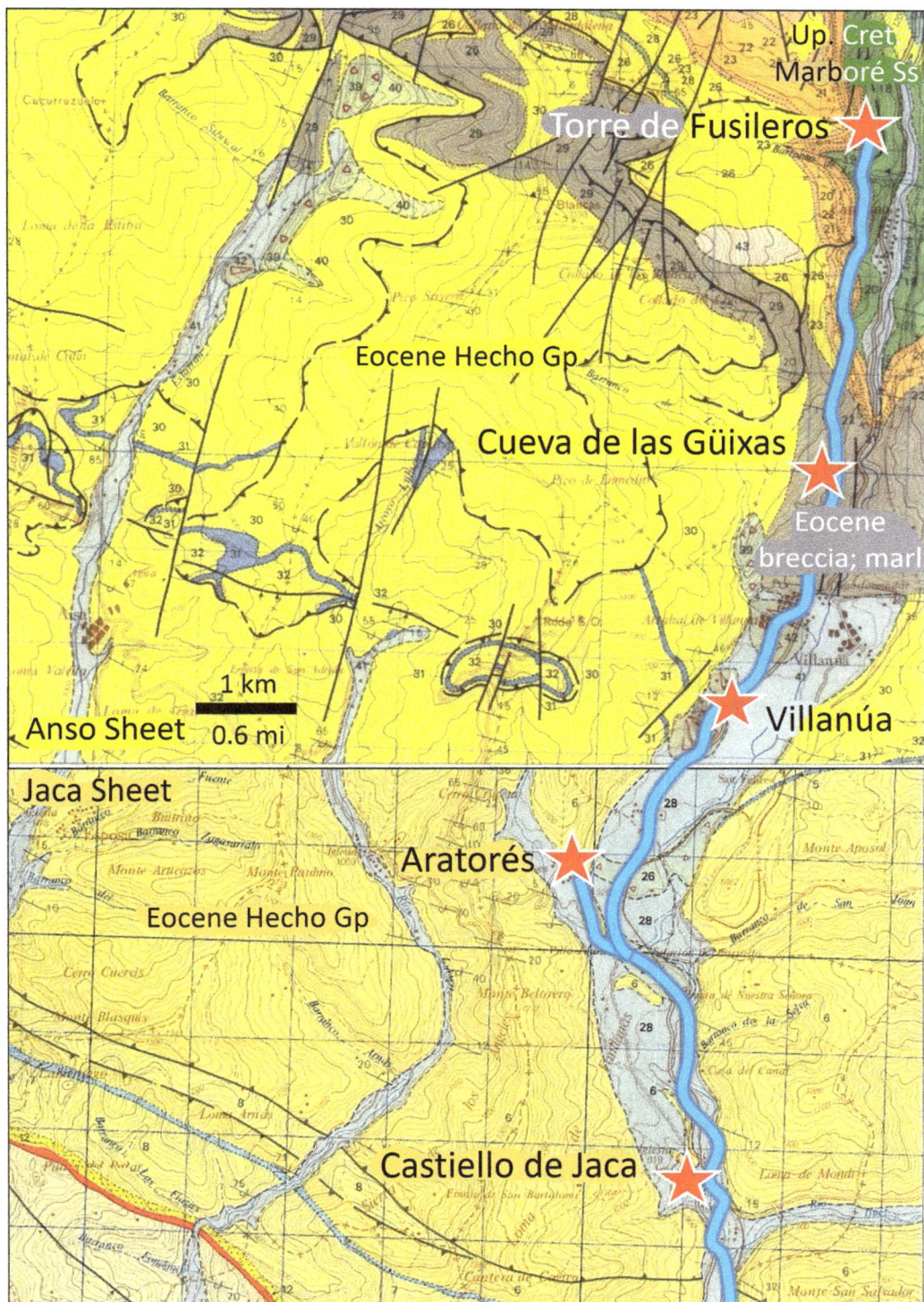

Geologic map of the Castiello de Jaca to Torre de Fusileros area. From Instituto Geológico y Minéro de España (IGME), Jaca and Anso Sheets. Source: © NC Geological Survey of Spain (IGME).

STOP 43 FLYSCH DEPOSITS, CASTIELLO DE JACA

Flysch deposits (deep marine turbidites) make ideal building stone due to their abundance and ease of extraction as slabs of rock. Owing to their varying thicknesses, these rocks have been used as natural bricks, tiles, and pavers for walls, roofs, and floors. Due to its complex history, it has been converted into a compact rock that is resistant to erosion. Traditionally these have been used uncut (unworked) except for corners, windows, and doors, which used lime mortar to hold the blocks together.

The homes and retaining walls in Castiello de Jaca are made of slabs of flysch.

Turbidite flagstone has been used in construction for centuries as a natural brick. Multiple generations of stonework are evident by the different shades of weathering.

Castiello de Jaca to Aratorés: *Return to N-330/E-7 and turn left (north); drive north on N-330 for about 3 km (1.9 mi) to the sign for Aratorés/Borau/Aisa and exit on the right; drive 1 km (0.6 mi) to parking on the right. This is **Stop 44, Aratorés and Pyrenean Glaciation** (42.656333, -0.563083) for a total of 4.1 km (2.5 mi; 5 min).*

As you drive north from Castiello de Jaca, at km 652 (42.631084, -0.548690) there are the most amazing chevron folds in thin bedded Hecho Group sandstones. There is no good pullout for a stop, but keep your eyes open for this remarkable outcrop. If you want to see it up close, drive another 400 m (0.25 mi) and pull off on a track on the right; turn around and drive back 230 m (0.1 mi) and pull over on the right shoulder. Walk south 130 m (430 ft) to the outcrop. It is best seen from across the highway. There is a lot of traffic; please exercise caution walking along the road.

Chevron folds are characterized by straight limbs and sharp hinges, creating a V-shaped appearance. As we saw at Collegats Gorge, they are indicative of compression and usually form in thin-bedded rock. In this case the chevron folds form in the Eocene Hecho Group containing alternating strong, brittle sandstone and weak, more ductile shale.

Chevron folds in Hecho Group thin-bedded sandstones and shales at km 652 just north of Castiello de Jaca. Google Street View northwest.

STOP 44 PYRENEAN GLACIATION, ARATORÉS

Earth's climate has not been stable over geological time. The most recent phase of cooling began during the Tertiary and climaxed during the Quaternary in several stages of glaciation, what we call Ice Ages. Climate cooling in the Pyrenees resulted in accumulation of thick bodies of ice on the high peaks, which moved slowly down the valleys under the influence of gravity. Deposits of rock carried by the glaciers created a landform unique to glacial landscapes: moraines. Moraines are any accumulation of unconsolidated rocks that was carried along by a glacier or ice sheet and deposited at the side (lateral moraines) or end (terminal moraines) of the ice flow.

The retreat of the glaciers at the end of the last Ice Age produced a succession of staggered terminal moraines as well as lateral moraines along the sides of the glaciers. Owing to the softness of turbidites and the ease of erosion, this led to U-shaped valleys with nearly vertical sides. From this we can deduce that the ice was several hundreds of meters thick during the last glaciations.

We are now in the Internal Sierras. Look up the valley to see the U-shaped canyon and lateral and terminal moraines.

View up the Río Aragón Valley. Note the U-shape of the valley that is a result of scouring by glaciers. Dotted line outlines the shape of the valley. Photo courtesy of Jsanchezes, https://commons.wikimedia.org/wiki/File :Valle_rio_aragon.jpg

*Aratorés to Villanúa: Return to N-330/E-7 and turn left (north); after 4 km (2.5 mi) take the offramp (sign says "Villanúa, area de descanso" (Villanúa rest area) and park by the interpretive display. This is **Stop 45, Villanúa, High Sierra Karsting** (42.674056, - 0.541583) for a total of 4.1 km (2.5 mi; 4 min).*

STOP 45 HIGH SIERRA KARST, VILLANÚA

You are crossing Paleocene limestones and dolomites of the Salarons Formation (Instituto Geológico y Minéro de España, 1989, Anso Sheet). All carbonates are prone to dissolving slowly in the very weak acid formed as rain falls through the atmosphere. When limestone dissolves it forms caverns, sinkholes, and underground rivers collectively known as "karst terrain." As you leave the Jaca Basin and enter the Pyrenean Axial Zone, you are passing through the highest elevation karst in Europe. The boundary between the Jaca Basin of the Southern Pyrenees and the Axial Zone is generally taken at the transition from Upper Cretaceous and younger sediments to Paleozoic rocks.

While in the area, check out Cueva de las Güixas, Av. de la Fuente, 1, 22870 Villanúa, Huesca (42.682611, -0.532197).

Phone: +34 97 437 8465.
Website: https://www.turismovillanua.net/descubrir/visita-la-cueva-las-guixas/
General Admission (2025):   9 €
Children 4–14 years: 7.50 €
Hours vary depending on the time of year. Check the website.

According to the Anso Sheet, the town of Villanúa lies on, and the Cueva de las Güixas is developed in, Paleocene calcareous megabreccias.

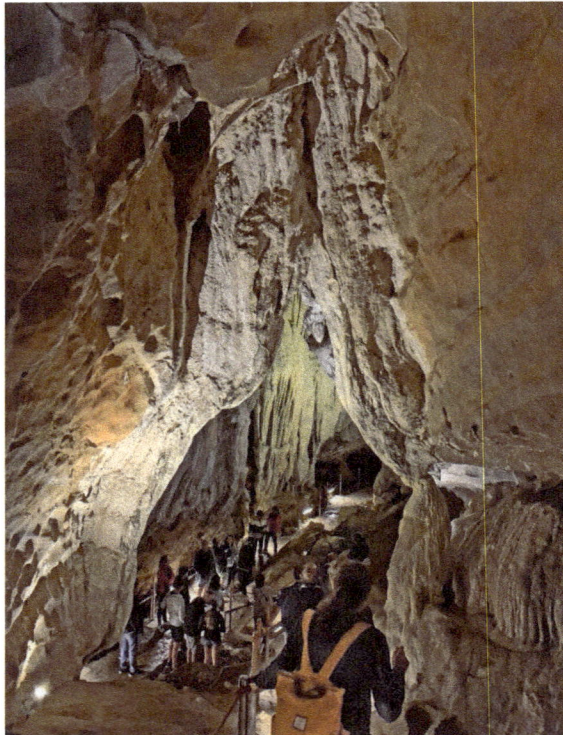

Cueva de las Güixas, Villanúa. Photo courtesy of Héctor Ochoa "Robot8A," https://commons.wikimedia.org /wiki/File:Cueva_de_las_G%C3%BCixas.jpeg

Villanúa to Torre de Fusileros: *Continue north on N-330/E-7 to Estacion de Canfranc; make a U-turn and return south to the rest area on the right. This is **Stop 46, Torre de Fusileros, Pyrenean Thrusting** (42.736361, -0.523278) for a total of 12.6 km (7.8 mi; 14 min).*

STOP 46 PYRENEAN THRUSTING, TORRE DE FUSILEROS

The Interpretive Panel at this stop is labeled "The Earth, Builder of Mountains."

Formation of the Pyrenees was the result of hundreds of km of shortening that squeezed and uplifted the crust between the colliding Iberian and European tectonic plates. This collision, the Pyrenean Orogeny, is characterized by the uplift, folding, and stacking of formations by thrust faults.

Marine sedimentation occurred between Iberia and Europe prior to Paleocene time: we see that in the marine rocks that were deposited. Compression began with the Alpine Orogeny, in late

Paleocene, which manifested as reverse and thrust faults affecting brittle units and thickening affecting the more ductile units. Horizontal shortening caused great displacement on the faults, thrusting older units over younger ones. The softer units deformed plastically (flow folding) such that they were often thickened rather than faulted.

A byproduct of compression of the softer units was development of Schistosity, which is seen in the rocks at Torre de Fusileros. Schistosity is a foliation (fine scale layering) in metamorphic rocks that is a result of the squeezing and alignment of platy minerals. In this area it is a result of high pressure and temperature due to both burial and compression.

Erosion of these materials during the Quaternary exposed the landscape seen today. From bottom to top (oldest to youngest) this stop is in southwest-dipping Upper Cretaceous calcareous sandstone with nodular limestone, Zuriza Marl, Tuca Blanca Sandstone, and Marboré Sandstone (Instituto Geológico y Minéro de España, 1989, Anso Sheet).

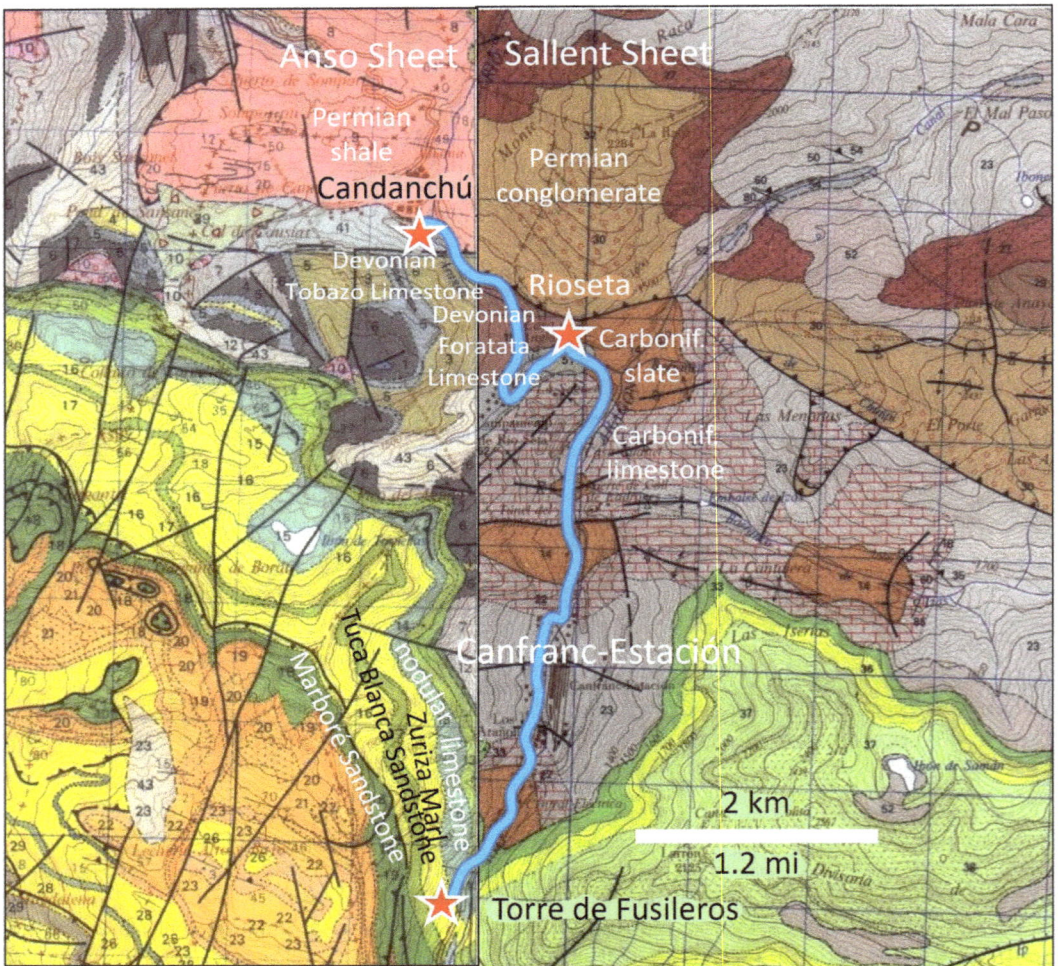

Geologic map of the Torre de Fusileros to Candanchu area. From Instituto Geológico y Minéro de España (IGME), Anso and Sallent Sheets and Rodríguez-Méndez et al., 2013. Source: © NC Geological Survey of Spain (IGME).

Torre de Fusileros to Rioseta: Continue north on N-330/E-7 for 0.5 km (0.3 mi) and bear right toward N-330a "Francia por Puerto Canfranc Estación";' at 5 km (3.1 mi; 6 min) pull into parking area on the right. This is **Stop 47, Pyrenean Axial Zone, Rioseta** *(42.776, -0.512333).*

STOP 47 PYRENEAN AXIAL ZONE, RIOSETA

As the rocks get older we leave behind the Jaca Basin and enter the Pyrenean Axial Zone. The boundary between the Jaca Basin of the Southern Pyrenees and the Axial Zone is generally taken at the transition from Upper Cretaceous and younger sediments (to the south) to Paleozoic rocks (to the north).

View north from the Rioseta stop. The day I was here it was raining and much of the scenery was obscured. Yet I could still see the upper and middle steps of the three-stepped cliffs, and also a small-scale thrust and buckle fold. Dotted line is a thrust fault; dashed line follows bedding.

This stop is in Middle Devonian Foratata Formation reef limestone and slate (Instituto Geológico y Minéro de España (IGME), Sallent Sheet). But that's not the purpose of the stop. We stopped to check out the Cretaceous-Paleocene three-stepped cliff unique to the Interior Sierra of Aragon.

The 1,000 m (3,280 ft) high cliffs are made of three distinct limestone formations deposited in differing marine environments over a period of 45 million years. At the base of the trio is the Upper Cretaceous Cañones Limestone, massive white to brown limestone containing coral, mollusk, and bivalve fossils and sitting unconformably above folded Devonian reef limestones. The Devonian limestone was folded during the Late Carboniferous–Permian Variscan mountain-building event, then underwent 200 Ma of erosion before the Cretaceous units were deposited.

The middle unit consists of a lower less resistant section, the brown Zuriza Marl, and an upper, buff, well-stratified Marboré Sandstone.

The upper unit is white, highly resistant Paleocene limestone. Note the small thrust and buckle fold in the uppermost cliff. As with chevron folds (a type of buckle fold), buckle folds form when a mechanically strong layer (like a limestone) is surrounded by a weak layer (like a shale). The weak layer flows while the strong layer buckles.

Pyrenean mountain-building folded the Upper Cretaceous–Paleocene section, and refolded the Devonian and older rocks.

Rioseta to Candanchú: *Continue north on N-330a for 2.3 km (1.4 mi) and turn left onto Calle Unica toward Candanchú; at the sign "Candanchú" turn left and pull into the parking area on the right. This is* **Stop 48, Candanchú, the Axial Zone** *(42.785583, - 0.524528) for a total of 2.6 km (1.6 mi; 3 min).*

Stop 48 Axial Zone Structures, Candanchú

This stop is on the Devonian Tobazo Limestone (Instituto Geológico y Minéro de España, 1989, Anso Sheet). We are now well and truly into the Axial Zone, the "heart of the Pyrenees" as indicated by the much older rocks. These rocks tell the story of a mountain range older than the Pyrenees, that of the Variscan (sometimes still referred to as Hercynian) Mountains. These layers reveal their distinct depositional environments and the various deformation events that affected them.

The Devonian limestones (360–420 million years ago) were deposited as reefs in a warm sea in the Southern Hemisphere. Compression during the Variscan Orogeny raised and folded these limestones to elevations approaching 6,000 m (20,000 ft). It is not obvious from the outcrop pattern on the geologic map, but the ski slopes are on the northeast flank of an enormous west-northwest-trending anticline that can be seen from the ski area parking lot.

At the end of Permian time these mountains were subjected to intense erosion and generated conglomerates and sandstones. At the same time volcanic rocks were erupting just east of here.

Folding in the Axial Zone is a result of superposition of Alpine and earlier Variscan folding events. The core of the Axial Zone here is characterized by an antiformal south-directed structure which is a result of the stacking and folding of imbricate thrust sheets of Variscan-age rocks (Rodríguez-Méndez et al., 2013).

Between these stops we enter France.

Candanchú to Peyrenere: *Return south to N-330a and turn left (north); at the French-Spanish border the road becomes Col du Somport; at 4.9 km (3.0 mi; 7 min) pull into the parking area on the left. This is* **Stop 49, Peyrenere, Deformed Permian Redbeds** *(42.803139, -0.545361).*

View southwest from the ski area parking lot, Candanchú. The ridge and slopes are part of a large anticline.

Stop 49 Deformed Permian Redbeds, Peyrenere

The red color of the 260 million year old Permian sandstones and shales seen here is the result of oxidation of iron minerals in a semi-arid environment. These redbeds were later folded during the Alpine Orogeny.

Looking west from this stop we see, at the highest elevations, Upper Cretaceous limestone. At lower levels and in the foreground are Carboniferous limestones and schists. The ridges to the right (north) contain Permian redbeds. Early Carboniferous (Mississippian) Culm Series sandstone and shale (weathered out in valley) and Permian dark shale and red sandstone of the Somport Series form the ridge to the east (Instituto Geológico y Minéro de España, 1989, Anso Sheet).

Red Permian conglomerate, sandstone, and shale can be seen in a roadcut a few meters downslope from this stop. Interbedded in the redbeds are orange volcanic ash layers that reveal active Permian volcanism that was occurring east of here.

Permian redbed sandstone and conglomerate of the Somport Series outcrop along the highway, Peyrenere stop.

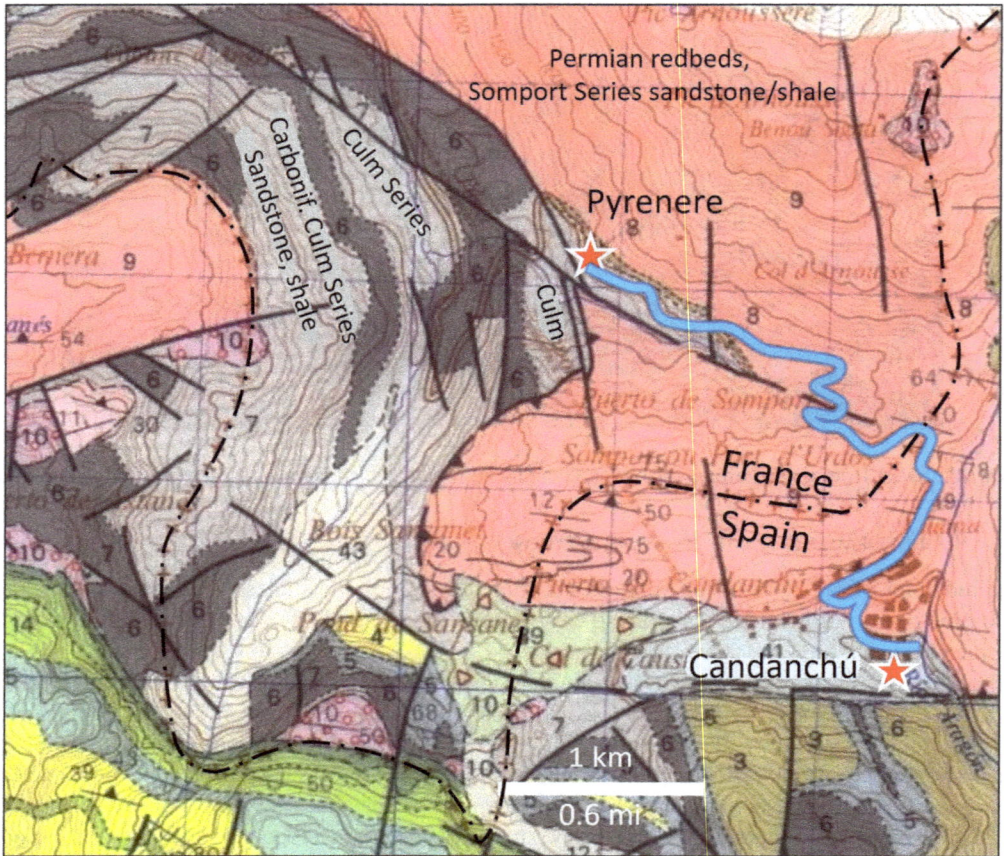

Geologic map of the Candanchú to Peyrenere and Puerto de Somport area. From Instituto Geológico y Minéro de España (IGME), Anso Sheet. Source: © NC Geological Survey of Spain (IGME).

Peyrenere to Fort du Portalet: *Continue north on Col du Somporte; turn right (north) onto N134; at 13 km (8.1 mi; 12 min) and pull over on either side of the road for **Stop 50, Fort du Portalet, Large-Scale Folding** (42.88175, -0.560278).*

STOP 50 LARGE-SCALE FOLDING, FORT DU PORTALET

This stop presents us with a view into the giant Fort du Portalet Anticline. The cliff-forming layers are Devonian limestone. The fold asymmetry is obvious: the south limb is much steeper than the north limb, so even though we are still in the Axial Zone, the fold verges to the south.

This might be a good place to mention the bidirectional nature of Pyrenees thrusting. Like many thrust belt mountain ranges, the Pyrenees are bounded by thrusts of opposite vergence. For the most part, the north flank of the mountain range has north-directed thrusts, whereas the south flank has primarily south-directed thrusts. On a smaller scale, the same thing can be said for any given zone. Although the dominant thrust vergence in a zone may be to the south, there may be north-directed thrusts as well. These are due to complex tectonic processes (e.g., inherited structure, rotation in imbricate thrust fans, triangle zones) within an overall zone of compression.

These rocks were folded during the Permian-Carboniferous Variscan Orogeny.

The younger surrounding rock is Early Carboniferous (Mississippian) turbidite sandstone and shale.

Fort du Portalet south-verging anticline from TransPyrenean stop. Dashed lines follow bedding. View northwest.

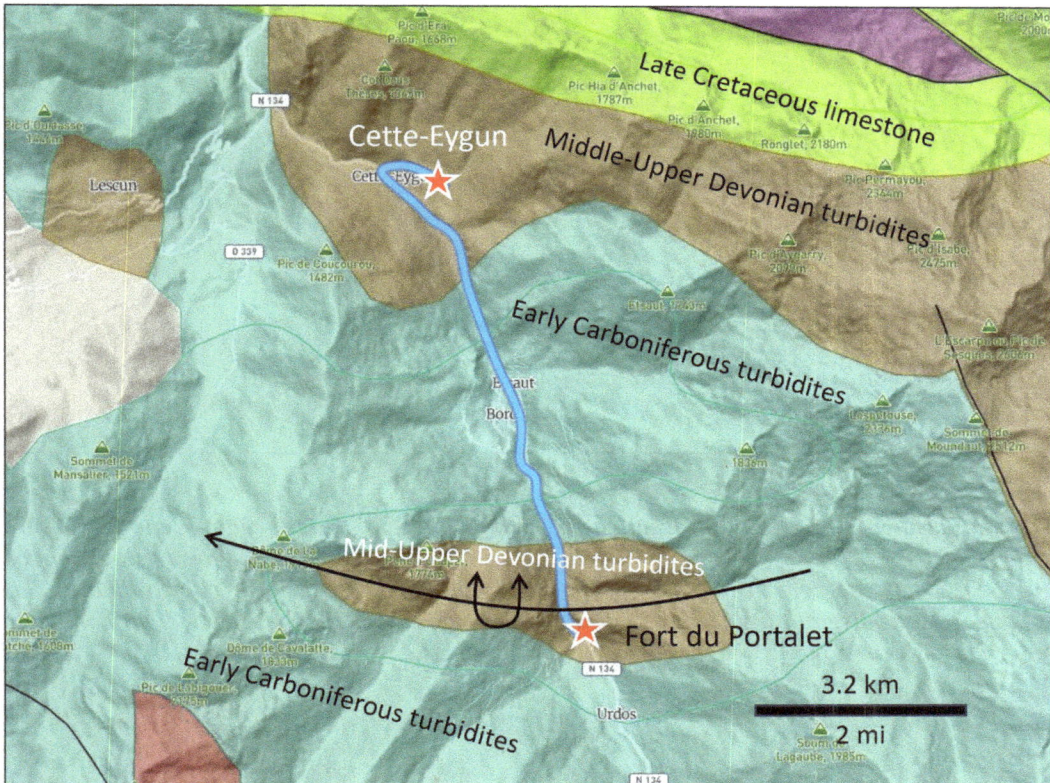

Geologic map of the Fort du Portalet to Cette-Eygun area. Base geology from MacroStrat, 2015.

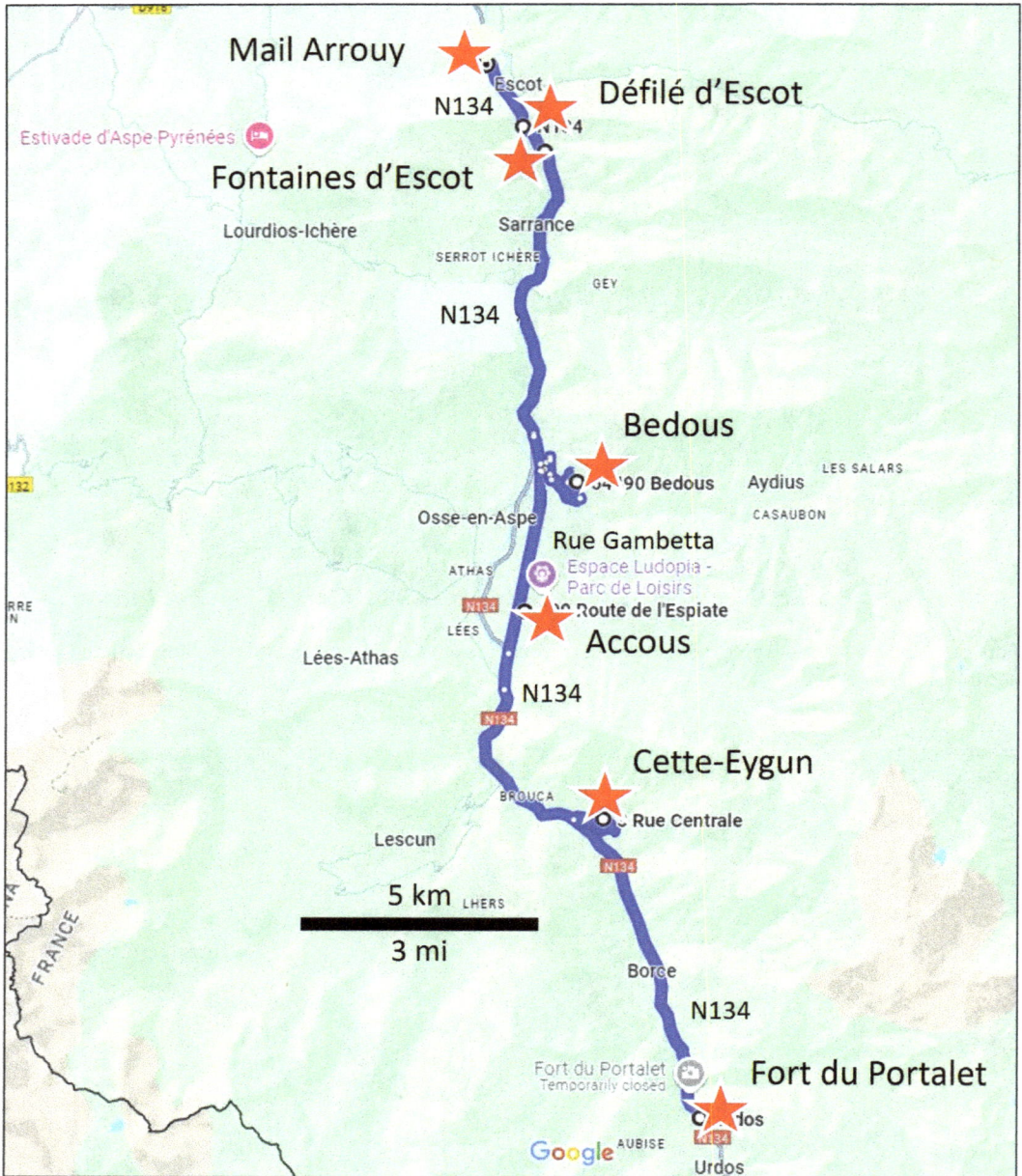

*Fort du Portalet to Cette-Eygun: Continue north on N134; at the north end of Cette-Eygun turn right (east) onto D539/Rese; continue onto Vge de Cette; at the top of the hill turn left and park on the right. Follow the signs to "Site Géologique" and walk about 100 m north to the trailhead. This is **Stop 51, Cette-Eygun, Cretaceous Seabed on Mountain Top** (42.938116, -0.584990) for a total of 9.2 km (5.7 mi; 13 min).*

STOP 51 SEABED ON A MOUNTAINTOP, CETTE-EYGUN

The theme of this stop is "from the seabed to the top of mountains." The limestones that today form the mountain peaks were originally deposited on an ocean floor that existed across all of the Pyrenees 90 million years ago.

Looking west we see Upper Cretaceous limestones folded into a gentle anticline. Below this cliff-forming layer is an unconformity, an old erosion surface. Below the unconformity are Carboniferous schists and gray Devonian limestone.

From this stop there are a couple of eight-hour, round-trip walks that take you to the angular unconformity. This erosion surface is missing 260 million years-worth of rocks. At the trailhead (and along the road up to this spot) there are good outcrops of near-vertical Carboniferous schist and slate.

View west from the Cette-Eygun interpretive panel.

At the trailhead: near-vertical slates that, 300 million years ago (give or take a few), were once horizontal mudstone at the bottom of a Carboniferous sea.

Cette-Eygun to Accous: *Continue north on N134 to roundabout at 6.9 km (4.3 mi); at roundabout take the 1st exit (north) onto D834/Rou Gambetta; turn right (east) on D637 to Accous; drive 150 m (0.1 mi) and turn left (north); follow signs to Route Geologique and park. Walk north to the trail-head.* ***This is Stop 52, Accous, Collision of Tectonic Plates*** *(42.977528, -0.605111) for a total of 8.0 km (5.0 mi; 11 min).*

STOP 52 COLLISION OF TECTONIC PLATES, ACCOUS

When we leave the Paleozoic rocks and enter the Mesozoic section, we have left the Axial Zone and entered the North Pyrenean Zone, a zone of dominantly north-directed thrusts and folds.

Photo is looking east at the boundary thrust separating the European from the Iberian tectonic plates. The Iberian Plate plunges beneath the European Plate. The sketch is from the TransPyrenean GeoDrive panel at Accous. Red triangles point to the boundary thrust.

At this stop you can actually look at the geologic boundary between the European (or Eurasian) Plate and the Iberian tectonic plate whose collision formed the Pyrenees. Look east to the Col d'Iseye. Just south of the pass, on a gentle slope, is the trace of the thrust fault that marks the boundary. The thrust puts Triassic shale, black Jurassic limestone and dolomite, and massive Lower Cretaceous limestone, folded into a large syncline on La Marere Peak, of the European Plate over massive Upper Cretaceous limestone of the Iberian Plate. This is where Europe is overriding Iberia. *Cette géologie est vraiment remarquable.*

Accous to Bedous: *Return to D834/Rou Gambetta and turn right (north); at 3.2 km make a sharp right onto Rue Notre-Dame; immediately turn left onto Chem. de Loueillet; in 366 m (0.2 mi) sharp left; in 127 m (0.1 mi) sharp right; in 930 m (0.6 mi) turn left; in 440 m (0.3 mi) bear right; drive to overlook and display at the curve and pull over on the right. This is* **Stop 53, Bedous, Geologic Time** *(43.002167, - 0.591944) for a total of 5.5 km (3.4 mi; 11 min).*

STOP 53 GETTING A FEEL FOR GEOLOGIC TIME, BEDOUS

Getting a grasp on geologic time is difficult: it is immense, way beyond our mere mortal experience.

Geology explains the landscape. Consider the millions of years it took for sediments to be deposited one grain at a time; the additional millions for the rocks to be buried, compressed, deformed, and raised up; then the millions of years they were subjected to erosion and slowly ground down to form the present-day landscape.

View up the Aspe River Valley. The present river floodplain, bounded by escarpments, is visible along with the glacial-carved U-shaped valley in the distance. View south.

Glaciers influenced this landscape over hundreds of thousands of years. The Bedous Valley, in the center of the mountains, was shaped by the Aspe Valley glacier. The moraines left behind are evidence of the glacier. The moraines are an accumulation of material of various size and composition, left behind as the glacier retreated, far from their place of origin. The moraines tell us that the glaciers did not extend much north of Bedous.

Rivers affected the landscape over thousands of years. The Aspe River and its floods provided the final touch in forming the landscape, carving the river floodplain and depositing alluvium and outwash fans in the 10,000 or so years since the last Ice Age.

*Bedous to Fontaines d'Escot: Return to N134 and turn right (north); turn into the parking area on the right and proceed to the displays. This is **Stop 54, Fontaines d'Escot Methane Seep** (43.064639, -0.600333) for a total of 18.7 km (11.6 mi; 21 min).*

STOP 54 METHANE SEEP, FONTAINES D'ESCOT

The theme of this stop is "gas in the rocks." The rocks that outcrop here are equivalent to the reservoir rock in the gas field at Lacq y de Meillon-San Fausto to the north. The fetid odor emitted when you break the rocks indicates that there is still some methane in the rocks.

Decomposing organic material in swamps form "swamp gas," which is mostly methane. Oil and gas form when organic material is buried before it can decompose. This becomes the "source rock," usually an organic-rich shale or coal. Continued burial under heat and pressure cook the organic material until either liquid or gaseous hydrocarbons are driven from the source rock. Think of cooking soup in a pot: when the pressure gets too high, it lifts the lid and the soup runs out. The pressure of cooking hydrocarbons forces the oil and gas to migrate through porous and permeable rocks till they encounter a reservoir rock in a trap. The reservoir rock, often a sandstone or limestone, has pore spaces to contain the oil or gas, and the trap keeps it from migrating farther or escaping to the surface. Traps can be domes or anticlinal folds, faults, or just impermeable layers.

The rocks at the surface here are mapped as Jurassic dolomite. The equivalent rocks at the Lacq gas field in the Aquitaine Basin to the north serve as reservoir rock, but also are a possible source rock (MacroStrat, 2015; Bahnan et al., 2021). As of my visit in 2025, the roadcut is mostly overgrown.

*Fontaines d'Escot to Défilé d'Escot: Return to N134 and turn right (north); drive 700 m (0.4 mi; 1 min) and pull into the parking area on the left. This is **Stop 55, Défilé d'Escot, Fossils in Schist and Limestone** (43.069306, -0.605861).*

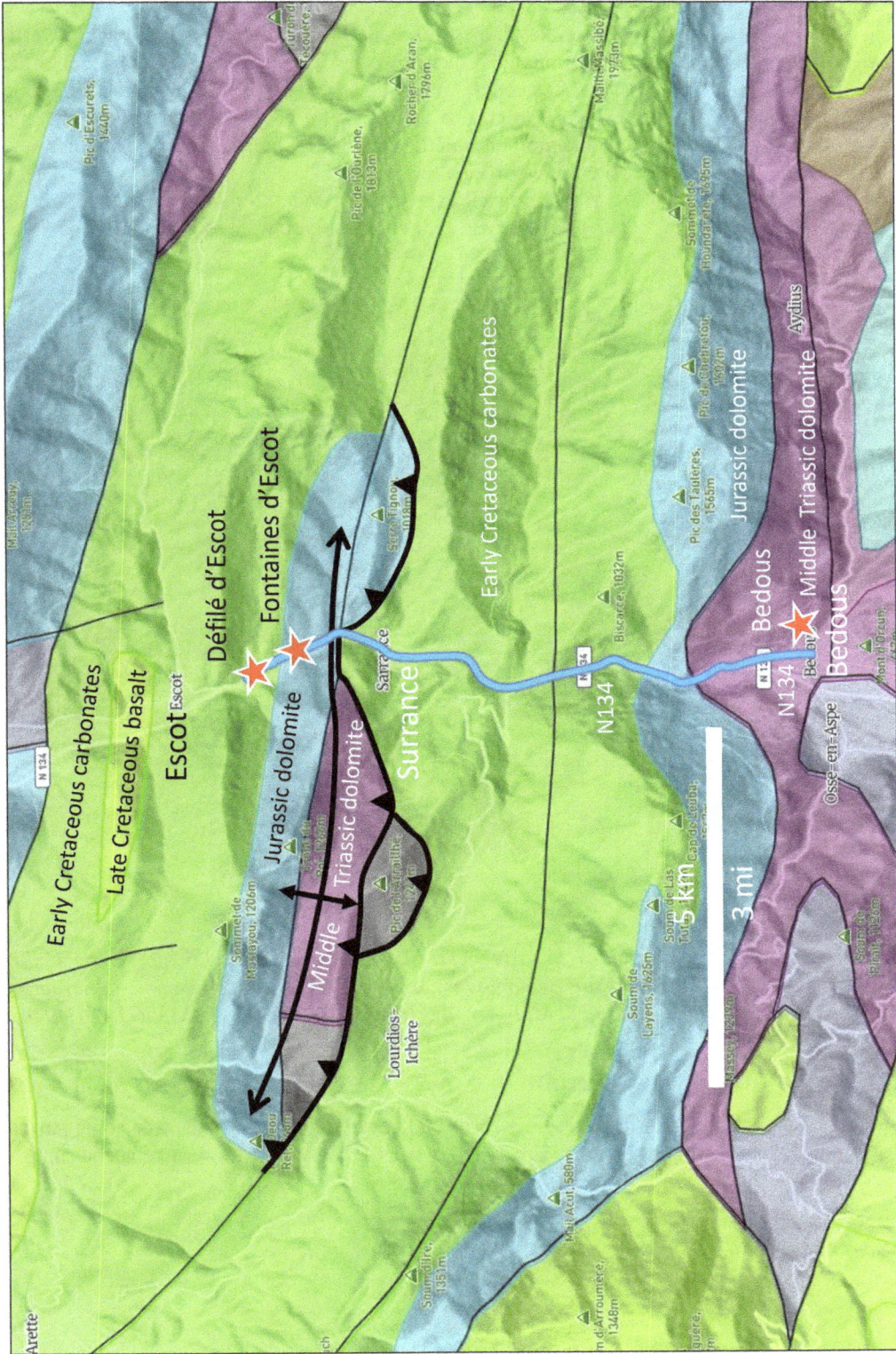

Geologic map of the Bedous to Escot area, including the Fontaines d'Escot and Défilé d'Escot. Base geology from MacroStrat, 2015.

STOP 55 FOSSILS IN SCHIST AND LIMESTONE, DÉFILÉ D'ESCOT

Erosion allows us to view the layers on the flank of an anticline. Here the layers are nearly verti-cal. Over a distance of 100 m (330 ft) along this roadcut we can see about five million years of the Earth's history, from 115 to 110 Ma (Cretaceous). What we see is a gradual transition from shale (oldest to the south) to marl to limestone (youngest to the north). Fossils, a clue to both age and environment of deposition, range from ammonites in the deep water shale to rudists (a mollusc) and sea urchins in the limestone deposited in warm, shallow water.

This is a good roadcut, and the rocks are well exposed, but to get to them you have to cross a busy highway and jump a low concrete barrier. Be very careful if you want to look for fossils here.

Near vertical beds at the Défilé d'Escot Cretaceous roadcut. View northeast.

Défilé d'Escot to Mail Arrouy: *Return to N134 and turn right (north); drive 1.6 km (1.0 mi; 2 min) and turn left (west) at the sign for "Site Géologique"; drive ~120 m (400 ft) on Soum de Bourdettes and park by the display. This is **Stop 56, Mail Arrouy, Jurassic Terraces** (43.081306, - 0.614944).*

STOP 56 JURASSIC TERRACES, MAIL ARROUY

The ridge to the northeast is a succession of scarps and covered zones, all of Jurassic age. At the base is an important thrust fault that puts Jurassic and Triassic rocks over Lower Cretaceous units.

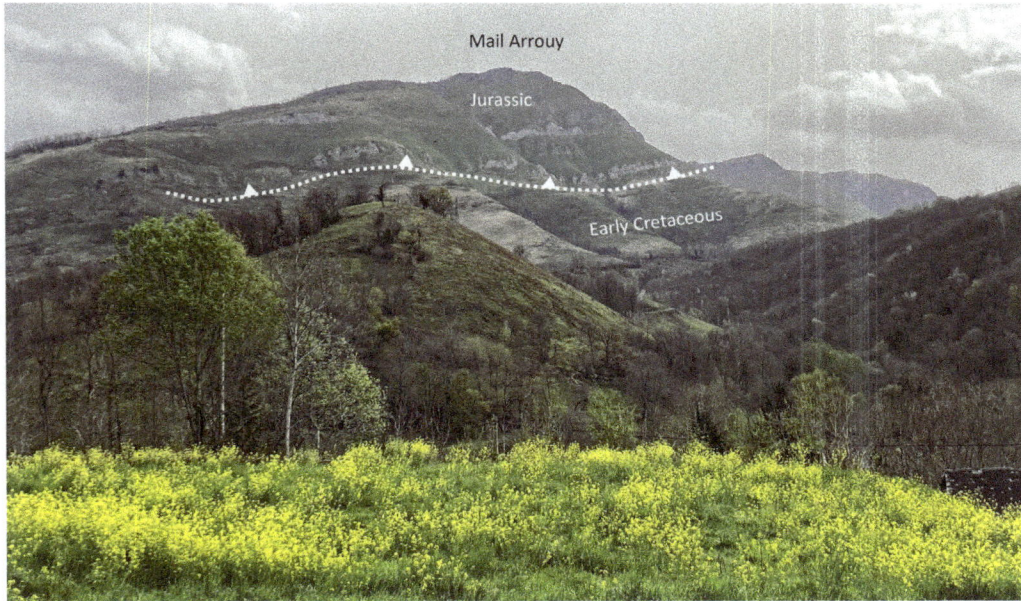

View east from Mail Arrouy stop to Mail Arrouy Peak. Dotted line is a thrust fault.

Mail Arrouy to Courrèges: *Return to N134 and turn left (north); as you enter Oloron Ste Marie, N134 becomes Rue Carrerot; turn right (east) at sign for Centre Ville onto Rue Alfred de Vigny; turn right (east) onto N134/Rue Sadi Carnot; turn right (southeast) onto N134/Rue Jeliotte towrd Pau and Gan; at the roundabout in Herrère de Bas take the 1st exit (south) onto D920 toward Laruns, Arudy, Herrère; drive southeast on D920/Av. Dossau; just before the railroad crossing there is a very small sign, mostly obscured by shrubs, for Site Géologique on the left. There is a small pullout on the left that leads to three tracks. Take the leftmost track maybe 20 m (60 ft) and park. Walk <10 m (<30 ft) to the old abandoned quarry. This is **Stop 57, Courrèges, Ancient Seafloor Volcanism** (43.159333, -0.523194) for a total of 22.7 km (14.1 mi; 25 min).*

STOP 57 ANCIENT SEAFLOOR VOLCANISM, COURRÈGES

At the close of Lower Cretaceous time this area was affected by extension. Faults extending deep into the crust lowered the pressure at depth and induced melting. The resulting magmas moved upward along these open faults and poured out on the ocean floor. Rapid cooling of the basaltic lava formed elongated blobs, or "pillow lavas," that can be seen in this quarry.

In this area the lava flowed downslope to the south. The faults bounded east-west depressions, or basins, in which deep marine black shales and marls were deposited over the seafloor lavas.

Courrèges to Bel Air: *Continue driving southeast on D920/Av. Dossau 1.9 km (1.2 mi) to D416/Av. de Pau; turn left (northeast) toward Stade; turn right (east) onto N134; drive another 3.8 km (2.4 mi) and pull into the large parking area on the right. This is **Stop 58, Bel Air, Overview of the Pyrenees** (43.162225, -0.451840) for a total of 7.2 km (4.5 mi; 8 min).*

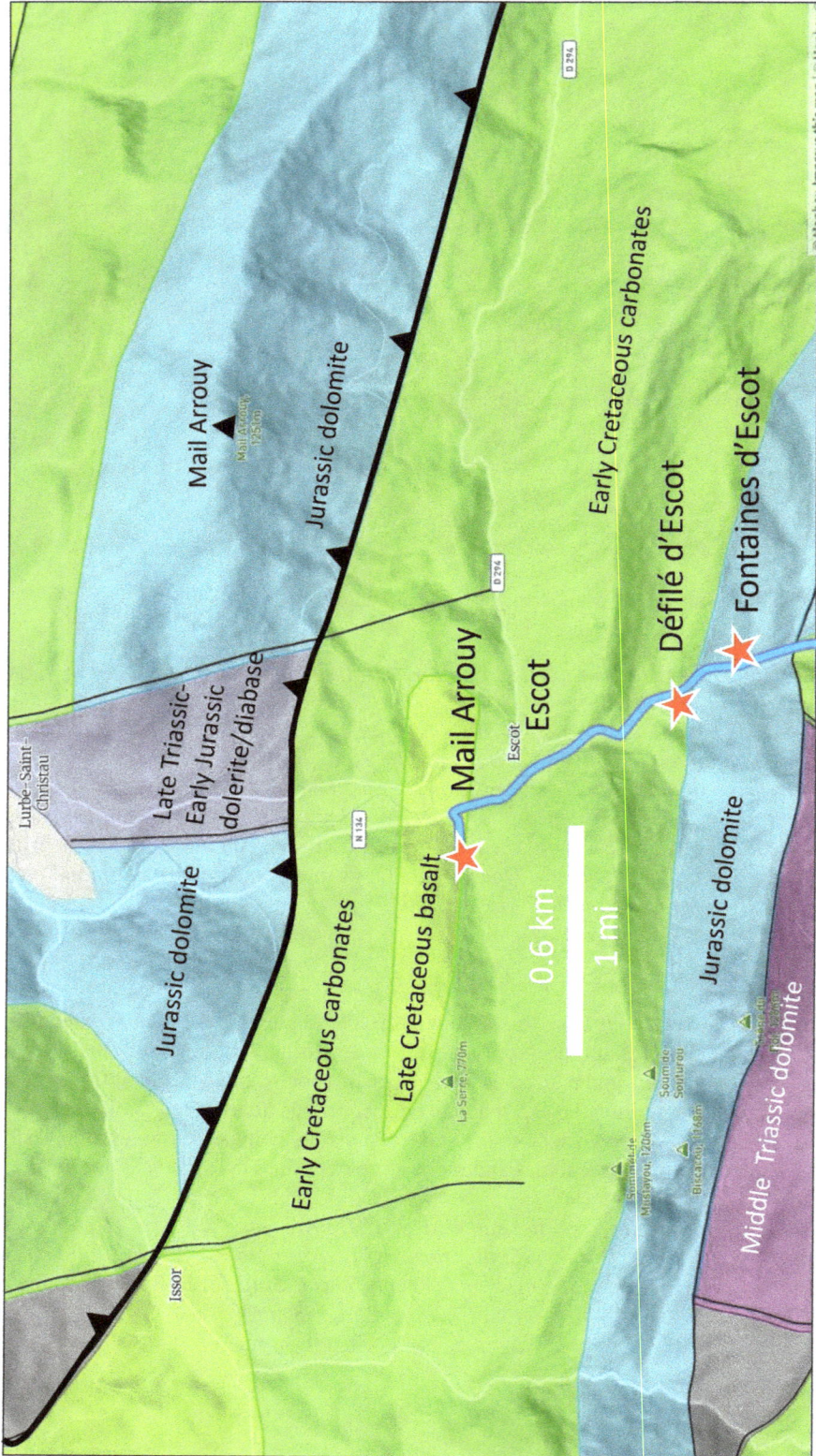

Geologic map of the Défilé d'Escot to Mail Arrouy stops. Base geology from MacroStrat, 2015.

Lower Cretaceous	red Sandstone and black Marl	
Jurassic	Black Dolomite	Black Dolomite Gp (Mano, Breche de Garlin, Lons, Cagnotte Fms)
	interbedded Limestone and black Marl	
	black fetid Dolomite gray Limestone	
	black Limestone	
	Marl buff Limestone	Anhydrite Fm.
Triassic	Variegated Shale	Keuper Gp.

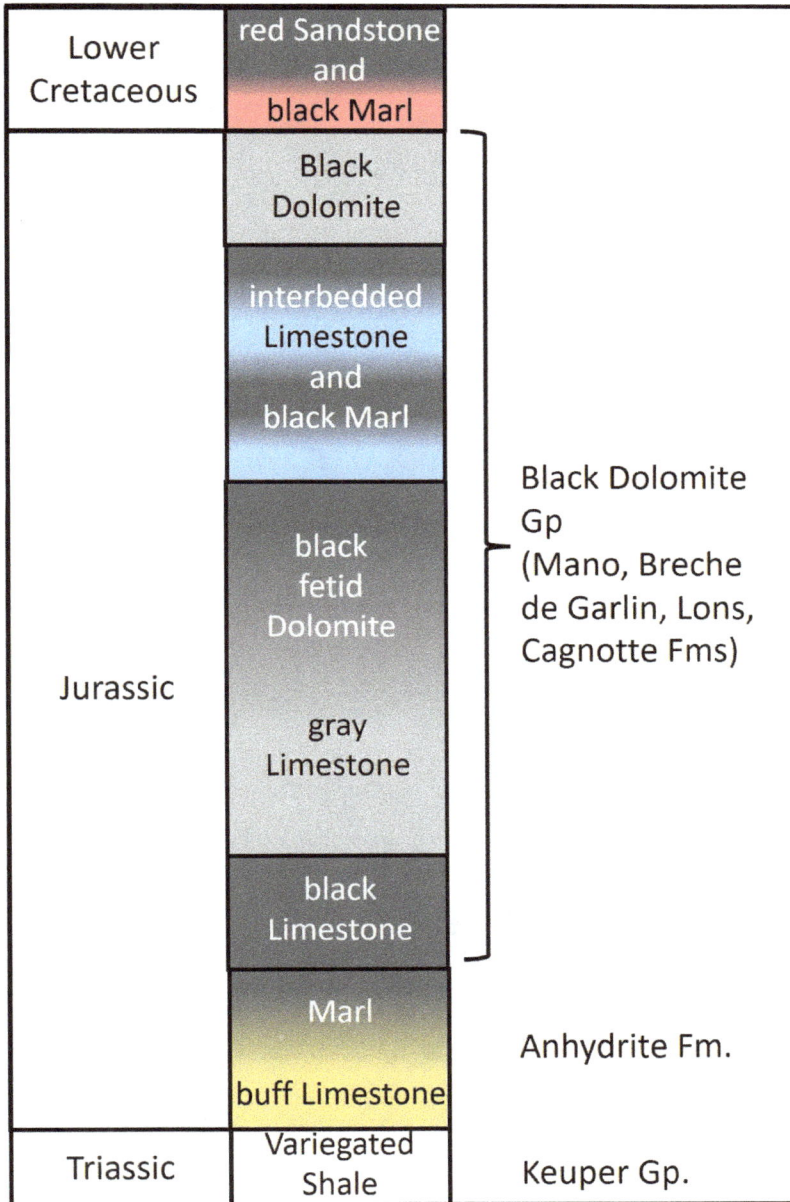

The stratigraphy at Mail Arrouy. Information derived from TransPyrenean GeoDrive panel; Rougier et al., 2016.

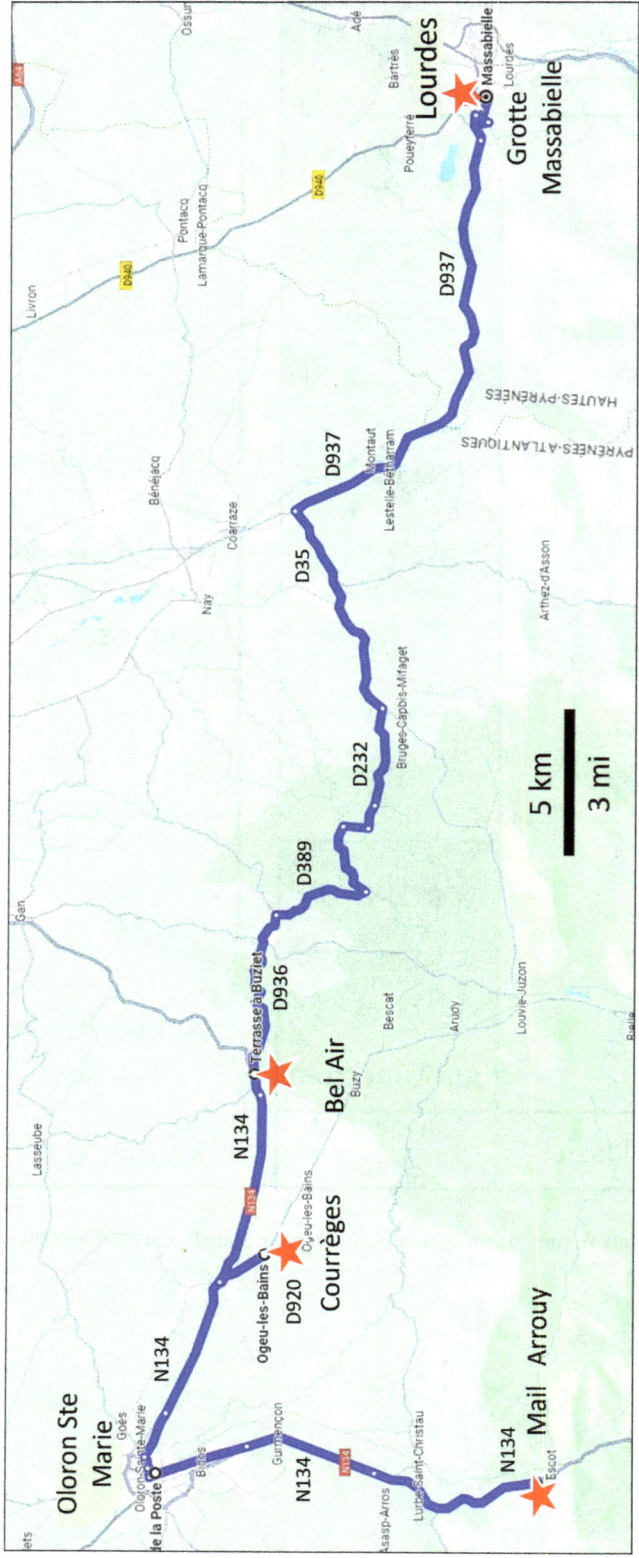

Pillow lava in an abandoned quarry at Courrèges.

STOP 58 OVERVIEW OF THE PYRENEES, BEL AIR

This is the last (or first, if you are going from north to south) of the TransPyrenean GeoDrive panels we will encounter. From this vantage point you can see three informal zones in the Pyrenees: in the far distance is the High Sierra (Axial Zone); in the middle distance is the Bernais Sierra; in the foreground is the Piedmont, or foothills zone, mostly covered by recent alluvium. Rivers carved the valleys, which were enlarged by Ice Age glaciers.

The High Sierra, around 3,000 m (9,840 ft) in elevation, consists of granites, sedimentary rock, and volcanic rocks of Paleozoic age. Occasionally these are covered by Upper Cretaceous rocks.

The Bearnais Sierra, whose peaks rarely exceed 2,000 m (6,560 ft), are mostly Mesozoic carbonates and shales that, affected by folding and faulting, underpin the high relief zone.

The Piedmont, averaging 500 m (1,640 ft) elevation, is in relatively soft, folded Cretaceous rocks. In places these are covered by Tertiary and Quaternary river terraces, alluvium, and glacial moraines, all derived from erosion of the Pyrenees.

Bel Air to Grotte Massabielle: Continue driving east on N134; almost immediately bear right onto D936 toward Rébénacq; stay on D936 to just past Rébénacq, then turn right (south) onto D389/ Rte de Lys; at the T intersection turn left (northeast) onto D287 toward Lys; drive 2.7 km (1.7 mi) and turn right (south) onto Chem. de Badie toward Lys; at the Stop sign in Lys turn left (east) onto D388/Chem. du Bourg; continue straight onto D232; in Bruges turn left (northeast) onto D35/Rue Albert de Saleza; continue on D35 to Igon and turn right (south) onto D937/Av. du Pic du Midi toward Lourdes; continue straight onto D13/Rue de Pau to Lourdes Centre; turn right (south) onto D13/Rte. de Batsurguere; turn right (south) to continue on D13 toward Omex and Ossen; turn left (east) onto 7 Rte de la Forêt; upon entering Lourdes park in spaces along the right side. This is Stop 59, Grotte Massabielle, Lourdes (43.097706, -0.058545) for a total of 43 km (26.7 mi; 52 min). Walk ~50 m (150 ft) north to the cave on the riverbank.

At the Bel Air stop you have a panoramic view of the High Pyrenees.

Stop 59 Grotte Massabielle, Lourdes

Lourdes is famous for its healing waters and for the multiple sightings of the Virgin Mary by Bernadette Soubirous at the Grotto of Massabielle in 1858. The name "Massabielle" comes from the local Bigorre dialect and means "ancient rock."

The Grotto lies at the base of a 27 m (89 ft) high limestone cliff on the bank of the Gave River. It is part of the Mont des Espélugues (the Mountain of Caves) because there are numerous caverns in the limestone. The opening where the Virgin appeared connects to a cavern system that continues up into the mountain and connects to other caves (Berti, Lourdes).

A spring in the Grotto, discovered in 1858, is said to have healing properties linked either to faith or to its chemical composition. Water from the spring in the cave is used by pilgrims and visitors for drinking and bathing. Each year five to six million visitors and pilgrims visit this site. The spring discharges groundwater from a fractured and karsted carbonate aquifer (Dobrzynski and Rossi, 2017).

The area around the Grotto is mainly Lower Cretaceous reef limestones, black marls, shaley marls, and Upper Cretaceous turbidites, all underlain by Jurassic limestones and dolomites. The French Geological Survey has mapped the limestone as Lower Cretaceous Toucasia Limestone (Urgonian facies; Garvemip Project). The Cretaceous and Jurassic rocks form a gentle syncline whose axis is covered by glacial material. Groundwater flows along the axis of the Batsurgure Syncline from west-southwest to east-northeast and discharges into the Gave de Pau River and its tributaries. In the area of the Grotto the river flows west along a steep escarpment, part of the 50 km (30 mi) long active scarp of the Lourdes Fault that extends from Lourdes west to Arette. The water from springs around Lourdes are all low in total dissolved solids and have a slightly alkaline pH, have low temperatures (11° to 13°C, or 52° to 55°F), and contain calcium bicarbonate, all typical for groundwater systems developed in carbonate-dominated bedrock (Dobrzynski and Rossi, 2017).

At Lourdes, deformation studies indicate a two-phase structural history beginning with gravitational spreading, followed by a compressive thrusting (Aerden, 1995).

The Upper Cretaceous limestone at Lourdes is riddled with caverns. These are just above Rte de la Forêt.

The Grotte de Massabielle at Lourdes, just below the Basilique Notre Dame du Rosaire.

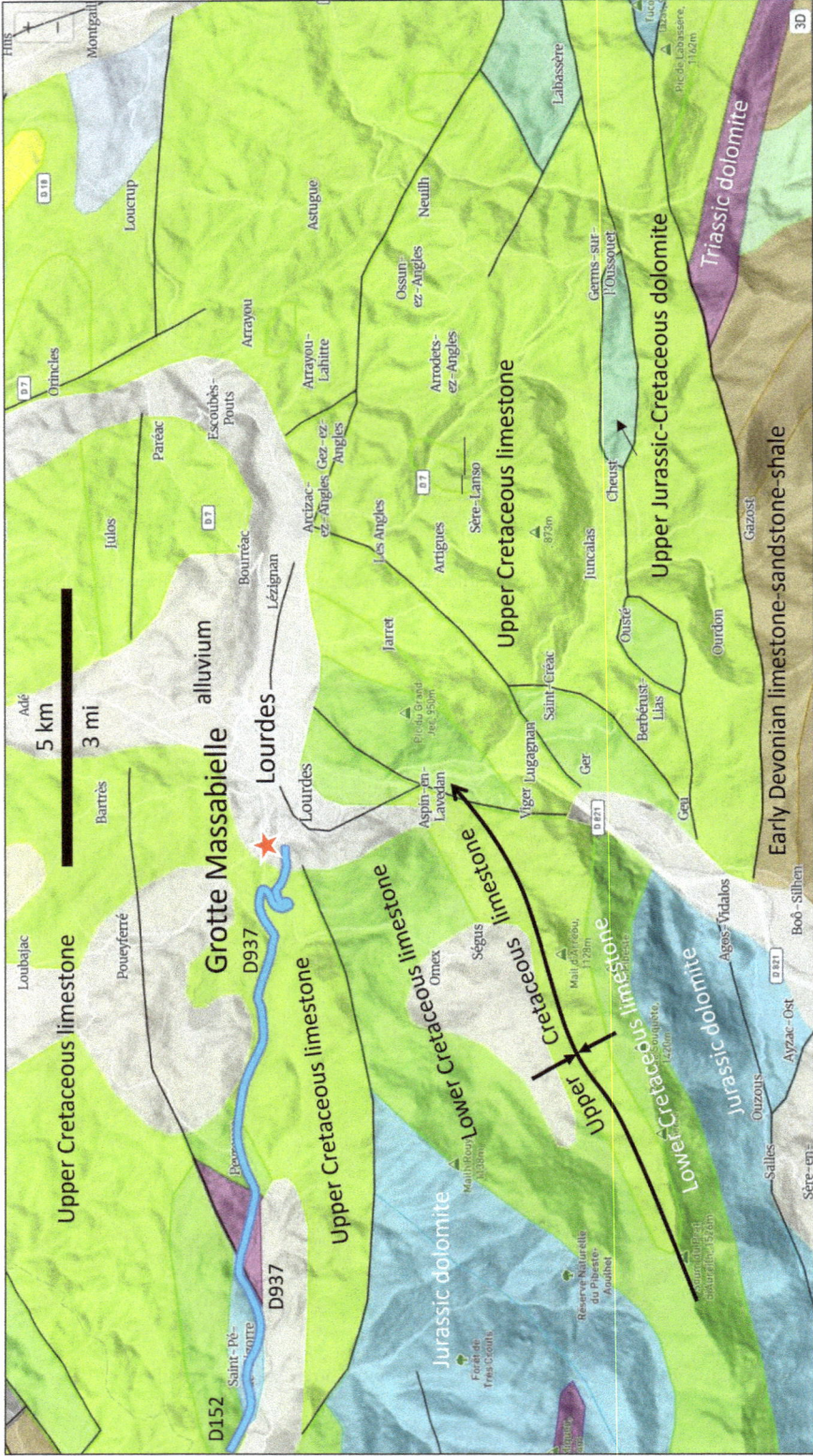

Geologic map of the Lourdes area. Base geology from MacroStrat, 2015.

Visit

The Grotte Massabielle is open every day year-round, and access is free.

Address: Avenue Mgr Théas, Sanctuaire ND de Lourdes, 65100 LOURDES
Parking: The car park closest to the grotto is located at the top of Boulevard de la Grotte, near
            Boulevard du Lapacca. Street parking is available but limited.
Websites: https://en.lourdes-infotourisme.com/explore/have-a-spiritual-experience/explore-the
            -sanctuary/the-unmissables/the-grotto-of-apparitions/ and https://www.lourdes-france
            .org/horaires/

Grotte Massabielle to Grotte de Gargas: *Return west on 7 Rte de la Forêt; turn right (north) onto D13/Batsurguere; turn right (east) onto Rue de Pau toward Lourdes Centre; in central Lourdes turn right (south) onto Bd. de la Grotte; turn left (east) onto Pl. Jeanne d'Arc; continue straight onto Bd. Du Lapacca; at the roundabout take the 3rd exit (north) onto Av. Victor Hugo; at the roundabout take the 1st exit (east) onto D937; at Montgaillard turn right (south) onto D935/Rue du 11 Novembre; in Pouzac turn left (east) onto D26/Rou de l'Adour; in Haut de la Côte turn left (northeast) onto D938/Rte de Toulouse; stay on D938 to Anères; turn right onto D626 toward Nestier/Nistos; bear left (east) onto D26; in Nestier continue straight onto D75; in 370 m (1,200 ft) continue straight onto D26; turn right (south) at the sign for Gorttes de Gargas/Nestploria; drive 1.1 km (0.6 mi) and park. This is* **Stop 60, Grotte de Gargas, Saint Gaudens** *(43.055409, 0.536347) for a total of 63.6 km (39.5 mi; 1 hr 13 min).*

Stop 60 Grotte de Gargas and Prehistoric Cave Art, Saint Gaudens

The Gargas Cave is famous for prehistoric cave art, in particular the negative hand prints and carved animal bones dated from 27,000 to 15,000 years before present. The hands are either red (limonite) or black (manganese oxide), where mineral powder mixed with animal fat was sprayed from the mouth around a hand pressed against the wall. Some have one or more fingers missing: it is speculated that this could be because of disease, frostbite, ritual amputation, or the bending of one or more fingers. Many figurative engravings were found in other parts of the cave. These depict horses, bison, stags, aurochs, ibex, mammoth, and birds. Human remains, as well as bones of cave bears and hyenas, have been found.

The cave was first described in writing in 1575, but it was not until 1906 that the first hand print was discovered. Grotte de Gargas was declared a historical monument in 1910, but it was not opened to the public until 1950.

The Grotte de Gargas is a karst feature developed in Jurassic limestone (according to the tour guide) or Early Cretaceous limestone (MacroStrat, 2015; French Geological Survey, Garvemip Project). Originally there were two large openings on the northwest-facing hillside. The lower was the entrance to the broad and low Gargas I, whereas the upper led to the narrow and taller Gargas II. The two caves are connected by a passage that was blocked by a clay plug that has since been excavated (Showcaves of France; Wikipedia, Caves of Gargas). The upper cave is developed along vertical fissures and has a high ceiling; the lower cave is developed along a horizontal contact and has a low ceiling. The horizontal contact has limestone over a clay surface. Bedding on the hillside appears to be dipping about 70° to the west.

The upper opening is now the main entrance and the lower opening is the exit.

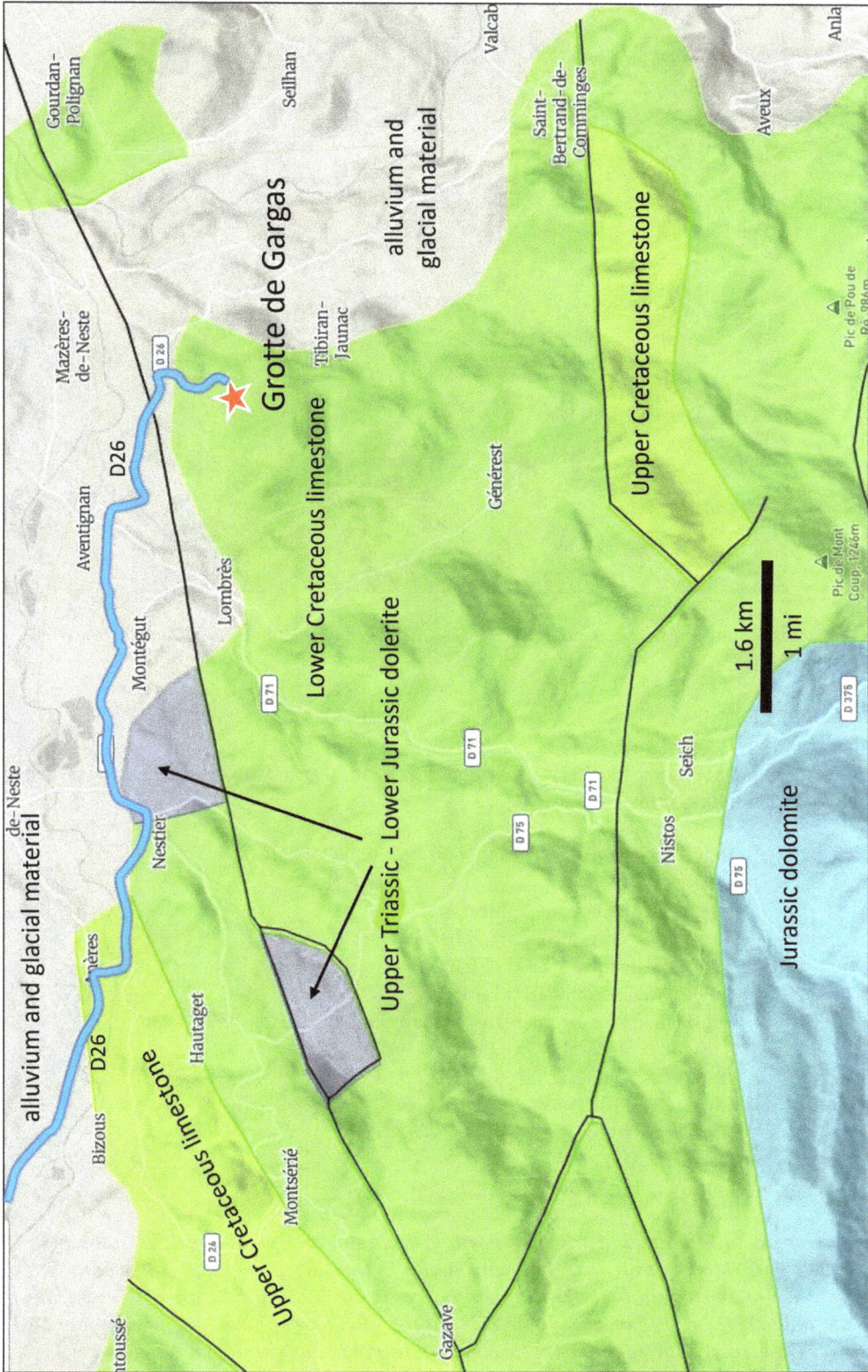

Geologic map of the Grotte de Gargas area. Base geology from MacroStrat, 2015.

Grotte de Gargas. Photo courtesy of Félix Régnault (1847–1908); https://commons.wikimedia.org/wiki/File
:Grotte_de_Gargas.jpg

VISIT

No special equipment is required. Wear good shoes and dress warm. Unfortunately, they no longer allow photography underground.

Address:	Route départementale 26, 65660 AVENTIGNAN
Phone:	+33-05 62 98 81 50
Email:	gargas.nestploria@nestploria.fr
Website:	https://www.grottesdegargas.fr/
Hours:	May – June daily 10:30 – 17:30
	July – August daily 10:00 – 18:00
	September – April Tuesday through Sunday 10:30 –17:30
	Closed 25 December – 1 January
Entrance Fee:	Adults 12.50 € (2025)
	Children 6–16 7.50 €
	Children <5 free
	Disabled 10.00 €
	Family (2+2) 33.00 €

Grotte de Gargas to Montcalibert: Return north to D26 and turn right (east); take the first left (northeast) toward Montréjeau; turn left (northwest) onto D72; turn right (east) onto Rte du Montréjeau; continue straight onto D34E; turn right onto D817 toward Centre Ville; take the 1st left (north) onto D34E; continue straight on D34; at the roundabout take the 2nd exit onto D817/ Av. de Saint Gaudens; at the roundabout take the 4th exit onto D645 ramp to A64 toward Tarbes/

*Foix/Toulouse; take the exit onto A64/E80 toward Foix/Toulouse; take Exit 20 toward Foix; at the roundabout take the 3rd exit (south) onto D117; at Saint Lizier turn left (east) onto D103; cross the Salat bridge and bear right onto D3 toward St. Girons and Centre Ville; turn left (north) onto Vignes de l'Evêcher; take the first right (east) onto La ville; turn left to stay on La ville; turn right (east) onto Rte de Montjoie; at the stop sign turn left (north) onto Carrère de Naout and toward Musée la Naïf; continue straight on Peyraute; pull over on the right at the intersection with Caoue. This is **Stop 61, Montcalibert Pyrenees View** (332 Caoue, or 43.0041818, 1.158998) for a total of 67.9 km (42.2 mi; 58 min).*

STOP 61 MONTCALIBERT PYRENEES VIEW

This is a scenery stop. We are in the French Piedmont. From here you get a panoramic view south to the Pyrenees. The peaks on the skyline, from east to west, are Mont-Roig (2,840 m or 9,320 ft), Mont Valier (2,786 m or 9,140 ft), and Tud de Maubèrme (2,880 m or 9,450 ft).

Bedrock at this stop is Jurassic dolomite.

View north to the Pyrenees from Tuc de Montcalibert.

*Montcalibert to Cascade d'Alzen: Return south on D218; at the roundabout turn left (east) onto D117 toward La Bastide-de-Sérou; about 3 km (1.8 mi) past La Bastide-de-Sérou turn right (south) onto D221; turn left (south) onto D21 toward Alzen and Cascade; drive another 1.1 km (0.7 mi) and pull over on the right at the sign for Cascade d'Alzen. This is **Stop 62, Cascade d'Alzen** (42.995144, 1.466741) for a total of 31.3 km (19.4 mi; 29 min). Walk <200 m (650 ft) on the trail to the falls. Listen for the birds!*

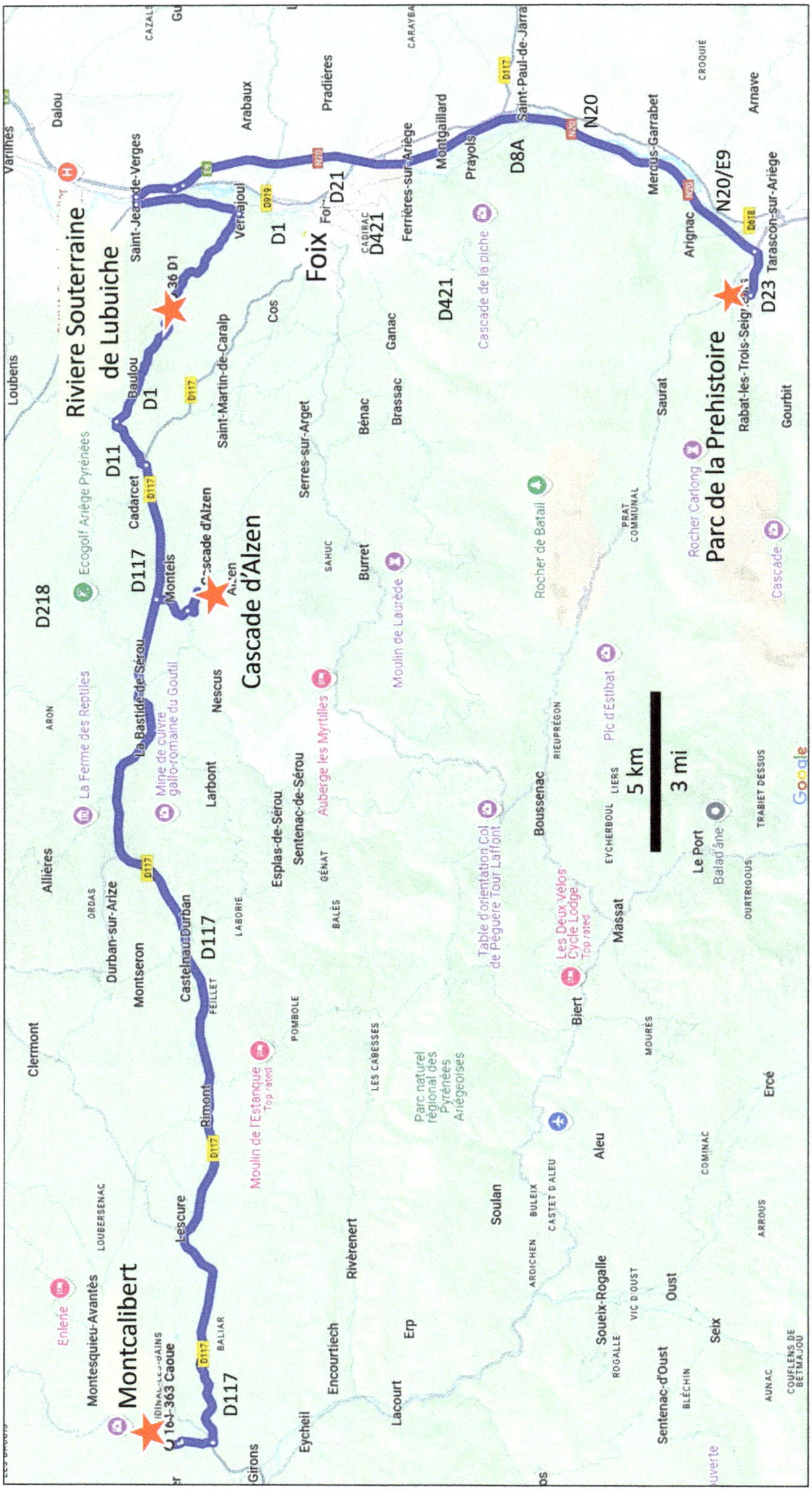
Montcalibert
Cascade d'Alzen
Riviere Souterraine de Lubuiche
Parc de la Prehistoire
Foix
D117
D1
D421
N20
N20/E9
D8A
D23
D218
D11
D21
5 km
3 mi

STOP 62 CASCADE D'ALZEN

According to MacroStrat (2015), the photogenic Alzen waterfall drops over Middle to Upper Triassic dolomite. This natural waterfall is located in the Arize Massif, northwest of the town of Alzen and is accessible by a short path. Starting at 612 m (2,008 ft) above sea level, Alzen Creek drops about 43 m (141 ft). The creek is a tributary to the Arize River.

This is one of the most peaceful walks I have had. There were no people and few car sounds. Listen for the birds.

Cascade d'Alzen.

Cascade d'Alzen to Rivière Souterraine de Lubuiche: *Return north to D117 and turn right (east) toward Foix; drive 4 km (2.5 mi) and turn left (northeast) onto D11 toward Loubens/Baulou; turn right (southeast) onto D1 toward Baulou/Riviere Souterraine; drive 4.3 km (2.7 mi) and pull into parking area on the left. This is* **Stop 63, Rivière Souterraine de Lubuiche** *(43.003367, 1.574121) for a total of 12.5 km (7.8 mi; 15 min).*

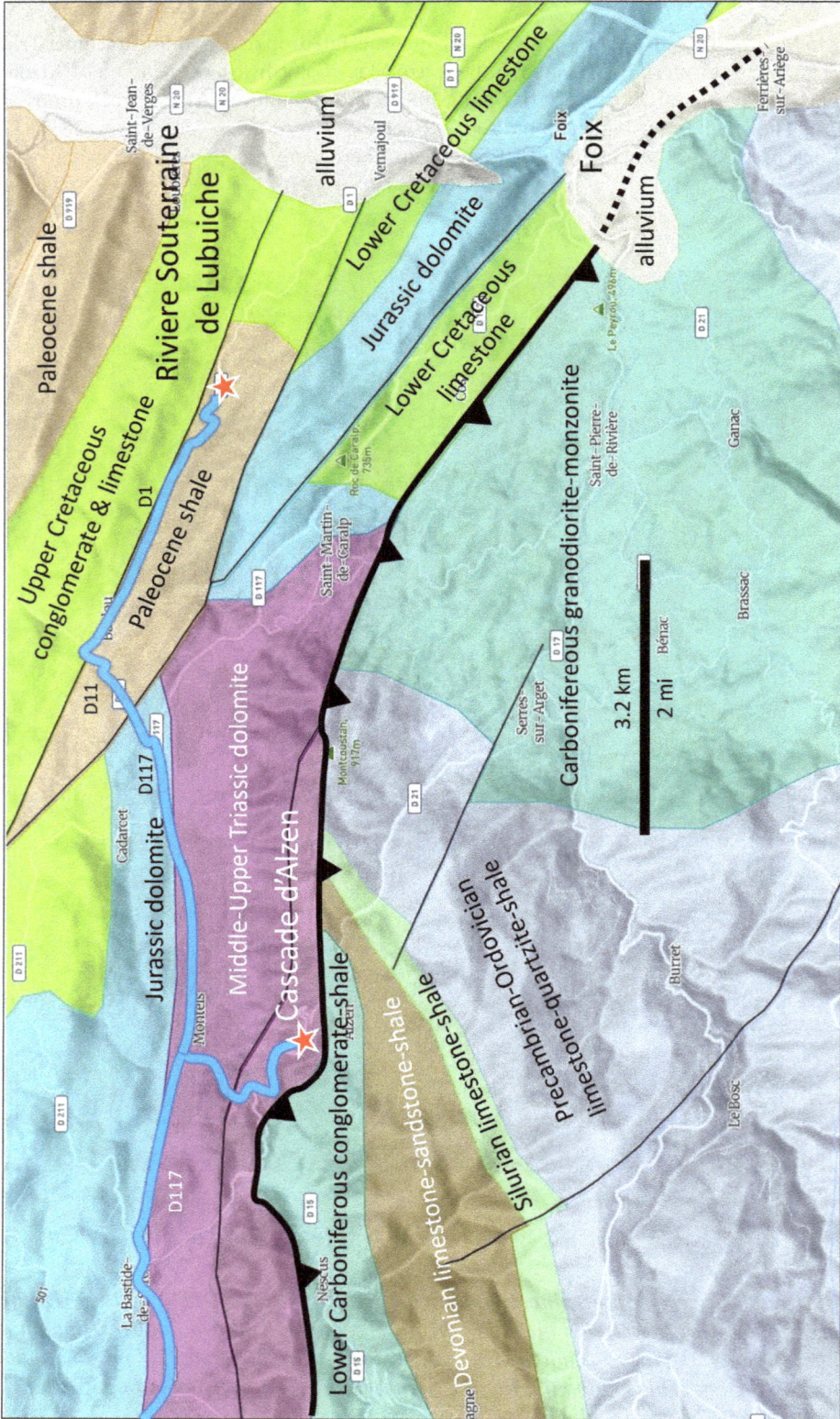

Geologic map of the Cascade d'Alzen to Rivière Souterraine area. Base geology from MacroStrat, 2015.

Stop 63 Rivière Souterraine de Lubuiche

Six km (3.7 mi) northwest of Foix on the D1, the Rivière Souterraine de Lubuiche, also known as the "Venice of Ariege," has been open to the public since 1938. The tour extends over 1,500 m (0.9 mi) and is approximately 60 m (200 ft) below the surface.

Discovered in 1908 by a local doctor, this is Europe's longest navigable underground river, with over 3,800 m (2.4 mi) of galleries explored. Visitors will see impressive stalagmites, stalactites, and even an underground waterfall: the tour ends at a waterfall known as the Cascade Salette, which spills into a shimmering turquoise pool.

Prehistoric flint weapons and reindeer antlers have been found in the cavern, indicating early humans lived, or at least visited, here.

Visit

Tour the underground river seated in a 12-person boat with guide. Trips last about 75 minutes (15 min in the boat; 60 min on land). Sweaters are recommended. Closed shoes are a must, as are warm clothes, as the cave is a constant 13°C (65°F). You must be able to navigate 240 steps during the tour.

Reservations by phone or online are strongly recommended.

Address: Labouiche, 09000 BAULOU FRANCE
Phone: +33 5 61 65 04 11
Email: riviere.labouiche@nordnet.fr
Website: https://www.labouiche.com
Hours: 1 April – 30 September: 10–11 and 14:00–16:30
           July-August: every day 9:30 – 17:00.
           October – 11 November: Saturdays, Sundays, public holidays, and school holidays from
           10:00 – 11:00 and 14:00–16:30; during the week 14:00 – 16:00
           Closed on Mondays

Please see the website for current entrance fees: https://www.labouiche.com/?g-path=%2Fcart%2Factivities%2FpMN9SEHFXVGqtL3mC76DljB5AyK2axUc81bo

Rivière Souterraine de Lubuiche to Parc de la Préhistoire: *Return to D1 and turn right (southeast); stay on D1 to Foix; turn left (south) onto D117; at the roundabout on the south side of Foix take the 1st exit (west) toward N20/Andorre; turn left to merge onto N20 south toward Andorre; at the roundabout take the 2nd exit (southwest) toward Parc de la Préhistoire; take D23 for 1.5 km (1 mi) and at the roundabout take the 1st exit and park on the left. This is **Stop 64, Parc de la Préhistoire** (42.854944, 1.57513) for a total of 23.7 km (14.7 mi; 24 min).*

Underground river of Labuiche. Photo courtesy of Tylwyth Eldar, https://commons.wikimedia.org/wiki/File :Rivi%C3%A8re_souterraine_de_Labouiche_07.jpg

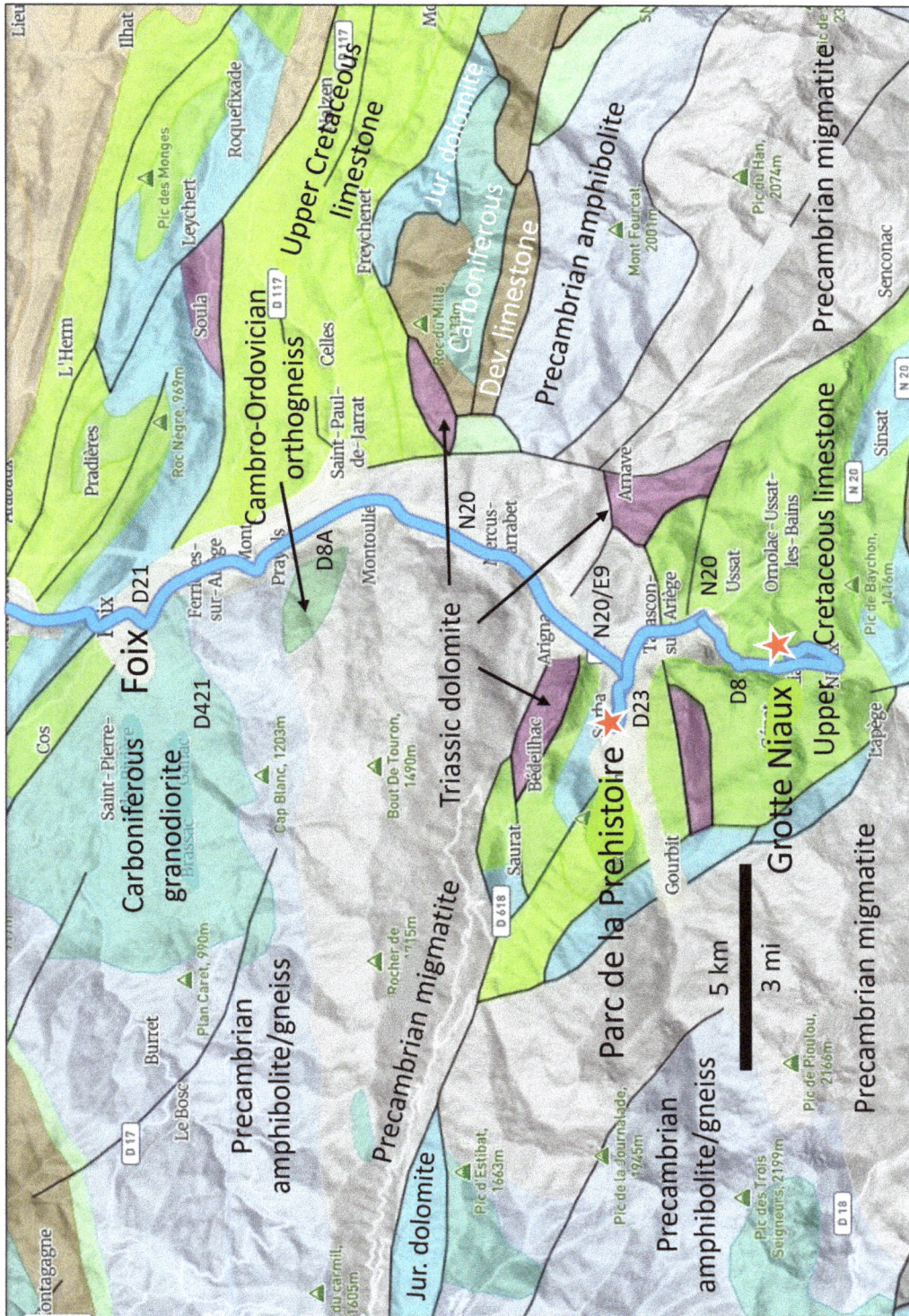

Geologic map of the Prayols-Cascade de la Piche area. Base geology from MacroStrat, 2015.

Stop 64 Parc de la Préhistoire, Surba

This is a Prehistory Museum and interpretive center in the village of Surba. It is meant to feel like you are entering a cave. There are model caves, lifelike reconstructions, and films.

Audioguides are available in French, English, Spanish, Catalan, German, and Dutch. These allow you to tour the prehistory of the French Pyrenees and Ariège at your own pace. Woolly mammoth, steppe bison, megaloceros deer, and cave lion are all brought to life. These prehistoric animals, long extinct, are presented in realistic settings. All the senses are involved, including the sense of smell! Covering an area of 500 m² (5,400 ft²), the "Giants of the Ice Age" exhibit presents a realistic depiction of the environments in which large prehistoric fauna lived during the Ice Age. Hunted for their meat, their fur, or their bones, these animals were admired, feared, and sources of inspiration for early humans. You enter the habitat of our ancestors and get to discover the environment they were in on a daily basis. A playful animated display, "the challenge of the giants," allows you to measure yourself against these prehistoric giants.

Megaloceros display. Photo courtesy of Tylwyth Eldar, https://commons.wikimedia.org/wiki/File:Parc_de_la _pr%C3%A9histoire_-_Megaloceros_01.jpg

Visit

Reservations are not required. This is a wonderful museum and outdoor park for the whole family. Displays include cave art, human history, and Ice Age mammals of the Pyrenees.

Address: Rte de Surba À Banat, 09400 Tarascon-sur-Ariège, France
Phone: +33561051010
Email: info@sites-touristiques-ariege.fr
Website: https://www.sites-touristiques-ariege.fr/parc-de-la-prehistoire/
            See website for latest entrance fees.

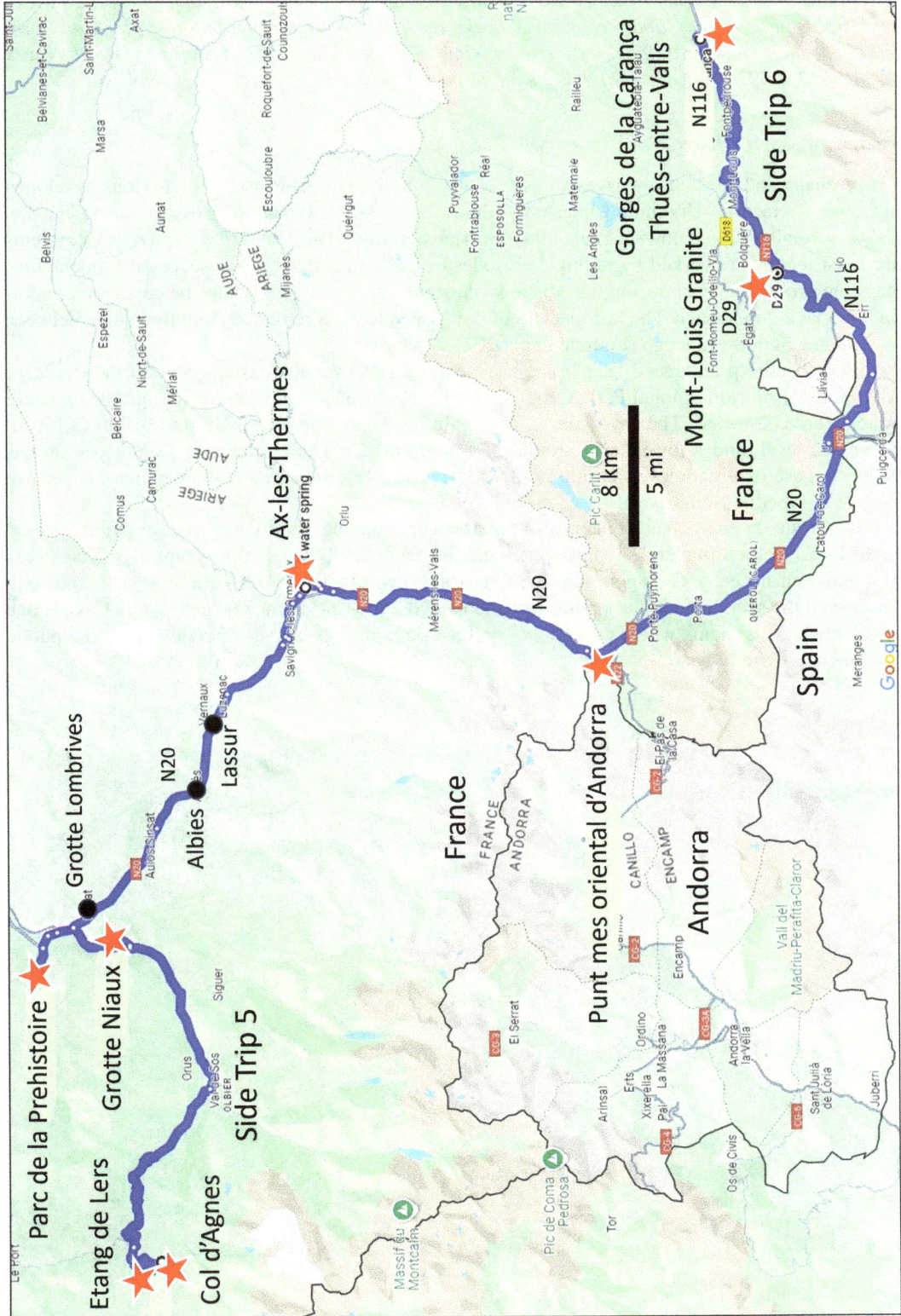

Side Trip 6

Gorges de la Carança
Thuès-entre-Valls

Mont-Louis Granite

France

N20

N116

Spain

Ax-les-Thermes

hot water spring

N20

N20

Grotte Lombrives

Lassur

Albies

N20

Parc de la Prehistoire

Grotte Niaux

Side Trip 5

Etang de Lers

Col d'Agnes

France

Punt mes oriental d'Andorra

Andorra

5 mi

8 km

*Parc de la Préhistoire to Grotte Niaux: Return east on D23; at the roundabout take the 1st exit onto N20 south; at the Sabart roundabout take the 1st exit (southwest) onto D8; in Niaux turn left (east) onto D56/La Pujade and drive to Niaux Cave parking. This is **Stop 65, Grotte Niaux** (42.819629, 1.593695) for a total of 8.9 km (5.6 mi; 12 min).*

STOP 65 GROTTE NIAUX AND PREHISTORIC CAVE ART

Grotte Niaux and Grotte Lombrives are part of a system of caves up to 15 km (9 mi) long developed in Lower Cretaceous Urgonian Limestone in the Cap de la Lesse Massif between the Ariège and Vicdessos valleys (Bakalowicz et al., 1984; Sorriaux et al., 2018). The Ariège Valley cave systems are karst features reworked by multiple episodes of glacial meltwater over successive glaciations. Karsting probably began during the Miocene-Pliocene (8.6 ± 2.1 Ma). Later the caves were modified by glacial meltwater. The last glaciers of the Ariège Icefield retreated from the valley between 19,000 and 18,000 years ago (Laurent et al., 2021).

Grotte Niaux (pronounced "neo") is a large natural cavern at the eastern edge of the Pyrénées Ariégeoises Natural Regional Park. Grotte de Lombrives is also in this mountain and the two cave systems are connected. The caves are on two main levels, an upper level around 650 m (2,130 ft) above sea level, and a lower level around 550 m (1,800 ft). The entrance to Lombrives cave is between these two main levels at 600 m (2,000 ft). The levels are connected by numerous vertical shafts (Wikipedia, Lombrives).

The Grotte de Niaux has a history of human occupation and vividly drawn cave paintings executed in a black-outline style typical of the period from 17,000 to 11,000 years ago. Niaux was first explored by Félix Garrigou, anthropologist and hydrologist, in 1869. In 1925 J. Mandeman discovered the section with black paintings and named it the Cartailhac Gallery. Niaux Cave is one of the few cave systems where exceptional prehistoric paintings can still be viewed by the public (Wikipedia, Cave of Niaux).

Entrance to Niuax Cave.

The Salon Noir ("Black Hall") panel in Niaux Cave contains Stone Age sketches in black and red of male and female bison, represented in the upper part of the panel, and of Przewalski's horses on the lower part of the wall. The panel also contains two goats. The panel is dated 13,000 years old.

Cave art in Niaux Cave includes surprisingly realistic depictions of horses and bison, local wildlife at the end of the last Ice Age.

Grotte Lombrives is 8 km (5.0 mi) long in Cap de la Lesse, a limestone mountain. Lombrives cave has two large chambers. The Cathedral Chamber, 80 m (260 ft) high, is on the standard tour. It is said that the Cathedral of Notre-Dame in Paris would fit inside this cavern. The larger Salle de l'Empire de Satan ("Chamber of the Empire of Satan") is at the end of the long cave tour, approximately 4 km (2.5 mi) from the entrance. Lombrives cave has a variety of stalactites, stalagmites, pools, cave pearls (cave pool oolites), helictites (hairlike or tubelike growths), and calcite crystals.

Excavations by Félix Régnault confirmed that humans occupied this cave during the Neolithic. The skeleton of a Bronze Age man has been found. Between the 12th and 14th centuries the "heretic" Cathars hid at the site, as did Huguenots when they were persecuted in the 1720s (Wikipedia, Lombrives).

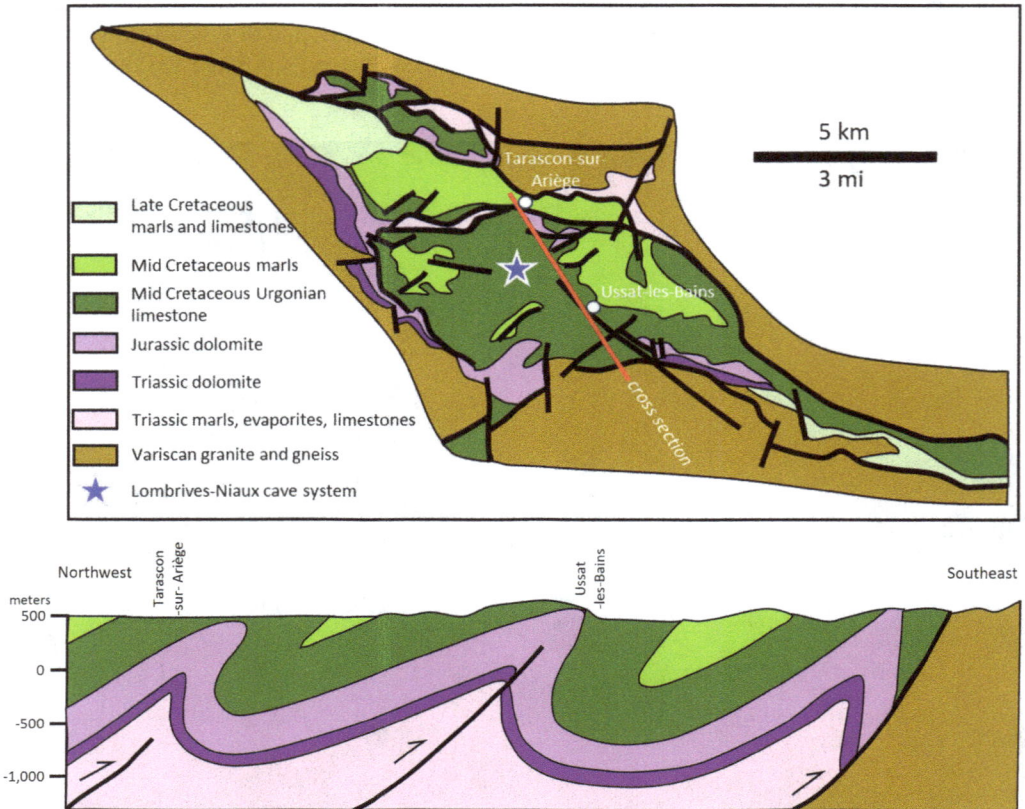

Geologic map and cross-section of the Grotte de Lombrives and Grotte Niaux area. Modified after Laurent et al., 2021.

Visit

You must book in advance online: https://sites-touristiques-ariege.oxygeno.fr/?lang=en.

Arrive at least 15 minutes early. The caves have uneven and wet floors so bring sturdy shoes. They are cool, so bring warm clothes. The Niaux Cave tour lasts ~ 1 hr 45 min.

Address: Grotte de Niaux, Route Nationale 20, 09400 Ussat-les-Bains (follow signs)
Phone: +33 561 05 10 10
Contact: https://www.sites-touristiques-ariege.fr/en/get-in-touch/
Website: https://en.pyrenees-ariegeoises.com/offers/niaux-cave-niaux-en-4309965/
Days: The cave is not open every day. See website for open days.
Fee: See website for current entry fees.

Side Trip 5, Grotte Niaux to Étang de Lers and Col d'Agnès: *Return to D8 and turn left (southwest); continue on D8 to Val-de-Sos; turn right (northwest) onto D18/Rte de Suc; follow the signs to Étang de Lers/Col d'Agnès and turn left (west) onto D8F; at 29.4 km (18.3 mi; 41 min) pull over on the right at roadcut. This is* **Stop ST5.1, Étang de Lers Peridotite** *(42.803164, 1.376058).*

If you prefer to skip Side Trip 5:

Grotte Niaux to Bassin de Ladres, Ax-les-Thermes: *Return north to N20 in Sabart; at the round-about take the 1st exit (south) onto N20; take the exit toward Ax-les-Thermes; at the roundabout take the 3rd exit (east) onto Rte d'Espagne; at the town center roundabout take the 2nd exit and park on the right. This is* **Stop 66.1, Bassin de Ladres** *(42.719863, 1.839711) for a total of 30.6 km (19.0 mi; 32 min).*

SIDE TRIP 5 ÉTANG DE LERS, TYPE LOCALITY FOR LHERZOLITE

Despite being in an area with Precambrian and Paleozoic outcrops, we are still in the North Pyrenean Zone, that is, we are still on the European Plate north of the North Pyrenean Thrust. This area comprises Variscan crystalline massifs and Jurassic and Cretaceous sediments, locally metamorphosed during the Pyrenean Orogeny (Satterfield et al., 2019).

Lherzolite is an ultrabasic (dark) igneous rock dominated by olivine with clinopyroxene and orthopyroxene in equal proportions. Lherzolites are a variety of peridotite, which is the main component of the Earth's upper mantle (Haldar, 2017; Strekeisen, 2006–2020). It is not at all common at the surface.

The Étang de Lers (Lers Pond, formerly Lherz Pond) is the type locality for lherzolite. It is the best known of 40 or so small mantle slices emplaced in the Northern Pyrenean Zone, a narrow zone just north of the North Pyrenean Fault. The North Pyrenean Fault is the main tectonic boundary between the European and Iberian plates in this area.

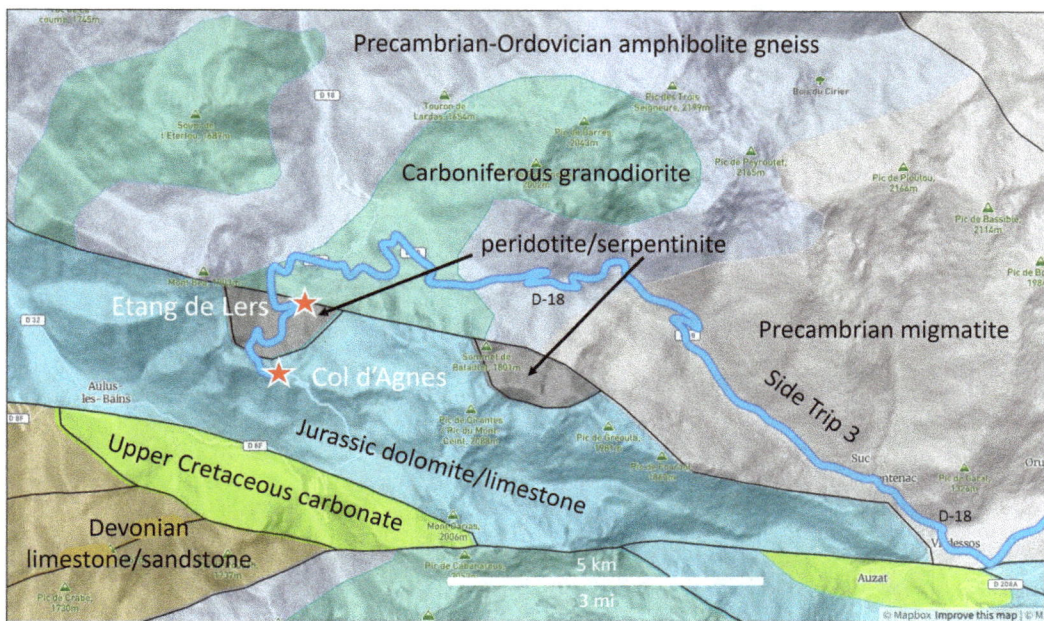

Geologic map of the Étang de Lers–Col d'Agnes area. Base geology from MacroStrat, 2015.

STOP ST5.1 ÉTANG DE LERS PERIDOTITE

The emplacement of the Lers peridotite was the result of middle Cretaceous counterclockwise rotation of the Iberian plate with respect to Europe. The stepwise rotation caused lithospheric thinning associated with successive opening of elongated pull-apart basins (Strekeisen, 2006–2020). These rocks were originally considered to be classic mantle peridotite, but recent work indicates they are depleted mantle harzburgites (Satterfield et al., 2019). Harzburgite is a variety of peridotite consisting of 40% to 90% olivine (a glassy green mineral; the gem variety is peridote) formed by partial melting of a more pyroxene-rich lherzolite.

If that is too technical, here is the important part. Peridotites are thought to be ancient subcontinental mantle that was uplifted and exposed on the seafloor during Early Cretaceous pre-Alpine extreme crustal stretching and thinning. Once on the seafloor, the mantle material was subject to submarine weathering. Submarine weathering is when rocks and minerals are broken down by seawater. This particular altered seafloor then got caught up in the collision between Iberia and Europe and was squeezed up, along with all the other rocks and sediments, as lenses and pods of Lherzolilte during the Pyrenean mountain-building event.

Roadcuts expose Lherzolite near Étang de Lers. Street View northwest.

Typical blue-green Lherzolite from near Étang de Lers. Photo courtesy of Asabengurtza, https://commons .wikimedia.org/wiki/File:Lherzolite_Lers.jpg

The rocks are well exposed in the hillsides and roadcuts around Étang de Lers. The rocks contain green pyroxene (a silicate mineral common in igneous and metamorphic rocks), often as cm-wide veins. Flow banding is indicated by deformed pyroxene veins. In places there are veins of black serpentinite (Satterfield et al., 2019).

Marble, some massive and some brecciated (shattered), hosts the peridotites. The road zig-zags across the peridotite-marble contact several times. The marble was deposited as mid-Cretaceous limestone in the Aulus Basin. The limestone was later subjected to low pressure–high temperature metamorphism as a result of being just above the hot mantle from 110 to 85 Ma. In the Lhers Massif, the last stage of peridotite exhumation is characterized by extensive fragmentation (creation of breccia) of the outer margin of the peridotite body and in the intermingled marble. Some of these breccias are tectonic, brought about by the mountain-building process itself, whereas others are probably a result of submarine debris flows (Lagabriel et al., 2016; Satterfield et al., 2019; Strekeisen, 2006–2020).

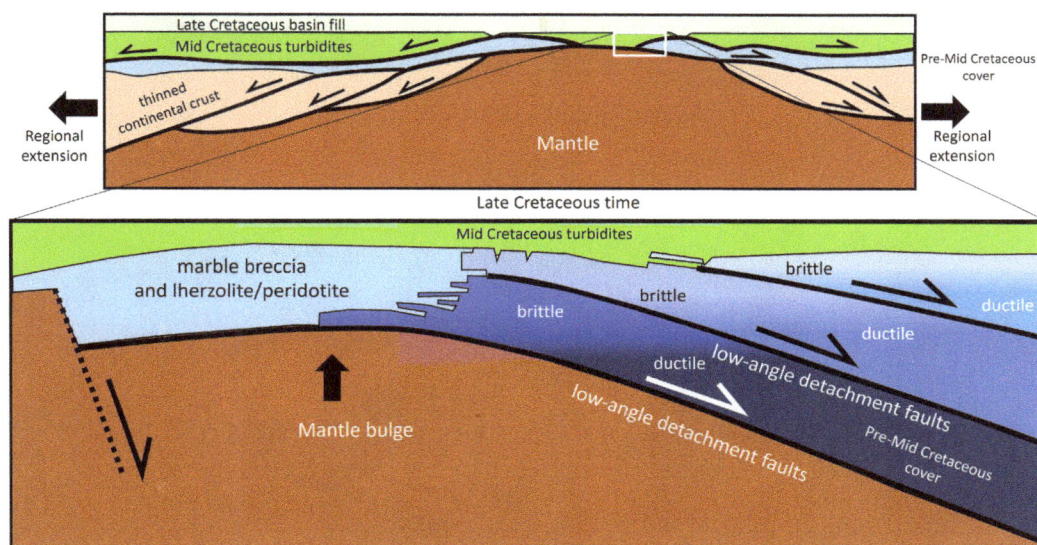

This schematic representation shows one way mantle rock (brown) can get to the seafloor in a setting of major low-angle detachment faults. These faults were active during crustal extension and mantle uplift prior to the Alpine Orogeny. This crustal evolution leads to juxtaposition of metamorphosed and ductile-deformed sediments (blue) directly on top of the uplifted and exhumed mantle (ie., no continental crust). Heavy black lines are the detachment faults; the dotted line is a late normal fault. Modified after Lagabriel et al., 2016.

*Side Trip 5, Étang de Lers to Col d'Agnès: Continue south on D8F for 2.4 km (1.5 mi; 4 min) to the pass and pull into parking area on the right. This is **Stop ST5.2, Col d'Agnès** (42.794087, 1.373830).*

Stop ST5.2 Col d'Agnès

This is a view stop. And the views are expansive, especially if you walk or drive half a km south of the pass.

Mapping here shows that massive Triassic and/or Jurassic layers of marble are separated by breccia intervals. Recumbent folds (folds laying over on their sides) have been mapped within these massive units. The marble layers are mapped adjacent to Triassic metasediments and doleritic tholeiites (a type of seafloor basalt) in this area.

Col d'Agnes, Street View south.

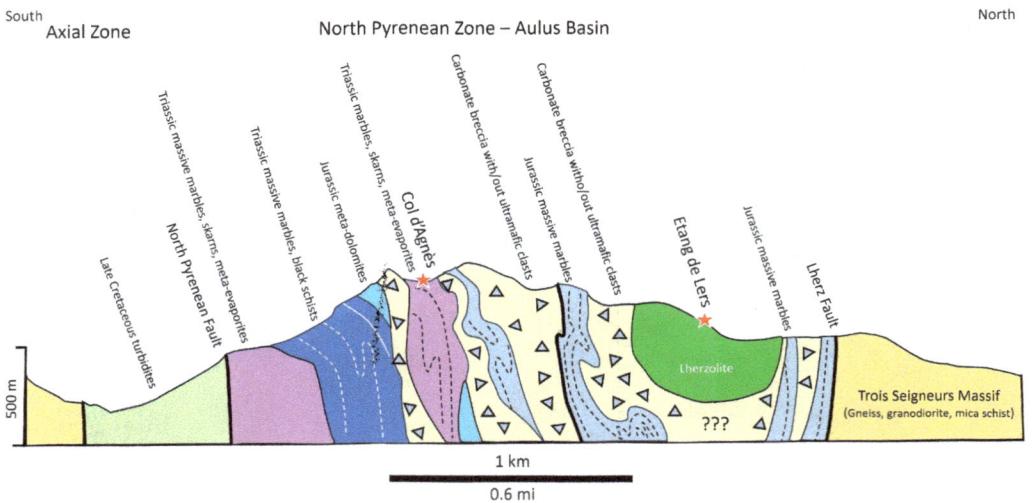

Cross-section of the Aulus Basin through the Étang de Lers and Col d'Agnès. Modified after Lagabrielle et al., 2016.

STOP 66 HOT SPRINGS, AX-LES-THERMES

The North Pyrenean Fault trends west-northwest in this area and separates Devonian schist (south) from Lower Cretaceous carbonates, mainly limestone (north of the fault). We cross it on N20 between Albiès and Lassur. As we cross the North Pyrenean Fault, we enter the Axial Zone. The geology changes from dominantly Mesozoic Age units to dominantly Paleozoic with some Precambrian rocks exposed. Precambrian-Ordovician schists trend east-west through the valley at Ax-les-Thermes. We are truly back in the Axial Zone of the Pyrenees.

Ax-les-Thermes (from Latin *Aquae* – water, and French *Thermes* – hot springs), at an elevation of 700 m (2,300 ft), is a spa town of 1,200 inhabitants (2015) known for its sulfurous hot springs that flow from the ground at 25° to 78°C (77° to 172°F). The springs, originally developed by the Romans, were claimed to treat rheumatism, skin diseases, and other ailments. In medieval times the springs were used to treat returning crusaders afflicted with leprosy. There are three main pools that are free to the public: The Bassin des Ladres, in the center of town, one behind the casino, and the

Source Eau Chaude in a square by the river. You can also enjoy the hot springs at private spas. A spa tourism industry developed in the early 1800s (Wikipedia, Ax-les Thermes; Kukavana).

Sir Charles Lyell, author of the first comprehensive geology text, visited Spain in the summer of 1830 to study the volcanos of Olot in Catalonia. On a visit to Ax-les Thermes he made observations on the Pyrenees that would be included in his second and third volumes (Vergili, 2007).

Today, the thermal springs are the property of SEMTTAX, Société d'Economie Mixte Thermale et Tourist Office of AX, and are operated by the Eurothermes Group.

Originally 16 springs fed the thermal baths. In 2010 two boreholes were drilled to deliver water at a temperature of 76°C and 68°C to the two private spas, the recreational spa center, and the Saint-Louis hospital. Residual heat from the springs is also used to heat municipal buildings, and some streets are cleared of snow using spent thermal water.

The thermal waters in Ax-les-Thermes contain sodium sulfide and relatively low amounts of dissolved minerals. Surface water infiltrates outcrops of the Ax-les-Thermes granite massif, descends to depths of around 5 km (16,400 ft), where the water reaches temperatures between 120° and 130°C. As the heated water rises over hundreds or thousands of years, it accumulates beneath schist layers. The hot springs are located along fractures that penetrate the schist. Sulfides in the water are thought to result from transformation of the mineral pyrite into pyrrhotite in a zone of increasing heat (Voirand et al., 2022). This transformation involves the loss of sulfur, which is then carried to the surface in the thermal waters.

Heat from the Ax-les-Thermes Granite, dated at 306 Ma (Late Carboniferous), metamorphosed its host rock during late stages of the Variscan Orogeny (Clariana, 2017). This heat has long since dissipated. The thermal waters at Ax-les-Thermes are heated simply by the Earth's thermal gradient, 25°–30°C/km (72°–87°F/mi).

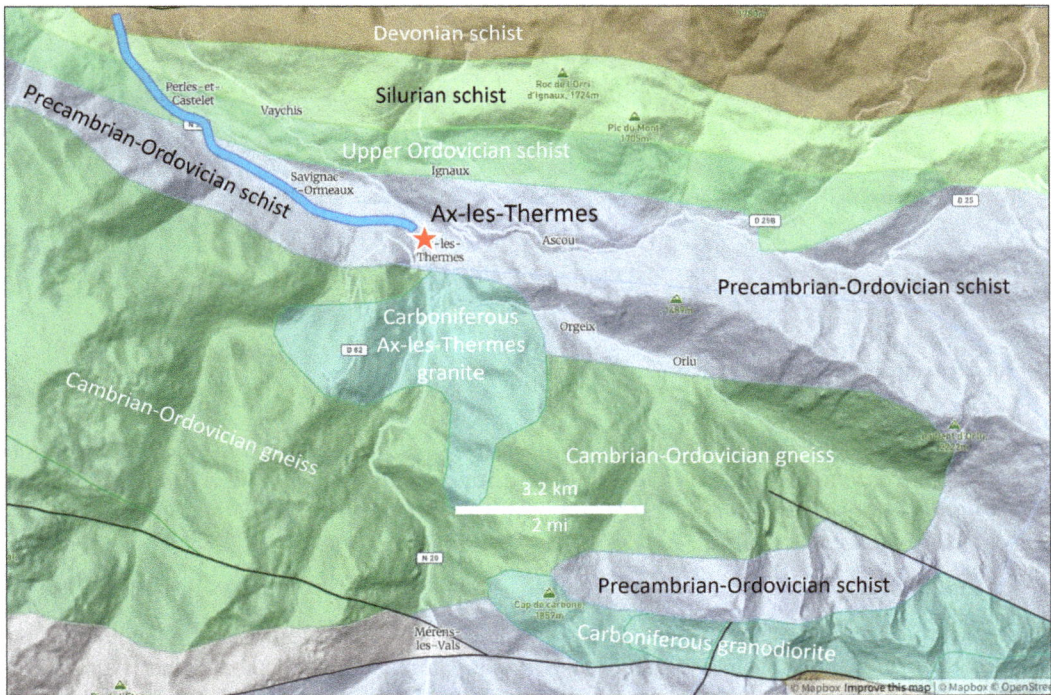

Geologic map of the Ax-les-Thermes area. Base geology from MacroStrat, 2015.

Side Trip 5, Col d'Agnès to Bassin des Ladres: *Return north on D8Fto D18; turn right (east) on D18 and drive to Val de Sos; turn left (east) onto D8/Rue Grand Rue; at the roundabout in Sabart take the 1st exit (south) onto N20; take the N20 exit toward Ax-les-Thermes; at the roundabout take the 3rd exit (east) onto Rte d'Espagne; at the town center roundabout take the 2nd exit and park on the right. This is **Stop 66.1, Bassin des Ladres** (42.719863, 1.839711) for a total of 59.3 km (36.8 mi; 68 min).*

STOP 66.1 BASSIN DES LADRES

The Bassin des Ladres (Lepers' Pond) is fed by hot springs at a temperature of 77°C. The ancient (and still in operation) hospital St. Louis is immediately adjacent to the pool. The pool was built by Roger IV, Count of Foix, in 1250 to treat soldiers returning from the Crusades with leprosy. The hospital, built during the reign of Saint Louis for the same purpose, dates to 1260.

While a small number of new cases are still diagnosed in metropolitan France each year, there is no indication that leprosy is currently a concern at the Bassin des Ladres.

Bassin des Ladres and the Hôpital St. Louis.

Bassin des Ladres to Source Eau Chaude: *Walk north on Pl Du Breilh and turn left (west) on Rue de l'Horloge; continue straight on Pl Roussel to the Source Eau Chaude. This is **Stop 66.2, Source Eau Chaude**, Ax-les-Thermes (42.719474, 1.837332) for a total of 260 m (850 ft; 3 min).*

STOP 66.2 SOURCE EAU CHAUDE

Soak your sore feet in the hot springs pool after exploring the town. Many exclaim the therapeutic, healing value of the hot, sulfurous water. To me, and apparently to many others, it just feels good to relax my feet in a hot water bath.

In addition to these two pools, there are a number of hot spring locations throughout town, including along the river walk on the L'Oriège.

Source Eau Chaude.

Source Eau Chaude to Easternmost Point Andorra: *Drive southeast on Blvd de L'Oriège; continue straight on Av. Durandeau; turn right (south) on Av. Gomma; Av. Gomma turns right and becomes Aouradu; continue straight on Bousiguette; at the roundabout take the 2nd exit (south) onto N20/E9; after about 18 km turn left (east) onto N22; continue on N22 to a pullout on the right at 22.7 km (14.1 mi; 27 min). This is* **Stop 67, Easternmost Point, Andorra** *(42.572608, 1.789120).*

STOP 67 EASTERNMOST POINT, ANDORRA

You are not actually in Andorra: you are in France about 260 m (850 ft) southeast of the easternmost point in Andorra. The easternmost point is in the bottom of a steep ravine. This is a view stop situated in the eastern High Pyrenees. You are standing on Precambrian-Ordovician schist. Just to the north is the contact with Cambrian-Ordovician muscovite gneiss.

The geology of Andorra is mainly Cambrian or Ordovician metasedimentary rocks, phyllite, quartzite, and slate of the Pyrenean Axial Zone. These rocks were metamorphosed during the Variscan Orogeny, then folded and thrust during the Pyrenean Orogeny. Muscovite gneiss and schist are found in the cores of antiforms in the northeast of the country. These folds are separated by nearly horizontal thrusts carrying nappes of metasediments. The Paleozoic metasediments found across most of Andorra have also been tightly folded. The Mt. Louis-Andorra Pluton, a composite granitic body, occurs in the southeast and extends into Spain. The base of the pluton is exposed in eastern Andorra (Wikipedia, Geology of Andorra).

Google Street View north into Andorra. The outcrops, center, are Cambrian-Ordovician muscovite gneiss.

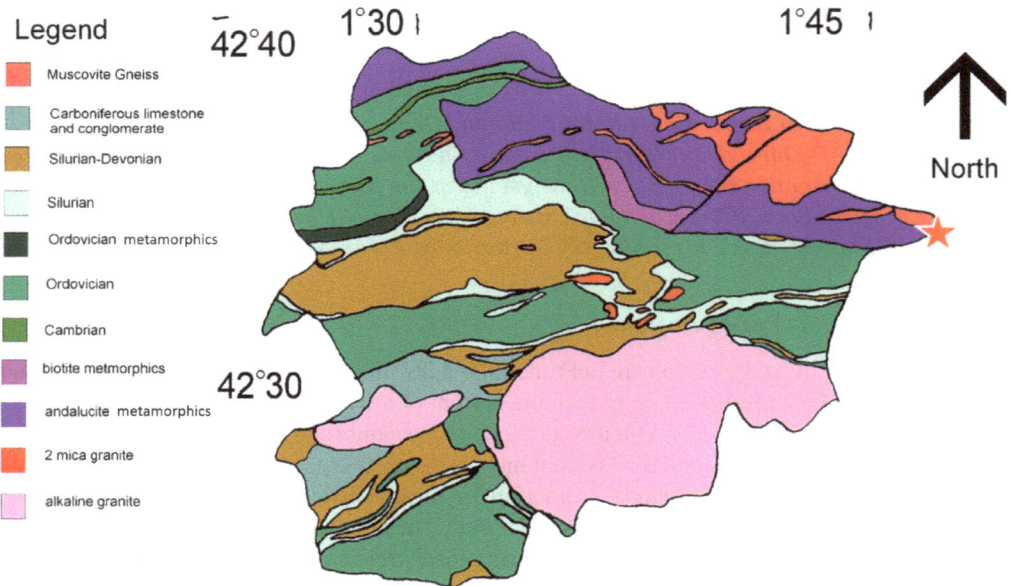

Geologic map of Andorra. Star indicates the easternmost point. Diagram courtesy of Graham Bartlett, https://commons.wikimedia.org/wiki/File:Andorra_Geology.PNG

Side Trip 6, Easternmost Point Andorra to Mont-Louis Granite: *Return to N320 and turn left; drive north to N20 and turn right (south); drive through the N20/E9 tunnel; at the south end of Ur turn left (east) onto D30 toward Caldégas; at the end of D30 turn left (east) onto D66; at Aire de Pont de Bou turn left (west) onto D29 to Font-Romeu; drive 230 m (0.15 mi) and pull over on the right past the guardrail. This is* ***Stop ST6.1, Mont-Louis Granite*** *(42.485549, 2.058331) for a total of 39.5 km (24.6 mi; 41 min).*

If you prefer to skip Side Trip 6:

Easternmost Point Andorra to Mirador de la Roca d'Alp: *Return to N320 and turn left; drive north to N20 and turn right (south); drive through the N20/E9 tunnel; just before the Spanish border at the roundabout take the 2nd exit toward Barcelone; at the next roundabout take the 1st exit onto N-154 toward Barcelone; at the Puidcerda roundabout continue straight (south) onto N-152; on leaving Puigcerda at the roundabout take the 2nd exit (south) onto N-260; continues straight onto C-162 toward La Molina; at the Alp roundabout take the 2nd exit to stay on C-162; in Das turn left (east) onto GI-404 toward La Molina; turn right (south) onto GI-400; drive 0.9 km (0.55 mi) and turn right at the sign for "Coll de la Creueta 1888 m." This is* **Stop 68, Mirador de la Roca d'Alp** *(42.360199, 1.880976) for a total of 40.3 km (25.0 mi; 43 min).*

Side Trip 6 Eastern Axial Zone

This side trip examines one of the Variscan-age intrusives in the eastern Axial Zone of the Pyrenees as well as a ductile deformation zone. Mylonite, a fine grained foliated rock formed by intense shearing in a ductile deformation zone, often with a banded appearance, is here associated with an Alpine reverse fault (like a thrust fault but nearly vertical) that was reactivated as a normal fault during late Tertiary to Quaternary extension. Displacement along the regional southwest-northeast Têt Fault separates the granitic Variscan Mont-Louis Massif in the hanging wall to the north from the Canigou-Carança Massif in the footwall to the south (Maurel et al., 2008; Milesi et al., 2022).

Stop ST6.1 Mont-Louis Granite

Granite is generally not that exciting. But this is the first granite we have seen up close in the Axial Zone of the Pyrenees.

The Pyrenean chain contains many granodioritic to granitic plutons of Variscan (Late Carboniferous, 303–305 Ma) age (Le Bayon and Cochelin, 2020). Most are located in the eastern Axial Zone and intruded and metamorphosed Cambrian to Lower Carboniferous sediments that now make up the basement rocks. The entire assemblage was uplifted, thrust south, and unroofed during the Alpine Orogeny (Gleizes et al., 1993; Milesi et al., 2022).

The Mont-Louis Granite is at the eastern end of the Mont-Louis-Andorra pluton and is zoned from quartz diorite in the core to granodiorite on the flanks (Gleizes et al., 1993). A pluton is a large body of igneous rock formed when magma cools and solidifies underground. Magnetic mapping indicates that magma emplacement was related to a northeast-southwest-trending fault system, probably ancestral to the 100+ km (60+ mi) long Têt Fault zone. A late stage of deformation saw right-lateral shear with a thrust component along northwest-southeast-striking faults that break the massif into three blocks. The eastern block (Mont-Louis Granite) represents the former base of the pluton, whereas the western block was the top (Bouchez and Gleizes, 1995). After Paleogene south-directed compression, these faults were reactivated during Oligo-Miocene to Recent extension with normal offset. This created uplifted blocks and downdropped basins and has been related to the opening of the Gulf of Lion (Maurel et al., 2008; Milesi et al., 2022).

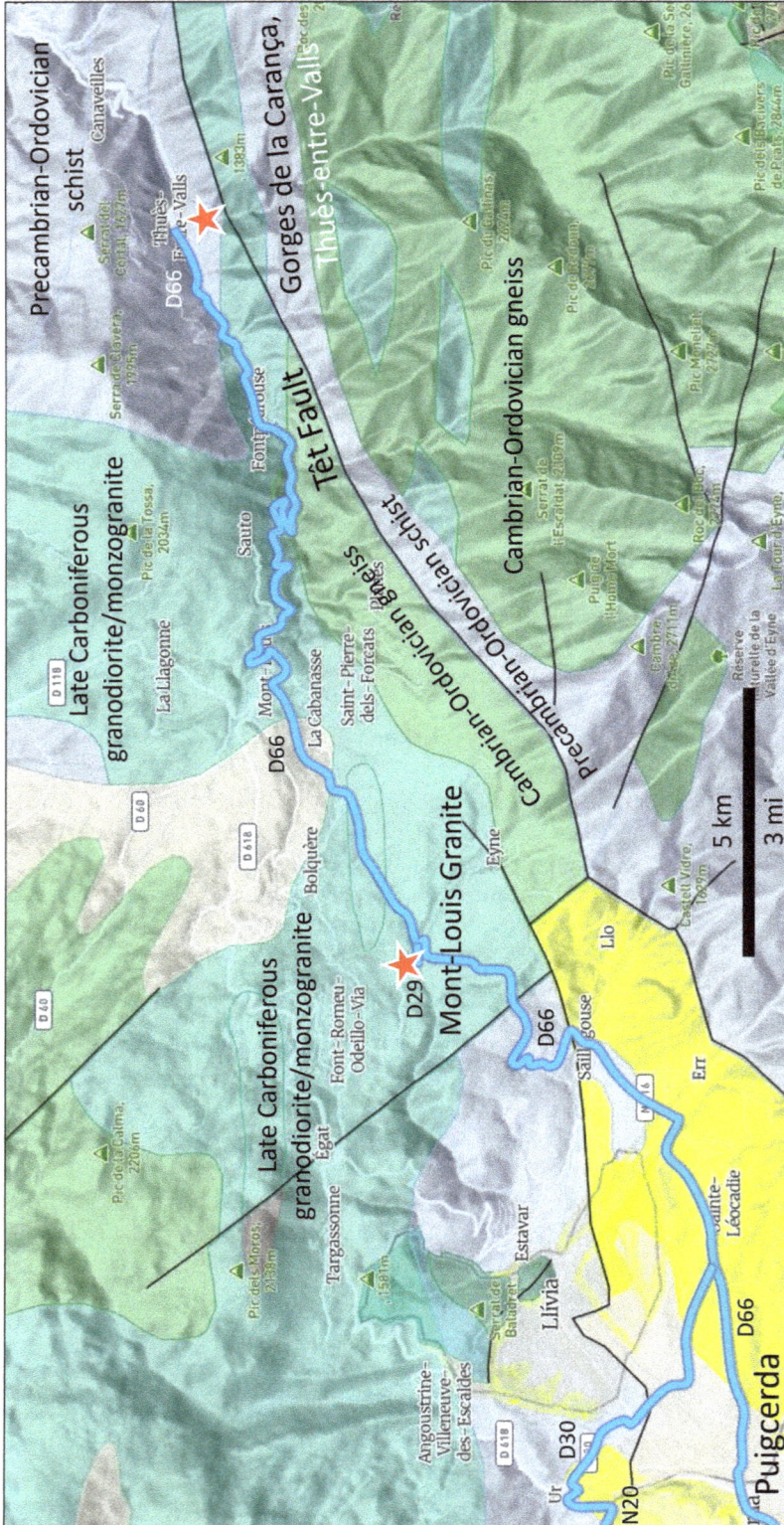

Geologic map of the Mont-Louis Granite and Thuès-entre-Valls Mylonite stops. Base geology from MacroStrat, 2015.

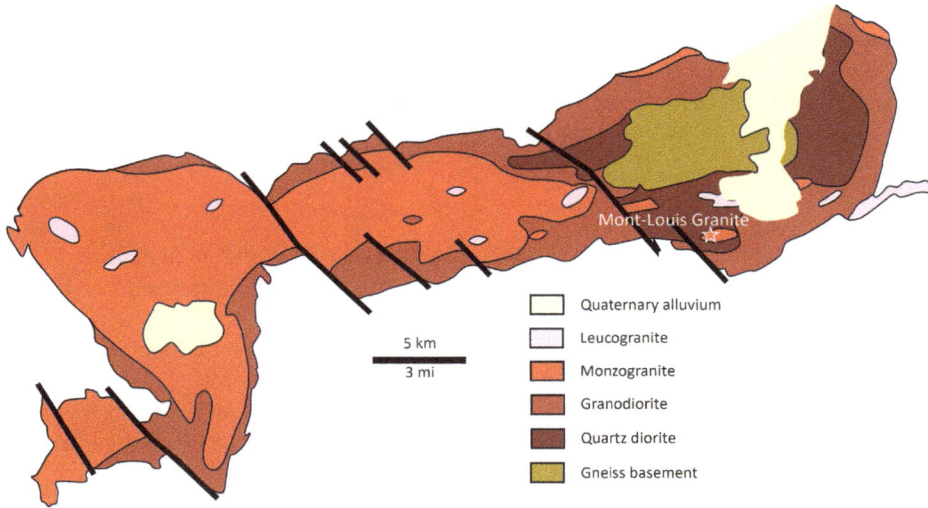

The Mont-Louis–Andorra Pluton showing compositional gradations and structural segmentation. The overall northeast trend appears to follow an ancestral Têt Fault. The stop is at the star. Modified after Gleizes et al., 1993.

At this location the Mont-Louis Granite exposes fresh granite with dark xenoliths and, in places, stretched xenoliths (Satterfield et al., 2019). Xenoliths (literally "foreign rocks") are bits of preexisting rocks that were engulfed by the granite as it moved upward to the surface; thus they are foreign to the granite. In this case the xenoliths were mafic igneous rocks (rocks that contained dark magnesium and iron minerals).

Mont-Louis granodiorite with deformed (elongated) mafic xenoliths (dark patches).

Side Trip 6, Mont-Louis Granite to Thuès-entre-Valls Mylonites: Return east on D29 to D66; turn left (northeast) onto D66; turn right (south) at sign to Thuès-entre-Valls/Gorges de la Carança Parking; follow the signs to Gorges de Carança Municipal Parking. Walk 250 m (820 ft) to the gorge. This is **Stop ST6.2, Thuès-entre-Valls Mylonites, Carança–Canigou Massif** *(42.523898, 2.221168) for a total of 19.8 km (12.3 mi; 26 min).*

STOP ST6.2 THUÈS-ENTRE-VALLS MYLONITES, CARANÇA–CANIGOU MASSIF

The Carança-Canigou gneiss Massif main uplift (cooling event) occurred from 32 to 18 Ma, during early Oligocene, and was followed by more moderate uplift from late Oligocene to early Miocene. This corresponds to normal offset on the Têt Fault and unroofing of the massif.

The Têt normal fault is a northeast-southwest-trending 100+ km (60+ mi) long crustal scale fault that cuts Palaeozoic magmatic and metamorphic rocks of the Mont-Louis, Canigou, and Carança massifs. The Têt Fault, dipping steeply to the north, is associated with at least two mylonitic-creating events that record south-vergent reverse offset.The youngest reverse offset dates to Pyrenean shortening around 60 Ma (Maurel et al., 2008).

A second stage of normal motion along the Têt fault is indicated between the middle Miocene and the late Pliocene, with associated vertical offset in the range of 150 m to several km (500 ft to miles). Since the end of Miocene, only the western Têt Fault has been active, causing the opening of the Cerdagne pull-apart basin by normal and right-lateral offset (Milesi et al., 2022).

The Gorges de la Carança is located at the village of Thuès-entre-Valls, in the Carança–Canigou Massif. From the car park there is a narrow path carved out of the cliffs. The first part of the path is in quartz-rich mylonites. The mylonite zone ends at the south end of the narrowest part of the gorge, some 250 m (820 ft) from the start. The gabbro-mylonite contact is just below the first footbridge.

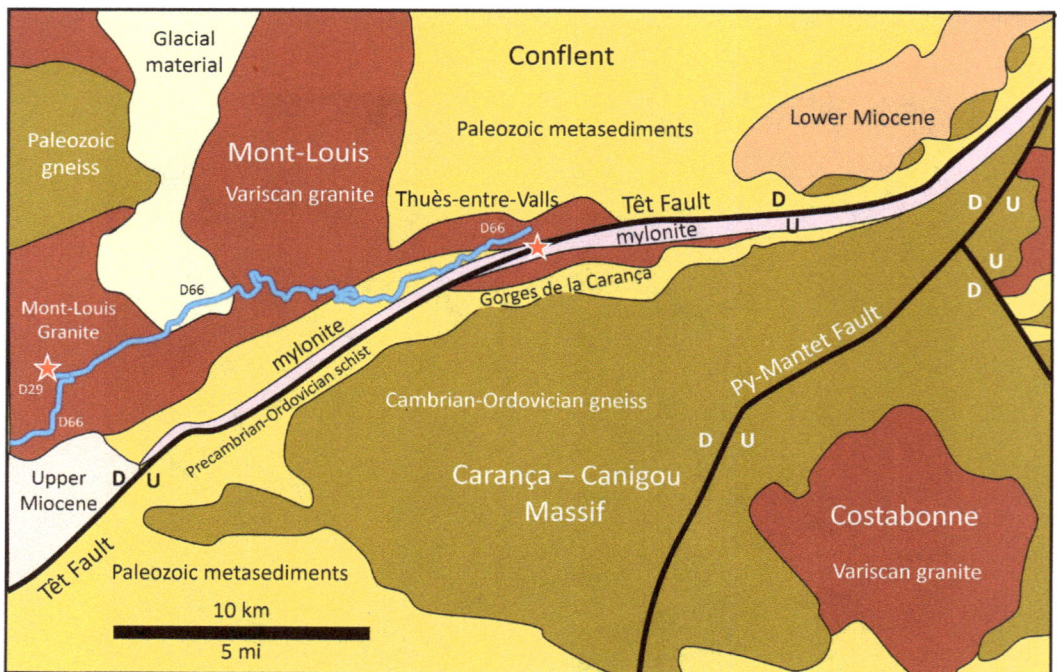

Geologic setting of the Têt Fault mylonites at Thuès-entre-Valls. Modified after Maurel et al., 2008.

Mylonite, Gorges de la Carança. Notice the sheared and elongated rock fabric.

Gorges de la Carança

Side Trip 6, Thuès-entre-Valls to Mirador de la Roca d'Alp: Return north to D66 and turn left (west); take D66 to the roundabout just before the Spanish border; take the 2nd exit onto N20 toward Puigcerda; cross the border and continue on N-152; at the next 2 roundabouts take the 2nd exit to continue on N-152; on leaving Puigcerda at the roundabout take the 2nd exit (south) onto N-260; continues straight onto C-162 toward La Molina; at the Alp roundabout take the 2nd exit to stay on C-162; in Das turn left (east) onto GI-404 toward La Molina; turn right (south) onto GI-400; drive 0.9 km (0.55 mi) and turn right at the sign for "Coll de la Creueta 1888 m." This is **Stop 68, Mirador de la Roca d'Alp** (42.360199, 1.880976) for a total of 48.8 km (30.3 mi; 57 min).

STOP 68 MIRADOR DE LA ROCA D'ALP

We have entered the Parc Natural del Cadí-Moixeró, a nature park in northern Catalonia. The park was established in 1983 and encompasses 41,060 hectares (101,500 acres) of mountainous terrain.

This view stop provides a magnificent 180° panorama of the eastern Pyrenees, with several peaks topping 2,900 m (9,500 ft). There is a tourist panel that names all the peaks and other landmarks.

View northwest from Mirador de la Roca d'Alp.

This overlook is underlain by highly faulted Devonian metasediments that originally were mainly limestones with minor shales. Carboniferous metasediments derived from shale, sandstone, and conglomerate crop out just to the south.

Mirador de la Roca d'Alp to Guardiola de Berguedà Thrust: *Return north to C-162 and turn left (southwest); take the ramp onto C-16 south; after 27.0 km (16.8 mi; 30 min) there is a large parking and rest area on the right. This is* ***Stop 69, Guardiola de Berguedà Thrust*** *(42.204386, 1.867202).*

STOP 69 GUARDIOLA DE BERGUEDÀ THRUST

Driving south from Guardiola de Berguedà you cross the trace of multiple south-directed thrust sheets carrying Cretaceous limestones and marls, but it is largely forested so you can't see much of the geology. But at this pullout/parking area you can clearly see Late Cretaceous limestones carried on the Cadí Thrust that have been emplaced over Paleocene conglomerate, sandstone, and coal.

The Mesozoic section indicates that we have returned to the South Pyrenean Zone.

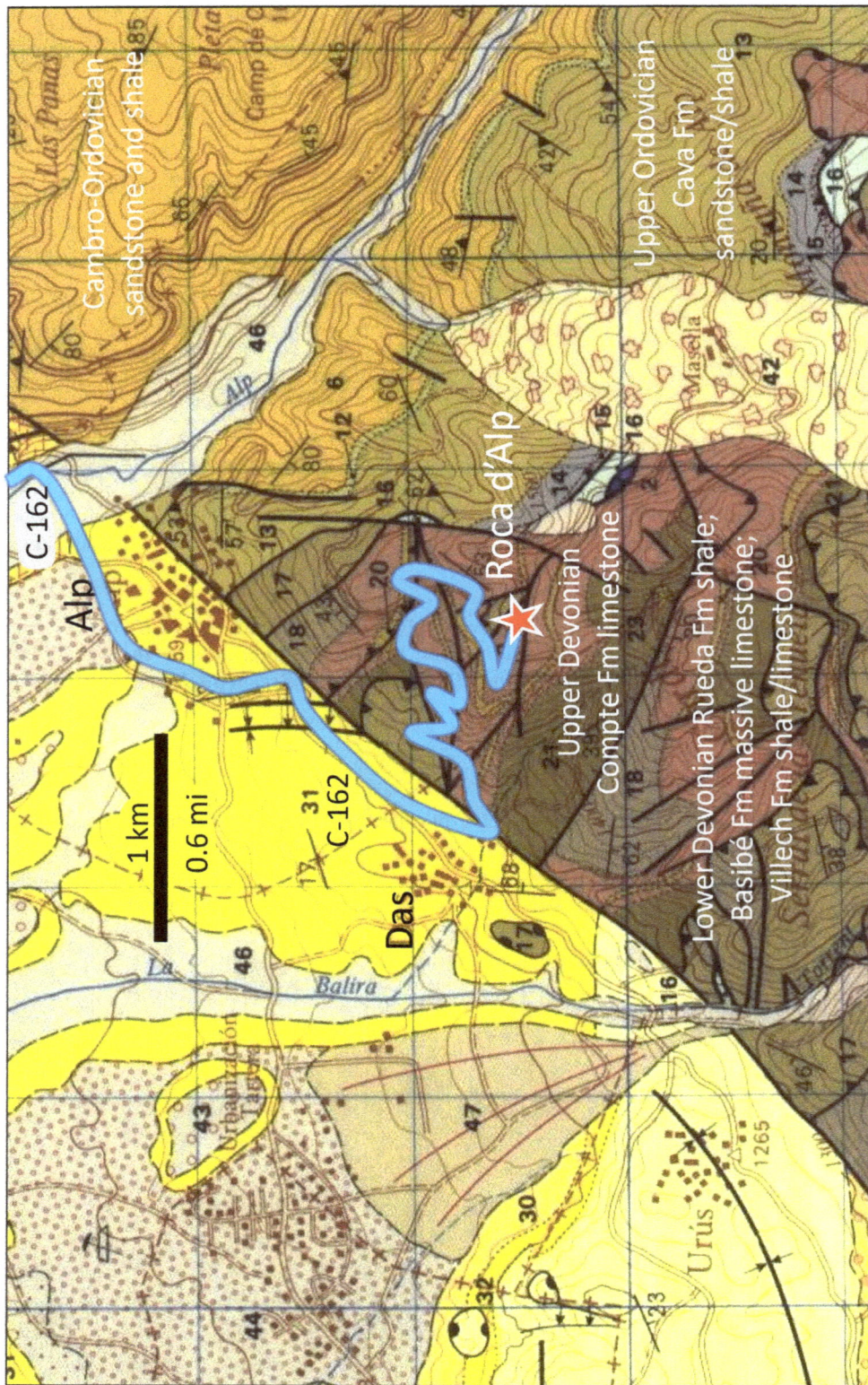

Cambro-Ordovician sandstone and shale

Upper Ordovician Cava Fm sandstone/shale

Roca d'Alp

Upper Devonian Compte Fm limestone

Lower Devonian Rueda Fm shale; Basibé Fm massive limestone; Villech Fm shale/limestone

C-162

Alp

1 km

0.6 mi

C-162

Das

La Balira

Urús

Geologic map, Roca d'Alp area. From Instituto Geológico y Minéro de España (IGME), Puigcerdà Sheet. Source: © NC Geological Survey of Spain (IGME).

South-directed thrusts carry Upper Cretaceous marls over Paleocene conglomerate and sandstone. Dotted line is roughly the trace of the thrust fault. View northwest.

Side Trip 7, Guardiola de Berguedà Thrust to Fumanya South Dinosaur Tracks: *Continue south on C-16; at 4.6 km (2.9 mi) make a sharp right (west) onto BV-4025 toward St. Josep/St. Corneli/ Figols; at 15.3 km (9.5 mi) make a right turn onto Cal Marxa and drive ~220 m (720 ft) and pull into the parking area on the left. This is **Side Trip 7, Fumanya South Dinosaur Track Site** (42.178839, 1.795054) for a total of 14.6 km (9.1 mi; 18 min).*

If you prefer to skip Side Trip 7:

Guardiola de Berguedà Thrust to Pont de la Baells: *Continue south on C-16; at 12.9 km (8 mi) take the exit (east) toward Ripoll and merge onto C-26; drive another 2.7 km (1.7 mi) and pull over on the right before the guardrails and bridge. This is **Stop 70, Pont de la Baells** (42.126760, 1.877304) for a total of 14.7 km (9.1 mi; 14 min).*

SIDE TRIP 7 FUMANYA SOUTH DINOSAUR TRACKS

The Fumanya dinosaur track site is part of Cadí-Moixeró National Park, and has been accepted as a UNESCO World Heritage site. The Fumanya sites (Fumanya South, Mina Esquirol, Fumanya North, Mina Tumí, and others) are designated as "Cultural Sites of National Interest," providing the maximum legal protection in Spain. Furthermore, since 1993 the Catalan Cultural Heritage Law prevents illegal fossil hunting in the area (Vila et al., 2008).

The site is fairly remote, is not well marked, and is little known or visited. The owner of the former coal quarry, Carbones de Berga SA, began building a Visitor Center but then stopped work. Preservation of the tracks is a serious problem, in that many tracks have been lost to weathering and erosion (Rius, 2022).

The Fumanya site has over 50 sauropod (primarily titanosaur) trackways containing nearly 3,000 footprints. This is a key location to understand the diversity of the last European dinosaurs during the latest period of the Cretaceous, just before their extinction.

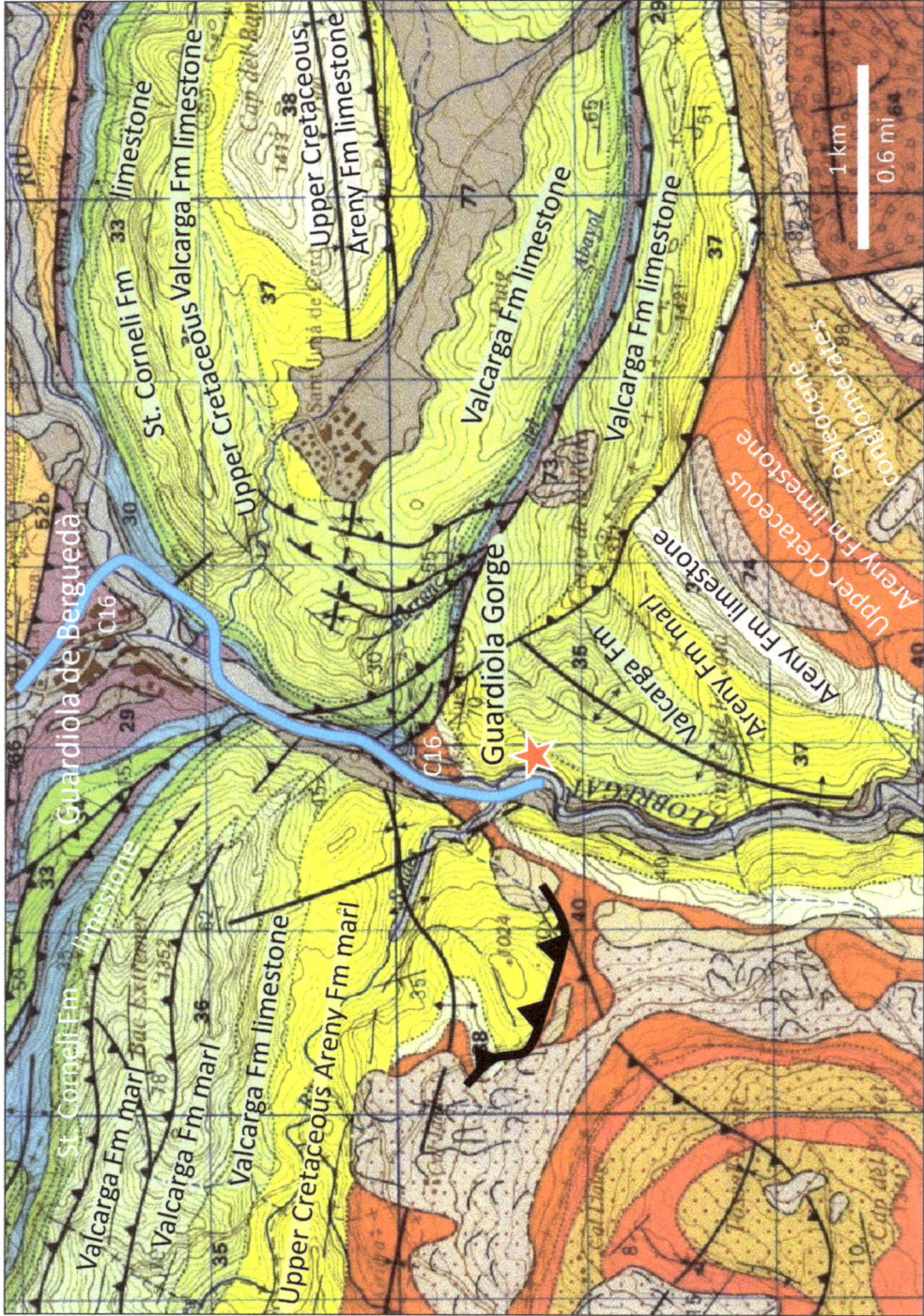

Geology of the Guardiola Gorge area. From Instituto Geológico y Minéro de España (IGME), La Pobla de Lillet Sheet. Source: © NC Geological Survey of Spain (IGME).

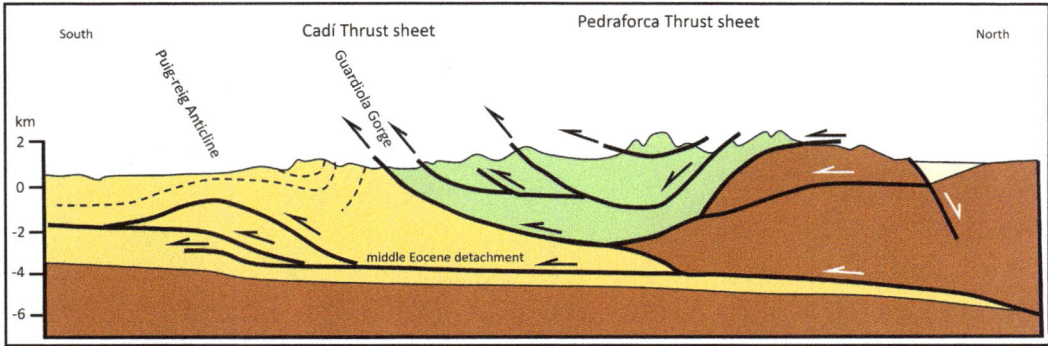

Cross-section through the frontal zone of the Pedraforca-Cadi thrust sheets. Modified after Sans and Vergés, 1995.

Sauropod tracks at the Fumanya South site. Photo courtesy of Jmarmi, https://commons.wikimedia.org/wiki/File:Diapositiva_Viladrich_1986.jpg

The Fumanya site was discovered in 1985 (Vila et al., 2008). Quarrying of lignite (low-grade coal) at Fumanya uncovered these extraordinary dinosaur traces, the world's largest outcrop of Maastrichtian (latest Cretaceous) sauropod tracks. The footprints were originally laid down on a calcareous lakebed (the St. Corneli Marl) that has since been uplifted and rotated to near-vertical. This deformation occurs in the lower Pedraforca Thrust sheet. The Pedraforca Thrust sheet moved south during early Eocene, resulting in folds that are oriented roughly east-west (Rius, 2022).

The tracks occur in continental strata of the Upper Cretaceous–Paleocene Tremp Formation. The Tremp Formation here is transitional from deltaic/lagoonal to river-dominated floodplain and lake environments. From base to top it includes the following zones:

(1) Cretaceous transitional Gray Garumnian (nonmarine marls, coals, limestones, and sandstones). This zone contains the coal that was being mined.
(2) Cretaceous fluvial (river-dominated) Lower Red Garumnian (mudstones, sandstones, and paleo-soils)
(3) Paleocene lacustrine (lake) Vallcebre Limestone
(4) Paleocene Upper Red Garumnian (mudstones, sandstones, conglomerates, and limestones).

The Fumanya South site is at the very base of unit (1), the Gray Garumnian.

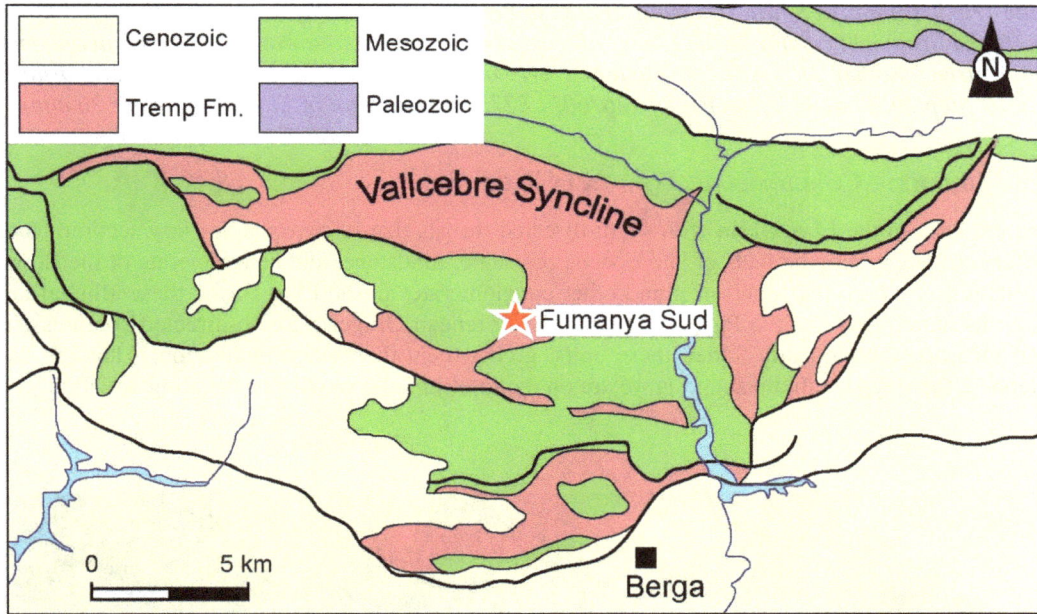

Geologic map of the South Fumanya site showing distribution of the Paleocene Tremp Formation. Map courtesy of Vila et al., 2013, https://commons.wikimedia.org/wiki/File:Tremp_Formation_-_Fumanya_Sud_Map.jpg

Titanosaurs were some of the heaviest creatures ever to walk the earth, weighing up to 100 tonnes (110 tons). Parallel trackways and the presence of multiple individuals provide evidence of social behavior in these large herbivores. Small trackways allow comparison between adult and baby titanosaurs (Vila et al., 2008). Footprints found at Fumanya make this site the largest (in terms of prints) in Europe and third in the world after the Lark Quarry, Australia, and Chaoyang, China (Rius, 2022).

The only disappointment with this site is that you can no longer walk up to the tracks, but have to view them from a path that is around 30 m (100 ft) away. Viewing is good when the pavement surface is in full sunlight, and even better if the light casts shadows on the track imprints (early to mid-afternoon). Otherwise, the contrast between tracks and background is fairly low.

Visit

Hours:
From Easter week to 14 July: 11:00 – 15:00.
From 15 July to 31 August: 11:00 – 15:00 daily except Tuesdays
From 1 September to 8 December: Weekends 11:00 – 15:00.
The last group will be admitted at 14:00.
Tickets can be purchased online (recommended) or at the reception area.
Email: m.cercs@diba.cat
Address: 08698 Fumanya, Barcelona, Spain
Phone: 938248187 or 938249025
Website: https://mmcercs.cat/producte/entrada-dinosaures-fumanya/

Contact the website for current entry fees.

*Side Trip 7, Fumanya South Dinosaur Tracks to Pont de la Baells: Return east on BV-4025 to the C-16 and turn right (south); at 19.1 km (11.8 mi) take the exit (east) toward Ripoll and merge onto C-26; drive another 2.7 km (1.7 mi) and pull over on the right before the guardrails and bridge. This is **Stop 70, Pont de la Baells** (42.126760, 1.877304) for a total of 21.9 km (13.6 mi; 26 min).*

STOP 70 EOCENE CONGLOMERATE, PONT DE LA BAELLS

We are back in the Ebro Basin with south-directed thrusts that involve the Tertiary section. Here we are in steeply dipping beds of late Eocene red shale, sandstone, and conglomerate of the Berga Formation. As with the Montserrat and other conglomerates in the Ebro Basin, these alluvial fan deposits herald uplift of the Pyrenees during the Pyrenean Orogeny. South-directed compression and thrusting subsequently folded these units, giving them their steep inclination. Although not mapped, the steep south dip suggests we are on the forelimb of an east-west-trending anticline.

Upper Eocene conglomerates of the Berga alluvial system.

Pont de la Baells to Balsareny Anticline: *Return west on C-26 to C-16 and turn left (south); C-16 merges with E-9 and continue south to Exit 71; take Exit 71 to Navas Sud and pull over on the right shoulder in the roundabout. This is **Stop 71, Balsareny Anticline** (41.888086, 1.875542) for a total of 30.5 km (19.0 mi; 20 min). Walk back north up the exit ramp to see the roadcut.*

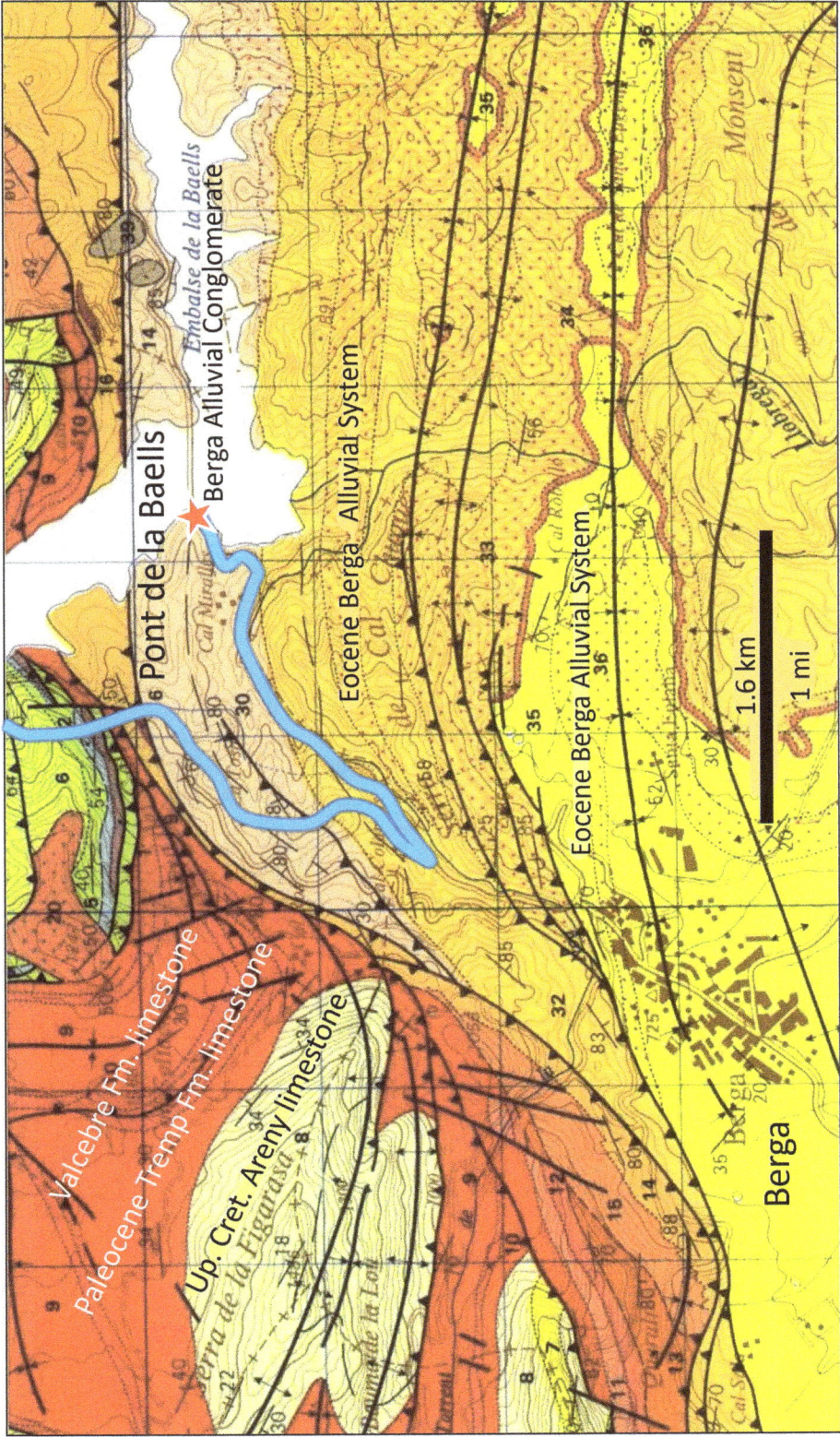

Geologic map of the Pont de la Baells area. From Instituto Geológico y Minéro de España (IGME), Berga Sheet. Source: © NC Geological Survey of Spain (IGME).

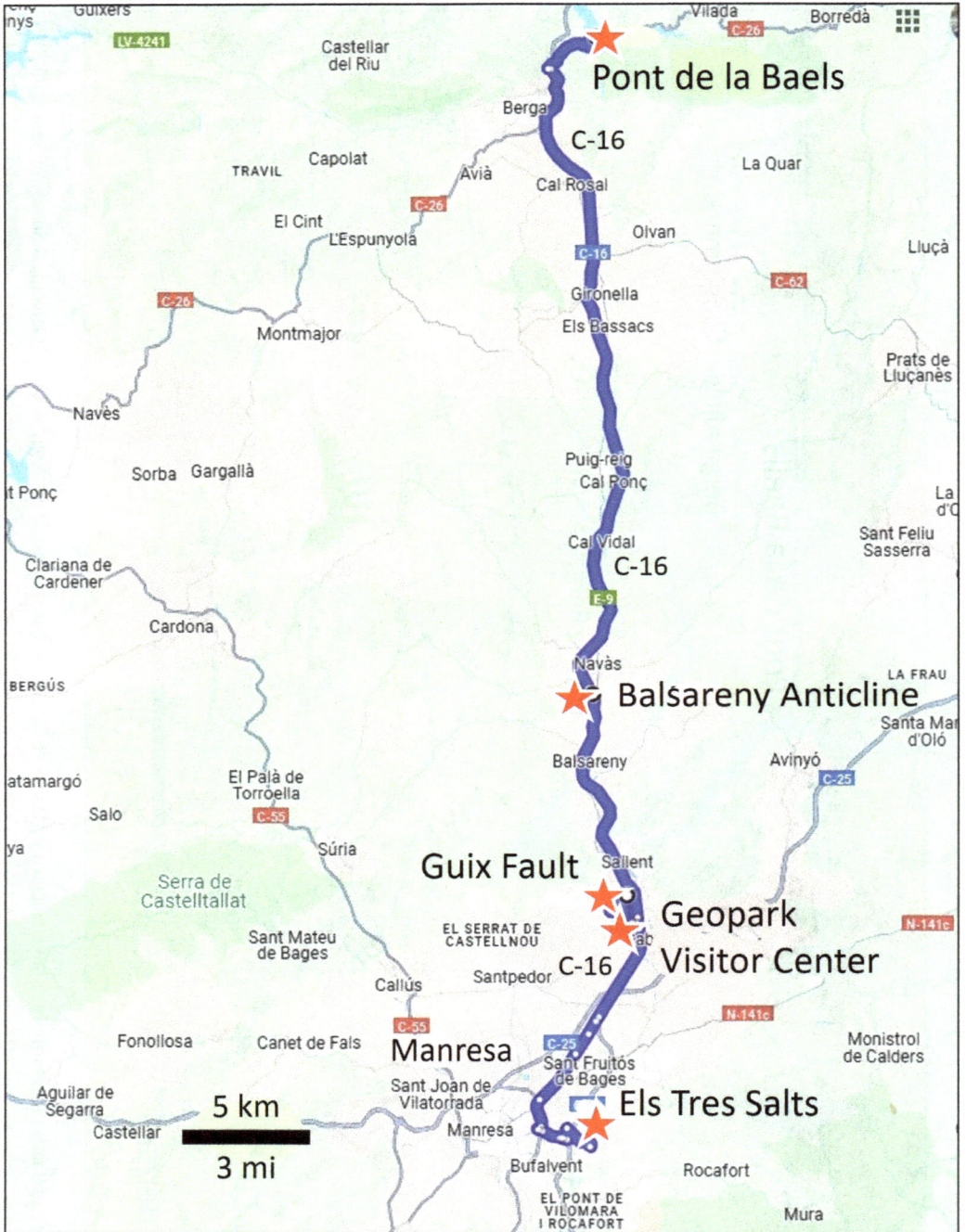

STOP 71 BALSARENY ANTICLINE

Once again we are in the Ebro foreland basin. The Balsareny Anticline is the eastern extension of the Suria salt-cored structure we saw earlier. As at Suria, this anticline developed in Eocene-Oligocene units of a thrust sheet on an Eocene evaporite (salt) detachment surface. Unlike at Suria, this is a single, simple, broad, and unfaulted anticline.

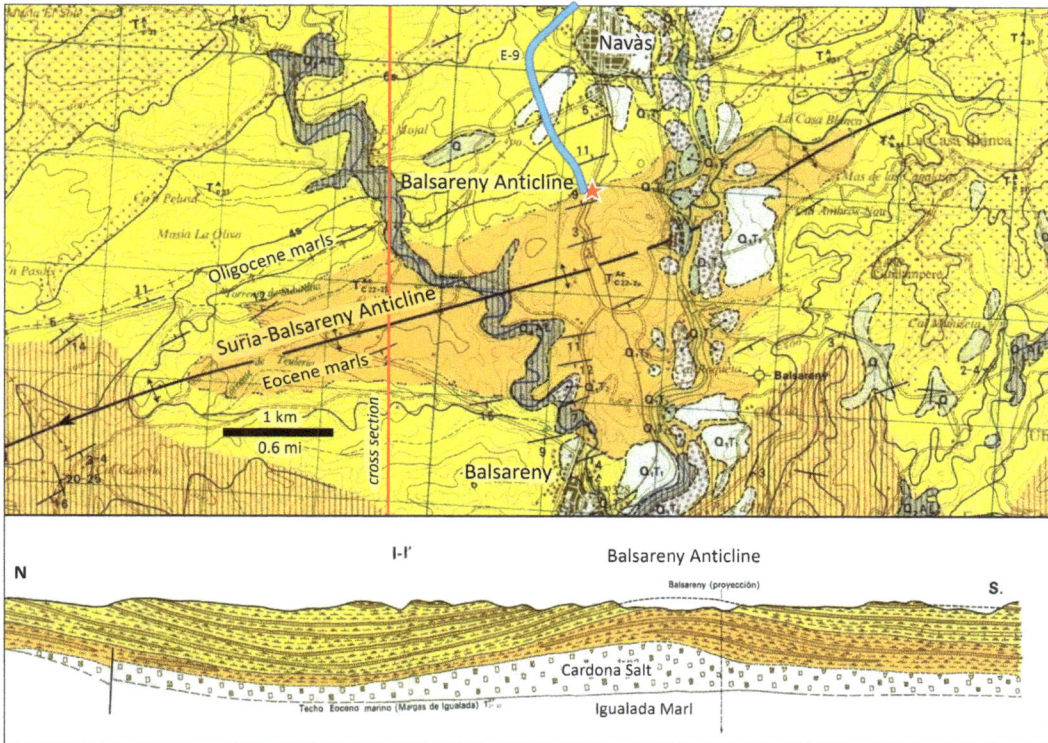

Geologic map and cross-section of the Suria-Balsareny Anticline. From Instituto Geológico y Minéro de España (IGME), Puigreig Sheet. Source: © NC Geological Survey of Spain (IGME).

At this stop the units are all dipping 3° to 9° north. As you drive a couple of km south of here on E-9 the units are all dipping 11° to 16° south. The steeper dips on the south limb indicates this fold is another south-verging anticline, a result of south-directed compression.

Balsareny Anticline to Guix Fault: *At the roundabout take the 2nd exit (south) onto the E-9/C-16 onramp; take Exit 62 Salent Dud and immediately turn right (west) and park; walk ~70 m (2,300 ft) south along the railroad tracks. This is **Stop 72, Guix Fault, Sallent** (41.81988, 1.8909) for a total of 8.6 km (5.3 mi; 6 min).*

STOP 72 GUIX FAULT, SALLENT

Continuing south in the Ebro Basin we come to the town of Sallent and the El Guix Anticline. This is another salt-cored anticline and thrust fault, the Guix Fault, that is the southernmost manifestation and latest phase of Pyrenean mountain building in Catalonia.

North flank of the Balsareny Anticline along the E-9 offramp. The north-dipping unit is red Eocene shale and marl.

The El Guix Anticline is located at the southern pinch-out of the Cardona Salt (Sans, 2003). The crest of this almost symmetrical fold is just south of the exposed thrust fault (Evans and Fischer, 2012). The name El Guix means "gypsum" in Catalan, and refers to the gypsum that outcrops near the axis of the anticline (Institució Catalana d'Història Natural, 2022). El Cogulló immediately west of this stop is a salt mountain, a large, man-made pile of salt.

Main Guix Thrust in railroad cut south of Sallent. The thrust here puts Eocene sandstone and shale over Eocene sandstone and shale. View looking west.

Guix backthrust, view northwest. This outcrop is only a few m north of the main thrust.

The Guix Thrust is particularly well exposed in a railroad cut on the east side of Cogulló Mountain. The fault trends northeast-southwest for about 10 km (6 mi), crossing the Llobregat River south of Sallent. The fault has its greatest offset and complexity in this area.

The Guix Fault cuts the El Guix Anticline just north of the anticlinal axis. The fault is a steep, south-directed thrust. The Cardona Salt is found roughly 400 m (1,300 ft) below the surface and is up to 300 m (1,000 ft) thick. The salt was mobilized into the anticline core during folding (Institució Catalana d'Història Natural, 2022).

The fold formed first, and was later cut by the thrust. Structural restoration through the El Guix Anticline indicates 5% shortening due to folding and thrusting and 16% due to internal layer parallel shortening (a result of e.g., pressure solution, porosity reduction, and/or bedding plane slip), which appears to be the main shortening mechanism in this area (Sans, 2003).

Guix Fault to Geopark Center: Return to the highway feeder road and turn right (south); go under the freeway and at the roundabout take the 5th exit (west); return under the freeway and get onto the onramp to C-16/E-9 south; take Exit 60 to Artés; at the end of the offramp turn right and drive to the parking lot for the Centre de visitants del Geoparc Mundial UNESCO de la Catalunya Central. *This is **Stop 73, Geopark Visitor Center** (41.80449, 1.89708) for a total of 2.5 km (1.6 mi; 3 min).*

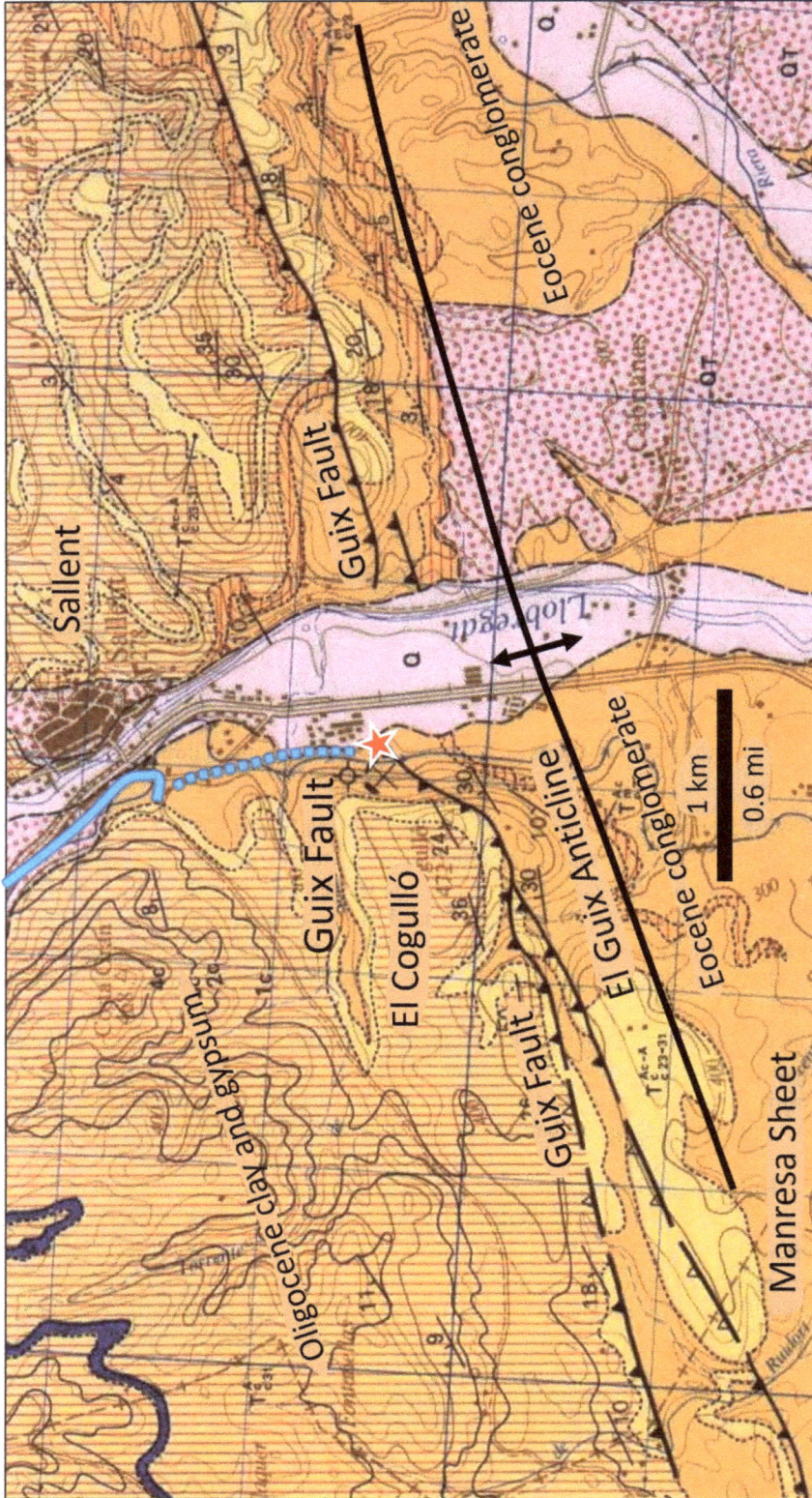

Geologic map of the Sallent area and Guix Fault. From Instituto Geológico y Minéro de España (IGME), Manresa Sheet. Source: © NC Geological Survey of Spain (IGME).

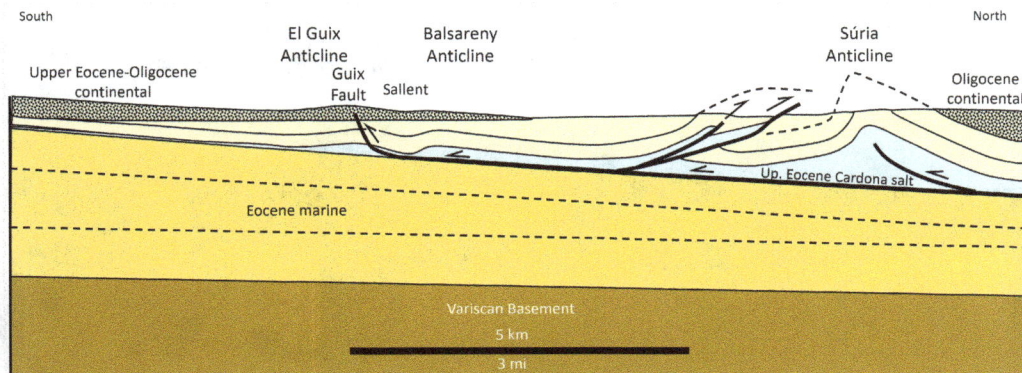

North-south cross-section through the Súria and Guix anticlines. Modified after Sans, 2003.

STOP 73 GEOPARK VISITOR CENTER

This is the main visitor center for all the central Catalan UNESCO Geoparks, including those we have visited: Parc Natural de la Muntanya de Montserrat, Muntanya de Sal de Cardona, and El Migmón. In addition to explanatory displays, there are samples of the various rocks one finds in central Catalonia as well as helpful maps and brochures. A docent is on-hand to answer questions.

The visitor center is open weekends and holidays from 9:30 to 14:00.

Geopark Center to Els Tres Salts: Return to the freeway onramp and turn right (south) to merge onto C-16; continue south onto E-9/C-16 to the northern outskirts of Manresa; take the exit to C-16C/Manresa; use the left lane to merge onto the Ronda Exterior de Manresa/C-55 south to Barcelona; in 1 km (0.6 mi) take the exit to Barcelona/Terrassa; at the roundabout take the 4th exit (east) onto Carretera de Viladordis; in 1 km (0.6 mi) bear right onto Carrer de Vilamajor; turn left (north) at the T intersection; in 0.5 km (0.3 mi) turn right (southeast) at the T intersection; pavement ends near here; cross the Llobregat River and continue to the right (south); drive about 550 m (1,800 ft) and park. Walk 30 m (100 ft) east to the river. This is Stop 74, Els Tres Salts del Llobregat (41.72559, 1.87616) for a total of 14.4 km (9.0 mi; 19 min).

STOP 74 ELS TRES SALTS DEL LLOBREGAT

This is a scenic stop with not much geology. Because sometimes you just need to enjoy the beauty. This is a popular spot for weekend outings and picnics, as it is less than an hour drive from Barcelona.

These small waterfalls on the Llobregat River, a couple of meters in height, are developed in slightly more erosion-resistant layers of Eocene sandstones and conglomerates.

Els Tres Salts to Serra de Collcardús: Return to Carretera de Viladordis and turn right (northeast) and merge onto C-16 (toll road) heading south; take Exit 33 toward Vacarisses/Olesa de Montserrat; make a slight left to merge onto C-58; in 1.2 km (0.7 mi) take the exit to Vacarisses/ Olesa de M; turn right (south) onto B-121/BV-1211; continue south on B-121 for 1.4 km (0.9 mi) and pull over on the right just past the guardrail. This is Stop 75, Serra de Collcardús Thrust (41.573675, 1.919462) for a total of 24.8 km (15.4 mi; 29 min). Carefully walk back northeast ~50 m (150 ft) inside the guardrail to see the outcrops and view the thrust.

Central Catalonia Geopark Visitor Center. Top: display panels. Bottom: rocks exhibit.

Els Tres Salts del Llobregat.

STOP 75 SERRA DE COLLCARDÚS THRUST

We are now at the northern margin of the Catalan Coastal Ranges. North-directed thrusting can be inferred from these roadcuts. As mentioned earlier, the Coast Ranges are the result of Eocene-Oligocene inversion of the Mesozoic Tethyan Basin (sediments deposited in the Tethys Ocean basin).

Here you see north-inclined Eocene Montserrat conglomerate, redbeds consisting of interbedded conglomerates and sandstones shed off the uplifting Catalan Coastal Ranges to the south as a fan delta into the Ebro Basin (Vergés and Burbank, 1996; López-Blanco, 2006). These are the same redbed fan deposits we see at Montserrat.

Across the valley to the south are near-vertical gray Triassic marine carbonates exposed in the E-9/C-16 roadcut. Interestingly, these Middle-Upper Triassic Muschelkalk carbonates are below (older than) the Keuper evaporite level that forms the main detachment surface for Pyrenean thrusting north of here. The detachment here appears to be in shale horizons of the Middle Triassic Buntsandstein redbeds.

Serra de Collcardús to Antiga Estació d'Olesa: *Continue south on B-121/BV-1211 for 1.1 km (0.7 mi; 2 min) and pull into the gap between guardrails on the right just past the second railroad overpass. This is **Stop 76, Antiga Estació d'Olesa Normal Fault** (41.566735, 1.922042).*

Stop 76 Normal Fault, Antiga Estació d'Olesa

This stop is near the old Olesa railway station. A normal fault juxtaposes near-vertical gray conglomerate (probably Middle Triassic Buntsandstein) against a red Paleozoic (Devonian?) slate (Sabadell Sheet), This fault, roughly parallel to the Vallès Fault (next stop), probably formed during late Oligocene-early Miocene rifting of the Valencia Trough to the east.

As you walk north on B-121 from here to the second overpass, you are going up-section from Middle Triassic redbeds into Middle-Upper Triassic carbonates.

*Antiga Estació d'Olesa to Les Ribes Blaves: Continue south on B-121/BV1211; at the stop sign continue straight onto B-120 to Olesa de M; follow the sign and turn left (south) to Ribes Blaves, then immediately turn left again into the parking area. This is **Stop 77, Les Ribes Blaves Cataclasite** (41.556680, 1.917116) for a total of 1.5 km (0.9 mi; 3 min).*

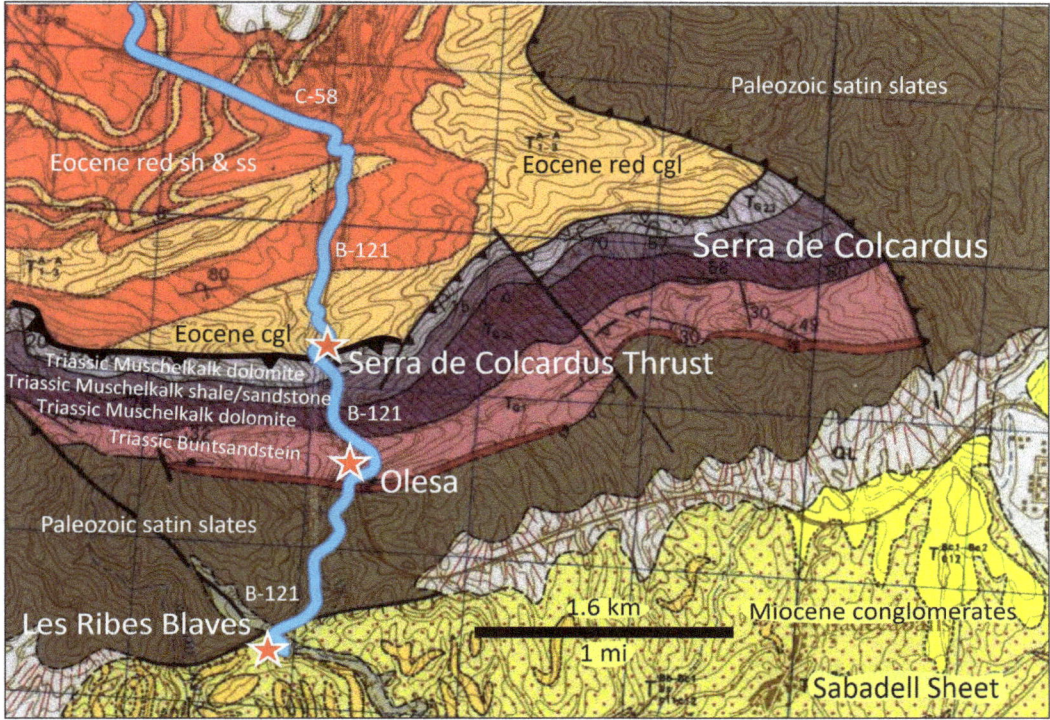

Geologic map of the Serra de Collcardús area. From Instituto Geológico y Minéro de España (IGME), Sabadell Sheet. Source: © NC Geological Survey of Spain (IGME).

Eocene redbeds in roadcut on BV-1211. Middle-Upper Triassic dolomite and sandstone-shale are exposed in the E-9 highway roadcuts in the background. The Serra de Collcardús Thrust lies in between. View south.

This normal fault puts red Devonian slates against gray, near-vertical conglomerates.

Stop 77 Les Ribes Blaves Cataclasite

Ribes Blaves (Blue Ribs, or Blue Shores) is listed by the Catalan Department of Territory and Sustainability as an area of international geological importance. The blue-gray badlands consist of ground up fault rock, or cataclasite, 50–100 m (165 to 330 ft) wide that is derived from crushing of Cambro-Ordovician satin slate. The east-northeast-trending Vallès Fault cuts across the area juxtaposing Paleozoic slate to the north against Miocene alluvial fan material to the south. The slate has been subject to low-grade metamorphism, has a penetrative foliation, and is strongly folded. The Miocene rocks in the southern part of the area are on the downdropped side of the Neogene normal fault. The conglomerates are interbedded with sandstones and clays. The conglomerate zones are graded from coarse at the base to finer upward. Pebbles are derived from metamorphic and calcareous rocks, sandstones, and older conglomerates. The conglomerates are interpreted as colluvial fans, sediments that accumulated at the base of slopes. It was eroded from a nearby source between 16 and 11.6 Ma.

The Vallès Fault was active during the Paleogene Alpine Orogeny as a transpressive structure that combined left-lateral offset and north-south shortening. It was later reactivated as a normal fault during extension associated with rifting of the Valencia Trough to the east (Mármol, 2014). The fault separates the Catalan Coastal Ranges from the Vallès-Penedès Basin to the north and west (Turisme Olesa de Montserrat, Les Ribes Blaves).

Les Ribes Blaves to Pedrera dels Ocells: *Continue south on B-120 to Olesa de Montserrat; at the traffic light turn left (southeast) onto Av. de Francesc de Macià; at the roundabout take the 2nd exit (straight) onto BV-1201/Ctra Martorell; continue southeast on BV-1201 to the C-243 roundabout; take the 3rd exit (south) onto the A-2 to Barcelona; take the E-15 exit to Barcelona; merge right onto C-1413a; at the traffic light turn left (east) onto BV-1468; continue on BV-1468 for 6.7 km (4.1 mi) and turn right (west) into the Ermita de Santa Creu d'Olorda; drive 0.3 km (0.2 mi) and pull over on either side of the road. This is **Stop 78, Devonian Slate, Pedrera dels Ocells, Serra Collserola** (41.415957, 2.057043) for a total of 30.5 km (18.9 mi; 36 min). Walk a few m to the old quarry.*

Ribes Blaves fault gouge along the Vallès Fault.

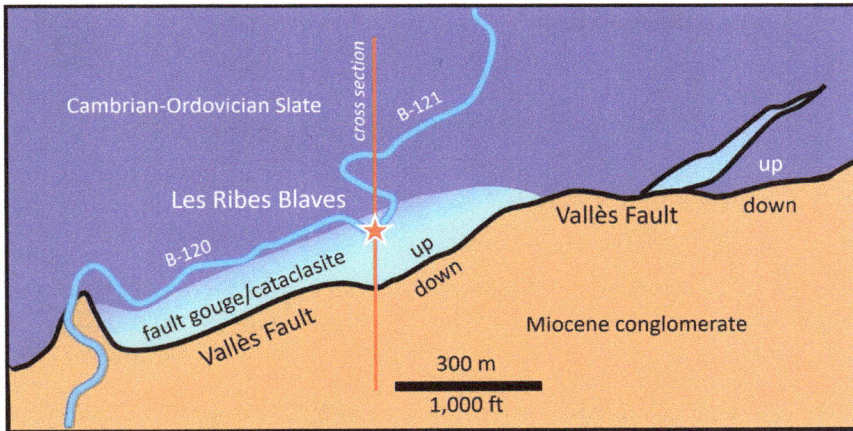

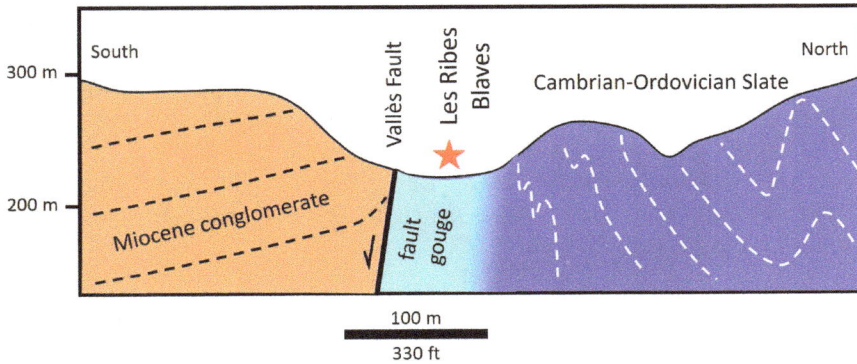

Geologic map and cross-section of the Les Ribes Blaves area. Modified after Mármol, 2014.

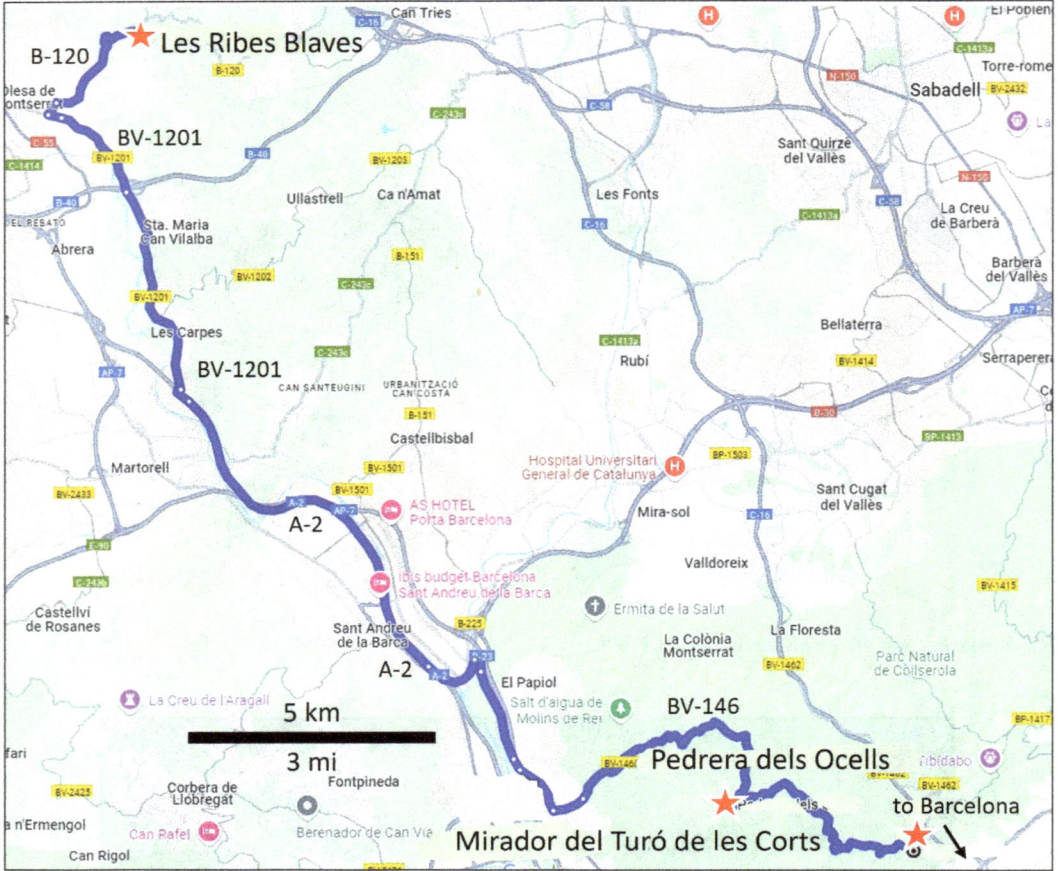

STOP 78 DEVONIAN SLATE, PEDRERA DELS OCELLS, SERRA COLLSEROLA

This stop is at an abandoned quarry that is now a picnic area in the Serra Collserola of the Catalan Coastal Ranges. The quarry was a source for black Devonian slate (360 to 410 Ma). The fact that we are back in truly old rocks, equivalent to the Axial Zone of the Pyrenees, suggests that these mountains too have been uplifted a lot.

The quarry is located in Collserola Natural Park. The black carbonaceous, pyritic slates were crushed to provide sulphate for grapevines. The unstable slopes caused the pit to be closed, but it later became a wetland for the observation and identification of birds. The public is restricted to a walkway.

The acidic runoff from the rocks collects in ponds in the quarry pit due to weathering of iron sulfides. This gives rise to sulfuric acid-rich waters loaded with metals.

Pedrera dels Ocells to Mirador del Turó de les Corts: *Continue east on BV-1468 for 5.7 km (3.5 mi; 9 min) and pull into the parking area on the right. This is Stop 79, Mirador del Turó de les Corts, Serra Collserola (41.407230, 2.101137).*

Pedrera dels Ocells and black carbonaceous Devonian slate.

STOP 79 MIRADOR DEL TURÓ DE LES CORTS, BARCELONA

This stop provides a scenic overview of Barcelona, at the margin between the Catalan Coastal Ranges and the offshore Valencia Trough. Most of the northeast-trending faults in this area are normal faults, down to the south. The rocks at this stop are brown Silurian slates and schists (410 to 440 Ma). They were metamorphosed during the Variscan Orogeny.

Return to Barcelona Airport and End: *Continue east on BV-1468; at the stop sign turn right (south) onto BV-1462 to Barcelona; at the traffic light turn right (southwest) onto Carrer del General Vives to the B-20; merge onto B-20/Ronda de Dalt; take Exit 54 (south) to B-22 toward C-31/Aeroport; take Exit 2 (south) to merge onto C-31 toward Terminal 1/Cargoparc; take Exit 189 (southeast) for B-22; follow the signs to Terminal 1/Car Rental for a total of 21 km (13.0 mi; 23 min).*

Here we end our traverse of the Pyrenees. The mountains formed in the collision zone between the Iberian and European tectonic plates. We have traveled from Barcelona to Jaca in the South Pyrenean Zone and Ebro foreland basin, crossed the width of the mountain range to Oloron St. Marie, traversed the northern Pyrenean Zone and southern Aquitaine Basin from Lourdes to Ax-les-Thermes, and once more crossed the Axial Zone heading south to the Catalan Coastal Ranges at Barcelona. We have seen Variscan structures, Alpine thrusting, salt-detachment folding, sediments that "grew" as structures developed, and alluvial fans that spilled from uplifting mountains. This trek has taken us through two UNESCO Global Geoparks (Sobrarbe Pirineos and Orígens), numerous national parks (Montserrat, Parc National des Pyrénées), natural parks (Parc Natural de la Muntanya de Montserrat, Muntanya de Sal de Cardona, and El Migmón Cadí Moixeró, Collserola), and World Heritage Sites (Fumanya, Monte Perdido, Pyrénées – Mont Perdu). We have seen hot springs, waterfalls, glacial valleys, and sawtooth peaks that merit the name Serra (Sierra). The Pyrenees are one of the great geologic treasures of the world.

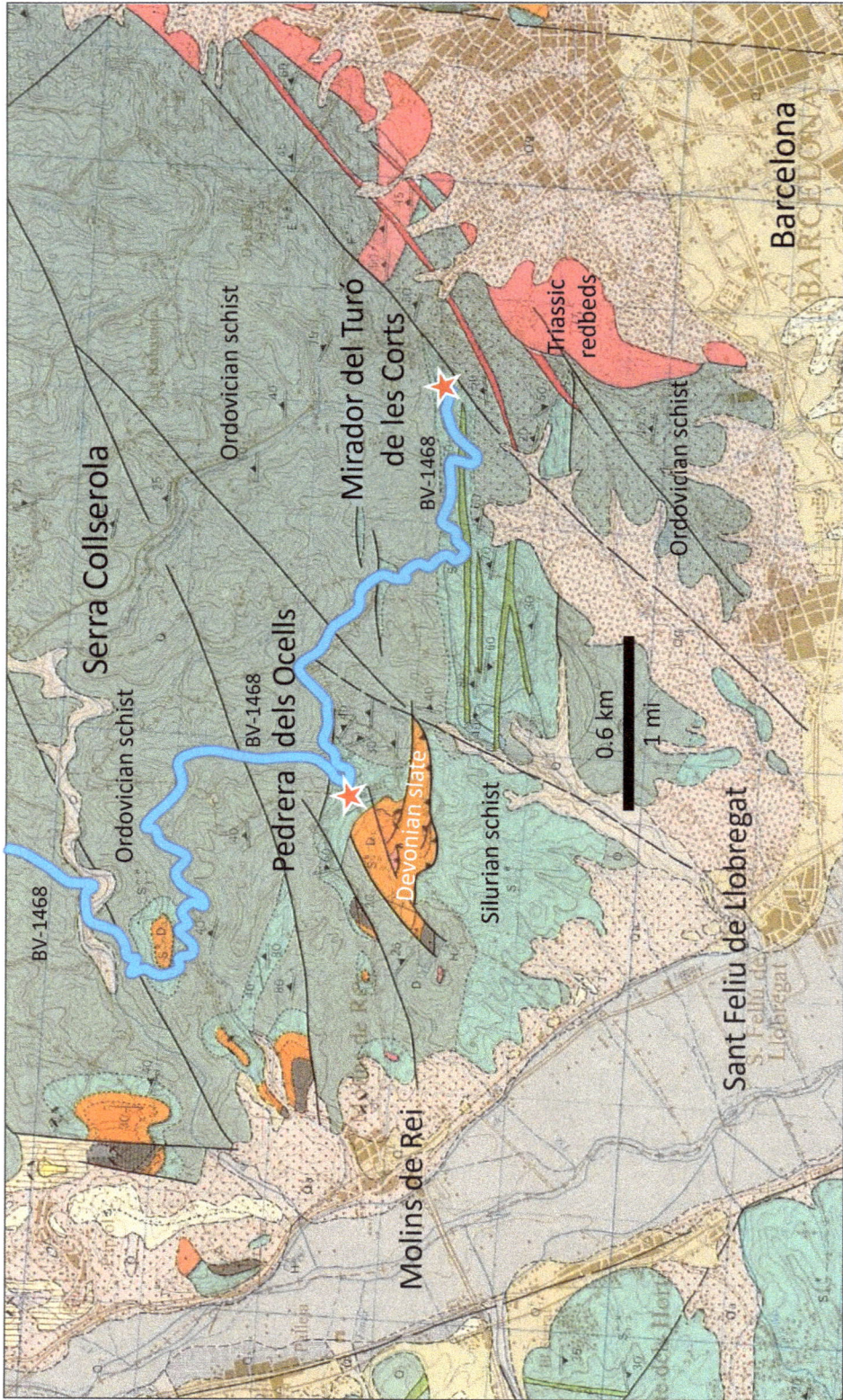

Geologic map of the Serra Collserola north of Barcelona. From Instituto Geológico Y Minéro de España (IGME), Hospitalet de Llobregat Sheet. Source: © NC Geological Survey of Spain (IGME).

Panoramic view south to Barcelona from the Mirador del Turó de les Corts.

REFERENCES

Aerden, D.G.A.M. 1995. The pyrite-type strain fringes from Lourdes (France): Indicators of Alpine thrust kinematics in the Pyrenees. *Journal of Structural Geology* v. 18 no. 1, p. 15–91.

All Pyrenees. 2018. Suria – a point of great geological Interest. Accessed 1 January 2023. https://all-andorra .com/suria/

Angrand, P., F. Mouthereau, E. Masini, and R. Asti. 2020. A reconstruction of Iberia accounting for Western Tethys–North Atlantic kinematics since the late-Permian–Triassic. *Solid Earth* v. 11, p. 1313–1332.

Ardèvol, L., J. Klimowitz, J. Malagón, and P.J.C. Nagtegaal. 2000. Depositional Sequence Response to Foreland Deformation in the Upper Cretaceous of the Southern Pyrenees, Spain. *American Association of Petroleum Geologists Bulletin* v. 84 no. 4, p. 566–587.

Arenas, C., H. Milán, G. Pardo, and A. Pocoví. 2001. Ebro Basin continental sedimentation associated with late compressional Pyrenean tectonics (north-eastern Iberia): Controls on basin margin fans and fluvial systems. *Basin Research* v. 13, p. 65–89.

Ariège Pyrenees. The Mines of the Biros Valley. Accessed 14 April 2024. https://www.ariege.com/en/discover -ariege/geology/mines-of-biros-valley#:~:text=The%20discovery%20of%20a%20seam,of%20Bulard %20and%20the%20Bentaillou

Atlas Obscura – Monastery of San Juan de la Peña. Accessed 19 January 2023. https://www.atlasobscura.com /places/monastery-of-san-juan-de-la-pena

Bahnan, A.E., J. Pironon, C. Carpentier, G. Barre, and E.C. Gaucher. 2021. The diagenetic history of the giant Lacq gas field, witness to the apto-albian riftingand the Pyrenean orogeny, revealed by fluid and basin modeling. *Marine and Petroleum Geology* v. 133, 19 p.

Bakalowicz, M., P. Sorriaux, and D.C. Ford. 1984. *Quaternary glacial events in the Pyrenees from U series dating of speleothems in the Niaux-Lombrives-Sabart caves, Ariège, France.* Norsk geogr Tidsskr, Oslo v. 38, p. 193–197.

Ball, J.E. 2020. *The structural and stratigraphic evolution of Gallego Gorge, Western External Sierras, Spain.* BSc Thesis, University of Aberdeen, 52 p.

Bauluz, B., A. Yuste, M.J. Mayayo, A.B. Rodríguez-Navarro, and J.M. González-López. 2012. Microtexture and genesis of clay minerals from a turbiditic sequence in a Southern Pyrenees foreland basin (Jaca basin, Eocene). *Clay Minerals* v. 47, p. 303–318.

Beamud, E., M. Garcés, L. Cabrera, J.A. Muñoz, and Y. Almar. 2003. A new middle to late Eocene continental chronostratigraphy from NE Spain. *Earth and Planetary Science Letters* v. 216, p. 501–514.

Bentham, P.A., D.W. Burbank, and C. Puigdefàbregas. 1992. Temporal and spatial controls on the alluvial architecture of an axial drainage system: late Eocene Escanilla Formation, southern Pyrenean foreland basin, Spain. *Basin Research* v. 4, p. 335–352.

Bentham, P., and D.W. Burbank. 1996. Chapter E13, chronology of eocene foreland basin evolution along the western oblique margin of the South-Central Pyrenees. In Friend, P.F., and C.J. Dabrio (eds), *Tertiary basins of Spain, The stratigraphic record of crustal kinematics.* Cambridge University Press, Cambridge, p. 144–152.

Berástegui, X., M. Losantos, J.A. Muños, and C. Puigdefàbregas. 1993. Tall geològic del Pirineu central 1:200 000. Map and explanations. Servei Geològic de Catalunya, Barcelona, 58 p.

Berti, Padre Paolo. Lourdes: Geology of the Grotto of Massabielle, Apparitions of Our Lady, St Bernadette. Accessed 20 January 2023. http://www.perfettaletizia.it/archivio/servizi/madonna_lourdes/english _massabielle.html

Bond, R. 2021. Cardona: Salt of the Earth. Metropolitan Barcelona. Accessed 19 January 2023. https://www .barcelona-metropolitan.com/travel/cardona-salt-mountain/

Bouchez, J.L., and G. Gleizes. 1995. Two-stage deformation of the Mont-Louis-Andorra granite pluton (Variscan Pyrenees) inferred from magnetic susceptibility anisotropy. *Journal of the Geological Society* v. 152 no. 4, p. 669–679.

Bosch, G.V., A. Teixell, M. Jolivet, P. Labaume, D. Stockli, M. Domènech, and P. Monié. 2016. Timing of eocene–miocene thrust activity in the Western Axial Zone and Chaînons Béarnais (west-central Pyrenees) revealed by multi-method thermochronology. *Comptes Rendus Geoscience* v. 348, p. 246–256.

Bourrouilh, R., L. Moen-Maurel, J. Muñoz, and A. Teixell. 2004. *Field trip guidebook B16, western pyrenees fold-and-thrust belt: Geodynamics, sedimentation and plate boundary reconstruction from rifting to inversion.* 32nd International Geological Congress, Italian Agency for the Environmental Protection and Technical Services (APAT), Rome, 64 p.

Burbank, D.W., J. Vergés, J. Anton Muñoz, and P. Bentham. 1992. Coeval hindward- and forward-imbricating thrusting in the south-central Pyrenees, Spain: Timing and rates of shortening and deposition. *Geological Society of America Bulletin* v. 104 no. 1, p. 3–17.

Burrel, L., A. Teixell, D. Gomez-Gras, and X. Coll. 2021. Basement-involved thrusting, salt migration and intramontane conglomerates: A case from the Southern Pyrenees. *EDP Sciences.* Accessed 19 January 2023. https://www.bsgf.fr or https://doi.org/10.1051/bsgf/2021013

Cantalejo, K., K.T. Pickering, C. McNiocaill, P. Bown, K. Johansen, and M. Grant. 2020. A revised age-model for the Eocene deep-marine siliciclastic systems, Aínsa Basin, Spanish Pyrenees. *Journal of the Geological Society*, v. 178. 18 p. Accessed 22 January 2025. https://www.researchgate.net/publication /343521338_A_revised_age-model_for_the_Eocene_deep-marine_siliciclastic_systems_Ainsa_Basin _Spanish_Pyrenees

Carola, E., O. Rerrer, O. Vidal-Royo, and J.A. Muñoz. 2017. Interpretation of salt-cored frontal structures in the Southern Pyrenees guided by analog modeling, surface and subsurface data. *Interpretation.* Accessed 19 January 2023. https://library.seg.org/doi/full/10.1190/INT-2016-0093.1

Carreras, J. 2022. Cardona Salt Mountain. Barcelona Field Studies Centre. Accessed 19 January 2023. https:// geographyfieldwork.com/CardonaSaltMountain.htm.

Carrigan, J.H., D.J. Anastasio, K.P. Kodama, and J.M. Parés. 2016. Fault-related fold kinematics recorded by terrestrial growth strata, Sant Llorenç de Morunys, Pyrenees Mountains, NE Spain. *Journal of Structural Geology*, v. 91, p. 161–176.

Carter, S.J., and P. Claughton. 2018. Mining, memory and the mountain: Iron mining on the Canigou Massif (Pyrénées-Orientales, France). *Journal of the Mining Heritage Trust of Ireland* v. 16, p. 21–32.

Catalonia Museums. Dinosaur Footprints at the Orcau 2 Site. Accessed 19 January 2023. https://visitmuseum .gencat.cat/en/museu-de-la-conca-della/extension/jaciment-de-petjades-d-orcau-2

Clariana, P. 2017. *Geological features of Variscan Pyrenees.* International Conodont Symposium 4, Valencia. Accessed 19 January 2023. https://www.researchgate.net/publication/319645568_Geological_features _of_Variscan_Pyrenees_International_Conodont_Symposium_4

Clariana, P., R. Soto, C. Ayala, A.M. Casas-Sainz, T. Román-Berdiel, B. Oliva-Urcia, E.L. Pueyo, C. Rey-Moral, F. Rubio, A. Margalef, S. Schamuells, N. Bach, and J. Martí. 2021. Basement and cover architecture in the Central Pyrenees constrained by gravity data. *International Journal of Earth Sciences* v. 111, p. 641–658.

Cofrade, G., P. Závada, O. Krýza, I. Cantarero, Òscar Gratacós, O. Ferrer, S. Adineh, P. Ramirez-Perez, E. Roca, and A. Travé. 2023a. The kinematics of a salt sheet recorded in an array of distorted intrasalt stringers (Les Avellanes Diapir – South-Central Pyrenees). *Journal of Structural Geology* v. 176, 24 p.

Cofrade, G., I. Cantarero, Òscar Gratacós, O. Ferrer, P. Ramirez-Perez, A. Travé, and E. Roca. 2023b. Allochthonous salt advance recorded by the adjacent *syn*-kinematic sedimentation: Example from the Les Avellanes diapir (South Central Pyrenees). *Global and Planetary Change* v. 220, 26 p.

Coll, X., M. Roigé, D. Gómez-Gras, A. Teixell, S. Boya, and N. Mestres. 2022. Interplay of multiple sediment routing systems revealed by combined sandstone petrography and heavy mineral analysis (HMA) in the South Pyrenean Foreland Basin. *Minerals* 12, 262, 29 p.

Connexion. 2020. Rare Gold Nugget Found in Pyrenees Hills. Accessed 14 April 2024. https://www.conne xionfrance.com/news/rare-gold-nugget-found-in-pyrenees-hills/399535#:~:text=The%20story%20of %20the%20prospector,river%20in%20the%20Pyrenees%20foothills.&text=Ken%20Degorgue%2C %2031%2C%20who%20works,joy%2C%E2%80%9D%20said%20Mr%20Lamarque

Dakin, N.C. 2016. *Geometry, architecture and the erosive nature of mass-transport deposits in deep-marine environments: Ainsa Basin (Pyrenees) and Buzzard Field (North Sea).* PhD dissertation, University College London, 426 p.

DeLisle, A. 2016. *Analysis of paleofluids in the Sant Corneli Anticline, South-Central Spanish Pyrenees.* Graduate Theses, Dissertations, and Problem Reports. 5467, West Virginia University, Morgantown. https://researchrepository.wvu.edu/etd/5467

de Saint Blanquat, M., F. Bajolet, A. Grand'Homme, A. Proietti, M. Zanti, A. Boutin, C. Clerc, Y. Lagabrielle, and P. Labaume. 2016. Cretaceous mantle exhumation in the central Pyrenees: New constraints from the peridotites in eastern Ariège (North Pyrenean zone, France). *Comptes Rendus Geoscience* v. 348, p. 268–278.

Diaz, J., J. Vergés, S. Chevrot, A. Antonio-Vigil, M. Ruiz, M. Sylvander, and J. Gallart. 2018. Mapping the crustal structure beneath the eastern Pyrenees. *Tectonophysics, Elsevier* v. 744, p. 296–309.

Dobrzynski, D., and D. Rossi. 2017. Geochemistry of trace elements in spring waters of the Lourdes Area (France). *Annales Societatis Geologorum Poloniae* v. 87, p. 199–212.

Edelmayer, F. 2018. The peoples, languages, and history of the pyrenees region. *Brewminate.* Accessed 19 January 2023. https://brewminate.com/the-peoples-languages-and-history-of-the-pyrenees-region/

Enge, H.D., S.J. Buckley, A. Rotevatn, and J.A. Howell. 2007. From outcrop to reservoir simulation model: Workflow and procedures. *Geosphere* v. 3 no. 6, p. 469–490.

Evans, M.A., and M. Fischer. 2012. On the distribution of fluids in folds: A review of controlling factors and processes. *Journal of Structural Geology* v. 44, p. 2–24.

Fernández, O., J.A. Muñoz, P. Arbués, and O. Falivene. 2012. 3D Structure and evolution of an oblique system of relaying folds: The Ainsa basin (Spanish Pyrenees). *Journal of the Geological Society* v. 169, p. 545–559.

Fernández, E. A geological race (Broto). Accessed 19 January 2023. https://www.geoparquepirineos.com/ contenidos.php?niv=&cla=_2OA1CD0KM&cla2=_2OB01HU8N&cla3=_3MD0KXOB9&tip=3&idi=3

Ford, M., E.A. Williams, A. Artoni, J. Vergés, and S. Hardy. 1997. Progressive evolution of a fault-related fold pair from growth strata geometries, Sant Llorenç de Morunys, SE Pyrenees. *Journal of Structural Geology* v. 19 no. 3–4, p. 413–441.

French Geological Survey, Garvemip Project. Accessed 30 June 2025. https://carto.picto-occitanie.fr/1/geolog ie.map

French Investment Property, Midi-Pyrénées History. Accessed 19 January 2023. https://www.regions-of -france.com/regions/midi_pyrenees/history

Friend, P.F., M.J. Lloyd, R. McElroy, J. Turner, A. van Gelder, and S.J. Vincent. 1996. Chapter El6, evolution of the central part of the northern Ebro Basin margin, as indicated by its Tertiary Fluvial Sedimentary Infill. In Friend, P.F., and C.J. Dabrio (eds), *Tertiary basins of Spain, the stratigraphic record of crustal kinematics.* Cambridge University Press, Cambridge, p. 166–172.

Gardiner, A., P. Corbett, and H. Lewis. 2003. Field excursion to the Southern Pyrenean and ebro foreland basins, Spain, 26th April to 4th May, 2003. Hariot Watt University, Edinburgh, 146 p.

Gleizes, G., A. Nédélec, J.-L. Bouchez, A. Autran, and P. Rochette. 1993. Magnetic susceptibility of the Mont-Louis Andorra Ilmenite-Type Granite (Pyrenees): A new tool for the petrographic characterization and regional mapping of zoned granite plutons. *Journal of Geophysical Research* v. 98 no. B3, p. 4317–4331.

Grool, A.R., M. Ford, J. Vergés, R.S. Huismans, F. Christophoul, and A. Dielforder. 2018. Insights into the crustal-scale dynamics of a doubly vergent orogen from a quantitative analysis of its forelands: A case study of the Eastern Pyrenees. *Tectonics* v. 37, p. 450–476.

Gutierrez, M. 2013. *Geomorphology.* Taylor and Francis CRC Press, Boca Raton, FL, 1014 p.

Gutmanis, J., L. Ardèvol I Oró, D. Díez-Canseco, L. Chebbihi, A. Awdal, and A. Cook. 2017. Fracture analysis of outcrop analogues to support modelling of the subseismic domain in carbonate reservoirs, south-central Pyrenees. In Ashton, M., S.J. Dee, and O.P. Wennberg (eds), *Subseismic-scale reservoir deformation.* Geological Society, London, Special Publications, 459. https://doi.org/10.1144/SP459.2

Haldar, S.K. 2017. Chapter 2 – Geology and geochemistry. In *Platinum-Nickel-Chromium Deposits* - Geology, *exploration and reserve base.* Elsevier, p. 37–61.

Hansson, S.V., Y. Grusson, M. Chimienti, A. Claustres, S. Jean, and G. Le Roux. 2019. Legacy Pb pollution in the contemporary environment and its potential bioavailability in three mountain catchments. *Science of the Total Environment* v. 671, p. 1227–1236.

Harman, A. 2013. Cardona Salt Mountain. Atlas Obscura. Accessed 19 January 2023. https://www.atlasobscura.com/places/cardona-salt-mountain

Hoffman, M.R. 2009. *Tectono-stratigraphic analysis of a deepwater growth basin, Ainsa Basin, Northern Spain.* Colorado School of Mines, MSc thesis, 203 p.

Hogan, P.J., and D.W. Burbank. 1996. Chapter E14, evolution of the Jaca piggyback basin and emergence of the External Sierra, Southern Pyrenees. In Friend, P.F., and C.J. Dabrio (eds), *Tertiary basins of Spain, the stratigraphic record of crustal kinematics.* Cambridge University Press, Cambridge, p. 153–160.

Huyghe, D., S. Castelltort, J. Serra-Kiel, P.-Y. Filleaudeau, L. Emmanuel, F. Mouthereau, and M. Renard. 2009. Sedimentology and stratigraphy of the middle Eocene Guara carbonate platform near Arguis, South-West Pyrenean foreland: Implications for basin physiography. *Geophysical Research Abstracts* v. 11, EGU2009-5439.

Institució Catalana d'Història Natural. 2022. El medi natural del Barges, Fault of El Guix (Sallent). Accessed 19 January 2023. https://elmedinaturaldelbages.cat/en/

Instituto Geográfico Nacional. Geology and Tectonics of the Catalan Coastal Range. Accessed 19 January 2023. https://www.ign.es/web/resources/sismologia/tproximos/sismotectonica/pag_sismotectonicas/costerocatalana_en.html

Instituto Geológico Y Minéro de España (IGME), 1991. Agüero Sheet 209, 1:50,000. Accessed 19 January 2023. http://info.igme.es/cartografiadigital/datos/magna50/jpgs/d2_G50/Editado_MAGNA50_209.jpg

Instituto Geológico Y Minéro de España (IGME), 1989. Anso Sheet 144, 1:50,000. Accessed 19 January 2023. https://info.igme.es/cartografiadigital/geologica/Magna50Hoja.aspx?Id=144&language=en

Instituto Geológico Y Minéro de España (IGME), 1991. Apies Sheet 248, 1:50,000. Accessed 19 January 2023. http://info.igme.es/cartografiadigital/datos/magna50/jpgs/d2_G50/Editado_MAGNA50_248.jpg

Instituto Geológico Y Minéro de España (IGME), 2000. Artesa de Segre Sheet 328, 1:50,000. Accessed 19 January 2023. http://info.igme.es/cartografiadigital/datos/magna50/jpgs/d3_G50/Editado_MAGNA50_328.jpg

Instituto Geológico Y Minéro de España (IGME), 1990. Barbastro Sheet 287, 1:50,000. Accessed 19 January 2023. http://info.igme.es/cartografiadigital/datos/magna50/jpgs/d2_G50/Editado_MAGNA50_287.jpg

Instituto Geológico Y Minéro de España (IGME), 1989. Berga Sheet 293, 1:50,000. Accessed 19 January 2023. http://info.igme.es/cartografiadigital/datos/magna50/jpgs/d2_G50/Editado_MAGNA50_293.jpg

Instituto Geológico Y Minéro de España (IGME), Bielsa Sheet 179, 1:50,000. Accessed 19 January 2023. http://info.igme.es/cartografiadigital/datos/magna50/jpgs/d1_G50/Editado_MAGNA50_179.jpg

Instituto Geológico Y Minéro de España (IGME), 1973. Calaf Sheet 362, 1:50,000. Accessed 19 January 2023. http://info.igme.es/cartografiadigital/datos/magna50/jpgs/d3_G50/Editado_MAGNA50_362.jpg

Instituto Geológico Y Minéro de España (IGME), 1999. Campo Sheet 212, 1:50,000. Accessed 19 January 2023. http://info.igme.es/cartografiadigital/datos/magna50/jpgs/d2_G50/Editado_MAGNA50_212.jpg

Instituto Geológico Y Minéro de España (IGME), 1973. Cardona Sheet 330, 1:50,000. Accessed 19 January 2023. http://info.igme.es/cartografiadigital/datos/magna50/jpgs/d3_G50/Editado_MAGNA50_330.jpg

Instituto Geológico Y Minéro de España (IGME), 1994. Fonz Sheet 288, 1:50,000. Accessed 19 January 2023. http://info.igme.es/cartografiadigital/datos/magna50/jpgs/d2_G50/Editado_MAGNA50_288.jpg

Instituto Geológico Y Minéro de España (IGME), 1973. Hospitalet de Llobregat Sheet 420, 1:50,000. Accessed 19 January 2023. http://info.igme.es/cartografiadigital/datos/magna50/jpgs/d4_G50/Editado_MAGNA50_420.jpg

Instituto Geológico Y Minéro de España (IGME), 1990. Jaca Sheet 176, 1:50,000. Accessed 19 January 2023. https://info.igme.es/cartografiadigital/geologica/Magna50Hoja.aspx?language=en&id=176

Instituto Geológico Y Minéro de España (IGME), 1991. La Pobla de Lillet Sheet 255, 1:50,000. Accessed 19 January 2023. http://info.igme.es/cartografiadigital/datos/magna50/jpgs/d2_G50/Editado_MAGNA50_255.jpg

Instituto Geológico Y Minéro de España (IGME), 1973. Manresa Sheet 363, 1:50,000. Accessed 19 January 2023. http://info.igme.es/cartografiadigital/datos/magna50/jpgs/d3_G50/Editado_MAGNA50_363.jpg

Instituto Geológico Y Minéro de España (IGME), Puigreig Sheet 331, 1:50,000. Accessed 19 January 2023. http://info.igme.es/cartografiadigital/datos/magna50/jpgs/d3_G50/Editado_MAGNA50_331.jpg

Instituto Geológico Y Minéro de España (IGME), 1975. Sabadell Sheet 392, 1:50,000. Accessed 19 January 2023. https://info.igme.es/cartografiadigital/geologica/Magna50Hoja.aspx?language=en&id=392

Instituto Geológico Y Minéro de España (IGME), 1987. Sallent Sheet 145, 1:50,000. Accessed 19 January 2023. http://info.igme.es/cartografiadigital/datos/magna50/jpgs/d1_G50/Editado_MAGNA50_145.jpg

Instituto Geológico Y Minéro de España (IGME), 1991. Yebra de Basa Sheet 210, 1:50,000. Accessed 19 January 2023. http://info.igme.es/cartografiadigital/datos/magna50/jpgs/d2_G50/Editado_MAGNA50_210.jpg

Kendall, J., J. Vergés, R. Koshnaw, and M. Louterbach. 2020. Petroleum tectonic comparison of fold and thrust belts: The Zagros of Iraq and Iran, the Pyrenees of Spain, the Sevier of Western USA and the Beni Sub-Andean of Bolivia. In Hammerstein, J.A., R. Di Cuia, M.A. Cottam, G. Zamora, and R.W.H. Butler (eds), *Fold and thrust belts: Structural style, evolution and exploration.* Geological Society, London, Special Publications, 490, p. 79–103.

Kucavana. 2020. Ax les Thermes: Free and Paid Hot Springs in France in the Middle of a Spa Town of skiers. Accessed 19 January 2023. https://www.kucavana.es/en/aguas-termales-en-francia/

Labaume, P., E. Mutti, and M. Seguret. 1987. Megaturbidites: A depositional model from the eocene of the SW-Pyrenean Foreland Basin, Spain. *Geo-Marine Letters* v. 7, p. 91–101.

Labaume, P., F. Meresse, M. Jolivet, A. Teixell, and A. Lahfid. 2016. Tectonothermal history of an exhumed thrust-sheet-top basin: An example from the south Pyrenean thrust belt. *Tectonics* v. 35, p. 1280–1313.

Labaume, P., and A. Teixell. 2018. 3D structure of subsurface thrusts in the eastern Jaca Basin, southern Pyrenees. Geologica Acta v. 16 no. 4, p. 477-498.

Lagabriel, Y., C. Clerc, A. Vauchez, A. Lahfid, P. Labaume, B. Azambre, S. Fourcade, and J-M. Dautria. 2016. Very high geothermal gradient during mantle exhumation recorded in mylonitic marbles and carbonate breccias from a Mesozoic Pyrenean palaeomargin (Lherz area, North Pyrenean Zone, France). *Comptes Rendus Geoscience v. 348,* 11 p.

La Moncloa, Social History of the Pyrenees. Accessed 20 January 2023. https://www.lamoncloa.gob.es/lang/en/espana/historyandculture/history/Paginas/index.aspx

Läuchli, C., M. Garcés, E. Beamud, L. Valero, L. Honegger, T. Adatte, J.E. Spangenberg, J. Clark, C. Puigdefàbregas, A. Fildani, E. de Kaenel, T. Hunger, A. Nowak, and S. Castelltort. 2021. Magnetostratigraphy and stable isotope stratigraphy of the middle-Eocene succession of the Ainsa basin (Spain): New age constraints and implications for sediment delivery to the deep waters. *Marine and Petroleum Geology* v. 132, 23 p.

Laurent, D., C. Durlet, G. Barre, P. Sorriaux, P. Audra, P. Cartigny, C. Carpentier, G. Paris, P.P. Collon, T. Rigaudier et al. 2021. Epigenic vs. hypogenic speleogenesis governed by H2S/CO2 hydrothermal input and Quaternary icefield dynamics (NE French Pyrenees). *Geomorphology* v. 387, p.107769.

Le Bayon, B., and B. Cochelin. 2020. *Anatomy of an extensional shear zone leading to the exhumation of the middle crust within the Canigou dome (Eastern Pyrenees, Axial Zone).* Elsevier. Accessed 19 January 2023. https://www.sciencedirect.com/science/article/pii/S019181412030420X

López -Blanco, M., M. Marzo, and J.A. Muñoz. 2003. Low-amplitude, synsedimentary folding of a deltaic complex: Roda Sandstone (lower Eocene), South-Pyrenean Foreland Basin. *Basin Research* v. 15, p. 73–95.

López-Blanco, M. 2006. Stratigraphic and tectonosedimentary development of the Eocene Sant Llorenç del Munt and Montserrat fan-delta complexes (Southeast Ebro basin margin, Northeast Spain). *Contributions to Science* v. 3 no. 2, p. 125–148.

López-Mir, B. 2013. *Extensional salt tectonics in the Cotiella post-rift basin (south-central Pyrenees): 3D structure and evolution.* PhD dissertation, Universitat Barcelona, 302 p.

López-Mir, B., J.A. Muñoz, and J. García-Senz. 2014. Extensional salt tectonics in the partially inverted Cotiella post-rift basin (south-central Pyrenees): structure and evolution. *International Journal of Earth Science.* v. 104, p. 419–434. Accessed 20 January 2023. https://www.researchgate.net/publication/268132251

López-Mir, B., J.A. Muñoz, and J. García-Senz. 2016. Geology of the Cotiella thrust sheet, Southern Pyrenees (Spain). *Journal of Maps* v. 12 no. S1, p. 323–327.

Lucha, P., F. Gutiérrez, and J. Guerrero. 2008. Environmental problems and geological implications derived from evaporite dissolution in the Barbastro salt anticline (NE Spain). *Environmental Geology* v. 53, p. 1045–1055.

MacroStrat. 2015. Accessed multiple times between 2022 and 2025, https://macrostrat.org/map/#/z=7.4/x=7 .3387/y=45.5845/bedrock/lines/

Maestro-Maideu, E., and J. Serra Roig. 1996. Chapter E12, the late eocene-early oligocene deposits of the NE Ebro Basin, West of the Segre River. In Friend, P.F., and C.J. Dabrio (eds), *Tertiary basins of Spain.* Cambridge University Press, Cambridge, p. 134–143.

Mármol, J.C. 2014. Geologia de Ribes Blaves – Riera de Sant Jaume: Una prova pilot d'integració de la informació geològica amb aplicacions de realitat augmentada. End of Degree Project, 30 p. Accessed 22 January 2023. https://ddd.uab.cat/pub/tfg/2014/125884/TFG_JordiCerezoMarmol.pdf

Martinius, A.W. 2012. Contrasting styles of siliciclastic tidal deposits in a developing thrust-sheet-top basins – The lower eocene of the Central Pyrenees (Spain). In Davis, R.A. Jr. and R.W. Dalrymple (eds.), *Principles of tidal sedimentology*, Springer, p. 473–506.

Maurel, O., P. Monie, R. Pik, N. Arnaud, M. Brunel, and M. Jolivet. 2008. The Meso-Cenozoic thermo-tectonic evolution of the Eastern Pyrenees: An $^{40}Ar/^{39}Ar$ fission track and (U–Th)/He thermochronological study of the Canigou and Mont-Louis massifs. *International Journal of Earth Science* v. 97, p. 565–584.

Melendez-Hevia, F., and E. Alvarez De Buergo. 1996. *Tertiary basins of Spain: Chapter 5, oil and gas resources of the Tertiary basins of Spain.* Cambridge University Press, Cambridge, p. 20–23.

Mencos, J., J.A. Muñoz, and S. Hardy. 2010. 3D kinematics of the Sant Corneli anticline: Insights from structural reconstruction and forward modelling. *Trabajos de Geología, Universidad de Oviedo* v. 30, p. 75–80.

Milesi, G., P. Monie, R. Soliva, P. Münch, P. Valla, S. Brichau, M. Bonno, C. Martin, and M. Bellanger. 2022. Deciphering the cenozoic exhumation history of the Eastern Pyrenees along a crustal-scale normal fault using low-temperature thermochronology. *Tectonics* v. 41 no. 4, 27 p.

Mindat, Salt Mountain, Cardona. Accessed 20 January 2023. https://www.mindat.org/loc-55717.html

Mining Heritage of Sobrarbe. Iron and Silver in the High Valleys of the Cinca River. Accessed 14 April 2024. chrome-extension://efaidnbmnnnibpcajpcglclefindmkaj/https://www.geoparquepirineos.com/descargas /mineria_ingles.pdf

Mochales, T., E.L. Pueyo, A.M. Casas, and A. Barnolas. 2010. Kinematic approach by means of AMS study in the Boltaña anticline (Southern Pyrenees). *Trabajos de Geología, Universidad de Oviedo* v. 30, p. 87–95.

Mochales, T., A. Barnolas, E.L. Pueyo, J. Serra-Kiel, A.M. Casas, J.M. Samsó, J. Ramajo, and J. Sanjuán. 2012. Chronostratigraphy of the Boltaña anticline and the Ainsa Basin (southern Pyrenees). *Geological Society of America Bulletin* v. 124 no. 7/8; p. 1229–1250.

Mochales López, T. 2011. *Chronostratigraphy, vertical axis rotations and AMS in the Boltaña anticline (Southern Pyrenees): Kinematic implications.* PhD dissertation, Universidad de Zaragoza, 222 p.

Mostaslavski, A. 2019. The history of the Pyrenees: chronology of major facts. Accessed 20 January 2023. https://all-andorra.com/history-of-the-pyrenees/

Mountain Bug, History of the Pyrenees. Accessed 20 January 2023. https://www.mountainbug.com/pyrenees -history/

Muñoz, J.A. 1992. Evolution of a continental collision belt: ECORS – Pyrenees crustal balanced cross – section. *In* McClay, K.R. (ed.): Thrust Tectonics. Chapman & Hall, London. p. 235-246.

Muñoz, J.-A., E. Beamud, O. Fernández, P. Arbués, J. Dinarès-Turell, and J. Poblet. 2013. The Ainsa fold and thrust oblique zone of the Central Pyrenees: Kinematics of a curved contractional system from paleomagnetic and structural data. *Tectonics* v. 32, p. 1142–1175.

Muñoz, J.A. 2017. Fault-related folds in the southern Pyrenees. *American Association of Petroleum Geologists Bulletin* v. 101 no. 4, p. 579–587.

Mutti, E., 1977. Distinctive thin-bedded turbidite facies and related depositional environments in the Eocene Hecho Group (South Central Pyrenees, Spain). Sedimentology, v. 24, p. 107- 131.

Mutti, E., M. Seguret, and M. Sgavetti. 1989. *Sedimentation and deformation in the tertiary sequences of the Southern Pyrenees.* American Association of Petroleum Geologists Mediterranean Basins Conference Guidebook Field Trip No. 7, Nice, Special Publication of the Institute of Geology, University of Parma.

Nichols, G. 2005. *Tertiary alluvial fans at the northern margin of the Ebro Basin: A review.* Geological Society, London, Special Publications no. 251, p. 187–206.

Novoa, E., J. Suppe, and J.H. Shaw. 2000. Inclined-shear restoration of growth folds. *American Association of Petroleum Geologists Bulletin* v. 84 no. 6, p. 787–804.

O'Brien, W. 2014. *Prehistoric copper mining in Europe: 5500-500 BC*. Oxford University Press, Oxford, Chap. 1055 France and the western Alps, p. 105–123.

Oms, O., J. Dinares-Turell, and E. Remacha. 2003. Magnetic stratigraphy from deep clastic turbidites: An example from the Eocene Hecho Group (Southern Pyrenees). *Studia Geophysica et Geodaetica* v. 47, p. 275–288.

P-O Life. 2024. A (very brief) history of the Pyrenees Orientales. Accessed 14 April 2024. https://anglophone-direct.com/history-of-the-pyrenees-orientales/

Payros, A., V. Pujalte, and X. Orue-Etxebarria. 1999. The South Pyrenean Eocene carbonate megabreccias revisited: New interpretation based on evidence from the Pamplona Basin. *Sedimentary Geology* v. 125, p. 165–194.

Pedley, A. 1994. *Eocene foreland basin carbonate facies, the External Sierras, Spanish Pyrenees*. PhD dissertation, Royal Holloway, University of London, p. 49–72.

Pickering, K.T., and J. Corregidor. 2005. Mass-transport complexes (MTCs) and tectonic control on basin-floor submarine fans, middle Eocene, south Spanish Pyrenees. Journal of Sedimentary Research v. 75, p. 761-783.

Pickering, K.T., J. Corregidor, and J.D. Clark. 2015. Architecture and stacking patterns of lower-slope and proximal basin-floor channelised submarine fans, Middle Eocene Ainsa System, Spanish Pyrenees: An integrated outcrop–subsurface study. *Earth-Science Reviews* v. 144, p. 47–81.

Pomar, L., E. Gili, A. Obrador, W.C. Ward. 2005. Facies architecture and high-resolution sequence stratigraphy of an Upper Cretaceous platform margin succession, Southern Central Pyrenees, Spain. *Sedimentary Geology* 175, p. 339–365.

Py-Saragaglia, V., et al. 2017. Late Holocene history of woodland dynamics and wood use in an ancient mining area of the Pyrenees (Ariege, France). *Quaternary International*. Accessed 14 April 2024. http://dx.doi.org/10.1016/j.quaint.2017.01.012

Rafini, S., and E. Mercier. 2002. Forward modelling of foreland basins progressive unconformities. *Sedimentary Geology* 146, p. 75–89.

Ramos, A., B. López-Mir, E.P. Wilson, P. Granado, and J.A. Muñoz. 2020. 3D reconstruction of syn-tectonic strata deposited during the inversion of salt-related structures: Insights from the Llert syncline (South-Central Pyrenees). *Geologica Acta* v. 18.20, p. 1–19.

Riba, O., 1973. Las discordancias sintectónicas del Alto Cardener (Prepirineo catalán), ensayo de interpretación evolutiva. *Acta Geológica Hispánica* v. 8 no. 3, p. 90–99.

Riba, O., 1976. Syntectonic unconformities of the Alto Cardener, Spanish Pyrenees: A genetic interpretation. *Sedimentary Geology* v. 15, p. 213–233.

Rius, A.M. 2022. Fumanya Dinosaurs. Accessed 20 January 2023. https://geographyfieldwork.com/FumanyaDinosaurs.htm

Roberti, K. 2008. *Revisiting the past, Leiden back in the Pyrenees*. LGV 75th anniversary Pyrenees Field Trip, June 11–17, 2008. University Leiden, 73 p. Accessed 20 January 2023. https://www.icgc.cat/content/download/74032/637621/version/5/file/IGC_LGV_2008_fieldtrip.pdf

Rodriguez, V. 2022. Barcelona. Britannica Online. Accessed 19 January 2023. https://www.britannica.com/place/Barcelona

Rodríguez-Méndez, L., J. Cuevas, and J.M. Tubía. 2013. Geological map of the Central Pyrenees between the Tena and Aragon valleys (Huesca). *Journal of Maps* v. 9 no. 4, p. 596–603.

Rougier, G., M. Ford, F. Christophoul, and A.-G. Bader. 2016. Stratigraphic and tectonic studies in the central Aquitaine Basin, northern Pyrenees: Constraints on the subsidence and deformation history of a retro-foreland basin. Comptes Rendus Geoscience v. 348, p. 224-235.

Roigé, M., D. Gómez-Gras, E. Remacha, R. Daza, and S. Boya. 2016. Tectonic control on sediment sources in the Jaca basin (middle and upper Eocene of the South-Central Pyrenees). *Comptes Rendus Geoscience* v. 348, p. 236–245.

Roigé, M., D. Gómez-Gras, E. Remacha, S. Boya, M. Viaplana-Muzas, and A. Teixell. 2017. Recycling an uplifted early foreland basin fill: An example from the Jaca basin (Southern Pyrenees, Spain). *Sedimentary Geology* v. 360, p. 1–21.

Roigé, M., D. Gómez-Gras, D. F. Stockli, A. Teixell, S. Boya and E. Remacha. 2019. Detrital zircon U-Pb insights into the timing and provenance of the South Pyrenean Jaca Basin. *Journal of the Geological Society*. v. 176 no. 6. p. 1182–1190. Accessed 20 January 2023. https://www.researchgate.net/publication/333226645_Detrital_zircon_U-Pb_insights_into_the_timing_and_provenance_of_the_South_Pyrenean_Jaca_Basin

Sans, M., J.A. Muñoz, and J. Vergés. 1996. Triangle zone and thrust wedge geometries related to evaporitic horizons (southern Pyrenees). *Bulletin of Canadian Petroleum Geology* v. 44 no. 2, p. 375–384.

Sans, M. 2003. From thrust tectonics to diapirism. The role of evaporites in the kinematic evolution of the eastern South Pyrenean front. *Geologica Acta* v. 1 no. 3, p. 239–259.

Sans, M., and J. Vergés. 1995. Fold development related to contractional salt tectonics: Southeastern Pyrenean thrust front, Spain. In Jackson, M.P.A., D.G. Roberts, and S. Snelson (eds.), *Salt tectonics: A global perspective: American Association of Petroleum Geologists Memoir* v. 65, p. 369–378.

Santolaria, P., C. Ayala, E.L. Pueyo, F.M. Rubio, R. Soto, P. Calvín, A. Luzón, A. Rodríguez-Pintó, C. Oliván, and M.M. Casas-Sainz. 2020. Structural and geophysical characterization of the western termination of the South Pyrenean Triangle Zone. *Tectonics Research Article* v. 39, 26 p. Accessed 21 January 2023. https://agupubs.onlinelibrary.wiley.com/doi/epdf/10.1029/2019TC005891

Saqalli, M., A. Hamrita, H. Mouri, H. Rejeb, M.L. El Hachimi, E.M. Benyassine, and N. Rebaï. 2022. The canary in the mine: Mediterranean mines as indicators of the hold on territories and resources: French Pyrenees, Moroccan Middle Atlas, and Tunisian coastal mountains. *Frontiers in Sustainable Cities*, 17 p.

Satterfield, D., H. Rollinson, and R. Suthren. 2019. Classic localities explained: The Eastern French Pyrenees – from Mountain Belt to Foreland Basin. *Geology Today* v. 2019, 35 p.

Saura, E. 2004. *Análisi Estructural de la Zona de les Nogueres (Pirineus Centrals)*. PhD dissertation, Universitat Autónoma de Barcelona, 355 p.

Saura, E., L. Ardèvol i Oró, A. Teixell, and J. Vergés. 2016. Rising and falling diapirs, shifting depocenters, and flap overturning in the Cretaceous Sopeira and Sant Gervàs subbasins (Ribagorça Basin, southern Pyrenees). *Tectonics Research Article*. Accessed 21 January 2023. https://agupubs.onlinelibrary.wiley.com/doi/epdf/10.1002/2015TC004001

Schellart, W. P. 2002. Alpine deformation at the western termination of the Axial Zone, Southern Pyrenees. In Rosenbaum, G. and G.S. Lister. 2002. Reconstruction of the evolution of the Alpine-Himalayan Orogen. *Journal of the Virtual Explorer* v. 8, p. 35–55.

Shackleton, J.R., M.L. Cooke, and A.J. Sussman. 2005. Evidence for temporally changing mechanical stratigraphy and effects on joint-network architecture. *Geology* v. 33 no. 2, p. 101–104.

Showcaves of France, Grottes Prehistorique de Gargas. Accessed 21 January 2023. https://www.showcaves.com/english/fr/showcaves/Gargas.html

Silva-Sánchez, N., and X.L. Armada. 2023. Environmental Impact of Roman Mining and Metallurgy and Its Correlation with the Archaeological Evidence: A European Perspective. *Environmental Archaeology*, 26 p.

Sorriaux, P., M. Delmas, M. Calvet, Y. Gunnell, N. Durand, and E. Pons-Branchu. 2018. Relations entre karst et glaciers depuis 450 ka dans les grottes de Niaux-Lombrives-Sabart (Pyrénées ariégeoises) Nouvelles datations U/Th dans la grotte de Niaux. *Karstologia* no. 67, p. 3–16.

Strekeisen, A. 2006–2020. Lherzolite (Type Locality). Accessed 21 January 2023. https://www.alexstrekeisen.it/english/pluto/lherzolite(tl).php

Suppe, J., F. Sàbat, J.A. Muñoz, J. Poblet, E. Roca, and J. Vergés. 1997. Bed-by-bed fold growth by kink-band migration: Sant Llorenç de Morunys, eastern Pyrenees. *Journal of Structural Geology* v. 19 no. 3–4, p. 443–461.

Tavani, S., J. Mencos, J. Bausà, and J.A. Muñoz. 2011. The fracture pattern of the Sant Corneli Bóixols oblique inversion anticline (Spanish Pyrenees). Journal of Structural Geology v. 33 no. 11, p. 1662-1680.

Teixell, A. and J.A. Muñoz. 2000. Evolución tectonosedimentaria del Pirineo meridional durante el Terciario: Una síntesis basada en la transversal del río Noguera Ribagorçana. Rev. Soc. Geol. España v. 13 no. 2, p. 251-264.

Teixell, A., P. Labaume, and Y. Lagabrielle. 2016. The crustal evolution of the west-central Pyrenees revisited: Inferences from a new kinematic scenario. Comptes Rendus Geoscience v. 348. p. 257-267.

Turisme Olesa de Montserrat, Les Ribes Blaves. Accessed 21 January 2023. https://www.turismeolesademontserrat.cat/es/pl24/visitar/llocs-d-olesa/id48/ribes-blaves.htm

Turner, J.P. 1992. Evolving alluvial stratigraphy and thrust front development in the West Jaca piggyback basin, Spanish Pyrenees. *Journal of the Geological Society* v. 149 no. 1, p. 51–63. https://doi.org/10.1144/gsjgs.149.1.0051

UNESCO, Sobrarbe Geopark. Accessed 21 January 2023. https://www.geoparquepirineos.com/contenidos.php?niv=1&cla=_2OA1CCRP9&cla2=_2OA1CGQUH&cla3=&tip=2&idi=3

van Hinsbergen, D.J.J., T.H. Torsvik, S.M. Schmid, L.C. Matenco, M. Maffione, R.L.M. Vissers, D. Gürer, and W. Spakman. 2020. Orogenic architecture of the Mediterranean region and kinematic reconstruction of its tectonic evolution since the Triassic. *Gondwana Research* v. 81, p. 79–229.

Vergés, J., J.A. Muñoz, and A. Martínez. 1992. South Pyrenean fold and thrust belt: The role of foreland evaporitic levels in thrust geometry. In McClay, K. (ed.), *Thrust tectonics*. Chapman and Hall, London, p. 255–264.

Vergés, J. 1993. *Estudi geològic del vessant sud del Pirineu oriental i central. Evolució cinemàtica en 3D.* PhD dissertation, Universitat de Barcelona, 203 p.

Vergés J., and D.W. Burbank. 1996. Chapter Ell Eocene-Oligocene thrusting and basin configuration in the eastern and central Pyrenees (Spain). In Friend, P.F., and C.J. Dabrio (eds), *Tertiary basins of Spain*. Cambridge University Press, Cambridge, p. 120–133.

Vergés, J. and J. Garcia-Senz. 2001. Mesozoic evolution and Cainozoic inversion of the Pyrenean Rift. In Ziegler, P.A., W. Cavazza, A.H.F. Robertson, and S. Crasquin-Soleau (eds), *Peri-Tethys Memoir 6: Peri-Tethyan Rift/Wrench Basins and Passive Margins*. Memoires du Museum National d'Histoire Naturelle, Paris v. 186, p. 187–212.

Vergés, J., M. Marzo, and J.A. Muñoz. 2002. Growth strata in foreland settings. *Sedimentary Geology* v. 146, p. 1–9.

Vergés, J. 2022. *South Pyrenean fold belt: From foreland shortening to inversion of iberian margin. Field trip, 15–16 June 2022.* In Structural Styles and Hydrocarbon Prospectivity in Fold Thrust Belt Settings: A Global Perspective. Barcelona, Spain. American Association of Petroleum Geologists.

Vergili, C. 2007. Charles Lyell and scientific thinking in geology. *Comptes Rendus Geoscience* v. 339, p. 572–584.

Vidal-Royo, O., J.A. Muñoz, S. Hardy, H. Koyi, N. Cardozo. 2013. Structural evolution of Pico del Águila anticline (External Sierras, southern Pyrenees) derived from sandbox, numerical and 3D structural modelling techniques. *Geologica Acta* v. 11 no. 1, p. 1–26.

Vila, B., O. Oms, J. Marmi, and À. Galobart. 2008. Tracking Fumanya Footprints (Maastrichtian, Pyrenees): historical and ichnological overview. *Oryctos* v. 8, p. 115–130.

Vila, B., J.J. Moratalla, V. Santos, R. Gaete, and À. Galobart. 2011. *New titanosaur trackways from southern Pyrenees: Orcau-2 locality (Late cretaceous) revisited.* Dinosaur Track Symposium 2011, Obernkirchen, p. 34–35.

Vila, B., O. Oms, V. Fondevilla, R. Gaete, À. Galobart, V. Riera, and J.I. Canudo. 2013. The Latest Succession of Dinosaur Tracksites in Europe: Hadrosaur Ichnology, Track Production and Palaeoenvironments. *PLoS One* v. 8 no. 9, p. 1–15.

Vinyoles, A., M. López-Blanco, M. Garcés, P. Arbués, L. Valero, E. Beamud, B. Oliva-Urcia, and P. Cabello. 2020. 10 Myr evolution of sedimentation rates in a deep marine to non-marine foreland basin system: tectonic and sedimentary controls (Eocene, Tremp-Jaca Basin, Southern Pyrenees, NE Spain). *Basin Research* v. 33 no. 1, p. 447–477.

Voirand, A., M. Saplairoles, C. Lamotte, P-E. Privat. 2022. Valorisation énergétique des eaux thermales de la commune d'Axles-Thermes, Rapport final v3, BRGM/RP- 71407 -FR, 24 p.

Wikipedia, Barcelona. Accessed 21 January 2023. https://en.wikipedia.org/wiki/Barcelona

Wikipedia, Cardona Salt Mountain. Accessed 21 January 2023. https://es.wikipedia.org/w/index.php?title =Monta%C3%B1a_de_sal_de_Cardona&oldid=127345294

Wikipedia, Cave of Niaux. Accessed 21 January 2023. https://en.wikipedia.org/wiki/Cave_of_Niaux

Wikipedia, Caves of Gargas. Accessed 21 January 2023. https://en.wikipedia.org/wiki/Caves_of_Gargas

Wikipedia, Geology of Andorra. Accessed 21 January 2023. https://en.wikipedia.org/wiki/Geology_of _Andorra

Wikipedia, Geology of Pyrenees. Accessed 21 January 2023. https://en.wikipedia.org/w/index.php?search =Geology+of+Pyrenees&title=Special%3ASearch&ns0=1

Wikipedia, Lombrives. Accessed 21 January 2023. https://en.wikipedia.org/wiki/Lombrives

Wikipedia, Mallos de Riglos. Accessed 21 January 2023. https://en.wikipedia.org/wiki/Mallos_de_Riglos

Wikipedia, Montserrat. Accessed 21 January 2023. https://en.wikipedia.org/wiki/Montserrat

Wikipedia, Montserrat Abbey. Accessed 21 January 2023. https://en.wikipedia.org/wiki/Santa_Maria_de _Montserrat_Abbey

Wikipedia, Ordesa y Monte Perdido National Park. Accessed 21 January 2023. https://en.wikipedia.org/wiki/ Ordesa_y_Monte_Perdido_National_Park

Wikipedia, Royal Monastery of San Juan de la Peña. Accessed 21 January 2023. https://en.wikipedia.org/wiki /Royal_Monastery_of_San_Juan_de_la_Pe%C3%B1a

Williams, E., M. Ford, and A. Artoni. 1998. *Alluvial gravel sedimentation in acontractional growth fold setting, Sant Llorenç de Morunys, southeastern Pyrenees.* Semantic Scholar.

3 Western Alps of Italy and France

Matterhorn (Monte Cervino) from Valtournenche. Photo courtesy of Cortomaltese, https://commons.wikime dia.org/wiki/File:Matterhorn_from_the_south.jpg

OVERVIEW

The majestic Alps were formed by the slow but inexorable closing of the Tethys Ocean between Africa and Europe. Developing over the past 90 million years or so, these mountains have inspired many with their beauty, and they have compelled geologists to explain ice ages and glaciers, continental collisions and fold-thrust belts. It is an area rich in history, both geologic and human, and is well worth exploring.

Our journey across the Western Alps begins in Turin, at the junction between the Alps, the Po Basin, and the Apennine Mountains. Starting on rocks that were originally part of the African and Apulian tectonic plates, we examine tectonic suture zones as well as vineyards developed in the rich soils of the region.

On entering the mountains proper we pass from the Po Basin to the Southern Alps, cross the Periadriatic/Canavese Line separating European from African rocks, and enter the Austro-Alpine Domain. We begin to see panels of rock at the tops of mountains that originated as ancient ocean-bottom sediments. How did they get there? Generations of geologists working in these mountains have determined that the ancient rocks were pushed upward and thrust over younger rocks by near-horizontal faults. The Austro-Alpine segment consists of slivers of the northern margin of the African Plate pushed north and west over Europe. We begin to encounter European Plate rocks in the Penninic Nappes, rocks derived from the southern margin of Europe that were also thrust north and west. The Helvetic (in Switzerland) and Dauphinois (in France) Nappes are more bits of the European tectonic plate pushed back upon itself. In most of Europe, the term "nappe" means the

 DOI: 10.1201/9781003435594-3

same thing as "thrust sheet" in North America. It refers to the package of rocks carried on a thrust fault.

At the northern and western margins of the mountains we encounter the Molasse Basin and the Central European Rifts (Bresse Graben, Valence Graben). These are basins in front of (in the foreland of) the Western Alps, filled with sediments derived from erosion of the mountains. Here we find the hot springs of Aix-les-Bains, sample the Vins de Savoie, and enter gorges cut through enormous leading-edge folds of the Alpine Front. Near Valence we turn east and cross the Alps once more.

Along our path we traverse three UNESCO Global Geoparks (Sesia Val Grande, Massif des Bauges, and Cottian Alps), numerous World Heritage Sites (Turin buildings, Castello di Razzano, Barolo wine region), and national parks (Vercors). We visit the iconic Matterhorn, the glaciers of Mont Blanc, ophiolites, ancient silver mines, hot springs and travertine springs, and massive landslides.

Itinerary

Begin – Turin
Stop 1 Turin and Its Building Stones
    Stop 1.1 Granito Rosa di Baveno, Palazzo Carignano
    Stop 1.2 Verde Cesana, Galeria dell'Industria Sub-alpina
    Stop 1.3 Sienite Della Balma, Monumento Filiberto
    Stop 1.4 Calcare di Gassino, Palazzo dell'Università
    Stop 1.5 Cumiana Orthogneiss, Vittorio Emanuele I Bridge and Po River
    Stop 1.6 Casotto Breccia and Marmo di Frabosa, Chiesa della Gran Madre di Dio
Side Trip 1 Castello di Razzano and the Wines of Piemonte
Stop 2 Baldissero Massif, the Ivrea–Verbano Zone, and the Canavese Line, Baldissero Canavese
Stop 3 Apulia Diorite of the Ivrea–Verbano Zone, Bienca
Side Trip 2 Ivrea–Verbano Zone and the Serie dei Laghi
    ST2.1 Insubric Line, Boundary between European and African Crust at Scopetta
    ST2.2 Mantle Peridotite at Balmuccia
    ST2.3 Mid-Crustal Diorite of the Mafic Complex, Isola
    ST2.4 Granulite Paragneiss, Croso della Gavala
    ST2.5 Kinzigite Formation Migmatite, Varallo
    ST2.6 Mid-Crustal Kinzigite Formation, Roccapietra
Stop 4 Schistes Lustrés of the Sesia–Lanzo Zone, Parco Avventura La Turna
Stop 5 Orthogneiss in the Micaschist Complex, Settimo Vittone
Stop 6 Eclogite Micaschist Complex, Quincinetto
Side Trip 3 Val del Lys
    ST3.1 Metagranodiorite and Dikes, Fontainemore
    ST3.2 Micaschist Complex at Cascata del Bouro
Stop 7 Eclogite Micaschist Complex, Arco Donnas
Stop 8 Rôche Moutonnée, Montjovet
Side Trip 4 Matterhorn, Valtournenche
Side Trip 5 Buthier Valley
    ST5.1 Valpelline Series Garnet Amphibolite, Moulin
    ST5.2 Valpelline Series Amphibolite, Bionaz
    ST5.3 Metapelite of the Valpelline Series, Oyace
    ST5.4 Dent Blanche Nappe – Tsaté Nappe Contact, Plan Debat
    ST5.5 Arolla Series Orthogneiss of the Gignod-Roisan Zone, Roisan

Key stops on the Western Alps geo-tour.

GEOLOGY OF THE PO BASIN

Our trip starts in Turin, within the Po Basin which lies in the hinterland of the Western Alps. The 50,000 km^2 (19, 305 mi^2) Po Basin includes northern Italy's Lombardy, Po River, and Veneto plains. The basin coincides with the Po Valley that extends 650 km (400 mi) east-west from the Western Alps to the Adriatic Sea. The basin is surrounded by thrusted terrain of the Western and Southern Alps to the west and north, respectively, and the northern Apennine Mountains to the south (Lindquist, 1999). In the east the Po Basin continues below the Adriatic. Across the basin, the structures are related either to the Alps (north-south to northeast fold trends) or to the Northern Apennines (northwest-southeast trends), with interference between the two trends showing the greatest complexity in the far western Po Basin. The Alps and Apennines also contributed to and controlled the sediment filling of the basin (Turrini et al., 2014). In fact, the plain is an in-filled system of ancient canyons (the "Apennine Foredeep") that has been filling with sediment since at least 7 Ma. Elevations in the valley range from 540 m (1,770 ft) in the Alpine foothills near Cuneo to sea level near Venice.

The climate is temperate: cool, damp winters with frequent fog and mist alternate with hot, humid summers. Rain, between 700 and 1,200 mm (30 to 45 in), occurs throughout the year. The Po Valley is one of the most important industrial and agricultural areas in Italy, let alone Europe. Hydroelectricity is generated by dams on the Po and its tributaries. The river's water is used to irrigate the region's grains, sugar beets, and fruits (including wine grapes), and for livestock (Wikipedia, Po Valley). Tourism is a big part of the economy.

The Po Basin is subdivided into the Tertiary Alpine Basin in the north, filled with relatively shallow-marine sediments, and the Apennine Basin in the southwest, filled with deep marine sediments. These were essentially separate basins until Messinian time (late Miocene, 5–7 Ma), when uplift of the Apennine Mountains caused the two basins to merge. The Tertiary Alpine Basin, in turn, is subdivided into the Tertiary Lombardy Basin to the north and the Tertiary Piedmont Basin in the southwest.

The Tertiary Piedmont Basin, the sub-basin that contains both the Turin and Monferrato hills, is an Alpine hinterland basin since it directly overlies deformed Alpine units. It is also considered a foreland basin to the Apennine Fold-Thrust Belt. The sediment fill of the Tertiary Piedmont Basin reaches up to 6,000 m (20,000 ft) of mainly river and delta sediments derived largely from the adjacent Western Alps (Mutti et al., 2002).

The Tertiary Piedmont Basin overlies a southward-backthrusted segment of the Alpine system largely created after the main Alpine mountain building event. The shallow Tertiary basin fill and deeper, deformed Alpine units have been affected by Apennine deformation since the Oligocene. The stratigraphy records multiple depositional systems, from alluvial fans to turbidites. All of these units were subjected to several phases of deformation. The youngest sediments are thickest in the north and west parts of the basin that are closer to the Alps, whereas older sequences come to the surface farther south (Mutti et al., 2002; Cloetingh et al., 2015).

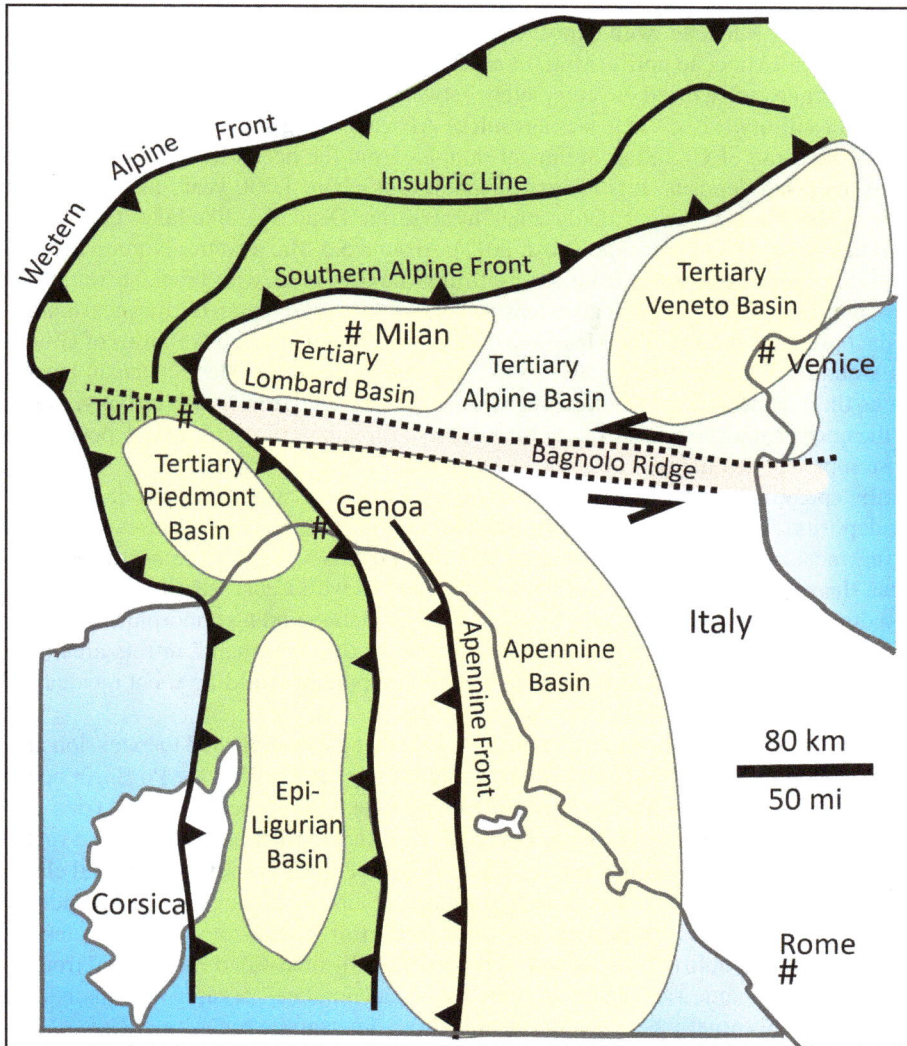

Oligo–Miocene basins of the Po Valley. Modified after DiBiase & Mutti, 2002.

The Tertiary basin fill rests unconformably on units deformed and metamorphosed during the Alpine Orogeny. From west to east the pre-Tertiary section is divided into domains consisting of slices of European crust with low-grade metamorphism (the Briançonnais Units) and slices of oceanic crust (ophiolites) and its sedimentary cover. The ophiolites and sediments were metamorphosed under high pressure–low temperature conditions (the Voltri Group), and low to very low-grade metamorphism (the Ligurian Units).

The sediments in the Tertiary Piedmont Basin consist dominantly of land-derived deposits. The basin has been subdivided into three sectors, each with a different sedimentary section and evolution. The western Langhe sector records a continuous deepening beginning in the early Oligocene and has the most complete Oligocene to Messinian section. The central sector (Alto Monferrato sector) was affected by early Miocene growth of a structural high capped by a carbonate platform that separated the eastern and the western sectors. The eastern Borbera–Scrivia sector contains an upper Eocene to early Miocene section that mirrors the process of exhumation of the metamorphic unit of the western Alps: the youngest rocks were eroded and deposited first, then progressively older units were uplifted, eroded, and deposited here. This sector experienced major subsidence during the Oligocene and early Miocene, with deposition of relatively deep-water turbidites. The entire basin experienced middle Miocene uplift (Mutti et al., 2002).

The Mediterranean Sea dried out completely between 5 and 6 million years ago in what is called the "Messinian salinity crisis." This was a result of Africa moving slowly north toward Europe and pinching off the Strait of Gibraltar. Sediment samples from the ocean floor show soil zones, plant fossils, and thick salt deposits. It is estimated that it took about 1,000 years for the sea to dry out completely in the hot, dry climate. Only a few hypersaline Dead Sea–like lakes existed 3 to 5 km (2 to 3 mi) below sea level (Prost and Prost, 2017). Around 5.5 Ma, a wetter climate resulted in the basin receiving more fresh water from rivers, filling the low spots with brackish water much like today's Caspian Sea. The drying left an extensive layer of salt and gypsum (evaporites) tens of meters thick on the bottom of the sea. The Messinian salinity crisis ended when the Strait of Gibraltar was breached around 5.3 Ma, allowing water from the Atlantic to fill the Mediterranean basin in what is known as the Zanclean flood (Wikipedia, Messinian Salinity Crisis). Interestingly, the Po Basin contains the northernmost evidence of the Messinian event consisting of chaotic deposits containing huge gypsum blocks floating in a fine-grained mud matrix.

Generally speaking, the rocks in the Po Basin are Mesozoic carbonates and dominantly clastic Cenozoic deposits, all sitting above a crystalline/metamorphic basement of Variscan (late Paleozoic) age. During most of Paleozoic time what is now the Po Basin was on the northwestern edge of Gondwana, the southern continent. Collision of Gondwana with Laurussia (the Paleozoic northern continent) during the Variscan Orogeny resulted in deformation and metamorphism of all preexisting Paleozoic rocks to the extent that they are effectively the "basement" in this area (Lindquist, 1999; Turrini et al., 2014). The Carboniferous Variscan mountain-building event melded the previous continents into the "supercontinent" Pangea.

In Upper Triassic–Lower Jurassic time this part of Pangea was subjected to extension and rifting that led to opening of the Tethys Ocean. During and shortly after rifting the Po Basin became part of the Adriatic (also called "Adria" or "Apulia") Microplate, part of the northern Africa carbonate-platform (think of the Bahamas Banks or Great Barrier Reef). These early Mesozoic rocks record the breakup of Pangea and a rising sea level. Thus, pre-rift Permo-Triassic continental clastic rocks (river and delta sandstones, siltstones, shales) grade upward into Triassic rift units, including marine sandstones, siltstones, shales, evaporites, and carbonates. Early Mesozoic shallow-water carbonate platforms developed on uplifted blocks and were separated by mud-filled extensional troughs, some of which contained organic-rich shales. Anoxic basins accumulated organic-rich muds in the Late Triassic and would eventually become excellent hydrocarbon source rocks.

Jurassic and Cretaceous strata reveal an upward transition from active rifting to a passive continental margin (that is, an area with no active mountain building) with shallow-water limestones to deeper water limestones and cherts. This indicates a slowly subsiding ocean basin. Extension ended in Cretaceous time, when the African Plate began to move north toward Europe, shrinking and ultimately closing the Tethys Ocean. From Late Cretaceous to Cenozoic time the European continental plate and adjacent Tethys oceanic crust was subducted below (plunged under) the African plate. Africa mostly overrode the Adriatic Microplate, but a good portion of that plate was preserved as a minimally deformed region under the central Adriatic Sea and under the Po Basin. The area

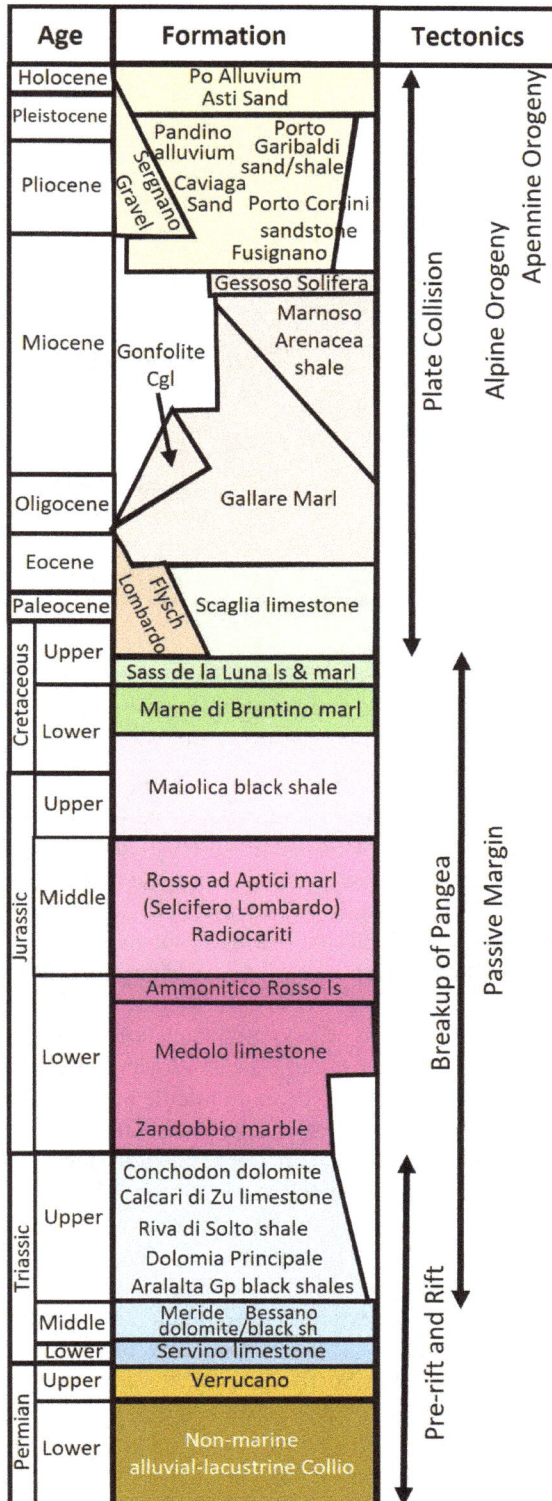

Stratigraphy of the Po Basin. Modified after Lindquist, 1999.

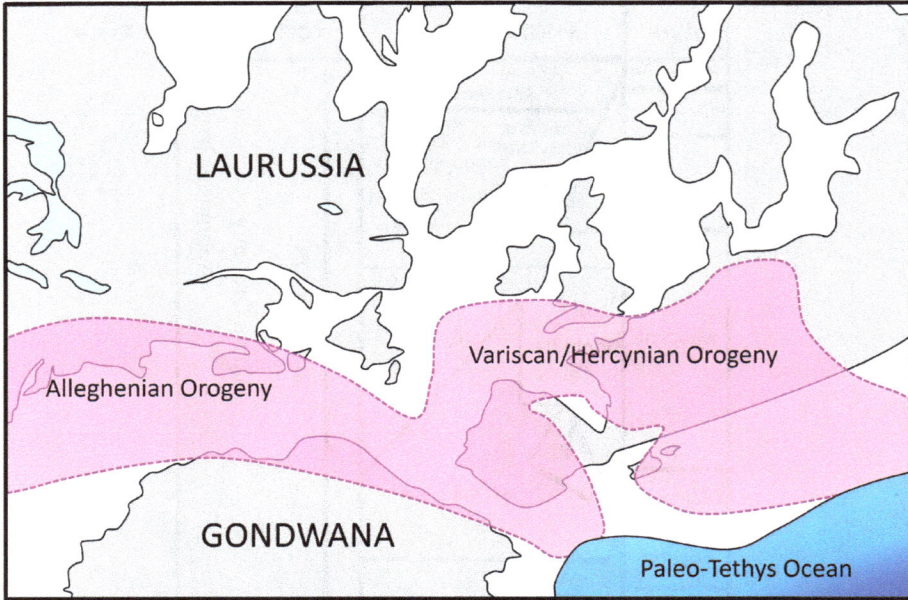

Location of the Variscan/Hercynian/Alleghanian mountain chains during middle Carboniferous time. Present-day coastlines are shown for reference. Capitalized names are the continents that collided during these orogenies. Modified after Woudloper, https://commons.wikimedia.org/wiki/File:Hercynides_EN.svg

evolved into a Late Cretaceous–Tertiary deep-water basin characterized by clastic turbidite and other marine deposits as a result of continued compression and sea level rise (Lindquist, 1999).

The earliest evidence for Alpine compression is recorded in the Upper Cretaceous. Late Cretaceous shortening in the Po Basin reactivated and inverted (reversed the offset direction on) some of the existing normal faults as thrusts (Turrini et al., 2014). By Oligocene time, most of the Tethys Ocean between Africa and Europe had been subducted, and the Adriatic continental microplate was being squeezed from several directions. A change in relative plate motion may have caused southward thrusting (backthrusting) in the Southern Alps. This could also be the result of rotation of an imbricate stack of thrusts, or of out-of-sequence thrusting. Also beginning in the Oligocene just south of the Po Basin was the earliest east-directed thrusting in the Apennines. This deformation would reach the southwestern Po Basin in middle Miocene time.

Late Miocene (Messinian) convergence of Europe and Africa resulted in separation of the Mediterranean Basin from open marine circulation. Lowering of sea level and ultimately drying of the ocean led to local unconformities and widespread evaporites being deposited. The Apennine Orogeny peaked in Pliocene time, although some deformation continued into the Pleistocene. Substantial subsidence north of the Apennine front resulted in deposition of at least 6 km (20,000 ft) of Pliocene sediment, filling the basin and establishing the Po River drainage system (Lindquist, 1999).

In addition to some pre-Apennine extension, Apennine foreland-related subsidence in the Piedmont Basin was enhanced by late-stage compression that, in places, is still active (Froitzheim, 2012; Cloetingh et al., 2015). Thus, the Po Basin is the northwesternmost buried sector of the Adriatic Microplate, is both a hinterland Alpine basin and foreland Apennine basin, and is one of the main hydrocarbon provinces of continental Europe (Turrini et al., 2014).

Older thrusts in the Po Basin are related to the Alps and are either south or southwest-directed; younger (mid-Miocene to Recent) thrusts are related to the Apennines and are generally northeast-directed (Lindquist, 1999; Froitzheim, 2012). The Sestri-Voltaggio Line is considered the boundary between the Alpine and the Apennine fold-thrust belts. The Monferrato and Turin hills are considered extensions of the Apennine Fold-Thrust Belt and contain thrust-related folds at the surface. South of the north-directed Monferrato thrust front are deep piggyback basins. Thrusting and basin development were contemporaneous, and are ongoing today (Giraudi, 2016; Dela Pierre et al., 2016).

North of the Monferrato and Turin hills thrust front, Tertiary sediments are preserved across the basin. Subsidence is linked to Apennine thrusting and has been ongoing since at least the Pleistocene (last 2 Ma; Giraudi, 2016).

Turin is developed on river and glacial sediments. It is located in the gap between the Turin Hills and the Western Alps. The Turin Hills are a north-thrusted series of folded Tertiary sediments related to the Tertiary Apennine Fold-Thrust Belt. The Western Alps are primarily west- to northwest-thrusted sediments and basement resulting from the Tertiary Alpine Orogeny.

Hydrocarbon Resources

The Po Basin has been the focus of petroleum exploration since the first well was drilled in 1890. Early exploration looked for onshore domes and anticlines with shallow, Tertiary-age reservoir rocks. Both oil and gas have been produced from a number of fields. The Villafortuna–Trecate field, 30 km west of Milan, was discovered in 1984 by Italy's ENI Group. It is the basin's largest oil producer, with 38.2 million m^3 (240 MMbbl) produced from a Triassic carbonate reservoir at depths between 3,000 and 6,000 m (10,000 to 20,000 ft). Miocene, Pliocene, and Pleistocene sandstones at depths between 1,000 and 3,000 m (3,000 to 10,000 ft) produce mostly biogenic gas, that is, gas produced at low temperatures by decomposition of organic matter (Turrini et al., 2014). The gas is part of the Porto Garibaldi petroleum system, a gas-generating system that contains approximately 453 billion m^3 (16 TCF) of ultimately recoverable Pliocene and Pleistocene gas.

Two other petroleum systems, with Triassic (Meride/Riva di Solto) and Miocene (Marnoso Arenacea) source rocks, contain approximately 159 million m^3 (1 billion bbl) of oil and thermogenic gas (derived from heating oil to high temperatures). The Malossa Field, discovered in 1973, produces gas condensate from the Triassic Dolomia Principale and the Jurassic Zandobbio Dolomite at a depth of 6 km (4 mi). The reservoirs are capped by the Cretaceous Marne di Bruntino Formation marl.

Most of the hydrocarbon generation and expulsion occurred during Late Neogene time, during the Alpine Orogeny (Lindquist, 1999).

The Piedmont region produced about 15,900 m^3 (100,000 bbl) of oil in 2018, and 5.37 million m^3 (190 MMCF) of gas in 2023, a small fraction of Italy's total production (Statista.com, 2019; Statista.com, 2024).

*Start – **Turin Airport to Palazzo Carignano:** From airport parking drive north on Lato Partenze; slight right onto SP2/Via Torino; continue straight onto RA10 and follow signs to Torino Centro; take exit toward Barriera di Milano/Centro/Collina; merge onto Corso Grosseto going east and immediately turn right (south) onto Corso Vercelli; turn left (southeast) onto Corso Novara; turn right (south) onto Corso Palermo; bear right onto Corso Regio Parco/Ponte delle Benne; turn left (south) onto Rondo Rivella and immediately turn right (southeast) onto Corso Maurizio and pull into the parking area on the right (45.072675, 7.691454) in Parcheggio dei Giardini Reali. Walk southwest ~700 m (0.4 mi; 10 min) to Via Accademia delle Scienze 5, Torino (45.068773, 7.686182). This is **Stop 1.1, Granito Rosa, Palazzo Carignano**, for a total of 17 km (10.5 mi; 21 min).*

BEGIN – TURIN

STOP 1 TURIN AND ITS BUILDING STONES

Turin (Torino in Italian) is a lovely city rich in history and architectural masterpieces. It is the capital city of Piedmont (Piemonte) and of the Metropolitan City of Turin (the administrative region around Turin). The city is located on the west bank of the Po River at an elevation of 240 m

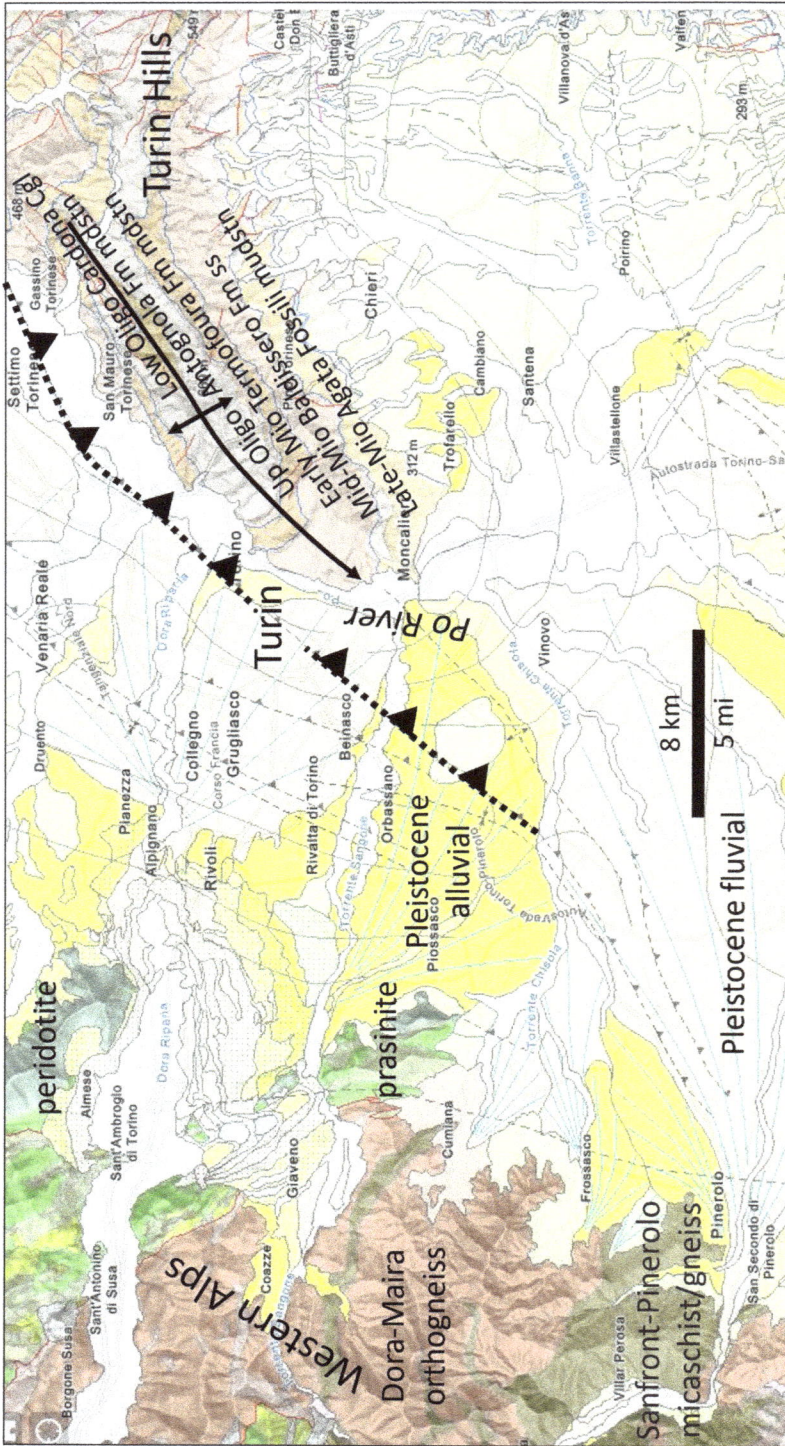

Geology of the Turin area. Map courtesy of Arpa GeoPiemonte, 2021.

(790 ft). The population of the city proper was 852,000 in 2021; the population of the urban area was estimated to be 1.7 million (Wikipedia, Turin).

Turin was founded around 2,400 years ago by the Taurini, a Celtic tribe. Originally "Taurasi," the name is derived from the Celtic word tau, or mountain. Gradually the name morphed into Italian, in which it means "little bull," and the bull is still on the city flag (Goodson, 2018).

During the Second Punic War, Hannibal and his army crossed the Alps in 218 BCE and destroyed much of Taurasi. It was later rebuilt as Castra Taurinorum. In 28 BCE it was conquered by Rome and became the military outpost of Augusta Taurinorum. The Romans also planted vineyards in the area.

With the collapse of the Western Roman Empire in the 5th century, the city was invaded by the Goths, Visigoths, Lombards, and Franks. The city became an earldom in the 8th century (World Guides, Turin History; Goodson, 2018).

In 1280 the French Savoy family conquered the city, and it would belong to the House of Savoy for the next 600 years. The city began its economic expansion in the 14th century under Amadeus VIII of Savoy, when it became the center of the merged provinces of Piedmont (Piemonte, The History of Turin).

The House of Savoy brought art, culture, and architecture to Turin. At that time rival cities tried to outdo each other in buildings and elaborate decoration. In 1498, Bishop Della Rovere commissioned the Cathedral of San Giovanni, a fine example of Renaissance architecture (World Guides, Turin History). Emanuele Filiberto, Duke of Savoy, made Turin the capital of his duchy in 1560. The Shroud of Turin, said to bear the likeness of Christ, was brought to the city during Filiberto's reign (Goodson, 2018). Relics of Savoyard rule are found in the palaces, boulevards, and squares of Turin. Turin became the capital of the Duchy of Savoy in 1563. In 1580, Charles Emmanuel I became duke and began to implement many architectural projects, extended the city south of its ancient walls, and was responsible for building the Piazza San Carolo, the Royal Palace, and the Palazzo Madama (World Guides, Turin History).

During the 17th century, Turin became a center for Baroque architecture in Europe. It is still regarded as one of the finest Baroque cities and attracts many tourists for this reason alone (Goodson, 2018).

In 1720, Sardinia was ceded by the Spanish throne to Victor Amadeus II, Duke of Savoy. The duke united it with his possessions on the mainland, including Savoy, Aosta, the Principality of Piedmont, and the County of Nice, creating the Kingdom of Sardinia.

In the 18th century, educational reform became a priority and a new university was built on the Via Po. In the late 1700s, Turin was occupied by the French under Napoleon. This period of occupation turned out to be short (World Guides, Turin History).

Turin has been called "the cradle of Italian liberty" for having been the birthplace and home of patriots (Garibaldi, Benso, and others) who contributed to the Risorgimento, or Italian unification. In the year 1861 the House of Savoy became the rulers of a united Italy with the crowning of Vittorio Emanuele II. Turin was the first capital of the Kingdom of Italy from 1861 to 1864, when the capital moved to Florence, then eventually to Rome (Wikipedia, Turin).

During Mussolini's time Turin stood out as an anti-fascist city. At the outbreak of World War II a resistance movement was organized there to fight the fascists.

Since the end of the war Turin has become a center of industry. Fiat (Fabbrica Italiana Automobili Torino), established in Turin in 1899, as well as Alfa Romeo and Lancia, made Turin one of the world's great automotive cities. The city is known for its art galleries, restaurants, churches, palaces, opera houses, piazzas, parks, theaters, libraries, and museums, among them the Egyptian Museum, second largest in the world. Turin is known for its Baroque, Rococo, Neoclassical, and Art Nouveau architecture. The historical center of Turin contains World Heritage buildings such as the Palazzo Reale and Residences of the Royal House of Savoy (Wikipedia, Turin; Goodson, 2018).

The province of Turin, consisting of Turin and 311 other cities, is the largest Metropolitan City in Italy (Wikipedia, Metropolitan City of Turin).

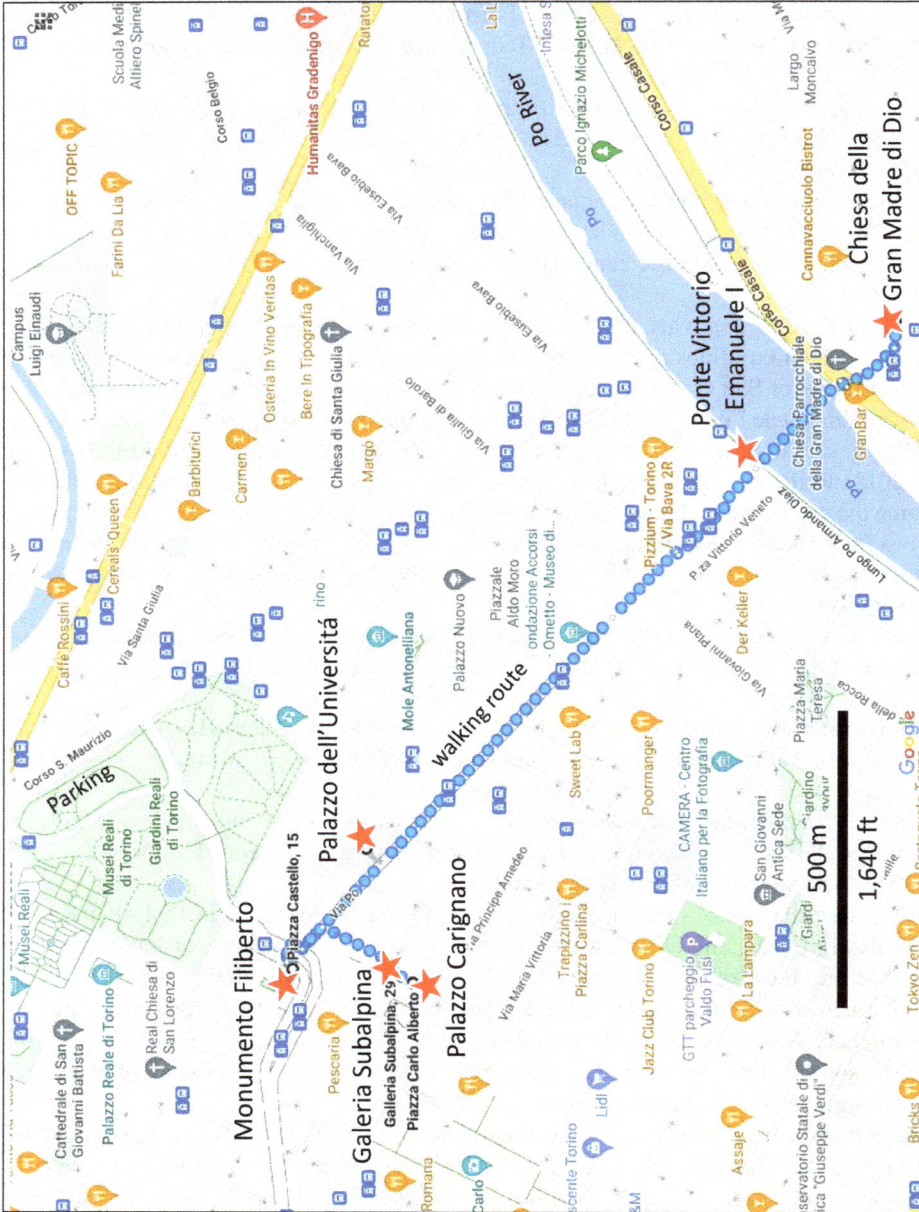

Walking map of typical building stone sites in Turin. Courtesy of Google Maps.

STOP 1.1 GRANITO ROSA DI BAVENO, PALAZZO CARIGNANO

Palazzo Carignano was originally built between 1679 and 1684 by Guarino Guarini for Prince Emmanuel Philibert, son of Thomas Francis, Prince of Carignano, and his French wife Marie de Bourbon. The building, built of brick in a typical Baroque style, has an elliptical main façade on the west side. Decorations over the windows of the *piano nobile* (noble floor) commemorate the campaign of the Carignano family and Carignan-Salières Regiment against the Iroquois in 1667. Palazzo Carignano served as the private residence of the Princes of Carignano.

A wing was added in 1864–71, doubling the size, and a new façade was added overlooking Piazza Carlo Alberto. The building was the birthplace of the first King of Italy, Victor Emmanuel II of the House of Savoy, in 1820. From 1848 to 1861 the palace held the House of Deputies of the Subalpine Parliament, and for a few years it held the parliament of the nascent Kingdom of Italy. It now houses the National Museum of the Italian Risorgimento and the management authority for Artistic and Historical Artifacts of Piedmont (Wikipedia, Palazzo Carignano; Gambino et al., 2017).

The 19th century façade is made of Granito Rosa di Baveno from quarries near Baveno in the Lago Maggiore area of northern Piemonte. These Permian granites are from the Southern Alpine Domain (Gambino et al., 2017).

The building is a UNESCO World Heritage Site.

Palazzo Carignano contains the Italian National Museum of the Risorgimento, the unification of Italy.

Palazzo Carignano to Galeria dell'Industria Subalpina: *Walk 40 m (135 ft; 1 min) northeast along Piazza Carlo Alberto to* ***Stop 1.2, Verde Cesana, Galeria dell'Industria Subalpina, 16, 10123 Torino*** *(45.069093, 7.686426).*

STOP 1.2 VERDE CESANA, GALERIA DELL'INDUSTRIA SUBALPINA

The Galleria dell'Industria Subalpina, opened in 1874, is a shopping mall in Turin, Italy. The structure was inspired by the typical 19th-century Parisian *passages*. Designed by Pietro Carrera, it is named for the Banca dell'Industria Subalpina, which financed its construction. Partly damaged during World War II, the gallery was rebuilt according to the original drawings.

The Galleria contains cafés, restaurants, a bookshop, art gallery, and other shops. The interior is open and spacious, a hall 50 m (165 ft) long and 14 m (46 ft) wide and lit by skylights 18 m (60 ft) above. The floor of the hall has a garden and distributed seating.

The Galleria features Verde Cesana (Verde Alpi Cesana), a green marble quarried at the Palestra di Roccia Quarry near Cesana Torinese in the Alps about 90 km (54 mi) west of Turin. This ornamental stone quarry was developed in carbonate-rich ophiolite of the Piemont-Liguria Zone of the Penninic Domain. The rock was formerly Jurassic-Cretaceous ocean floor and overlying carbonate sediment (Barale et al., 2020).

The Galleria dell' Industria Subalpina is a shopping mall with a wide variety of amazing stonework.

The main floor is an abstract mosaic of angular marble fragments; the shop walls are different varieties of gray, pink, and green marble.

Galeria dell'Industria Subalpina to Monumento Filiberto: Walk northeast on Via Carlo Alberto; turn left (northwest) on Via Po and walk to Piazza Costello for a total of 250 m (0.2 mi; 3 min). This is Stop 1.3, Sienite Della Balma, Monumento Filiberto (45.070463, 7.686450).

STOP 1.3 SIENITE DELLA BALMA, MONUMENTO FILIBERTO

The Filiberto Monument is built entirely of Sienite Della Balma, a rock quarried from the Balma pluton ~20 km (12 mi) north-northwest of Biela in the Austro-Alpine Domain. (A pluton is a large body of magma that cooled deep in the crust.) Sienite Della Balma is variously described as a syenite or monzonite, two closely related granitic rocks. It is a medium to coarsely crystalline, brown-gray Oligocene rock characterized by brown orthoclase and white plagioclase (Gambino et al., 2017; International Granites, Sienita della Balma).

The Monumento Emanuele Filiberto on Piazza Castello.

Monumento Filiberto syenite.

Monumento Filiberto to Palazzo dell'Università: *Walk 250 m (0.2 mi; 3 min) southeast on Via Po to **Stop 1.4, Calcare di Gassino, Palazzo dell'Università** (45.069358, 7.688809) on the left.*

Stop 1.4 Calcare di Gassino, Palazzo dell'Università

The University of Turin is one of the oldest in Europe, having been established in 1404 by Prince Ludovico of Savoy. The rector's residence, the Rector Palazzo, is an ornate structure built in the Neoclassical style with columns and statues. Famous alumni featured on the walls include Gianni Agnelli (head of FIAT), Primo Levi (20th-century Italian chemist, author, and Holocaust survivor), Desiderius Erasmus (16th-century Dutch philosopher and theologian), Umberto Eco (20th-century Italian literary critic, novelist, philosopher), Amadeo Avogadro (19th-century Italian physicist), and multiple Italian prime ministers, presidents, and Nobel Prize winners. It is associated with the Turin Museum of Natural History (36 Via Giolitti) and its well-known geology and paleontology galleries.

The columns of the Rector Palazzo are made of Gassino Stone. The Calcare di Gassino (Gassino Limestone) is an Eocene nummulitic limestone (nummulites are a variety of foraminifera, single-celled floating marine animals) quarried from the core of an anticline in the Turin Hills about 15 km (9 mi) east-northeast of central Turin (Gambino et al., 2017). The quarry near Gassino Torinese is now abandoned.

Internal courtyard, Palazzo dell'università.

Palazzo dell'Università to Vitorrio Emanuele I Bridge: *Walk southeast on Via Po to the Po River, the Ponte Vittorio Emanuele I, and **Stop 1.5, Cumiana Orthogneiss, Vitorrio Emanuele I Bridge** (45.063885, 7.696837) for a total of 0.9 km (0.6 mi; 10 min).*

Stop 1.5 Cumiana Orthogneiss, Vittorio Emanuele I Bridge and Po River

The Po is the longest river in Italy, flowing east across northern Italy for 652 km (405 mi). Rising in the Cottian Alps west of Turin, the headwaters are at a spring seeping from a stony hillside at Pian del Re, at the head of the Val Po under the northwest face of Monviso. The Po ends at a delta on the Adriatic coast near Venice (Wikipedia, Po River).

This beautiful bridge crossing the Po is the Ponte Vittorio Emanuele I. It is the main bridge in Turin, as it connects the historic Piazza Vittorio Veneto with the Chiesa della Gran Madre di Dio (Church of the Great Mother of God) on the east bank of the Po. It has been called the bridge of the Gran Madre, the bridge of Piazza Vittorio, or the Murazzi bridge.

During the French occupation of Turin in the early 1800s, Napoleon ordered the building of a solid and massive bridge to replace the existing, unstable stone bridge that dated back to 1404. The foundation stone of the bridge was laid in November 1810: embedded in the central column of the bridge were 88 coins and medals of the Napoleonic campaigns and a silver ingot. Designed by Claude-Yves La Ramée Pertinchamp, the bridge was built by the French engineer Charles Mallet and the Piedmontese engineer Pellegrini. The bridge has five arches, is 150 m (492 ft) long, and 12.9 m (42 ft) wide. It was completed in 1813. A year later the French occupation ended and the House of Savoy returned to the city. It was named after King Vittorio Emanuele I.

Ponte Vittorio Emanuele I is the oldest existing bridge in Turin. There have been essentially no upgrades to the bridge since its completion, the main change being the laying of tram tracks in 1876 and replacement of the old stone parapets with the current ones of cast iron. Modern lighting was added for the 2002 Winter Olympics. The bridge has withstood numerous floods of the Po (IGoToWorld.com).

Vittorio Emanuele I Bridge, view southeast to the Chiesa della Gran Madre di Dio.

The Vittorio Emanuele I Bridge is a masonry bridge clad in Cumiana Stone quarried near Cumiana, about 40 km (24 mi) southwest of Turin (Gambino et al., 2017). The rock is quarried from the pre-Triassic Dora Maira orthogneiss, which is part of the Inner Crystalline Massif (as at the Perosa Quarry stop). Orthogneiss is a high-temperature metamorphic rock derived from an igneous precursor. This rock forms the metamorphic basement of the Paleozoic continental margin of Europe (Barale et al., 2020).

Vitorrio Emanuele I Bridge to Chiesa della Gran Madre di Dio: *Walk 220 m (720 ft) southeast on Via Po and you are at* ***Stop 1.6, Chiesa della Gran Madre di Dio*** *(45.062507, 7.698687).*

Stop 1.6 Casotto Breccia and Marmo di Frabosa, Chiesa della Gran Madre di Dio

The only church in the city owned by the municipality, the Chiesa della Gran Madre di Dio is one of the most important Catholic places of worship in Turin.

In 1814 King Vittorio Emanuele I decided to build the Chiesa della Gran Madre di Dio to commemorate the defeat of the French and return of the House of Savoy. The church was completed in 1831 (Wikipedia, Ponte Vittorio Emanuele I).

The Neoclassic-style church is located on the east bank of the Po River, facing the Ponte Vittorio Emanuele I and the Piazza Vittorio Veneto on the west bank. The architect was Ferdinando Bonsignore.

The tympanum of the church states: ORDO POPVLVSQVE TAVRINVS OB ADVENTVM REGIS, translated as "The Nobility and the Population of Turin on the Return of the King." The church is styled after the Pantheon in Rome (Wikipedia, Gran Madre di Dio).

At the end of a wide staircase is a pronaos, or vestibule with six columns. The columns and pillars have Corinthian capitals that, with the pronaos, give the structure a Neoclassical aspect. The columns are carved from Malanaggio Stone, a metadiorite quarried from the Dora Maira intrustive of the Inner Crystalline Massif near Malanaggio, about 60 km (36 mi) southwest of Turin (Barale et al., 2020).

On the pediment there is a marble tympanum carved in high relief by Francesco Somaini di Maroggia, representing the Virgin with Child receiving homage from the Turinese.

On the sides of the portal are two niches containing statues of saints:

- St. Mark the Evangelist on the right, with the Gospel in his hand and a foot above the lion, the work of Giuseppe Chialli (1828), and
- San Carlo Borromeo on the left, by Giuseppe Pagliani (1829)

At the base of the staircase, each on a large base placed at the side of the structure, are two statues, one representing Faith (or Justice) and the other, Religion, both created by the Carrarese sculptor Carlo Chelli in 1828. Faith, on the left, is represented by a female figure who holds in her right hand an open book; in the left she raises a chalice to the sky. To her right is a small angel. Religion, on the right, is depicted by a woman looking toward the horizon. A young man kneeling beside her is holding two white stone tablets and a cross.

A statue of Vittorio Emanuele I of Savoy, in the center at the foot of the staircase, rises almost 10 m (33 ft) high. It is the work of Giuseppe Gaggini, professor at the Albertina Academy in Turin, completed in 1869. An inscription on the base states "Vittorio Emanuele I King of Sardinia, Returned to his People on May 20, 1814." A second inscription on the base is translated "Vittorio Emanuele II, King of Italy. This Statue Given to the People of Turin in the Year 1869."

The interior of the church is a single circular nave; the main altar is on the east in a semicircular apse flanked by two large red breccia columns, and is opposed by two similar ones that flank the access door on the west.

Most people find the most stunning characteristic of the church is its dome, a Piedmontese Neoclassical masterpiece. The structure is made entirely of concrete and ends with a circular oculus that provides light to the interior. If you are an admirer of beautiful rock, one of the most stunning aspects of the interior might be the eight large columns made from Casotto Breccia (a marble that had been previously used by Vittone for the majestic decorations of the chapel in Casotto Castle).

Vittorio Emanuele I had the church decorated entirely with marble from Piedmont, most of which came from the Monregalese area east of Cuneo. Marmo di Frabosa (Frabosa Marble, Internal Briançonnais Domain) is known for its easy workability due to the fine and homogeneous grain.

Marmo di Frabosa was used in statues and external and internal decorations such as capitals in white and internal slabs in green varieties for the Gran Madre di Dio. Nero Nuvolato di Miroglio, a black marble, was used for internal decoration of the Ossario dei Caduti of the Gran Madre di Dio (Storta et al., 2024).

In the church basement is a crypt designed by Giovanni Ricci as a Memorial to the Fallen of World War I.

A number of legends are associated with the church. It is said to be built on the site where, in ancient times, there was a temple dedicated to the Egyptian goddess Isis, also known as the "Great Mother."

One of the two statues on either side of the staircase, the one of a woman holding an open book in her right hand and raising a chalice in her left hand, is said to be the Madonna herself, holding the Holy Grail. If properly deciphered, she is said to be indicating the location of the chalice, either somewhere in the city or this church (Wikipedia, Gran Madre di Dio). She may have been carved from Grappiolo, a white marble from Garessio, Bardiglio.

In 1969 the church was featured in the film *The Italian Job*, when Michael Caine and his gang drove their Mini Coopers down the front staircase (Lonely Planet, Gran Madre di Dio).

Address: Piazza Gran Madre di Dio, 4, 10131 Torino (TO)
Phone: +39 0118193572
Email: parrocchiagranmadre@gmail.com
Hours: 7:30 to 19:00 daily.

Chiesa della Gran Madre di Dio.

The striking statue of Faith (or Justice), Chiesa della Gran Madre di Dio. Photo courtesy of Augusto De Luca, https://commons.wikimedia.org/wiki/File:Augusto_De_Luca_-_foto_della_Gran_Madre_di_Dio,_Torino .jpg

Side Trip 1, Turin to Castello di Razzano (tolls): *Head south on Corso Moncalieri; turn right onto Corso Sicilia; at the roundabout take the 1st exit (southwest) onto Viale Dogali; at the roundabout take the 1st exit (west) onto Ponte Franco Balbis; cross the Po River and turn left onto Corso Achille Mario Dogliotti; continue straight onto Corso Unità d'Italia; at the roundabout take the 2nd exit (straight) onto Corso Trieste; keep left onto A55/A6; take the exit onto E70 to Genova (east); take the Asti Est exit to E74/A33 to Cuneo (toll); use the right lane to take the ramp onto SS706 (north); at the T turn left (west), then left (north) again onto SP457var; turn left (northwest) onto SP52; bear right on SP15 to Alfiano Natta; in Natta turn left (west) onto SP13; in 0.4 mi turn left (south) onto Via S. Carlo; drive 0.7 mi to Castello on the right. This is* ***Side Trip 1, Castello di Razzano*** *(Via San Carlo 72, 15021 Alfiano Natta; 45.0414, 8.1995) for a total of 79.6 km (49.5 mi; 62 min).*

The magnificent columns inside the Gran Madre di Dio are carved from red Casotto Breccia.

If you prefer to skip side Trip 1:

Turin to Baldissero (tolls): *Head northeast on Corso Casale; turn left (north) onto Ponte Regina Margherita (sign for Centro; Aeropuerto); continue straight onto Corso Tortona; continue straight onto Corso Novara; continue straight onto Corso Vigevano; at Piazza Generale Baldissera take the 1st exit (northeast) onto Corso Venezia; continue straight onto RA10/Raccordo Autostradale Torino – Caselle; take Exit 1 to Tangenziale; bear right onto the ramp for E66, then merge onto A55 going northeast; continue straight on A5/E612; take the exit to San Giorgio Canavese; at the roundabout take the 2nd exit (west) onto SP53; at the roundabout take the 1st exit (north) onto SP41; turn left (west) onto SP41/Via Circonvallazione; turn left to continue on SP41; SP41 turns left and becomes SP di Aglié; sharp left onto SP222 and immediate right (north) onto Via Molino; turn left (west) on Via Pradone; turn right (north) onto SP61/Via Rimesa; turn right (east) onto Via Vidracco and park. This is* **Stop 2, Baldissero Massif** *(45.41307, 7.74400), on the west side of SP61/Via Pramarzo for a total of 48.5 km (30.7 mi; 47 min). Carefully walk ~50 m (150 ft) north on SP61 to the roadcut.*

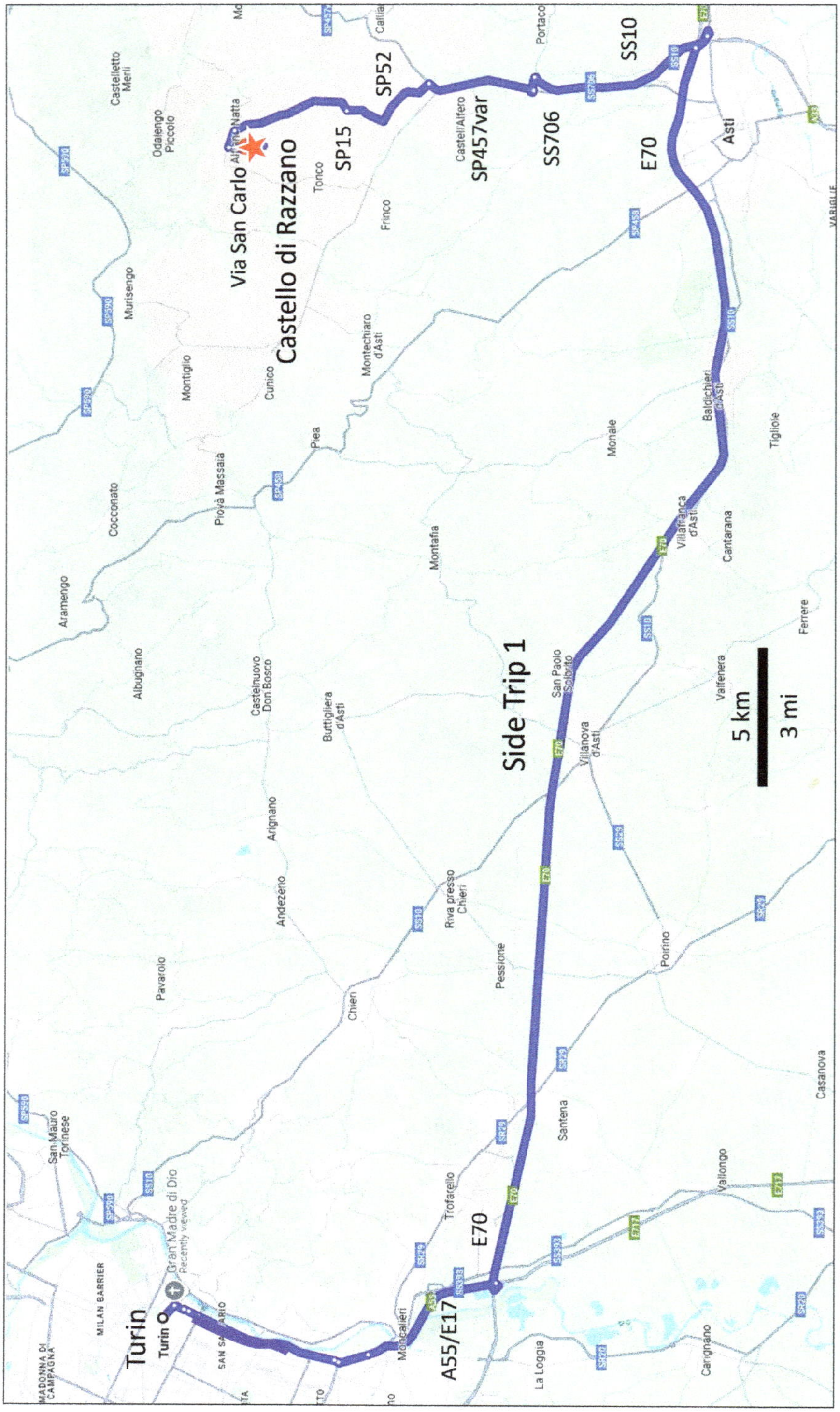
Via San Carlo
Castello di Razzano
SP15
SP52
SP457var
SS706
E70
SS10
Asti
Side Trip 1
5 km
3 mi
Turin
A55/E17
E70

SIDE TRIP 1 CASTELLO DI RAZZANO AND THE WINES OF PIEMONTE

This side trip takes us to the Asti area and gently rolling hills of the Monferrato region. These hills are the northwest extension of the Northern Apennine fold-thrust belt. Deep piggyback basins, subsiding in Plio-Quaternary time, lie behind (southwest of) the Monferrato thrust fronts. Subsidence was coincident with thrust tectonics. The Asti Basin, south of the central Monferrato front, subsided until early Pliocene, after which it has been slowly uplifting. Evidence suggests that, even after the end of the compressional tectonics, the Monferrato hills continue to uplift (Giraudi, 2016).

The Piedmont Basin is filled with upper Eocene to latest Miocene sediments that overlie Alpine basement deformed in response to the collision between the European and Adriatic plates. The Cenozoic basin fill is exposed in the Torino-Monferrato hills along the northern rim of the basin. Late Eocene–Oligocene strata are continental and shallow marine deposits, grading upward to upper Oligocene–lower Miocene turbidites that indicate a deepening basin. Uplift of the southern part of the basin and tilting to the north has occurred since middle Miocene. During late Miocene, northward thrusting created the Turin Hills. The hills were thrust north over the Po Basin as part of the late Neogene–Quaternary Padane Thrust at the leading edge of the Apennine Fold-Thrust Belt.

The Messinian section in the Piemonte Basin begins with outer shelf to slope mudstones of the Sant'Agata Fossili Marl. This is followed, at the basin margins, by evaporites of the Primary Lower Gypsum (PLG). The PLG is followed by chaotic evaporites of the Valle Versa Chaotic Complex (VVC). The Valle Versa is followed, in turn, by river, delta, and lake deposits. Deposition of Pliocene marine clay and marl of the Argille Azzurre Formation marks the end of the salinity crisis at 5.33 Ma.

Much of the Valle Versa Chaotic Complex is interpreted as large-scale landslide or slump deposits triggered by Messinian earthquakes. The stratigraphic relationships are difficult to see because of extensive weathering.

Since we are in the Piedmont Basin, it is incumbent on us to sample some of that which makes Italy, and the Piemonte in particular, such a pleasure to experience.

PIEDMONT WINES

Piemonte (Piedmont in English) is considered a top wine region in Italy. The Piemonte district lies between the western Alps and northern Apennines. When enthusiasts think of Piemonte, they think of Barolo and Barbaresco, famous Nebbiolo grape-based wines (Puckette, 2021).

History

This is the land the Ancient Greeks called Oenotrua ("land of vines"). It was heavily cultivated by the Romans. In the 14th century the Italian agricultural writer Pietro de Crescentius wrote *Liber Ruralium Commodorum* that specifically mentions Piemonte wines. He noted that the Piedmontese make "Greek style" sweet wines by twisting the stems of the grape clusters and letting them hang longer on the vine to dry and increase the sugar content (Wikipedia, Piemonte Wine).

Many Piemontese winemakers and land owners played important roles in the Risorgimento (Italian unification) of the 19th century. Giuseppe Garibaldi was a winemaker who, in the 1850s, introduced the Bordeaux grape to control the spread of oidium (powdery mildew) that was ravaging the area's vineyards. Camillo Benso, a wealthy vineyard owner, founded the political newspaper *Il Risorgimento*. At the time Austria controlled Lombardy, Emilia, and the Veneto. In an echo of the American Revolution, one of the early sparks of the Italian revolt against Austria was a response to Austria's doubling the tariffs on Piemontese wines imported into Austria. The tax cut the Piemontese wine industry off from its main export market. In 1846, King Charles Albert of Sardinia addressed the Piemontese winegrowers at a meeting of the Agricultural Congress, criticizing Austrian oppression. By 1848, Piemonte was at war with Austria in one of the early events that eventually led to unification of the Italian Peninsula (Wikipedia, Piemonte Wine).

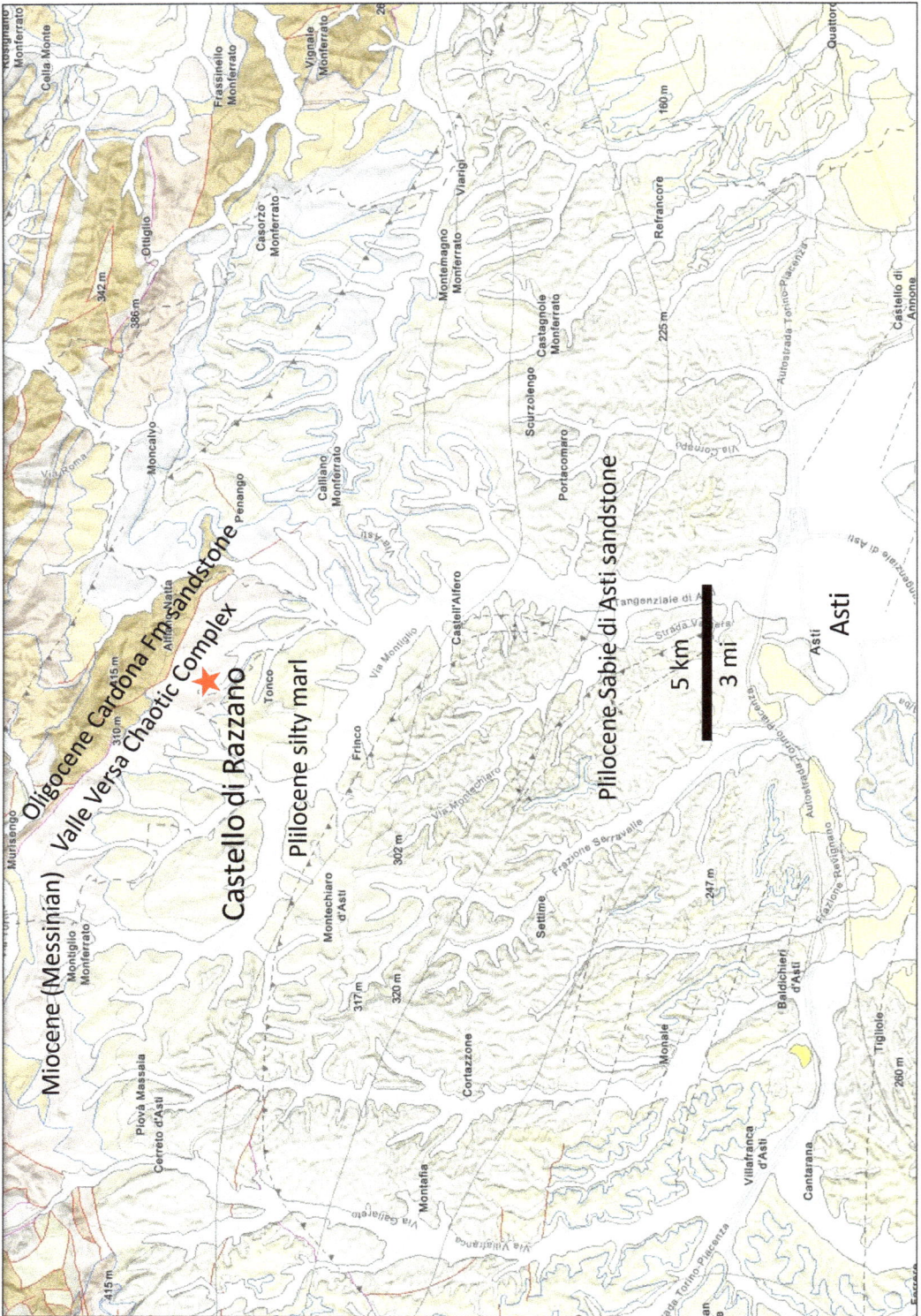

Geologic map, Asti to Castello di Razzano area. Map courtesy of Arpa GeoPiemonte, 2021.

Miocene (Messinian)

Oligocene Cardona Fm. sandstone

Valle Versa Chaotic Complex

Castello di Razzano

Pliocene silty marl

Pliocene Sabie di Asti sandstone

Asti

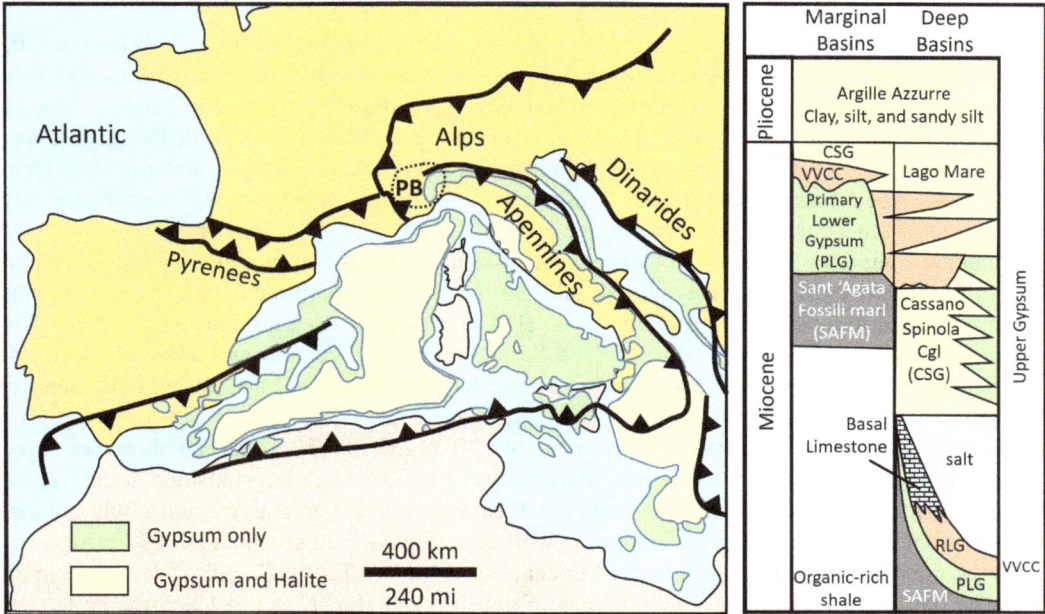

Left: Distribution of Messinian evaporites in Mediterranean Basin. PB = Piemonte Basin. Right: Piemonte Basin stratigraphy. RLG = Resedimented Lower Gypsum; VVCC = Valle Versa chaotic complex. Modified after Dela Pierre et al, 2016.

Landscape and vineyards around the Castello di Razzano. View west.

Terroir

Terroir refers to the totality of the environment that gives a wine its unique flavor. This includes the rocks and soil, the slope steepness and direction, and the climate (temperature and amount of sun, rain, fog). The region produces a stunning range of flavors, ranging from bold tannic reds to delicate peach-scented Moscatos, as a result of over a dozen grape varieties grown in different soils and microclimates. The soils are a product of the underlying bedrock and alluvial material shed from the Alps and Apennines. Some hillsides have sandier soils, some are rich in lime from underlying limestone, and others are a mixture of silt and clay.

Piemonte's wineries extend from Genoa to Turin, and yet intense agriculture leaves only about 30% of the land for planting vineyards, usually on the warmer, south-facing slopes. The regions of Piemonte are divided into five main zones: (1) Canavese, the area around Turin; (2) Colline Novaresi, in Novara province; (3) Coste della Sesia, comprising Vercelli and adjoining land; (4) Langhe, the hills around Alba; and (5) Monferrato, the lands surrounding Asti and Alessandria (Buckley, 2020).

Castello di Razzano is located in Monferrato at 230 to 250 m (750 to 820 ft) above sea level. Bedrock is a middle Pliocene marine silty marl (limy clay) and calcareous sandstone of the Argille Azzurre Formation, around 3.2 million years old. Weathering of this unit has created a light-colored soil that is enriched in lime and low in humus. This gives the Barbera grape an intense flavor.

To the north and slightly higher is the Miocene Valle Versa Chaotic Complex, deposited in an evaporite basin and containing large amounts of gypsum and clay. North of Cardona the land is higher, up to 400 m (1,310 ft) above sea level, and hillier. Here bedrock is upper Eocene to middle Oligocene marine sandstone and conglomerate of the Cardona Formation (28 to 33 million years old). There are also scattered outcrops of lower Miocene (23 million year old) marls. Fossils indicate a tropical to temperate climate.

Part of the geology display, Castello di Razzano museum.

Climate

Piemonte is at a similar latitude as the Bordeaux region in France, but it has a colder winter and a dryer overall climate because it is in the rain shadow of the Alps. During the night, cold Alpine air sinks into the valleys. In the morning warm, moist air moving in from the Mediterranean causes lowland fog. The south-facing slopes are ideal for sun-loving grapes like Nebbiolo and Barbera. The hillsides that face the Alps are better for Dolcetto and Moscato grapes that prefer cool, dry air. In the cooler areas, expect lighter tasting, higher acid wines (Mowery, 2019; Buckley, 2020; Puckette, 2021). Vineyards are typically planted at altitudes between 150 and 400 m (490–1150 ft).

Wines

DOCG and DOC are Italian quality designations for wine. The Piemonte region has 42 Denominazione di Origine Controllata (DOC) and 17 Denominazione di Origine Controllata e Garantita (DOCG). Piemonte produces more DOCG wines by volume than any other Italian wine region: nearly 84% of wine production is designated DOCG (Wikipedia, Piemonte Wine).

Wines from the northern Piemonte, or Alto Piemonte, are some of the oldest in the region. They include Ghemme, Gattinara, Lessona, and Bramaterra. These Nebbiolo-based wines from the sub-Alpine zone northeast of Turin were once popular both in Italy and abroad. Enjoyed by the nobility and commoners alike during the 19th century, the wines were popular throughout the Austro-Hungarian Empire. Since then these wines have become less popular, and the vineyards almost went out of existence. Now, however, some vineyards are trying to revive these wines. The wines are made from a blend of Nebbiolo, locally called Spanna, and other native grapes such as Vespolina, Croatina, and Uva Rara. The quality of wine from this region is considered high (Mowery, 2019).

Red Wines of Piedmont

Nebbiolo and Barbera are by far the most famous grape varieties from the region, and several wines are made from each of these grapes.

The Nebbiolo grape gets its name from the Piemontese word "nebia," meaning fog, and it is widely acknowledged as the finest of Piemonte's grapes. Nebbiolo is grown in clay-rich soils across the region. Nebbiolo has a distinct red cherry, tar, and roses bouquet. It is one of the highest tannin grapes in the world, which means Nebbiolo wines are dry and can be bitter when young, but they age well. Nebbiolo wines are best around the 10–15 year mark. As the wine matures the color becomes more rusty red, and gives the smooth feeling of drinking velvet (Wikipedia, Piemonte Wine; Mowery, 2019; Buckley, 2020; Puckette, 2021). Wines made from this grape include Barolo and Barbaresco (Puckette, 2021). Barolo and Barbaresco are the best-known wines from this region, yet they make up just 3% of the region's production.

Barolo is located southwest of the city of Alba in the Apennines. The only vineyards with Barolo DOCG status are on south-facing slopes. The wine is a pale brick red, has a bold taste with notable tannin (slightly bitter, makes your tongue feel dry) and slightly higher (13%) alcohol content. Barolo wines are aged in barrels for at least 18 months, and they are released only after three years or more.

In 1980, the wines of the Barolo region became one of the first Italian wines to receive DOCG status. Barolo is called the "king of wines and wine of kings," and has hints of ripe fruit, spices, licorice, and leather. Barolo has contributed, along with Barbaresco, to the region becoming a UNESCO World Heritage Site. South and southeast of Turin the lime-rich Tortonian Marl (7 to 11 Ma) contributes to wines of delicacy, perfume, and fruitiness. Other vineyards are developed on loose sandstone and limestone soils, called white-yellow marl, resulting in fuller bodied wines.

Barbaresco is located northeast of the city of Alba in the Apennines. The soils in Barbaresco are mostly derived from limestone, which results in less tannins in the wine. The weather has less of a daily range, which produces grapes that ripen sooner and have thinner skins. Thus, Barbaresco tends to have less tannin, a paler color, and fewer aromatic compounds. Barbaresco wines are usually lighter tasting and less dry than Barolo. Barbaresco is a softer, more approachable Nebbiolo-based wine, often with notes of roses, cherry, and tobacco. The Barbaresco DOCG regulation stipulates wines with minimum alcohol content of 12.5% and two years minimum aging in the winery for standard labels and four years minimum for "riservas" wines.

Another popular red wine grape is Barbera. It is grown in nearly every major winemaking region of the Piedmont. Wine made with the Barbera grape tends to be fruity, with high acidity. It can be delicate, having less tannin than wine made from Nebbiolo grapes (Wikipedia, Piemonte Wine). The wine has a dark ruby red color and tastes of black cherry, anise, coffee, smoke, and dried herbs. The tannins are distinctly softer than Nebbiolo grapes. Barbera is the most planted red grape in Piemonte. It is also the most consumed wine in the province (Mowery, 2019; Buckley, 2020; Puckette, 2021).

Barbera d'Asti DOCG and Barbera d'Alba Denominazione di Origine Controllata (DOC) are the most well-known appellations in Piedmont.

Dolcetto means "little sweet one," but the grape is neither. Wines made with Dolcetto grapes are dark red with flavors of blackberry, licorice, and tar. The wines don't age well because they have low acidity, but they have plenty of tannin, so they are dry. Dolcetto wine designated "Superiore" has 13% alcohol and has been aged longer, which helps smooth out the tannins (Puckette, 2021).

Typically found south and southeast of Turin, Dolcetto's fruitiness, moderate-to-low acidity, deep color, and affordable price point make it popular. It is produced mainly in the provinces of Asti and Cuneo. Dolcetto d'Alba Superiore requires a minimum of 12 months aging (Mowery, 2019).

White Wines

Although they're not as famous as the reds, Piemonte also produces notable white wines. These include Moscato Bianco, Cortese, Arneis, and Chardonnay.

Moscato Bianco, also known as Muscat Blanc à Petits Grains, is an ancient grape with intense aromas of roses, mandarin orange, melon, pear, cotton candy, and lychee. Two main styles are found in Piemonte: spumante and frizzante. Asti Spumante is a fully sparkling sweet wine with about 9% alcohol. Moscato d'Asti is a lightly bubbling (frizzante) sweet wine with about 5% alcohol (Buckley, 2020; Puckette, 2021). These wines come mainly from around Asti and Alba, southeast of Turin.

The top white grapes of the region are Cortese and Arneis. The former grows in the appellation of Cortese di Gavi DOCG, while the latter is at its best in Roero DOCG.

Cortese grapes, locally called "Gavi," create wines that are dry and known for their refreshing citrus overtones. Gavi wine is generally enjoyed young to take advantage of its crisp acidity and white peach, almond, and floral qualities.

Arneis, meaning "rascal" in Piemontese, is a medium-bodied fruity, dry wine that often has notes of bitter almond, flowers, peach, apple, and hazelnut, or pears and apricots (Wikipedia, Piemonte Wine). The vineyards are planted on sandy soils from an ancient seashore. These wines are described as fresh and grassy and somewhat similar to Sauvignon Blanc (Mowery, 2019; Buckley, 2020; Puckette, 2021).

Piemonte Chardonnays are worth tasting, as they're quite different than the southern varieties (Buckley, 2020).

CASTELLO DI RAZZANO

I selected this winery at random because it allowed tours by as few as a single person. My intent is not to promote any particular winery but to provide an example of a vineyard experience. There are many vineyard tours, and a quick Google search turns up dozens of local wineries.

In 1927 Ernesto Olearo and his wife Clementina Razzano began the first small winery at the current Cà di Corte estate with a few hectares of vineyards. Thus began their production and marketing of wines.

The company grew substantially in the 1940s as a result of the strong entrepreneurial ability of Eugenio Olearo, son of the founders and grandfather of the current generation. Eugenio established the distribution business by creating commercial relationships between Piedmont, Liguria, Valle d'Aosta, and Lombardy. The company expanded with the purchase of new land and properties, including Tenuta Castello di Razzano in 1968 from the Caligaris family, and Tenuta Campasso in 1975 from the Toso family, both located a few hundred meters from the main Tenuta Cà di Corte.

A turning point came in the 1980s with the entry into the company of Augusto Olearo, Eugenio's son. Following his oenology studies at the Alba school, Augusto completely changed the company's philosophy. Up to that time the business model was based on marketing of large quantities of wine and planting numerous hectares of vineyards. Augusto desired quality wines.

In 2002 Augusto Olearo began a complete renovation of the Castello di Razzano; after seven years the building was returned to its original splendor, with the opening of the Museo ArteVino Razzano.

Augusto's sons, Riccardo and Federico, joined the company in 2006. Currently, Castello di Razzano includes, in addition to the historic production of wine, the production of extra virgin olive oil, a charming country hotel, a restaurant, and the Museo ArteVino Razzano.

Castello di Razzano.

There are currently 30 ha (74 ac) under cultivation. They grow all their own grapes except the Ruche.

They offer 23 wines, including Spumante (sparkling), white, Rose, Young Red, Aged Red, Sweet, and Grappe.

Spumanti Metodo Classico (Sparkling)
    Lunadoro, Pinot Nero 100%
    Privilegio, Pinot Nero 70% and Chardonnay 30%
White
    Costa al Sole, Chardonnay 100%
    Sanspirit, Sauvignon Blanc 100%
    Desiderio, Sauvignon Blanc 100%
Rose
    Bellaria, Pinot Nero 100%
Young Reds
    Pianaccio, Grignolino 100%
    Briccone, Ruche 100%
    Serra del Bosco, Nebbiolo 100%
    La Leona, Barbera 100%
Aged Reds
    Campasso, Barbera 100%
    Beneficio, Barbera 100%
    Eugenea, Barbera 100%
    Valentino Caligaris, Barbera 100%
    Nero di Razzano, Bebbiolo 100%
    Cuntra', Merlot 100%
    Pian dei Tigli, Merlot 70%, Cabernet Sauvignon 30%
Sweet
    Calicantus, Moscato 100%
    Oro di Razzano, Moscato 100%
Aromatic
    China', Barbera Aromatizzato
    Grappa, Barbera
    Grappa, Moscato

Prices range from €10 to €30/bottle. Also available are olive oil (from their 10 ha of olives) and vinegar.

The Museo ArteVino Razzano was opened in October 2009 and is located inside the Castello di Razzano. The project was conceived by Augusto Olearo and was a collaboration with the artist Silvio Manzotti, originally from Milan but a proud transplant to Monferrato.

Guests can discover the exhibition spaces at their own pace and level of interest. The tour starts with a map describing the environments, spaces, and installations. Guests follow a historical path that captures the changes over time using both small and large details. The tour begins with the geology of the region, including rock types and fossils, and snippets of Roman, Gothic, medieval, and Baroque artifacts. An old farmhouse has been recreated, and a blacksmith and cooper exhibit describes how barrels were made. Old instruments for analyzing wines are shown, an old cellar houses a wine press and various tools used in the production of wine, and the actual aging cellars (still in use) are seen. The tour ends at the *Infernot*, an ancient escape route through a UNESCO World Heritage tunnel that today houses their Riserve.

Wine press and old tools, Castello di Razzano wine museum.

Visit

Visits are by reservation only. Book your visit at https://www.castellodirazzano.it/en/cantina/degus-tazioni/ .

No minimum number of people are required. The self-guided tour lasts 20 to 30 minutes and is followed by a tasting. The tour goes through the winery museum and wine cellars. There are "small," "medium," and "large" tastings. They also offer a breakfast and picnic option.

Address: Strada Gessi 2, Frazione Casarello, 15021 Alfiano Natta (AL)
Tel: +39 0141 922535
Email: info@castellodirazzano.it
Website: https://www.castellodirazzano.it/en/
Hours: variable, depending on season and tour. Please consult their website.

We now leave the Po Basin and enter the foothills of the Southern Alps, specifically the Canavese Zone.

Castello di Razzano wine tasting.

GEOLOGY OF THE WESTERN ALPS

In our discussion of the Po Basin we covered the basics of the collision of the European and African tectonic plates that led to the Alpine Orogeny. Here we extend the discussion to cover the Western Alps.

The Western Alps form an arc that extends north-northwest from the Mediterranean coast near Nice, bracketing the French-Italian border, and curves to the northeast starting between Valence, France, and Turin, Italy. From the Valence–Grenoble area the Western Alps extend to the area between Geneva and Aosta, whence begin the Central Alps.

The Alpine chain is a result of the collision and subduction of the European tectonic plate beneath the Adriatic Microplate, a northern extension of the African plate. Between Europe and Africa was the Alpine Tethys Ocean and a few smaller basins. As a result of this collision a number of crustal slices were compressed, pushed north, and stacked one on another (Harris, Outline of the Geology of the Alps).

Alpine Domains

To the uninitiated North American geologist (me, when I began this exploration) there are a bewildering number of Alpine zones and new terms: Helvetic-Dauphinois Nappes, Piemonte–Liguria and Valais oceans, the Adriatic Microcontinent, flysch. Some of these are simply European terms for common concepts: flysch is referred to in North America as deep marine turbidite deposits. Others are National artifacts: the Dauphinois Domain in France is the same as the Helvetic Zone in Switzerland.

Over the years the Alps have been subdivided by various workers into a number of geographic, tectonic, and lithologic zones. Commonly accepted subdivisions (domains, zones, or supergroups) of the Western Alps are, from north to south and west to east:

- Undeformed European continental crust.
- Slightly deformed European continental crust of the Jura Mountains. This is the most external domain, the last to be incorporated in the Alpine chain (from 11 to 3 Ma). It is characterized by thrusting that is detached in Triassic evaporites. This décollement horizon runs under the Molasse Basin and is rooted in the Helvetic Domain (Stampfli, 2001).
- The Molasse Basin, Bresse Graben, and Valence Basin, all foreland or rift basins filled with material eroded from the adjacent Alps. A graben is a rift valley or long, fault-bounded basin.
- The Helvetic and Dauphinois Nappes consisting of the former southern margin of the European Plate that has been thrust to the north and west. A "nappe" is a sheet of rock (a "thrust sheet") that has moved horizontally on a thrust fault up and over younger strata. These nappes are the most external and lowermost of the superunits. This domain contains the External Crystalline Massifs (e.g., Mont Blanc), comprising Permo-Carboniferous granitic and older metamorphosed sedimentary rocks that represent Variscan European basement. Mesozoic sediments in the Helvetic-Dauphinois Nappes are generally not metamorphosed (Stampfli, 2001; Froitzheim, 2012).
- The Penninic Nappes consist of the former southern continental margin of Europe, ophiolites (former oceanic crust) of the Piemonte–Liguria and Valais oceans, and parts of the European Briançonnais and African-Adriatic Microplates. These were all thrust north and west over the Helvetic Nappes. Thus, there are both oceanic- and continent-derived Penninic nappes. The Penninic nappes occur in a continuous zone in the Western Alps and in isolated outliers along the northwestern boundary of the Alps. High-level metamorphism in the Penninic Nappes is primarily Tertiary and ranges from greenschist facies to blueschist and eclogite facies (Froitzheim, 2012; Piana et al., 2017). In the Western Alps the Penninic Nappes have been further subdivided into the Briançonnais Zone (west), the Liguro-Piemontais Zone composed of ophiolites and oceanic sediments (central), and the Inner Crystalline Massifs such as the Gran Paradiso and Dora Maira intrusions (east; Stampfli, 2001; Faure and Chen, 2018).
- The Austro-Alpine Domain consists of slivers of the Adriatic Microplate and African Plate thrust north and west over the Penninic Nappes. This is the uppermost zone, that is, it sits above the other nappes (Froitzheim, 2012). Locally is is characterized by large-scale backfolding (e.g., folds originally leaning to the west that were refolded and now verge to the east), backthrusting, and high pressure–low temperature metamorphism (Stampfli, 2001).
- The Periadriatic Line is a Tertiary fault system that separates north-directed thrusting from south-directed thrusting. The Canavese Line, the western branch of the Periadriatic Line, is a mylonite zone 10–300 m (33–1,000 ft) thick. The Canavese Line has been interpreted as a suture between the European and Adriatic plates, albeit modified by later events (Faure and Chen, 2018).
- The Southern Alps, consisting of slightly deformed Adriatic Microplate and African continent, thrust south over the Po Basin. The South Alpine Domain contains the southern part of the Adriatic Microplate. The lower crust of the Adriatic Microplate outcrops in the Ivrea Zone (Stampfli, 2001). There is no Cretaceous thrusting and little Alpine metamorphism in the South Alpine Domain. South-directed thrusting is Tertiary in age. The most external (southernmost) thrusts at the margin of the Po Basin are still active today (Froitzheim, 2012).
- The Po Basin, a hinterland basin to the Alps and a foreland basin to the Apennine Fold-Thrust Belt (Harris, Outline of the Geology of the Alps).

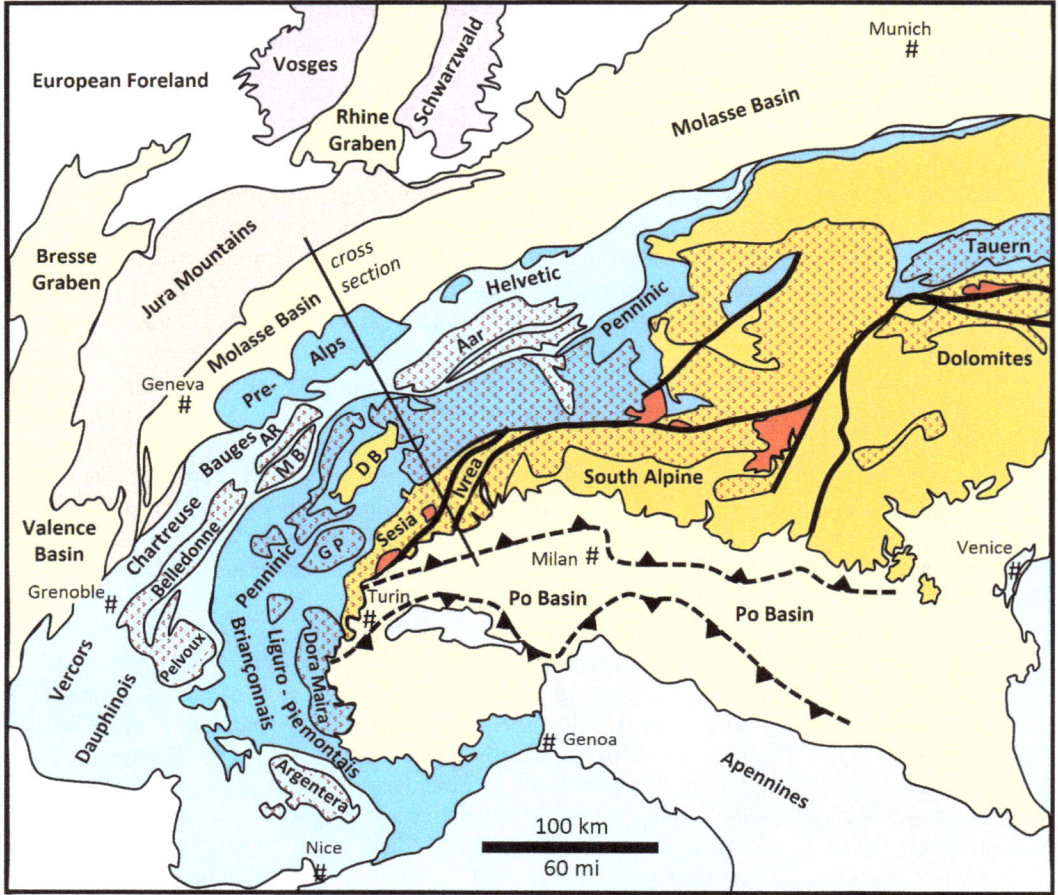

Structural domains of the western Alps. AR = Aiguilles Rouges; MB = Mont Blanc; DB = Dent Blanche; GP = Gran Paradiso. Pattern indicates crystalline massifs (granite, gneiss, schist). Modified after Pfiffner, 2016; Faure and Chen, 2018; Kuderski, 2020.

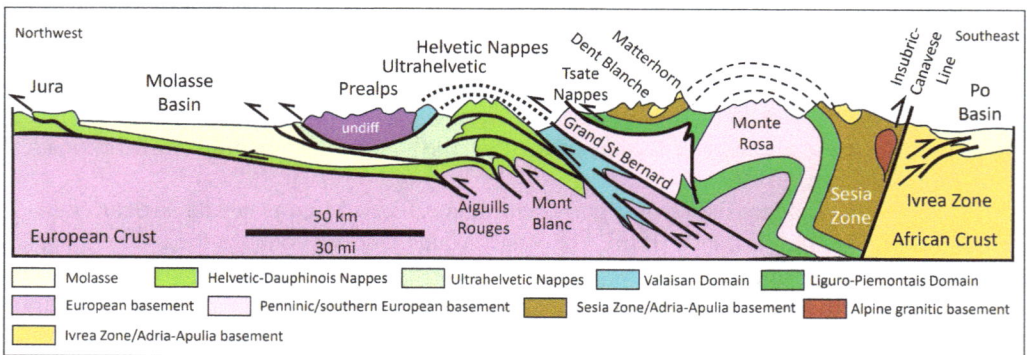

Cross section through the Western Alps. Location is shown on the previous figure. Modified after Faure and Chen, 2018.

Geologic History

The Variscan (formerly the Hercynian) Orogeny, in which the supercontinent Pangea was formed by the collision of Gondwana (southern continent) and Laurussia (northern continent), ended in Late Carboniferous time. The Paleo-Tethys Ocean was east and south of the terranes that now form the Alps.

Over time erosion wore down the Variscan Mountains that had formed during the continent-continent collision. During the Permian, the main deposits in Europe were sandstone and conglomerate, all eroded from the Variscan Mountains. At that time, crustal extension formed basins along the axis of the mountain range. Extension and rifting were accompanied by silica-rich (felsic) volcanism. A rising Triassic sea level flooded the continental margin of Pangea with shallow seas in which limestone and, locally, evaporites were deposited.

Alpine history begins in the Late Triassic with the breakup of the megacontinent Pangea and the birth of the east-west-oriented Tethys Ocean. The Tethys separated the continent of Laurasia in the north (mainly North America–Eurasia; not to be confused with the older, Paleozoic continent Laurussia) and Gondwana in the south (mainly Africa–South America–Australia). Extension and rifting continued during the Jurassic and lasted until Early Cretaceous time. This created minor land masses separated by relatively small oceanic basins. These include, from the northwest to the southeast, Europe, the Valais Ocean, the Briançonnais Terrane–Iberian Plate, the Piemont–Liguria Ocean, the Adriatic Microplate, the Tethys Ocean, and Africa (Kindler, 2014).

The Piemont–Liguria Ocean was a narrow western extension of the Tethys Ocean. A piece of continental crust connected to the northern African plate, the Adriatic Microplate, lay between Africa and Europe. Continued high sea levels during the Jurassic and Cretaceous encouraged the deposition of massive limestone layers.

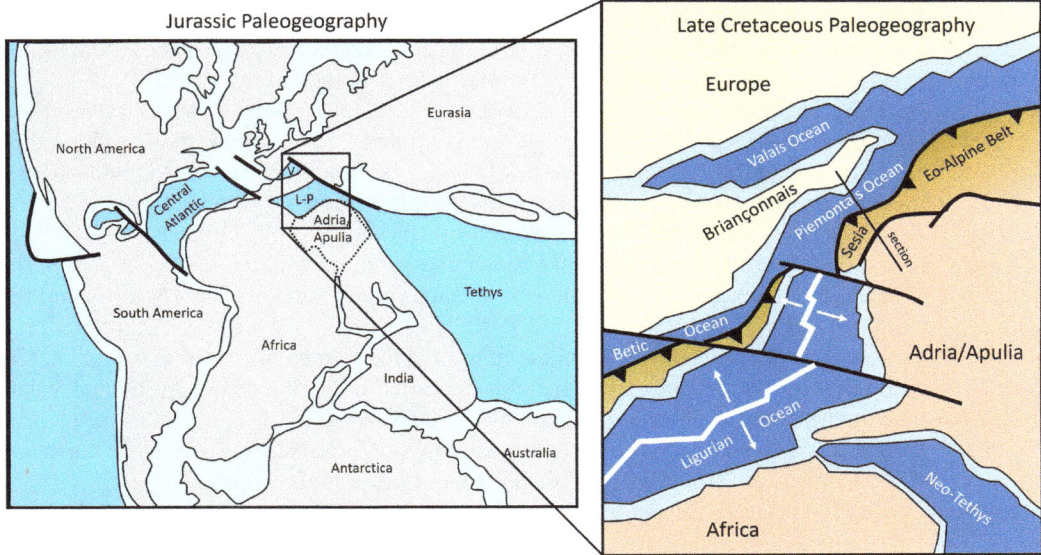

Left: Jurassic paleogeography. V = Valais Ocean; L-P = Liguro-Piemontais Ocean. Modified after Faure and Chen, 2018. Right: Late Cretaceous paleogeography, Western Alps. Modified after Festa et al., 2020.

In Late Jurassic–Early Cretaceous time the Iberian (and Briançonnais) microcontinents rifted away from the southwestern part of the European Plate, forming the Valais Ocean between them and Europe. The Briançonnais Microcontinent (also called a microplate, terrane, or continental ribbon) was a narrow strip of continental crust that was part of the European continental margin before the opening of the Valais Ocean. It was the eastern estension of the Iberian block.

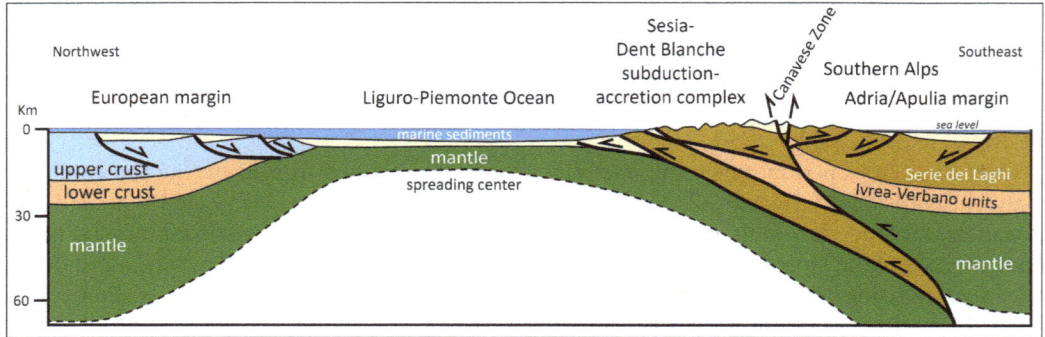

Late Cretaceous paleogeography, Western Alps. Modified after Festa et al., 2020.

At the end of the Jurassic the African-Adriatic plates began to move north and converge with the European plate. An oceanic trench formed at the northern margin of the Adriatic Microplate as the European Plate plunged down (was subducted) beneath Africa. This trench filled with deep marine sediments.

During the Cretaceous, Africa continued moving toward Europe, compressing, folding, and ultimately thrusting the sediments in the ocean basins and trenches upward and to the north and west. The Tethys/Piemont–Liguria Ocean between Africa and Europe continued to close as oceanic crust was driven beneath the Adriatic Microplate. The southern margin of the European plate, dragged down many km, became extensively metamorphosed. The Adriatic/African Plate moved northward over 1,000 km (600 mi) during the Cretaceous, or about 1.3 cm/yr (0.5 in/yr). Deformation of the sediments as they were forced up from the depths was associated with an east-west string of volcanic islands, an island arc. Volcanic rocks produced in these island arcs are found today among the ophiolites (oceanic crust) in the Penninic Nappes (Wikipedia, Geology of the Alps).

In the Late Cretaceous the first continent-continent collision took place as the northern part of the Adriatic Microplate collided with Europe. Because continents are less dense than seafloor, they don't subduct easily and tend to crumple upward. This was the first phase of the Alpine Orogeny. Eclogite, high-grade metamorphic rock in the Austro-Alpine Nappes, formed during the early Alpine Orogeny. This metamorphism is unrelated to the later, main, Alpine Orogeny.

Fragments of the Piemont–Liguria Ocean are now found in the Penninic Nappes. The part of the Adriatic Microplate that deformed at this time is now found in the Austro-Alpine Nappes and the Southern Alps.

Cretaceous limestone deposited on the southern margin of the European continent would later be incorporated into the Helvetic Nappes. Mud and clay deposited in the Piemont–Liguria and Valais oceans were metamorphosed to slates found in the Penninic Nappes.

The upper crust of the Adriatic Microplate (the later Austro-Alpine Nappes) was thrust north and west over the European crust. This was the main collisional phase in the Alps.

Convergence and shortening formed great folds and detached extensive sheets of rock that rose out of what had been the Tethys Ocean. The detached sheets were pushed north and west, often breaking and sliding one over the other on regional, near-horizontal thrust faults (the detachment surfaces). The dominant vergence (direction the folds lean toward) in these units is to the north and west (except in the Southern Alps, where folds lean to the south). Even crystalline basement rocks, such as granites intruded during the Carboniferous Variscan Orogeny, became detached and were thrust to the north and west.

During the Paleocene the Piemont–Liguria oceanic crust had plunged completely beneath the Adriatic Microplate and the Briançonnais Microcontinent had just arrived at the subduction zone.

Now the Briançonnais Microcontinent and Valais Ocean subducted beneath the Adriatic Microplate. They stayed at depths around 70 km (45 mi) during the Eocene, undergoing high-grade (eclogite facies) metamorphism. This material would later be incorporated in the Penninic Nappes.

The continent-continent collision reached a climax in the early Oligocene. The subducting slab broke off and sank into the mantle. The shallower parts of the slab rebounded upward, leading to uplift of the now thickened continental crust. The central zones of the Alps rose. Meanwhile, the thrust front of the Penninic and Austro-Alpine nappes continued moving west and north. The northwest-directed shortening caused another detachment to form beneath the Penninic and Austro-Alpine nappes. This thrusted material became the Helvetic and Dauphinois nappes in Switzerland and France, respectively (Wikipedia, Geology of the Alps). Continuing compression and shortening in late Miocene and Pliocene time led to folding and thrusting in the Jura Mountains, the northern-most external domain related to the Alpine Orogeny.

Subsiding basins formed both north and south of the Alps due to the weight of the stack of thrusts. Eroded sediment (first flysch and later molasse) accumulated in these basins from early Oligocene to the late Miocene (Kindler, 2014). "Flysch" is a European term for deep marine turbidity deposits; "molasse" is a European term for a mix of deposits (generally shallow marine or onshore conglomerate, sandstone, and shale) formed by erosion of actively rising mountains.

The Molasse Basin lies between the Alps and the Jura Mountains. It is the northern and western foreland basin of the Alps formed during the Oligocene and Miocene as a result of flexure of the European crust under the load of Alpine thrusts. It no longer accumulates sediments, as large parts of it are now being uplifted and eroded (Froitzheim, 2012). The Molasse Basin extends ~700 km (435 mi) in a northeast direction from Chambéry, France, to Linz, Austria. It is about 30 km (19 mi) wide in the west, and close to 200 km (120 mi) wide in Bavaria. The Molasse Basin is the deformed part of what was originally the North Alpine foreland basin, but that later became a piggyback basin when the Jura Mountains were formed. The Molasse Basin contains 1 to 3 km (3,280 to 9,840 ft) of Mesozoic sediments. An erosional unconformity separates predominantly Mesozoic carbonates from overlying late Eocene to late Miocene molasse deposits derived from erosion of the Alps (Kindler, 2014).

Along the southwestern side of the Western Alps the mountains abut the Chaînes Provençales, a set of east-west hills and ridges that are the far eastern extension of the Pyrenees Mountains. North of these chains, the Alps are bordered by the Tertiary Valence Basin and Rhône-Bresse Graben.

Cenozoic sediments in the Valence–Rhône–Bresse system reach a maximum thickness of 2,300 m (7,550 ft). They are part of the European Cenozoic Rift (or Central European Rift), a feature crossing the continent from the Mediterranean to the North Sea and representing a failed rift system. The Rhône–Bresse Graben began to subside during the middle Eocene, at the same time as the Molasse Basin. The graben is separated in the south from the Valence Basin by the Lyon Sill, a narrow ridge. Successive phases of subsidence eventually breached the Lyon Sill, creating communication with the Valence Basin (Sissingh, 1998). These basins are filled with Middle Eocene to Pliocene river and lake deposits, evaporites (gypsum, salt), conglomerates, marginal marine sandstones, and shallow marine limestones. Extension and subsidence of the Bresse Graben continued during the Early Pleistocene, ending around 1.4 Ma as the basin began to uplift and erode (Schlunegger and Mosar, 2010).

As detailed earlier, south of the Alps is the Po Basin. It is the hinterland basin of the Alps, in that it lies in back of the north-transported Alps. Filled mainly with Cenozoic sediments, it is actively subsiding and receiving sediments today (Froitzheim, 2012).

There were at least five major glacial periods over the past two million years. The last great glacial advance ended between 10,000 and 14,000 years ago. Besides leaving large granite boulders (glacial erratics) scattered across the landscape, the glaciers also left polished and striated (scratched) bedrock, and piles of rock pushed ahead and to the side of glaciers known as moraines (Wikipedia, Geology of the Alps). They also sculpted the peaks into horns and carved bowls (cirques) that contain glacial lakes (tarns).

There is evidence that the Alps are still being uplifted up to 2.5 mm/year (1 in/yr) in some areas. In addition to some ongoing tectonic uplift, this is probably largely because of glacial rebound after removal of the weight of 1 km (0.6 mi) of ice after the last Ice Age ended around 10,000 years ago. Thus, the famous Alpine landscape of craggy peaks and broad valleys is a relatively recent development.

ALPINE CONTRIBUTIONS TO THE CONCEPT OF THRUST FAULTING

The idea of collisional mountain building is older than plate tectonics. The Swiss geologist Émile Argand had already proposed this concept at the beginning of the 20th century. An early supporter of Alfred Wegener's theory of continental drift, Argand viewed continental collisions as the best explanation for the formation of the Alps. At that time, a driving mechanism for plate tectonics had not yet been recognized, but concepts such as ophiolitic sutures, Austro-Alpine overthrusting, and backthrusting had already been proposed (Faure and Chen, 2018).

As is so often the case in science, thrust faults were recognized by widely separated workers at around the same time. Arnold Escher von der Linth, Albert Heim, and Marcel Alexandre Bertrand were working in the Alps on the Glarus Thrust; Charles Lapworth, Ben Peach, and John Horne worked on the Moine Thrust in the Northwest Highlands of Scotland; Alfred Elis Törnebohm worked in the Scandinavian Caledonides, and R. G. McConnell recognized thrusting in the Canadian Rockies. The realization that older strata could be faulted over younger strata was arrived at more or less independently by each of these groups of geologists during the 1880s (Wikipedia, Thrust Fault).

In the Swiss Alps the Glarus Thrust was discovered and intensely debated during the 19th century. Arnold Escher and his student Albert Heim proposed that it was a "double fold" structure. Their early concept was that horizontal transport was accomplished by folding. And yet it was Heim's descriptions in his 1878 text on mountain building that led Marcel Bertrand, in 1884, to reinterpret the Glarus structure as a north-directed thrust sheet. Bertrand was a French geologist and instructor at the École Nationale Supérieure des Mines. As with many new concepts, his idea was not taken seriously until years later, in the 1890s.

In the mid 18th century there was an ongoing debate between Roderick Impey Murchison (and his student, Archibald Geikie) and James Nicol as to the stratigraphy of the Northwest Highlands. Murchison and Geikie argued for a normal stratigraphic succession (older rocks at the base, younger on top), while Nicol argued that he was mapping older rocks above younger rocks. Key observations were made independently by Charles Callaway and Charles Lapworth. Heim's *Untersuchungen uber den Mechanismus der Gebirgbildung* (Studies on the Process of Mountainbuilding, 1878) was read by the Scottish geologists, who applied the concepts to their observations. In 1883 Callaway described the Glencoul Thrust as having an older metamorphic basement over younger Cambrian sediments. Lapworth recognized a similar relationship on the Scottish north coast at Ben Arnaboll, where he also noted that the shearing along the thrust surface had altered the rock to mylonite, a fine-grained recrystallized fault rock. Two Scottish geologists, Ben Peach and John Horne, mapped the thrusts and provided the first coherent description of the Moine Thrust (Butler, Thrust Tectonics). Geikie, in a paper in *Nature* in 1884, coined the term *thrust* to describe this type of near-horizontal fault that accommodates crustal shortening.

Hats off to the geologists, geophysicists, and engineers who figured this out by walking and examining every bit of their respective countrysides over the past 150 years. Putting it all together and coming up with a coherent explanation for the complex geology of this region was truly a monumental accomplishment.

Castello to Baldiserro: *Return north on Via San Carlo; turn left (west) on SP13; slight right onto SP12; turn left (west) onto SP590/Via Casale; at the roundabout in Abate take the 1st exit (north) onto SP94; at the roundabout in Verolingo take the 2nd exit (north) toward Torino/Milano; continue*

*straight on Raccordo Est di Chivasso; at the roundabout take the 3rd exit (northwest) onto Via S. Francesco; at the roundabout take the 3rd exit (west); at the roundabout take the 2nd exit (west) onto Via Cene; at the roundabout take the 2nd exit (west) onto SP82; at the roundabout take the 1st exit (north) onto Circonvallazione di Montanaro; at the roundabout take the 1st exit (west) onto SP82DIR1; merge onto SP82; turn left (west) onto SP40; turn right (north) on SP40/Via Ivrea; at San Giorgio Canavese turn left (west) onto SP53/Circonvallazione; at the roundabout take the 1st exit (north) onto SP41; SP41 turns left and becomes SP di Aglie; shrp left (west) on SP222; immediate right (north) onto SP61/Via Rimesa; right on Via Vidracco and park on the left. Walk 120 m north to the roadcut at the curve. Mind the traffic This is **Stop 2, Baldissero Massif** (45.41307, 7.74400), on the west side of SP61/Via Pramarzo for a total of of 73.0 km (45.3 mi; 1 hr 16 min). Carefully walk ~50 m (150 ft) north on SP61 to the roadcut.*

Stop 2 Baldissero Massif, the Ivrea–Verbano Zone, and Canavese Line, Baldissero Canavese

We are still on the ancient continental crust of Adria, part of northern Gondwana. This stop is on Baldissero peridotite, part of the Ivrea–Verbano zone of the South Alpine Domain (Faure and Chen, 2018). The peridotite represents part of the lower crust of the Adriatic Microplate. The Ivrea–Verbano Zone contains, besides amphibolite, high-grade metasediments (gneiss of the Kinzigite

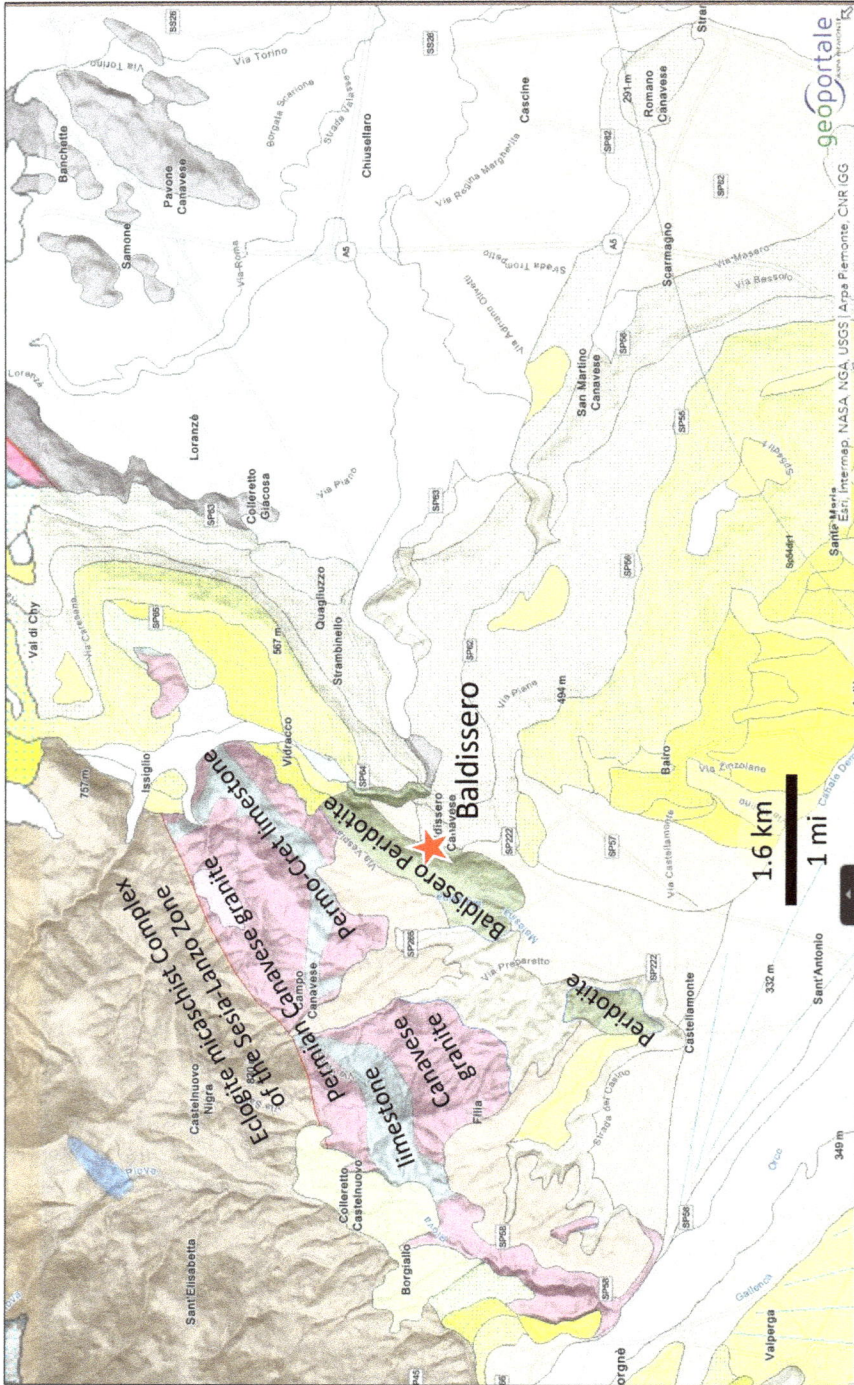

Geologic map of the Baldissero Massif area. From Arpa GeoPiemonte, 2021.

Formation) and mafic and ultramafic bodies that were intruded by Lower Permian gabbro and diorite (the Mafic Complex; Festa et al., 2020). Amphibolite is a dense, dark metamorphic rock primarily composed of hornblende (a type of amphibole) and plagioclase feldspar. Mafic and ultramafic rocks are dark, rich in magnesium and iron minerals, and they are commonly derived from oceanic crust. They include gabbros, basalts, and diorites, among others.

Baldissero massif.

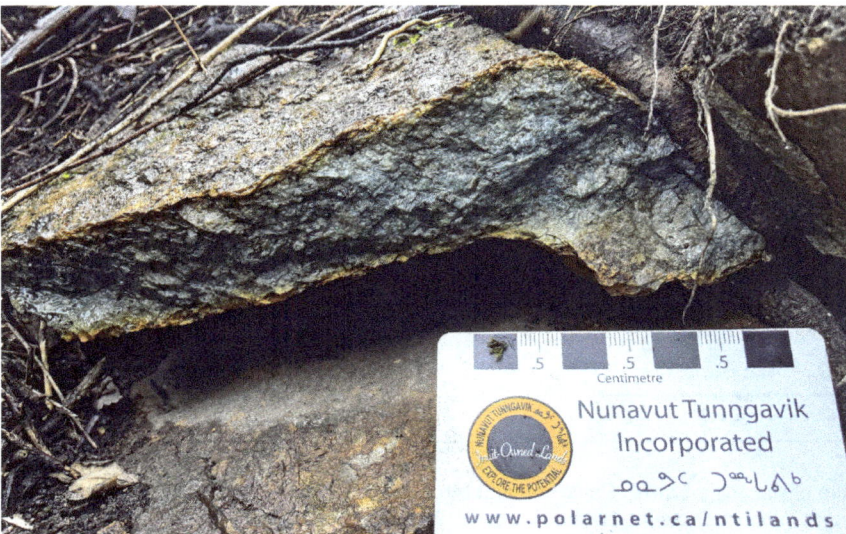

Baldissero peridotite. This dark blue-green rock weathers to a light orange-brown.

The Ivrea–Verbano Zone is bounded on the northwest by the Internal Canavese Line. This line is a left-lateral mylonite zone developed during low-grade Alpine metamorphism (prehnite-pumpellyite facies). Small peridotite massifs, such as the Baldissero Massif, outcrop near the Internal Canavese Line (Faure and Chen, 2018).

The Canavese Zone is only a few kilometers wide and about 40 km (24 mi) long. The Canavese Zone is a remnant of the distal passive margin of the Adriatic Microplate. It was stretched and thinned during Jurassic opening of the Tethys Ocean. Breakup of Pangea and the Adriatic Microplate margin was strongly influenced by the preexisting Proto-Canavese Shear Zone. The Canavese Zone is bounded on the northwest by the External Canavese Line, developed in Oligocene–early Miocene time, and corresponding to the southwestern extension of the Insubric or Periadriatic Line. It separates the Canavese Zone from the Sesia–Lanzo Zone in the Austro-Alpine Domain. The Sesia–Lanzo Zone, part of the northwest-verging Alpine Fold-Thrust Belt, represents Adriatic continental crust metamorphosed to blueschist-eclogite facies during the Alpine Orogeny (Festa et al., 2020).

This part of the Alps consists of tectonic slivers of the Adriatic Microplate that were subducted to great depth, metamorphosed to varying degrees, then squeezed back to the surface.

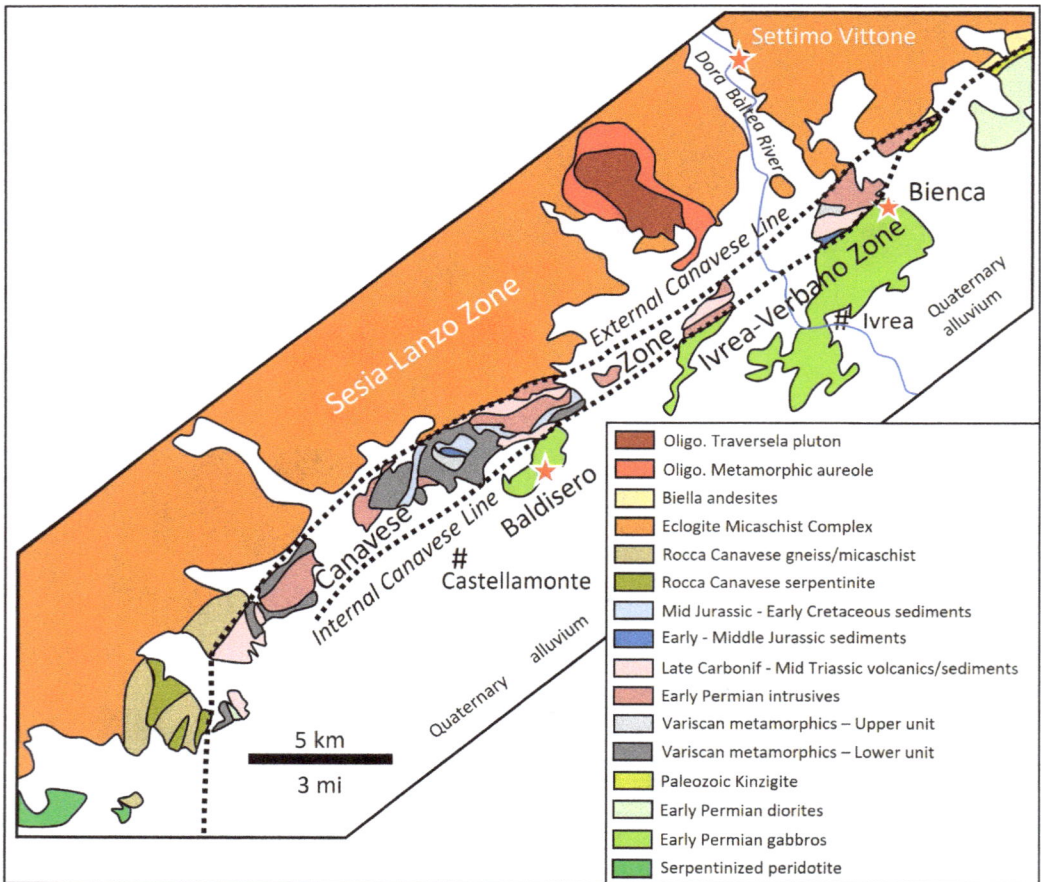

Geologic map of the Sesia-Lanzo, Canavese, and Ivrea-Verbano zones. Dotted lines are tectonic boundaries. Modified after Compagnoni et al., 2014; Festa et al., 2020.

Baldissero Massif to Bienca: *Drive south on Via Vidracco; continue straight on Piazza San Martino; turn left (south) onto Via Molino; turn left (east) onto SP222/Via Bettolino; continue straight onto Via Ponte Preti; continue straight onto SS565; at the roundabout take the 1st exit (northeast) onto Via Castellamonte; turn left (north) onto Via Pinchia; turn right (northeast) onto Via Miniere; at the roundabout take the 2nd exit (north) onto Corso Costantino Nigra; cross the Dora Bàltea River into Ivrea and drive east on Corso Re Umberto; turn left (north) onto Corso Botta; at the roundabout take the 2nd exit onto SS26 to Aosta, then turn right (northeast) onto SP75/Viale Monte Stella; continue straight onto SP74; continue on SP74 to* **Stop 3, Apulia Diorite, Bienca** *(45.505920, 7.898410) for a total of 20.2 km (12.5 mi; 24 min). Pull onto the small road on the left just north of the intersection. The outcrop is at the intersection.*

Stop 3 Apulia Diorite of the Ivrea–Verbano Zone, Bienca

This is a rare outcrop in an otherwise heavily forested area. It appears to have been smoothed, and in places actually polished, by glaciers that have left their striations running north to south across the rock.

We are in the Ivrea–Verbano Zone. This outcrop has been variously described as Apulia/Adriatic diorite and gabbro/mafic granulite of the Mafic Complex, Ivrea–Verbano Zone. All of these terms refer to a hard dark rock that was once the deep, lower part of Apulian continental crust (Faure and Chen, 2018; Arpa GeoPiemonte, 2021).

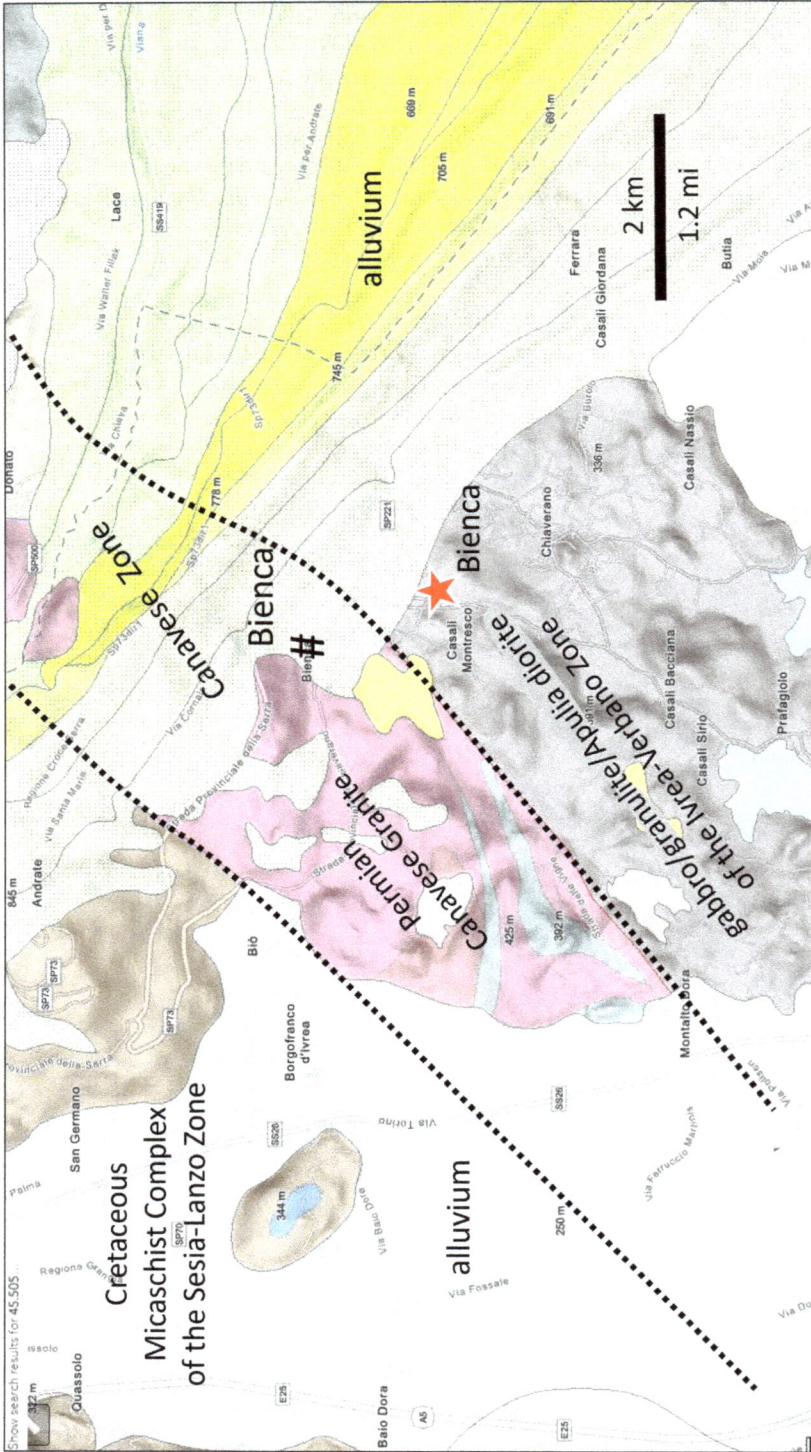

Geologic map of the Bienca area. Dotted lines are tectonic boundaries. From Arpa GeoPiemonte, 2021.

Adriatic Microplate diorite/gabbro outcrop at Bienca. Notice the glacial striations.

Side Trip 2, Bienca to Scopetta: *Return south on SP74; continue straight on SP75; at Lago Serio turn left (south) onto Via Lago S. Michele; turn left (southeast) onto Via Alessandro Favero; turn left (east) onto Via Cascinette; at the roundabout take the 1st exit (south) onto Via Strusiglia; at the roundabout take the 3rd exit (east) onto Corso Vercelli; at the roundabout take the 1st exit onto SP78; bear left onto SP79; in Albiano d'Ivrea continue straight onto SP80; take the A4 ramp to A26 (toll road); merge onto E25 east; at La Mandriotta take the exit onto A4/E64 to Milan; take the exit onto A26 north to Gravellona; take exit Romagnano Sesia-Ghemme north to Romagnano; continue north on SP299 to Scopetta and turn left (east) into the parking area. This is **Stop ST2.1, Insubric Line, Scopetta** (45.806607, 8.123754) for a total of 98.9 km (65.5 mi; 1 hr 35 min). Park and walk ~180 m (600 ft) east to the bridge over the Sesia River.*

If you prefer not to go on Side Trip 2:

Bienca to Parco Avventura: *Return south on SP74; continue straight on SP75; turn right (north) onto Via Lago Sirio; continue straight on Via S. Pietro; turn right (north) onto Via Giovanni Bosco; turn right (north) onto SS26; at the sign for Falesia on the right turn right (east) onto Via Vittorio Emanuele; at the sign for "Palestra di Roccia" turn left (north) and drive to the parking area on the right. This is **Stop 4, Schistes Lustrés, Parco Avventura La Turna** (45.537512, 7.839575) for a total of 12.9 km (8.0 mi; 17 min). Note that this is a bit roundabout because the more direct route was closed at the time of writing.*

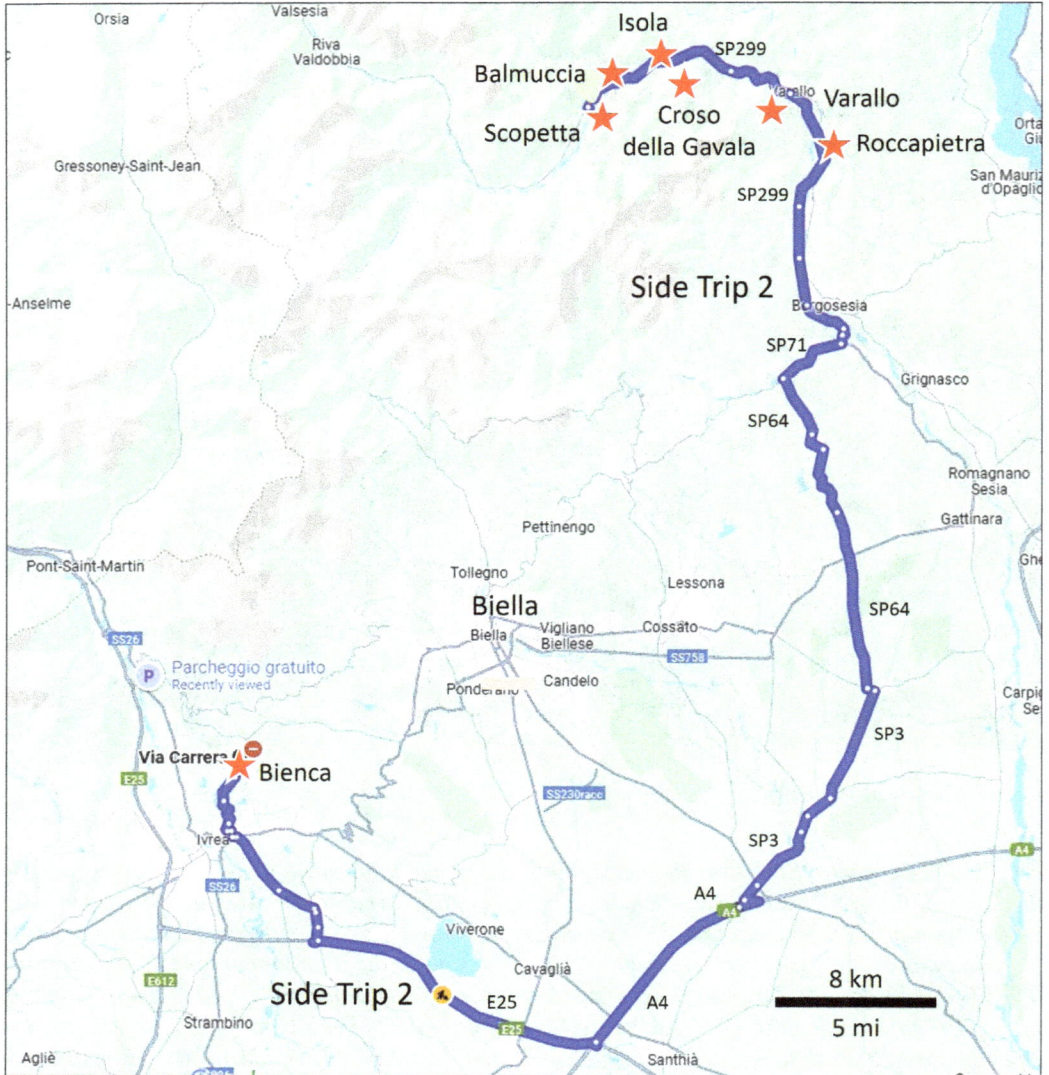

Sesia Val Grande UNESCO Global Geopark

The Sesia Val Grande UNESCO Global Geopark, about 2202 km² (850 mi²), was established in 2015 and extends from the high Alps to the Po Plain. It contains the largest wilderness area in Italy, as well as Europe. It protects both habitats and endangered animals and plants. The geopark offers visitors the opportunity to observe the effects of geologic processes, supports the concept of plate tectonics and mountain-building, and reveals some of the processes that led to development of the Austro-Alpine and Southern Alpine domains.

The park straddles the Canavese Line, the southwestern extension of the Insubric Line, which is regarded as the tectonic boundary that separates European from African continental crust. North of the Insubric Line there are north-vergent nappes of the Central and Western Alps (Austro-Alpine and Penninic domains of the European continental crust); south of the line are south-vergent folds and thrusts of the South Alpine Domain that developed on Adriatic continental crust.

Much of the geopark is in the South Alpine Domain. As we saw between Baldissero and Bienca, the Canavese Line trends southwest-northeast in this area and juxtaposes the Sesia–Lanzo Zone of the Austro-Alpine Domain against the Ivrea–Verbano Zone of the South Alpine Domain. Geosites are scattered across the park with interpretive displays for the interested geotourist (UNESCO, Sesia Val Grande Geopark).

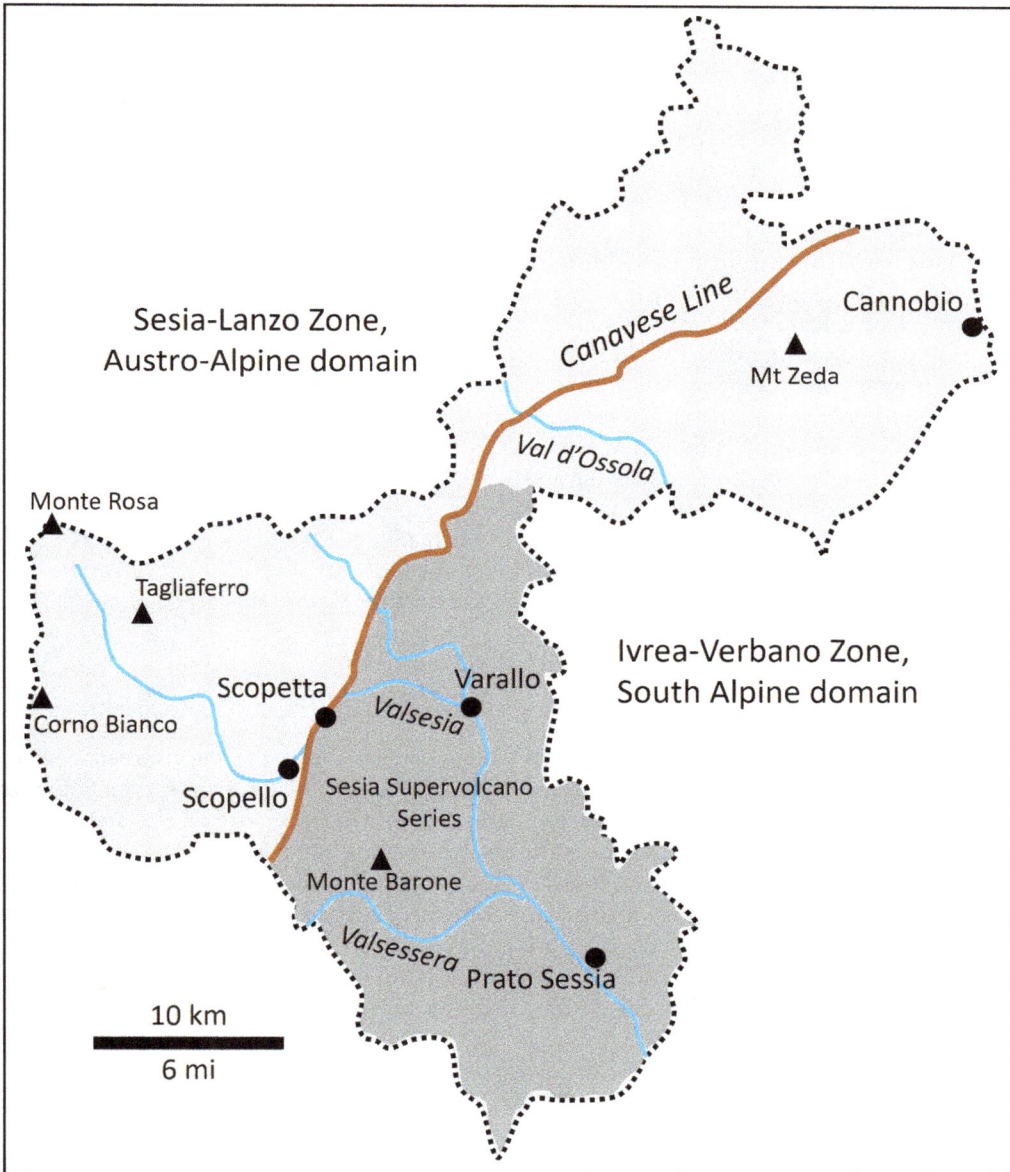

Location and key features of the Sesia-Val Grande UNESCO Global Geopark.

The Ivrea–Verbano Zone exposes late Paleozoic Variscan granulites that formed the lower crust of Adria. These high-temperature granulites were tilted to near-vertical along the Canavese line. Paleozoic diorite, high temperature gneiss (kinzigite) and micaschist, the basement rocks of the Adriatic Microplate, can be seen in outcrops extending from Ivrea to Milan.

SIDE TRIP 2 IVREA–VERBANO ZONE AND THE SERIE DEI LAGHI

Side trip 2 provides an opportunity to see mantle and deep crustal rocks of the Ivrea–Verbano Zone. The Ivrea–Verbano Zone is bounded to the southeast by amphibolite facies metamorphic rocks and granites of the Serie dei Laghi. The Ivrea–Verbano Zone and Serie dei Laghi (Lake Series) are the deep- and middle-upper crustal components, respectively, of the pre-Alpine crust of northwest Italy. This excursion is based on a technical field guide prepared by Mazzucchelli et al., 2014.

The Ivrea–Verbano Zone contains two major units, the Kinzigite Formation and the Mafic Complex. The Kinzigite Formation consists of high-grade (high temperature–medium pressure amphibolite and granulite facies) gneiss originally deposited as shale and sandstone. The Mafic Complex is dominated by gabbro, but contains lenses of mantle peridotite. This Permian-age gabbroic pluton intruded the deep crustal Kinzigite Formation.

Because the Sesia magmatic system has been tilted on its side, the entire system is exposed from the near-surface caldera to about 25 km (15.5 mi) depth. Intrusion of the Mafic Complex into the deep crust caused melting of the more silica-rich crustal rocks, upward movement of mixed and siliceous magmas, and extensive volcanism. Hybrid mafic-silicic melts were produced by partial melting of the upper crust and mixing this melt with a substantial mantle-derived magma.

These units are exposed at the surface as a result of Mesozoic crustal thinning and associated vertical uplift, followed by tectonic stacking and tilting of thrust wedges during the Alpine Orogeny.

We will cross the magmatic system, from the deepest exposures of the Mafic Complex up to mid-crustal levels. Starting at the Insubric line, we will see mantle peridotite near Balmuccia, visit the middle and upper mafic complex and the contact with the roof rock Kinzigite Fm at Varallo. Mid-crustal mingling of mafic and silicic magmas can be seen at the base of the large granitic pluton at Roccapietra.

ST2.1 INSUBRIC LINE, BOUNDARY BETWEEN EUROPEAN AND AFRICAN CRUST, AT SCOPETTA

Mylonite of the Insubric Line, a major regional fault system, is exposed in the riverbed. Mylonite is a fine-grained metamorphic fault rock caused by the breakdown and recrystallization of the constituent minerals in ductile shear zones. Ductile shear zones are broad, elongated zones where rocks have deformed by plastic flow, leading to significant deformation without fracturing.

The Insubric Line here separates plutonic and high-temperature, medium-pressure metamorphic rocks (amphibolite-granulite facies) of the Ivrea–Verbano Zone on the east from eclogite facies (high temperature-high pressure) mica schist of the Sesia–Lanzo Zone/Cretaceous basement of the Austro-Alpine Domain to the west (Mazzucchelli et al., 2014).

This section of the valley is controlled by and lines up with the fault zone that runs through it.

Side Trip 2, Scopetta to Balmuccia: *Return north on SP299 3.1 km (1.9 mi; 4 min) to* ***Stop ST2.2, Mantle Peridotite, Balmuccia*** *(45.820200, 8.153153) and pull into the parking area on the right. Take the path to outcrops along the river.*

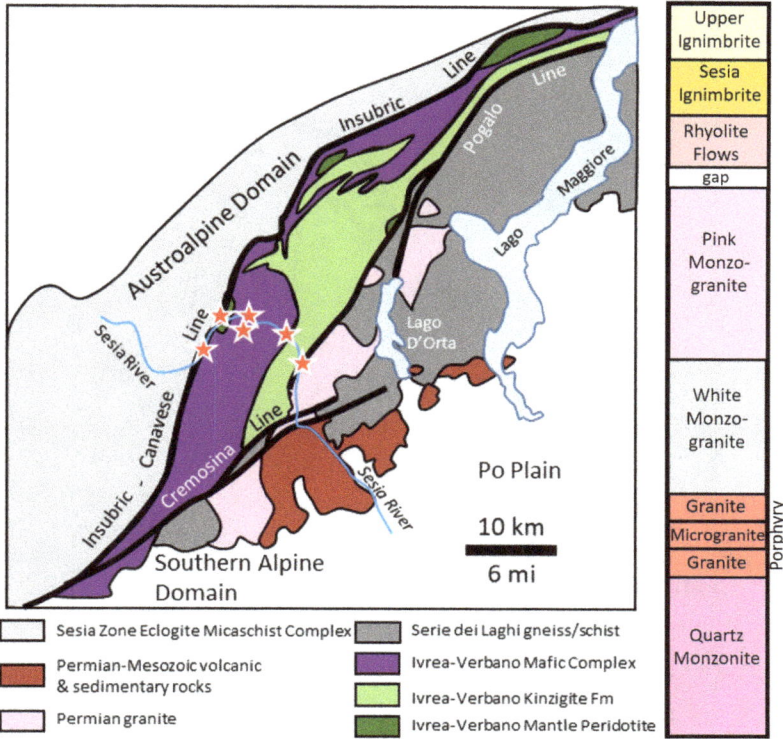

Left: Geologic map of the Ivrea–Verbano zone. Heavy lines are tectonic boundaries. Stars represent stops. Modified after Sinigoi et al., 2010; Mazzucchelli et al., 2014; Tavazzani et al., 2020. Right: Stratigrahy of the Sesia volcanic system. Modified after Tavazzani et al., 2020.

Light-colored foliated mylonite occurs along the Insubric Line at Scopetta.

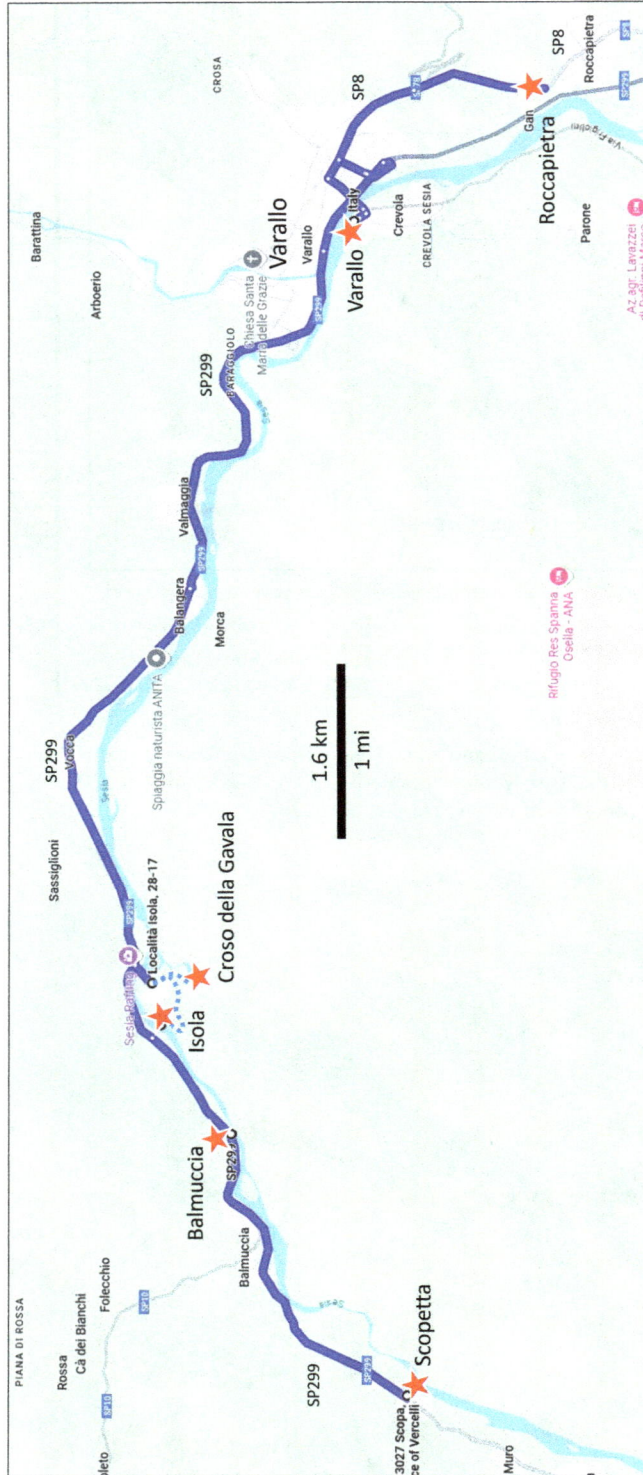

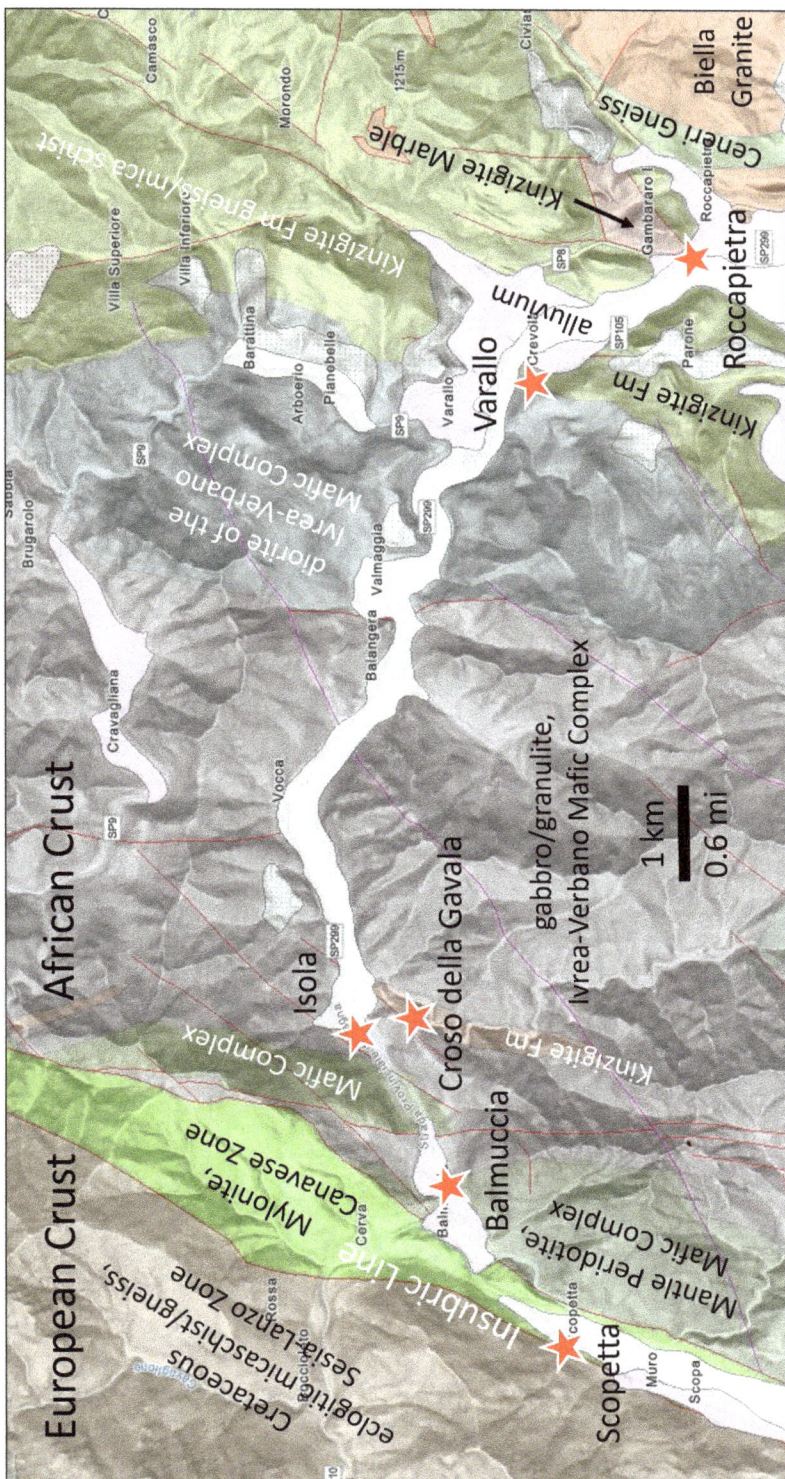

Geologic map of the Varallo to Scopetta area. From Arpa GeoPiemonte, 2021.

ST2.2 Mantle Peridotite at Balmuccia

The largest of the Ivrea–Verbano peridotite bodies are, from south to north, the Baldissero, Balmuccia, and Finero massifs. These bodies are aligned along the western, deepest margin of the Mafic Complex. Minor peridotites occur at various stratigraphic levels in the Ivrea–Verbano Zone and within the metasediments of the Kinzigite Formation along high temperature shear zones.

The association of mantle peridotites with high temperature shear zones suggests that they were not emplaced by mantle diapirism, that is, they do not represent an undisturbed mantle-crust transition that has just been uplifted to the surface. Rather, peridotite emplacement appears to predate the Permian mafic intrusions. The consistant association with high temperature extensional shear zones, and interfingering of lenses of peridotite with Kinzigite Formation, suggests that peridotites were emplaced early. They may be part of the accretionary wedge, the jumble of rocks caught up in a really old Ordovician subduction zone.

The peridotite at Balmuccia consists of dunnite. Dunnite is a yellow-green ultramafic intrusive igneous rock composed almost entirely of olivine, a glassy green mineral that is a key component of the upper Mantle. Tabular dunite bodies occurring in ophiolites are thought to be the result of deep mantle melts infiltrating shallower mantle levels. Most ophiolitic dunites occur at relatively shallow mantle levels. Ophiolitic dunites are usually related to mid-ocean ridges, possibly to seamounts, or subduction-related magmas (Mazzucchelli et al., 2014).

This stop is worth a look because it is considered one of the best-preserved mantle peridotites in the world.

Balmuccia peridotite is a beautiful dark green stone.

Side Trip 2, Balmuccia to Isola: *Continue east on SP299 to Località Isola; turn right (west) and drive to* **Stop ST2.3, Mid-Crustal Diorite, Isola** *parking area (45.826490, 8.170647), for a total of 2.0 km (1.2 mi; 2 min). Park and walk 620 m (2,000 ft; 10 min) west to the large outcrops on the south bank of the Sesia River (45.825572, 8.165497).*

ST2.3 Mid-Crustal Diorite of the Mafic Complex, Isola

A large outcrop along the Sesia River exposes mid-crustal gabbros/peridotites of the Mafic Complex. You can see the contact between the Balmuccia Peridotite (dark rock) and diorite-gabbro of the Mafic Complex (lighter rock).

Deformation seen here includes stretching foliation, isoclinal folds (fold where the sides are parallel to each other), and boudinage (sausage-shaped blobs caused by stretching and fracturing of a rigid rock layer encased within a more ductile layer) (Mazzucchelli et al., 2014).

Google Maps image of Isola and stops ST2.3 and ST2.4. ©2025 Airbus, Maxar Technologies.

Contact (dotted line) between Balmuccia Peridotite and Diorite-Gabbro, Isola.

Isoclinal folding in the Mafic Complex diorite.

Side Trip 2, Isola to Croso della Gavala: *Return almost to the parking area and at the trail intersection turn right and walk ~300 m (1,000 ft; 6 min) south and west to a bridge crossing the stream. This is* ***Stop ST2.4, Granulitic Paragneiss, Croso della Gavala*** *(45.826339, 8.170247).*

ST2.4 Granulitic Paragneiss, Croso della Gavala

Magma intrusions can engulf the crustal rocks they are moving upward through. The gabbro intrusion known as the Mafic Complex enveloped large volumes of deep crustal rock. The banded granulite/gneiss in the riverbed has been transformed by heat and partial melting to a dense gneiss containing white feldspar, red-brown garnet, and some quartz and other minerals.

The rock in the riverbed is a granulite/gneiss, a high-grade metamorphic rock that has experienced high-temperature, moderate-pressure metamorphism. This rock has a strong parallel fabric, is mostly feldspar, and has abundant garnet. The original rock was probably shale of the Kinzigite Formation; here it occurs in a narrow, roughly north-south band surrounded by norite, a member of the gabbro family that is part of the Ivrea–Verbano Zone Mafic Complex.

Granulitic paragneisses composed of metamorphosed shales, sandstones, and carbonates are found as septa (elongated zones) within the Mafic Complex. Septa are characterized by extreme length-to-width ratios, with some only a few meters thick extending for kilometers. The septa are thought to form due to weakening of the roof rock above a magma chamber, followed by detachment of a sliver of roof rock that sinks into and becomes incorporated in the underlying magma. From that point the septa moves with the magma (Sinigoi et al., 2010).

Banded gneiss in the riverbed near the contact with Mafic Complex gabbros.

It is worth pointing out the difference between orthogneiss and paragneiss here. Orthogneiss is a high-temperature metamorphic rock derived from an igneous precursor, whereas a paragneiss is derived from a sedimentary rock.

About 20 m (60 ft) away is the entrance to an abandoned nickel mine. Iron-nickel sulfide (pentlandite) occurs in a cumulate pyroxenite, a rock formed when pyroxene crystals accumulate on the bottom of a magma chamber. The mine was active up until World War II (Mazzucchelli et al., 2014).

Side Trip 2, Croso della Gavala to Varallo: *Continue driving east on SP299 to Varallo; take the exit to Varallo/Civiasco; turn left (northwest) onto Via Brigate Garibaldi; immediately turn left (southwest) onto Via Frigiolini; cross the Sesia River and turn right (north) toward kayak club and park. This is **Stop 2.5, Kinzigite Formation, Varallo** (45.810714, 8.257005) for a total of 10.2 km (6.4 mi; 12 min). Take the steps above the small playground and turn right on the grassy path alongside the club building. Walk ~160 m (530 ft; 5 min) to just below the weir on the Sesia River.*

ST2.5 KINZIGITE FORMATION MIGMATITE, VARALLO

We are about 15 km (9 mi) deep in the crust at the contact between the gabbro of the Mafic Complex and the rock it intruded. Heat from the cooling Mafic Complex has partially melted the crustal rocks, causing the light and dark minerals to segregate into bands.

This is one of the best exposures of the contact between the Mafic Complex and its roof rock, the Kinzigite Formation. Deformed amphibolite facies migmatite of the Kinzigite Formation is in contact with garnet-bearing diorite of the Mafic Complex about 200 m (660 ft) upstream (Mazzucchelli et al., 2014).

Migmatite, meaning "mixed rock," contains two intermingled rocks, usually a granite injected into a metamorphic rock to form a paragneiss.

Migmatite of the Kinzigite Formation in the riverbed at Varallo.

Side Trip 2, Varallo to Roccapietra: Return northeast on Via Frigiolini; turn right (east) onto Via Brigate Garibaldi; at the south end of Varallo continue straight onto SP8/Via Fratelli Varalli; at the entrance to Roccapietra pull over and park on the right. Cross to the roadcut by the steps. This is **Stop 2.6, Mid-Crustal Kinzigite, Roccapietra** *(45.795990, 8.272073) for a total of 2.9 km (1.8 mi; 5 min). There is a narrow shoulder and you are on a curve. Be mindful of traffic!*

ST2.6 MID-CRUSTAL KINZIGITE FORMATION, ROCCAPIETRA

This roadcut exposes the Kinzigite Formation where it is intruded by a dike of fine-grained granite that is probably related to the Early Permian Roccapietra Pluton. Peak magmatic activity occurred over a ~10 m.y. interval, from 288 to 278 Ma. The nearby Roccapietra Pluton intrudes the "Serie dei Laghi" that outcrops to the south and east of this stop (Mazzucchelli et al., 2014).

The Ivrea–Verbano Zone and the Serie dei Laghi (collectively termed the dei Laghi Massif) represent lower continental crust (Ivrea–Verbano Zone) and middle continental crust (Serie dei Laghi). Uplift of this zone probably occurred during the Alpine Orogeny. The dei Laghi Massif represents the westernmost, highest-grade metamorphic and deepest buried part of the South Alpine Domain. It is bounded on the west by the Insubric Line, active during the Alpine Orogeny, which separates this zone from the Austro-Alpine Domain to the west (James, 2002). The Serie dei Laghi consists of amphibolite facies (high-grade) metasediments, metagranites, and post-metamorphic Permian granites. Metamorphism occurred long before the Alpine Orogeny, around 340 Ma (Early Carboniferous, or Mississippian), probably related to the Variscan Orogeny. The contact between the Ivrea–Verbano Zone and Serie dei Laghi occurs along what has been called the Cossato–Mergozzo–Brissago Line, a shear zone that is not well expressed in this area (Laura et al., 2005).

Roadcut at Roccapietra.

Roccapietra Kinzigite paragneiss.

Side Trip 2, Roccapietra to Parco Avventura: *Continue south on SP8 to Via al Sesia; turn right (west) on Via al Sesia and drive under the highway and turn left (south); turn right onto the ramp to Borgosesia and merge onto SP299; drive south on SP299 to Bornate; at the roundabout take the 1st exit (west) onto SP71; at the roundabout in Azoglio take the 3rd exit (south) onto Via Piamonte; continue straight on SP64; at the roundabout south of San Maurizio take the 1st exit (southwest) onto SP142; at the roundabout in San Giacomo take the 2nd exit (south) onto SS758; at the roundabout in Filatura Fontanella take the 2nd exit (west) onto SS758; at the roundabout on the west side of Biela take the 2nd exit (south) onto SS758; at the roundabout in Mongando take the 2nd exit (south) onto SP338; at the next roundabout take the 1st exit (west) onto SP419; south of Andrate turn left (south) onto SP73; turn right (north) onto SS26; at the sign for Falesia on the right turn right (east) onto Via Vittorio Emanuele; at the sign for "Palestra di Roccia" turn left (north) and drive to the parking area on the right. This is **Stop 4, Schistes Lustrés, Parco Avventura La Turna** (45.537512, 7.839575) for a total of 77.5 km (48.2 mi; 1 hr 15 min).*

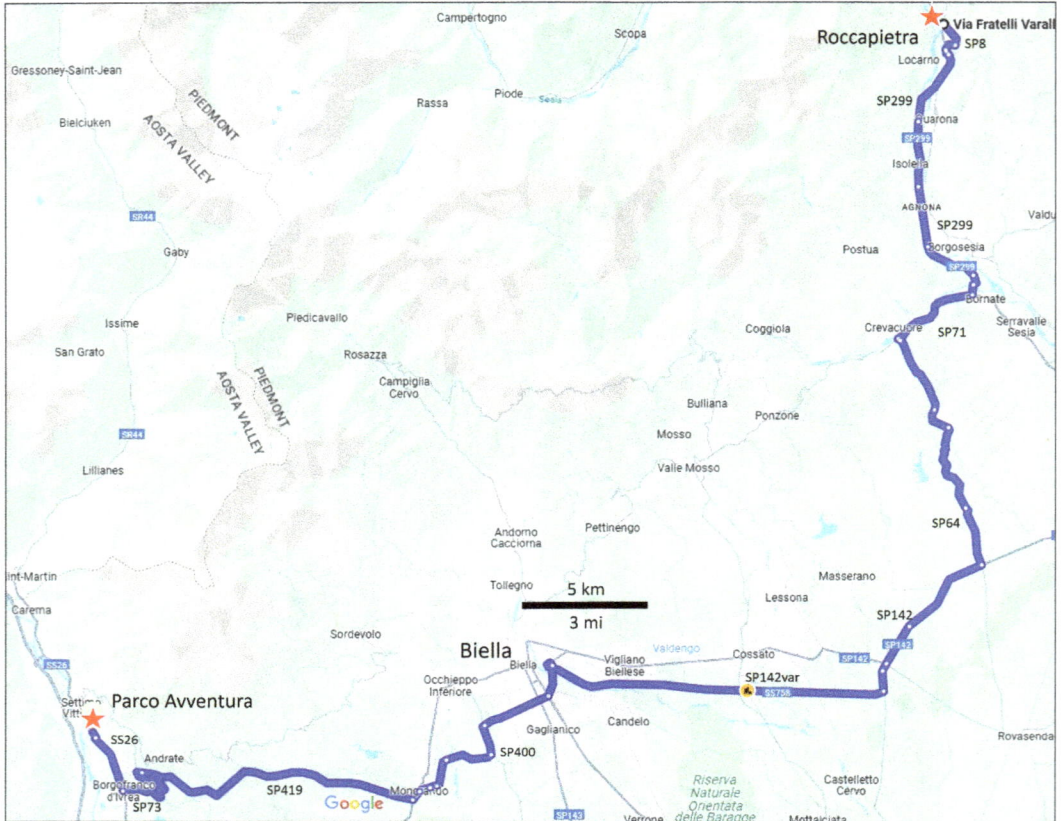

STOP 4 SCHISTES LUSTRÉS OF THE SESIA-LANZO ZONE, PARCO AVVENTURA LA TURNA

We have now crossed the Canavese Zone and entered the Sesia (or Sesia–Lanzo) Zone, some-times called the Sesia Massif. The outcrops are part of the "Schistes Lustrés" (glimmering shales), high pressure micaschists. These rocks, originally Late Cretaceous (85–68 Ma) shale or mudstone, were subducted beneath Adriatic Microplate and metamorphosed to blueschist-eclogite facies dur-ing the Alpine Orogeny, then quickly uplifted (Piana et al., 2017; Faure and Chen, 2018; Arpa GeoPiemonte, 2021).

Eclogite, an ultrahigh pressure metamorphic rock consisting largely of pyroxene and garnet, occurs throughout the Sesia zone, but the nature of the zone remains disputed. Many workers con-sider the zone to be the deepest and most metamorphosed part of the Austro-Alpine Domain in which the Sesia Zone was originally part of the Adriatic continental margin. Other workers con-sider the Sesia Zone to be the easternmost sector of the Inner Crystalline Massif, that is, part of the European continental crust (Faure and Chen, 2018).

The Sesia Zone is among the best-preserved continental crust that has been subjected to high pressure metamorphism. These tectonic slivers share a common subduction history that started in the Late Cretaceous and was followed by extension, decompression, and tectonic uplift and erosion in the mid-Tertiary (Regis et al., 2015).

These schists are but one of three units found in this zone. The units are defined on the basis of metamorphic grade and dominant rock type. The Schistes Lustrés here are part of the Eclogite Micaschist Complex, but there is also the Gneiss Minuti Complex and the Second Diorite–Kinzigite Complex (Compagnoni et al., 2014). Kinzigite, as we saw near Varallo, is a gneiss that is derived from shale that has been subject to high-grade metamorphism.

View north to outcrops of mica schist of the Micaschist Complex at Parco Avventura.

The outcrops on the far side of the park are part of the Eclogite Micaschist Complex in the Sesia–Lanzo Zone. The Argentera Quarry just above the park (45.540472, 7.844519) is mining a jadeite-bearing orthogneiss ("verde argento"), also from the Micaschist Complex.

This adventure park is a favorite of rock climbers.

The mica schist at Parco Avventura is highly foliated (consists of thin layers like sheets of paper) and sparkles with muscovite mica.

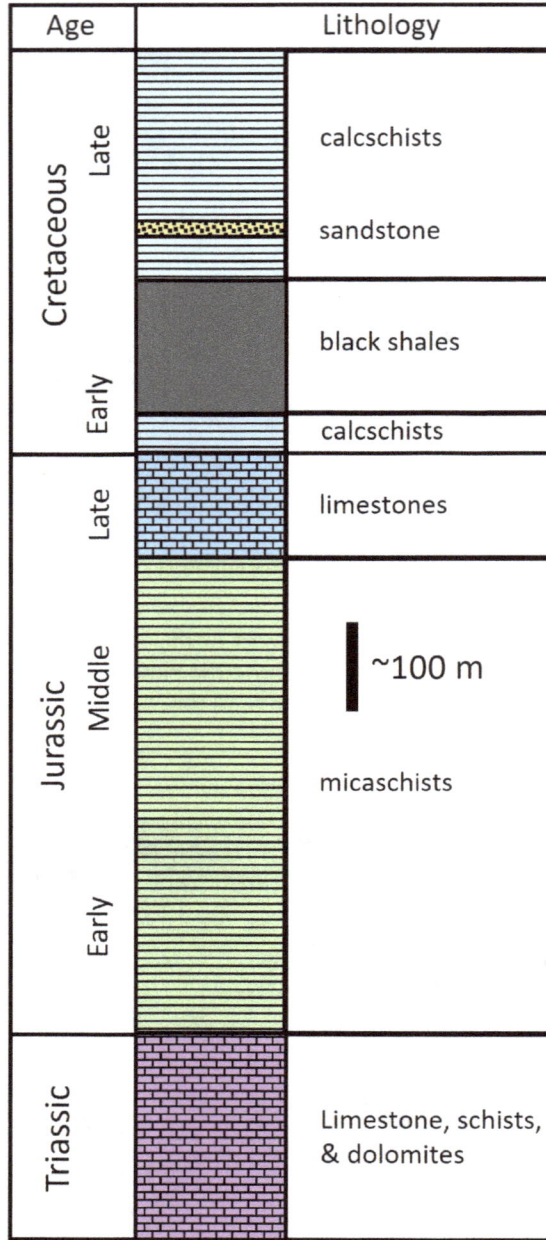

Stratigraphy of the Schistes Lustrés near Parco Avventura. Modified after Faure and Chen, 2018.

Parco Avventura to Settimo Vitone: *Return southeast on Via Vittorio Emanuelle; turn right (west) to SS26; turn right (northwest) on SS26; drive a total of 2.9 km (1.8 mi; 4 min) slow down and pull over on the right immediately after the sign for Aosta and in front of the small shrine. This is **Stop 5, Orthogneiss, Settimo Vitone** (45.555383, 7.827653). Walk back south along the shoulder for ~60 m (200 ft) to the outcrop.*

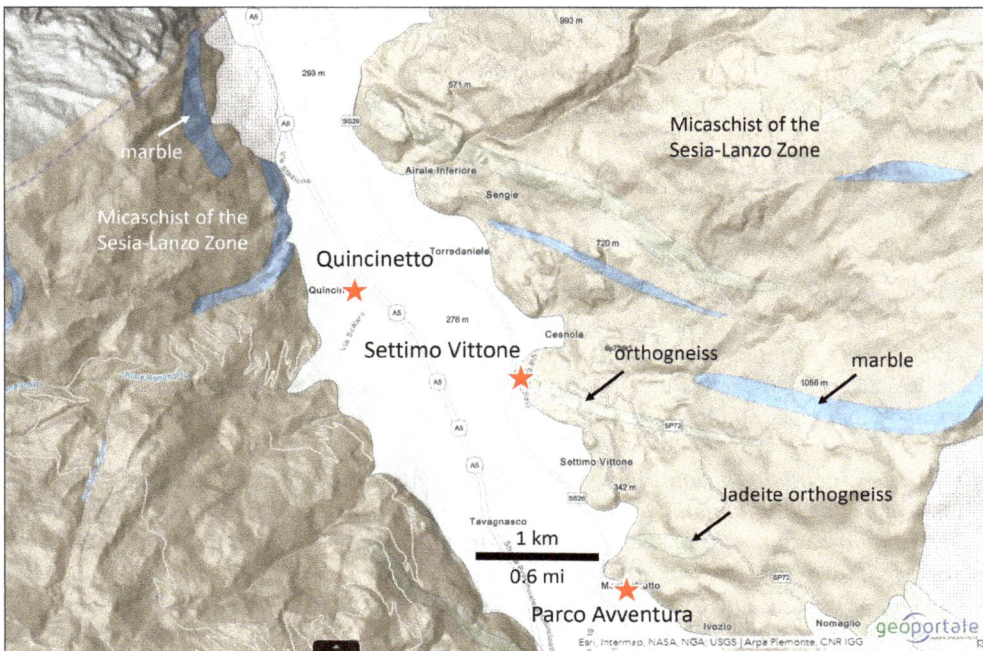

Geologic map of the Parco Avventura to Quincinetto area. From Arpa GeoPiemonte, 2021.

STOP 5 ORTHOGNEISS IN THE MICASCHIST COMPLEX, SETTIMO VITTONE

This stop allows us to see an orthgneiss stringer or septa within the Micaschist Complex. Again, orthgneiss is a high-temperature metamorphic rock derived from an igneous precursor. Here we can find garnet-bearing mafic boudins (Faure & Chen, 2018).

Outcrop of garnet-bearing orthogneiss (light band at base) in the Micaschist Complex, Sesia–Lanzo Zone. View north.

Settimo Vittone to Quincinetto: *Continue north on SS26 to the roundabout; take the 2nd exit (west) to A5/Torino/Aosta and cross the Dora Bàltea River; at the roundabout take the 1st exit; then turn left (southwest) toward Lago Europa; turn right (northwest) onto Lago Europa; turn left (west) onto Via Piamonte; drive to* **Stop 6, Eclogite, Quincinetto** *(45.56320, 7.80815) for a total of 2.3 km (1.5 mi; 3 min). Since the street is very narrow, park on Largo Europa before you turn onto Via Piemonte and walk the last few meters.*

STOP 6 ECLOGITE OF THE MICASCHIST COMPLEX, QUINCINETTO

Examine the handsome gray building stone used for walls on the home on the north side of the road. The Late Cretaceous rock is part of the eclogite Micaschist Complex (Compagnoni et al., 2014). Eclogite is a metamorphic rock containing garnet hosted in a matrix of sodium-rich pyroxene (omphacite). It can be derived from either the mantle or the crust, and it represents extremely high pressure and moderate to high temperature metamorphism.

Side Trip 3, Quincinetto to Fontainemore: *Return south on Largo Europa; turn left (east) onto Via 4 Novembre; at the roundabout take the 3rd exit (under the freeway); at the roundabout take the 2nd exit onto SS26 north; at the Pont Saint Martin roundabout continue straight onto SR44/ Via Resistenza; turn right (southeast) to continue on SR44/Via Gressoney; drive northeast on SR44 to the sign for Versa and Theilly on the left; pull into the parking area on the left before the bridge. This is* **Stop ST3.1, Metagranodiorite, Fontainemore** *(45.643225, 7.855720) for a total of 13.0 km (8.0 mi; 18 min). View from the bridge, or walk to the outcrops below the bridge.*

Eclogite mica schist used as building stone on Angelo's Cafe, Quincinetto.

If you prefer not to go on Side Trip 3:

Quincinetto to Arco Donnas: *Return south on Largo Europa; turn left (east) onto Via 4 Novembre; at the roundabout take the 3rd exit (under the freeway); at the roundabout take the 2nd exit onto SS26 north; at 7.8 km (4.8 mi) at the roundabout take the 2nd exit (west) onto Localitá Antiche Industrie and park in the large parking lot on the right. This is **Stop 7, Eclogite Micaschist, Arco Donnas** (45.602177, 7.759622) for a total of 7.9 km (4.9 mi; 10 min). Walk east on the path ~100 m (330 ft) to the arch.*

SIDE TRIP 3 VAL DEL LYS

The picturesque Val del Lys (Valley of the River Lys) is carved into the Eclogite Micaschist Complex of the Sesia Zone. It is part of the Adriatic-derived "Internal Complex" of Giuntoli and Engi (2016).

ST3.1 METAGRANODIORITE AND DIKES, FONTAINEMORE

Exposed in the streambed is Permian granodiorite that has been deformed and metamorphosed by high temperatures and pressures to eclogite facies. The metagranodiorite has been intruded by undeformed Oligocene (34 to 30 Ma) andesite dikes (Compagnoni et al., 2014). In the Western Alps, such post-metamorphic intrusions crosscut the Alpine nappes. The magmatism was probably generated by partial mantle melting with some crustal contribution.

Side Trip 3, Fontainemore to Cascata del Bouro: *Continue north on SR44 to the signs for Lago Vargno, Colombit, Rifugio della Barma, Chuchal and others; turn right (east) and cross the River Lys; after the sign for Barme turn right (south) and drive toward Coumarial and Rifugio della Barma; bear right toward Lillianes/Champvieil; drive to where rocks block the road and park. **This is Stop ST3.2, Micaschist Complex, Cascata del Bouro** (45.639959, 7.857586) for a total of 5.1 km (3.1 mi; 10 min). [There was a way to get here going south on SR44 but the road is closed due to landslides].*

The Permian granodiorite (light gray) in the riverbed has been deformed and metamorphosed to eclogite facies. It was then intruded by undeformed andesite dikes (dark bands).

ST3.2 MICASCHIST COMPLEX AT CASCATA DEL BOURO

It's hard to make a glittering mica schist look bad, but this stop makes it look positively beautiful. A scenic waterfall cuts the Eclogite Micaschist Complex of the Sesia Zone both above and below the road (Halama et al., 2018). Below the road is a lovely aquamarine pool.

Side Trip 3, Cascata del Bouro to Arco Donnas: Return north to SR44 and turn left (south); in Pont-Saint-Martin turn right (northwest) onto Via Chanoux; continue straight onto Via Nazional Donnas; at the roundabout take the 1st exit (west) onto SS26; at the roundabout take the 2nd exit to Antiche Industrie and park on the right. This is **Stop 7, Eclogite Micaschist, Arco Donnas** *(45.602177, 7.759622) for a total of 16.5 km (10.2 mi; 25 min).*

STOP 7 ECLOGITE MICASCHIST COMPLEX, ARCO DONNAS

A section of a Roman road is cut into the Sesia high-pressure orthogneiss and mica schist (Faure and Chen, 2018; Giuntoli and Engi, 2016). The Ecolgite Micaschist Complex was later overprinted by a late-Alpine greenschist facies (low-to-moderate grade) metamorphism.

At this site a 220 m (720 ft) section of a Roman road is preserved. The road was built to link Rome to the Rhone Valley in Gaul on the north side of the Alps (present-day France). The road and 4 m (13 ft) high arch were carved by slaves in order to avoid rock falls. The arch also served as a mile marker: the number 36 (XXXVI) indicates the distance in miles between Donnas and Augusta Praetoria (Aosta). The roadbed contains grooves cut by carriages. Ponte del Diavolo, the stone bridge in nearby Pont Saint Martin, is also part of this Roman road and was used for over 1,000 years to cross the Lys River.

Cascata del Bouro. There are falls both above and below the road (top, center). Modified from photo by Paola Marsala, Google Maps.

Arco Donnas, view east. Google Maps photo.

Folding in outcrops exposed along the Roman road near Arco Donnas. The Roman road was cut into the Eclogitic Micaschist Complex of the Sesia Zone.

Northwest of here is the Fortress of Bard: you can see it when looking north as you drive by the hydroelectric plant 1.1 km north of this stop. Originally built in 25 BCE, after the Romans defeated the local Celtic tribes, the fortress was destroyed in 1800 by French troops under Napoleon. It was rebuilt in 1827 by Carlo Felice of Savoy. Today the Fortress of Bard houses the Museum of the Alps. The fortress is built on a spur that is part of a complex fold structure where the three main complexes of the Sesia Zone merge (Compagnoni et al., 2014). These are the Schistes Lustrés/Eclogite Micaschist Complex, the Gneiss Minuti Complex, and the Diorite-Kinzigite Complex we discussed back at Parco Avventura.

Arco Donnas to Montjovet: *Continue driving west and north on SS26 toward Aosta; at the signs for Estaod/Chenal/Ciseran/Balmas take a sharp right (southeast) onto a minor road to Saint Germain and the Château de Montjovet, and drive c. 460 m (1,500 ft) south and park on a track that leaves the road on the right. Walk 1 minute northwest. This is **Stop 8, Rôche Moutonnée, Montjovet** (45.727571, 7.668253) for a total of 19.1 km (11.9 mi; 23 min).*

THE THEORY OF ICE AGES AND GLACIATION

We are now entering parts of the Alps that have been seriously affected by glaciation. Like the concept of thrust faulting, the theory of Ice Ages and an explanation of glaciation was to a large extent worked out right here in the Alps.

The concept of a worldwide Ice Age seemed to be popping up everywhere in the mid-1700s. The Swedish mining engineer Daniel Tilas was, in 1742, the first person to suggest drifting sea ice was a cause of the erratic boulders scattered across Scandinavia and the Baltic. In the same year Pierre Martel, an engineer and geographer in Geneva, visited the Chamonix area of the Swiss Alps. Two years later he published an account of his journey, reporting that the local people attributed the distribution of large boulders to glaciers that had once extended much farther. In 1795 the Scottish naturalist James Hutton explained erratic boulders in the Alps as a result of the action of glaciers. In 1815 Jean-Pierre Perraudin explained erratic boulders in the Val de Bagnes in the Swiss Valais as being due to glaciers. In 1818 the Swedish botanist Göran Wahlenberg published his theory of regional glaciation that had covered all of Scandinavia.

A few years later, the Danish-Norwegian geologist Jens Esmark argued for a sequence of world-wide ice ages. In 1824 Esmark published his ideas that changes in climate caused glaciations. Robert Jameson at the University of Edinburgh heard of Esmark's ideas and speculated that glaciers had once covered Scotland (Wikipedia, Ice Age).

The Swiss-German geologist Jean de Charpentier had been studying glaciers since the Glacier de Gietroz disaster. In 1818 a lake dammed by the glacier burst through the ice and flooded the areas downstream. After studying the Rhone Valley and the huge blocks of granite scattered from the Alps to the Jura, Charpentier supported the theory proposed in 1821 by his friend Ignaz Venetz that these "erratic" boulders could have been moved only by the action of glaciers. The glaciers must have been derived from the Alps, since the boulders were identical to rocks found on Alpine peaks (Norman, 1991a).

In 1829 Venetz, a Swiss civil engineer, put his thoughts on paper. He explained the distribution of erratic boulders in the Alps, the nearby Jura Mountains, and the North German Plain as being due to glaciers that had once covered the area. He read his paper before the Swiss Society for Natural Research (Schweizerische Naturforschende Gesellschaft) and got a mostly skeptical reception. His friend Jean de Charpentier reworked Venetz's idea into a theory of Alpine glaciation (Wikipedia, Ice Age).

Charpentier further explained the distribution of the erratic blocks and his theory of glaciers in a paper read before the Swiss Society for Natural Research in 1834. He maintained his position in spite of a hostile reception, and he invited his critics to see the evidence for themselves (Norman, 1991b).

Slowly the idea spread. The German botanist Karl Schimper studied moss growing on erratic boulders in Bavaria. Wondering where such large stones had come from, in 1835 he traveled to the Bavarian Alps. Schimper concluded that ice must have carried the boulders. Schimper then coined the term "ice age" (*Eiszeit*) for the glacial epoch.

Schimper spent the summer of 1836 near Bex in the Swiss Alps with Venetz, Charpentier, and his former university friend Louis Agassiz. Schimper, Charpentier, and Venetz convinced Agassiz that there had been a time when glaciers flowed all the way down the valleys.

Agassiz returned to his position at the University of Neuchâtel a convert. During the winter of 1836/37, Agassiz and Schimper worked on a theory of multiple glaciations. They did extensive field-work in order to persuade their critics. Together they drew on the works of Venetz, Charpentier, and on their own observations. Agassiz visited England, and with William Buckland, the only English naturalist who shared his ideas, toured the British Isles in search of glacial phenomena. He became convinced that his ice age theory was correct.

In 1837 Agassiz presented their joint conclusions before a meeting of the Swiss Society for Natural Research. He proposed that ancient glaciers had flowed out from the Alps, and that ice sheets had covered the plains and mountains of Europe, Asia, and North America, in fact the entire Northern Hemisphere, in a prolonged ice age. He was the first to propose in a scientific forum that the Earth had been subject to a past ice age. Again, the audience was skeptical (Wikipedia, Ice Age).

Agassiz published his "Study on Glaciers" (Études sur les glaciers) in 1840. Charpentier published his "Essay on the glaciers and on the erratic terrain of the Rhone Basin" (Essai sur les glaciers et sur le terrain erratique du bassin du Rhône) in 1841 (Norman, 1991a; Wikipedia, Ice Age). And yet it took decades for this concept to be fully accepted.

By the time James Croll published "Climate and Time in Their Geological Relations; A Theory of Secular Changes of the Earth's Climate" in *Nature* (1875), a consensus had developed that there had been several ice ages as a result of changing climate patterns.

The evidence for ice ages and glaciation comes from rock scouring, polishing, and scratching; glacial moraines, drumlins, eskers, kettles, U-shaped and hanging valleys; and the deposition of till, glacial outwash, and glacial erratics. At Montjovet we see one particular line of evidence of glaciation, the rôche moutonnée.

STOP 8 RÔCHE MOUTONNÉE, MONTJOVET

A "rôche moutonnée" is formed by moving ice. The name, meaning "rock wig," was coined in 1787 by the alpine explorer/geologist Horace Bénédict de Saussure, who thought these whale-back ridges resembled the wigs that men wore at the time, wigs that were slicked down with mutton grease (Cairngorms Nat Park, Rôche Moutonnée, Dulnain Bridge). In this area the ice moved from north to south, grinding down the bedrock into asymmetric ridges that have a gentle slope facing the oncoming ice, and a steep slope in the lee of the ice. The ridges and grooves scoured into the bedrock by the moving ice indicate the direction of ice movement. This rock was polished by the advancing Valle d'Aosta glacier. The retreating glacier left a number of erratic boulders of porphyritic granite from the Mont Blanc Massif (Compagnoni et al., 2014).

The outcrop is a rôche moutonnée developed in foliated and crenulated Jurassic serpentinite and talc schist (Macrostrat, 2015). The serpentinite is a result of metamorphic recrystallization of an originally basaltic ophiolite (Compagnoni et al., 2014). Thus, this is likely a remnant of Liguro-Piemontais oceanic crust.

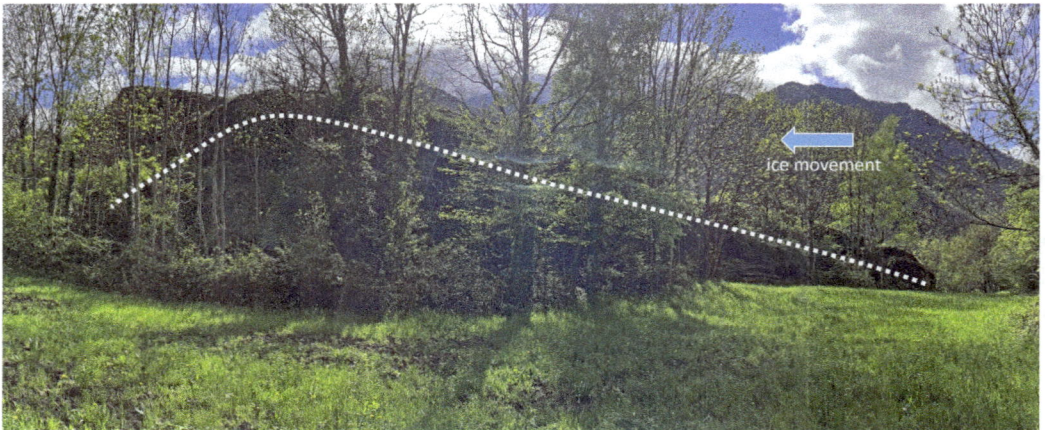

It's a bit hard to see looking into the sun, but this rôche moutonnée (dotted outline) was developed in serpentinite. Ice movement here was from right to left (north to south).

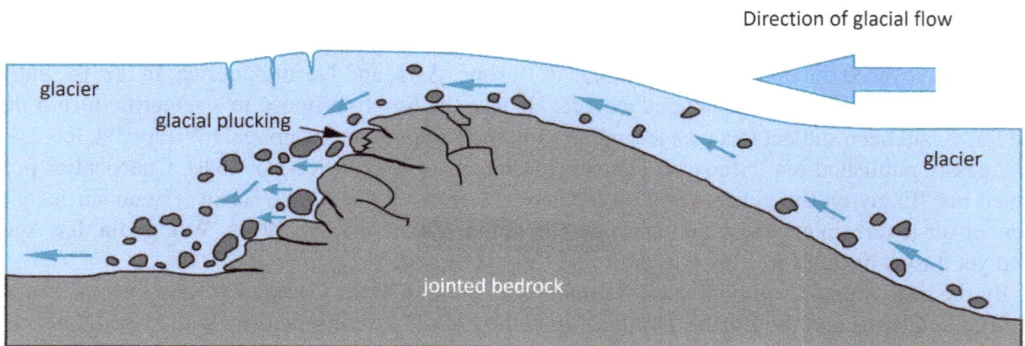

Relationship of a moving glacier to the rôche moutonnée. The direction of ice movement is from left to right. From US National Park Service, https://www.nps.gov/media/photo/view.htm?id=0ca840be-e574-4667-9c38-7883dcc5674e

Side Trip 4, Montjovet to Side Trip 4, Matterhorn: *Return north to SS26 and drive north to Châtillon; at the roundabout take the 1st exit (north) onto SR46/Valtournenche; drive to the Breuil Cervinia parking area on the right. This is* **Stop ST4, Matterhorn, Valtournenche** *(45.925924, 7.620100) for a total of 32.1 km (19.9 mi; 40 min).*

If you prefer not to go on Side Trip 4 but rather Side Trip 5:

Side Trip 5, Montjovet to Side Trip 5.1, Moulin: *Return north to SS26 and drive north and west to Châtillon; turn left (south) onto Via Pellissier; at the roundabout take the 2nd exit (east) onto Via Chanoux; turn right and enter the E25 toll road to Aosta [you can avoid tolls by staying on SS26]; take the exit to Aosta/E27/Gran Saint Bernard and continue on E27; bear right at the sign for SR28/Bionaz; at the roundabout continue straight on Frazione Signayes; continue straight on Frazione chez Roncoz; bear right onto Rhinz; stay on the main road to Valpelline and Oyace; continue on the main road to Bionaz; before entering the tunnel pull over and park on the right. This is* **Stop ST5.1, Valpelline Series, Moulin** *(45.868606, 7.411449) for a total of 53.1 km (33 mi; 53 min). Walk back south ~ 75 m (250 ft) to the outcrop at (45.868833, 7.410839).*

If you prefer not to go on Side Trip 4 or Side Trip 5:

Montjovet to Courmayeur/Ermitage: *Return north to SS26; turn right (north) at the sign to Torino/ Aosta/Châtillon; immediate right (west) onto Località Panorama; continue straight on Località Soleil; turn left (south) onto the E25/A5 toll road toward Aosta; west of Aosta take the Aosta Ovest/ St. Pierre exit; keep left to continue to the SR47 toll road; at the roundabout take the 2nd exit (west) onto SS26; at Chenoz continue straight to stay on SS26dir; immediately after the sign for*

Courmayeur turn right (east); continue east through the roundabout and turn left (north) onto Via Circonvallazione; Via Circonvallazione turns slight left and becomes Via le Monte Bianco; at the roundabout take the 2nd exit (north) onto Str. Sergio Viotto; turn right (east) onto Str. Adolphe Rey; continue straight on Strada Grand Ru; turn left (northeast) on Strada del Villair; turn right (southeast) on Str. per Plan Gorret; turn left to stay on Str. per Plan Gorret and drive to the end of the road at parking for the Ermitage Restaurant on the left. This is **Stop 9.1, Mont Blanc Massif, Courmayeur/Ermitage** *(45.796003, 6.982320) for a total of 68.6 km (42.6 mi; 54 min).*

Side Trip 4 Matterhorn, Valtournenche

From this stop you get a fine view of the Matterhorn, assuming the weather is good. The iconic peak straddles the border between Switzerland and Italy. The near-symmetric, pyramid-shaped peak is located in the Pennine Alps. At 4,478 m (14,692 ft), it is one of the highest peaks in Europe. The name comes from the German *Matt*, ("alpine meadow"), and "horn" for the horn-shaped peak. The French and Italian names Cervin and Cervino are derived from the Latin *Mons Silvanus*, where *mons* is mount and *silva* means forest; over time this became corrupted to *Selvin* and then *Servin*. Horace Bénédict de Saussure changed the first letter from an *s* to a *c*, thinking the word derived from the Italian *cervo*, or "deer."

The Matterhorn has been part of the Swiss Federal Inventory of Natural Monuments since 1983 (Wikipedia, Matterhorn).

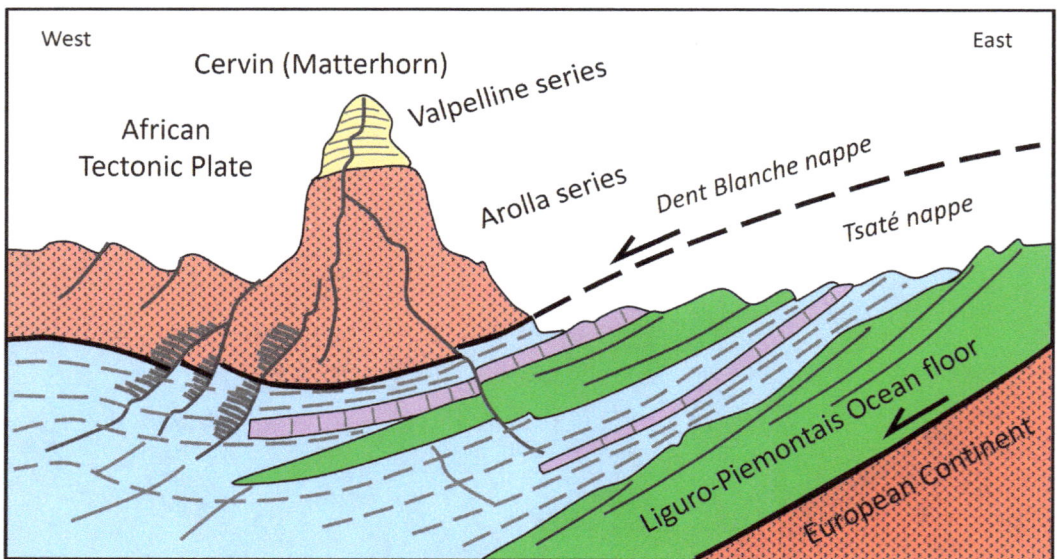

Cross section through the Matterhorn. Modified after a diagram that appears on several websites without citation.

Geology

Apart from the base of the mountain, the Matterhorn is composed of Paleozoic (Variscan) Valpelline Series gneiss belonging to the Dent Blanche Nappe, an isolated part of the Austro-Alpine Domain, lying over ophiolites and sedimentary rocks of the Penninic Nappes. The Dent Blanche Nappe is divided into the Arolla Series (elevation 3,400 to 4,200 m, or 11,200 to 13,800 ft) and the Valpelline Series (the summit). We will see the Valpeline and Arolla series in outcrop in the Buthier Valley. The Valpelline Series represents a slice of pre-Alpine Adriatic/African continental crust characterized by Permian high temperature metamorphism and partial melting. It contains metapelite (metamorphosed shale), amphibolite, and gneiss, among others. The Arolla Series consists of Permian granite, diorite, gabbro, and orthgneiss with some metasediments. The Austro-Alpine nappes in general are derived from the Adriatic Microplate (Wikipedia, Matterhorn; Faure and Chen, 2018).

Most of the base of the mountain lies in the Tsaté Nappe, a remnant of Liguro-Piemontais oceanic crust (ophiolites) and its sedimentary cover (Faure and Chen, 2018). (The Tsaté Nappe is a thrust sheet within the larger stacked Penninic Nappes system.) The ophiolites are the dark rocks on the north and east slopes. They formed between mid-Jurassic (~170 Ma) and mid-Cretaceous (~100 Ma), and were subsequently metamorphosed during the Alpine Orogeny.

The mountain's shape is a result of multiple glaciers diverging from the peak. At the head of each glacier is an eroded cirque, an amphitheater-like valley formed by glacial erosion.

This stop in the valley and the ridges to the east are all Jurassic (201 to 145 Ma) serpentinite and talc schist, part of the Liguro-Piemontais ophiolite sequence of the Penninic Nappes. The ridges to the west are Permian amphibolite, diorite, and hornblende gneiss and Devonian granites of the Austro-Alpine Domain (ISPRA, 2012a; Macrostrat, 2015).

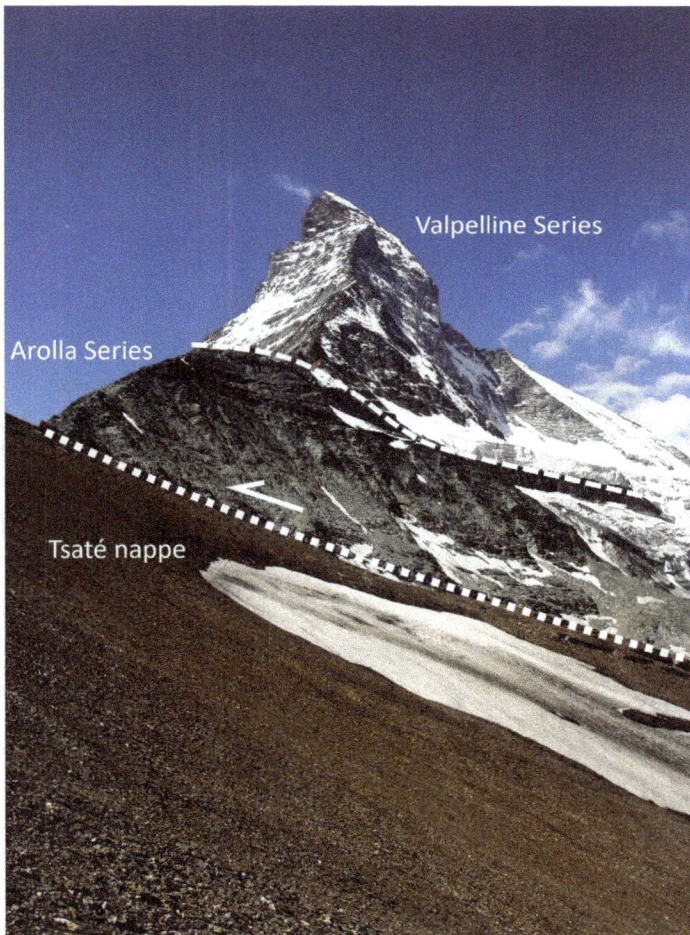

Matterhorn looking southwest from the trail to Hörnlihütte. The lower part is sedimentary rock (brown); the middle section is greenschist from the oceanic crust. The peak itself is gneiss from the crust of the African plate. The dashed contact is probably a thrust as well. Photo courtesy of Sunna, https://commons.wikimedia.org/wiki/File:Matterhorn002.jpg

Climbing History

The Matterhorn remained unclimbed after most of the other Alpine peaks had been conquered, not because of its technical difficulty but because of the fear it inspired in early mountaineers. The first recorded attempts were in 1857 and 1858. Climbing it became the object of international competition. The first ascent of the Matterhorn was on 14 July 1865 from Zermatt by a party led by Edward Whymper; it ended when four of its seven members fell to their deaths during the descent. Just three days later, on 17 July 1865, Jean-Antoine Carrel and Jean-Baptiste Bich reached the summit from the Italian side.

The west face, the highest of the Matterhorn's four faces, was finally climbed in 1962. It is estimated that over 500 alpinists have died on the Matterhorn, making it one of the deadliest peaks in the world (Wikipedia, Matterhorn).

Matterhorn from Valtournenche. Photo courtesy of yab994, https://commons.wikimedia.org/wiki/File :Matterhorn_from_Val_Tournenche.jpg

*Side Trip 4, Matterhorn to Side Trip 5.1, Moulin (tolls): Return south on SR46 to Châtillon; at the roundabout take the 3rd exit (east) onto SS26; turn right (south) on Via Pellisier; at the roundabout take the 2nd exit (east) onto Via Chanoux; turn right and enter the E25 toll road to Aosta [you can avoid tolls by staying on SS26]; take the exit to Aosta/E27/Gran Saint Bernard and continue on E27; bear right at the sign for SR28/Bionaz; at the roundabout continue straight on Frazione Signayes; continue straight on Frazione chez Roncoz; bear right onto Rhinz; stay on the main road to Valpelline and Oyace; continue on the main road to Bionaz; before entering the tunnel pull over and park on the right. This is **Stop ST5.1, Valpelline Series, Moulin** (45.868606, 7.411449) for a total of 73.9 km (45.9 mi; 1 hr 19 min). Walk back south ~ 75 m (250 ft) to the outcrop at (45.868833, 7.410839).*

If you prefer not to go to Side Trip Five:

*Side Trip 4, Matterhorn to Courmayeur/Mont Blanc (tolls): Return south on SR46 to Châtillon; at the roundabout take the 3rd exit (east) to SS26 and drive to Via Robert Pellissier; sharp right (south) onto Via Pellissier; and the roundabout take the 2nd exit (east) onto Via Emilio Chanoux; turn right (south) at the sign to E25/A5 toll road to Aosta [you can avoid tolls by staying on SS26]; take the ramp onto the E25 west to Aosta; stay on E25 to the Courmayeur exit; exit north onto SS26dir; immediately after the sign for Courmayeur turn right (east); continue east through the roundabout and turn left (north) onto Via Circonvallazione; Via Circonvallazione turns slight left and becomes Via le Monte Bianco; at the roundabout take the 2nd exit (north) onto Str. Sergio Viotto; turn right (east) onto Str. Adolphe Rey; continue straight on Strada Grand Ru; turn left (northeast) on Strada del Villair; turn right (southeast) on Str. per Plan Gorret; turn left to stay on Str. per Plan Gorret and drive to the end of the road at parking for the Ermitage Restaurant on the left. This is **Stop 9.1, Mont Blanc Massif, Courmayeur/Ermitage** (45.796003, 6.982320) for a total of 89.5 km (55.6 mi; 1 hr 22 min).*

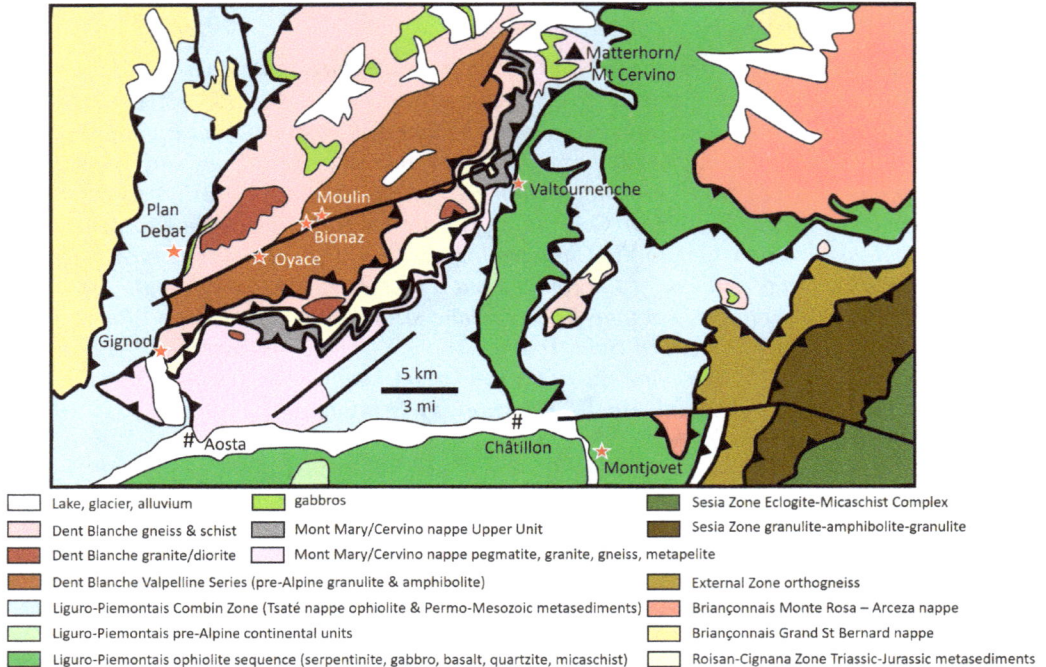

Geologic map of the Aosta Valley stops. Heavy lines are normal faults; heavy lines with teeth are thrust faults (teeth are on the thrusted sheet). Modified after Manzotti and Ballèvre, 2016; ISPRA, 2012a, 2012b; MacroStrat, 2015.

SIDE TRIP 5 BUTHIER VALLEY

This side trip is mostly to examine the Valpelline series, the rocks that form the peak of the Matterhorn. Valpelline series, part of the Dent Blanche Nappe of the Austro-Alpine Domain, represents the pre-Permian basement of the Adriatic Microplate. As mentioned previously, the Austro-Alpine Domain consists of slivers of the Adriatic and African plates that were thrust north and west over the Penninic Nappes.

We cross the Roisan–Cignana Shear Zone at Gignod on our way up the Buthier Valley. This shear zone is a 25 km (15.5 mi) long southwest-northeast-trending high strain zone. It extends northeast from Gignod in the Buthier Valley to the Lago Cignana area. It is best exposed along the ridge between the Valpelline and Valtournenche valleys. Northeast of Lago Cignana it becomes difficult to map.

The Roisan–Cignana Shear Zone records several Alpine deformation phases and is characterized by mylonites derived from the Dent Blanche Nappe and of the Mont Mary–Cervino Nappe basements, from pre-Alpine metasediments, and from Mesozoic cover of the Roisan Zone. The Roisan Zone occurs within the Roisan–Cignana Shear Zone; its thickness ranges from 30 to 500 m. It contains bands and pods of dolomite up to 100 m (328 ft) wide, embedded in mylonites. Triassic dolomites are preserved as small pods at the Col de St. Barthélémy and at Roisan village, and as large lenses at Cima Bianca.

The deformation in the Roisan–Cignana Shear Zone includes foliation, northwest-southeast folds, and a prominent northwest-southeast stretching lineation. Deformation indicators reveal a northwest-directed sense of movement.

The latest deformation produced large open northeast-southwest-trending folds, hundreds of meters in amplitude, with axial planes (the imaginary plane that bisects the fold along the fold hinge) inclined to the northwest, and moderate to pervasive crenulation or chevron folding. Crenulation folds are small-scale folds in the planar fabric of a schist or slate. The fold hinge, or hingeline, is the line of maximum curvature on a fold such as an anticline or syncline.

Temperature estimates for Alpine metamorphism in the Roisan Zone indicate a range from 450° to 520°C. Mineral equilibrium temperatures suggest pressures from 13 to 17 kbar (189,000 to 247,000 psi; Manzotti and Ballèvre, 2016). This is at the transition from Blueschist to Eclogite facies metamorphism, which suggests that the metamorphism occurred in a subduction (high pressure, moderate temperature) setting.

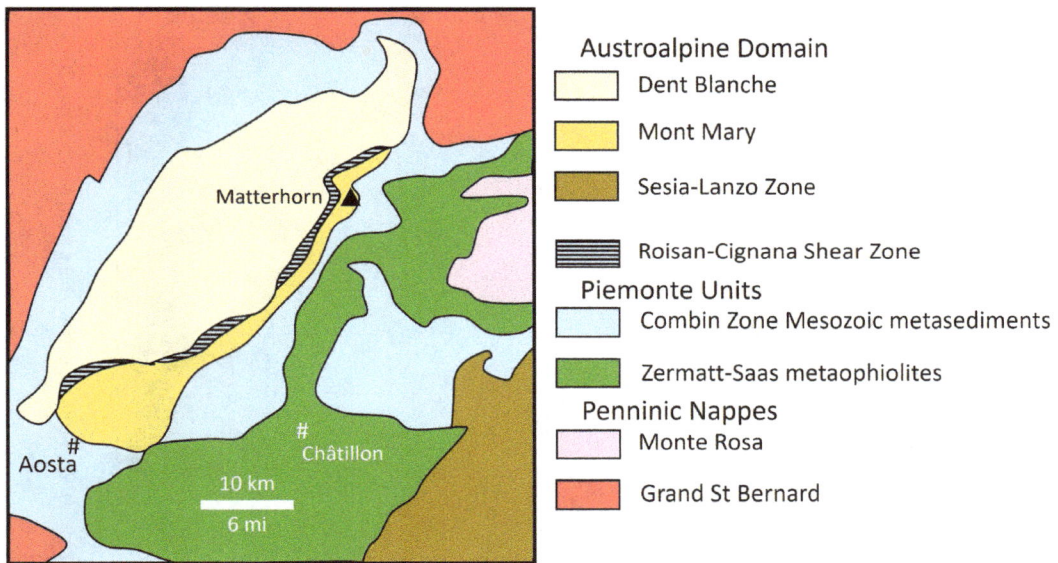

Map of the various nappes mentioned in the text. Modified after Manzotti, 2011.

ST5.1 VALPELLINE SERIES GARNET AMPHIBOLITE, MOULIN

In Moulin we encounter outcrops of the Valpelline Series, Dent Blanche Nappe of the Austro-Alpine Domain. These are high-grade garnet-bearing amphibolites interbedded with garnet-biotite gneiss. The amphibolite has garnet crystals commonly up to 5 cm (2 in). Be kind to the next visitor and do not remove them.

A gently folded pegmatitic dike cuts the outcrop. The axial plane of the pegmatite folds is parallel to foliation in the garnet-biotite gneiss. Pegmatite is a light-colored igneous rock with exceptionally large crystals, typically quartz, feldspar, and mica.

A ductile shear zone cuts this outcrop at a low angle to layering. This shear zone is pre-Alpine age, as high-temperature minerals are stable in the shear zone. The present orientation is the result of rotation during the Alpine Orogeny. If these layers are rotated back to horizontal, the displacement along the shear zone is seen to be the result of Permian extension (Manzotti and Ballèvre, 2016).

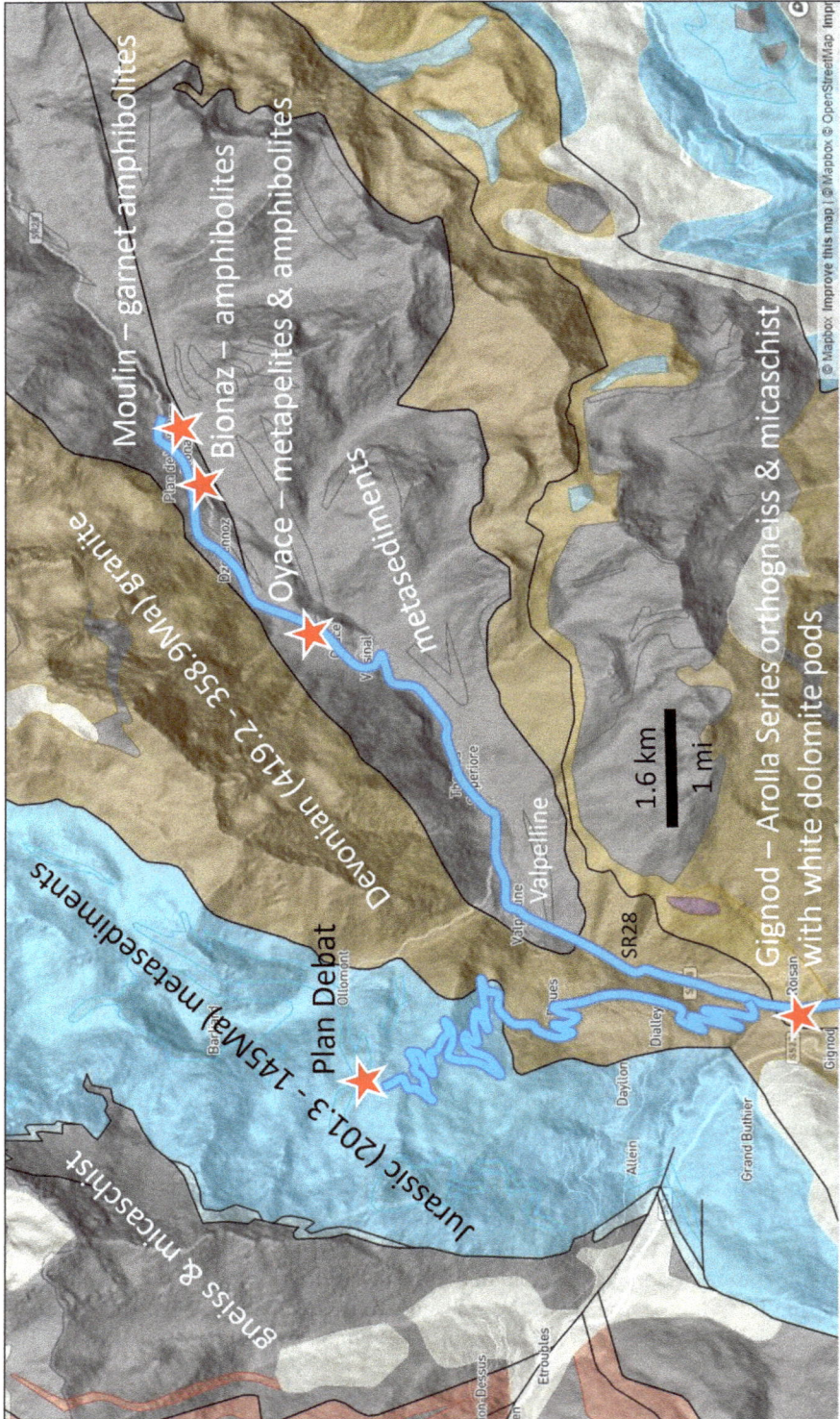

Geologic map of the Buthier Valley stops. Geology from ISPRA, 2012a, 2012b; Monte Cervino; base from Macrostrat, 2015.

Ductile shear zone and pegmatite, Moulin. Geology derived from Manzotti and Ballèvre, 2016.

Side Trip 5, Moulin to Bionaz: *Return southwest 650 m (0.4 mi; 2 min) on Dzovennoz to* **Stop ST5.2, Valpelline Series, Bionaz** *(45.867287, 7.404687). Park on the north side of the road and walk south 60 m (200 ft) to the outcrops.*

ST5.2 Valpelline Series Amphibolite, Bionaz

The rocks here are finely crystalline amphibolites of the Valpelline Series, Dent Blanche Nappe. Veins are thought to be the result of partial melting. The veins are generally parallel to pre-Alpine foliation. The amphibolites have a stretching lineation inclined to the north-northeast.

If you continue to walk south-southeast, you will encounter Valpellline slate and schist (metapelite) (Manzotti and Ballèvre, 2016).

Amphibolite outcrop (foreground), Bionaz. View east.

Side Trip 5, Bionaz to Oyace: *Return southwest 3.4 km (2.1 mi; 5 min) on Dzovennoz/Chentre/* *Balmes to **Stop ST5.3, Valpelline Series, Oyace** (45.850379, 7.381703). Stop close to St. Michele* *church.*

ST5.3 METAPELITE OF THE VALPELLINE SERIES, OYACE

This outcrop exposes high-grade metamorphosed mudstone (metapelite) of the Valpelline Series. In the outcrop you can see:

- Metapelite (former ocean floor muds)
- Amphibolite bodies interlayered with the metapelite. Foliation is indicated by light-colored plagioclase feldspar-rich layers and by the alignment of black hornblende crystals.
- Pegmatite veins that cut across the foliation.
- Broken rock (cataclasite) occurs along fault planes that dip to the southeast. Movement indicators suggest northwest displacement of the upper thrust sheet with respect to the footwall (Manzotti and Ballèvre, 2016).

Side Trip 5, Oyace to Plan Debat: *Continue southwest on Balmes/Via Frazione Pied de Ville; just* *south of Valpelline turn right (northwest) onto Frazione Arliod; in 0.6 km (0.4 mi) turn right; con-* *tinue straight on La Cretaz/Planavill; turn right on Frazione La Chenal; turn left onto La Cretaz;* *left on La Chenal; stay on main road (Coudrey); straight on Frazione la Cerise; turn right onto* *Frazione Chatellair; continue to the parking area on the right where the pavement ends. This is* ***Stop ST5.4, Dent Blanche and Tsaté Nappes, Plan Debat** (45.843487, 7.287732) for a total of 22.9* *km (14.3 mi; 40 min).*

Outcrop of red metapelite (slate), amphibolite, and pegmatite veins of the Valpelline Series.

Oyace is in a beautiful setting. I missed this view because it was raining when I was here. Photo courtesy of Patafisik, https://commons.wikimedia.org/wiki/File:Oyace_DSCN0262.JPG

ST5.4 DENT BLANCHE NAPPE – TSATÉ NAPPE CONTACT, PLAN DEBAT

This is a view stop. From northwest to northeast (left to right), the main summits and passes around Conca di By, the cirque at the head of the valley, include Mont Vélan, Grand Combin, Mont Avril, Fenêtre de Durand (at the contact between Tsaté Unit and Dent Blanche Nappe), Monte Gelé, and Monte Berrio. Within the Conca di By is a prominent ridge, extending from the M. Berrio to the Punta Cornet. Along this ridge is the best place to see the contact between the Combin Unit (the Tsaté Nappe) and the overlying Dent Blanche Nappe.

The view of the high Alps from here is spectacular at any time, but late afternoon summer sun provides an excellent view of the Dent Blanche – Tsaté contact at the Col Cornet – Monte Berrio ridge.

Mapping north and east of Plan Debat indicates that:

1. Continental rocks of the Dent Blanche Nappe (Austro-Alpine Domain) occur over oceanic rocks of the Tsaté Nappe of the Penninic Domain. The contact is interpreted as a thrust represented by an abrupt discontinuity between ocean-derived material below the thrust and continent-derived material above.
2. The Tsaté Nappe (Combin Zone) comprises ocean-derived sediments and ophiolites, remnants of the Liguro-Piemontais oceanic crust, that are pervasively foliated and have a northwest-southeast stretching lineation.
3. The Arolla Series of the Dent Blanche Nappe consists of Permian granitoids that intrude a high-grade amphibolite facies unit. The Berrio Gabbro, a layered gabbro complex, occurs near the contact between the Dent Blanche and Tsaté nappes. (A layered gabbro complex is an igneous intrusion that formed as a result of magma cooling and crystallizing slowly in a magma chamber, creating distinct, repetitive layers with different mineral concentrations.)
4. Alpine deformation of the Arolla series has created several stacked units that include undeformed granitoids in the upper units and ductile-deformed orthogneisses and metagabbros in the lower units.

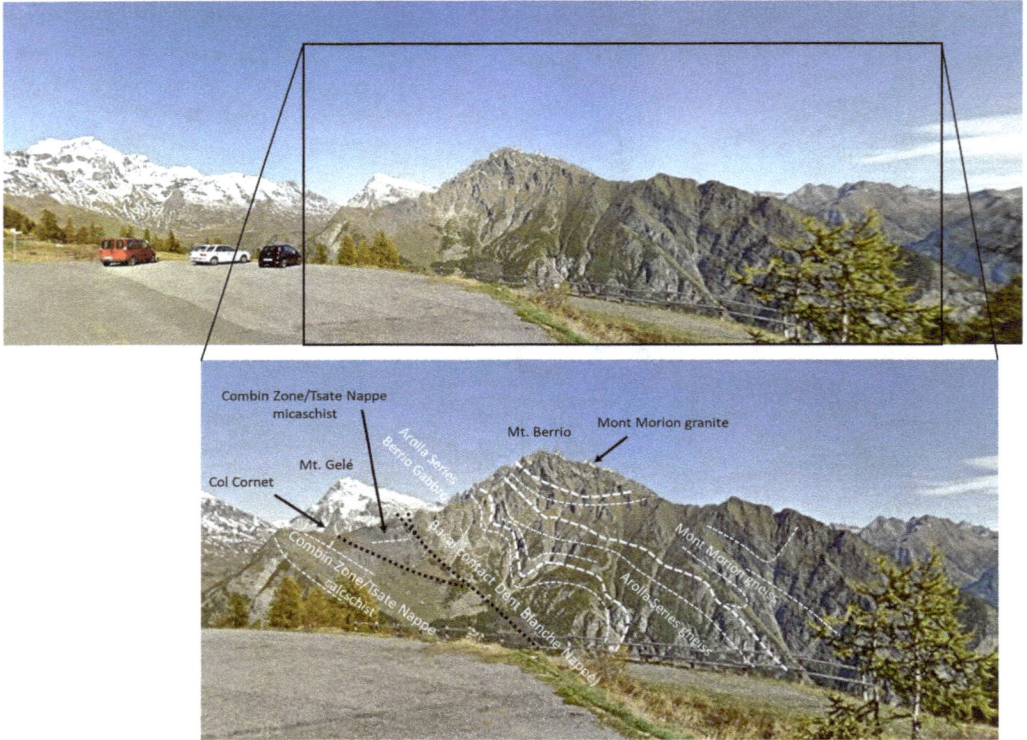

Above: Plan Debat, Google Street View northeast. Again, I use a Street View because it was raining when I was there. Below: Photo of Fenêtre de Durand (at the contact between Tsaté unit and Dent Blanche nappe), Monte Gelé, and Monte Berrio. Dashed lines follow bedding; dotted lines are faults. Geology derived from Manzotti and Ballèvre, 2016.

Side Trip 5, Plan Debat to Gignod–Roisan Zone: *Return south on Frazione Chatellair; turn left on Torrent la Cerise; continue straight on Frazione la Cerise; continue straight on Coudrey; merge onto La Chenal; continue straight on La Cretaz; turn right on Frazione la Chenal; continue straight on Frazione Rhins; at the sign for Chez Roux/Rovin/Gignod turn right onto Frazione Rovin and drive 100 m to a small pullout on the right at the intersection of the road to Rovin (45.785675, 7.304775). This is* **Stop ST5.5, Arolla Series, Gignod–Roisan Zone** *for a total of 19.2 km (11.9 mi; 34 min). Walk 100 m (330 ft) down to the outcrop at the intersection (45.784917, 7.305126). Watch for traffic!*

ST5.5 AROLLA SERIES ORTHOGNEISS OF THE GIGNOD–ROISAN ZONE, ROISAN

At this stop the Arolla series orthgneiss outcrops along with mica schist and dolomite lenses. This is the unit that forms the base of the Matterhorn. The unit may be overturned here. The outcrop is also in the Roisan–Cignana Shear Zone, as suggested by the pods of dolomite in the mica schist.

Starting at the north end and moving south, one can see:

- Orthogneiss of the Arolla series.
- Graphitic mica schist (a schist containing graphite and mica) with quartz-calcite veins. Grey marble layers, 5 to 10 cm thick, occur inside the micaschist.
- White dolomite pods (up to 20 × 10 m) occur within the micaschist. The dolomite pods are not foliated. Foliation in the mica schist wraps the dolomite pods and is locally crenulated.

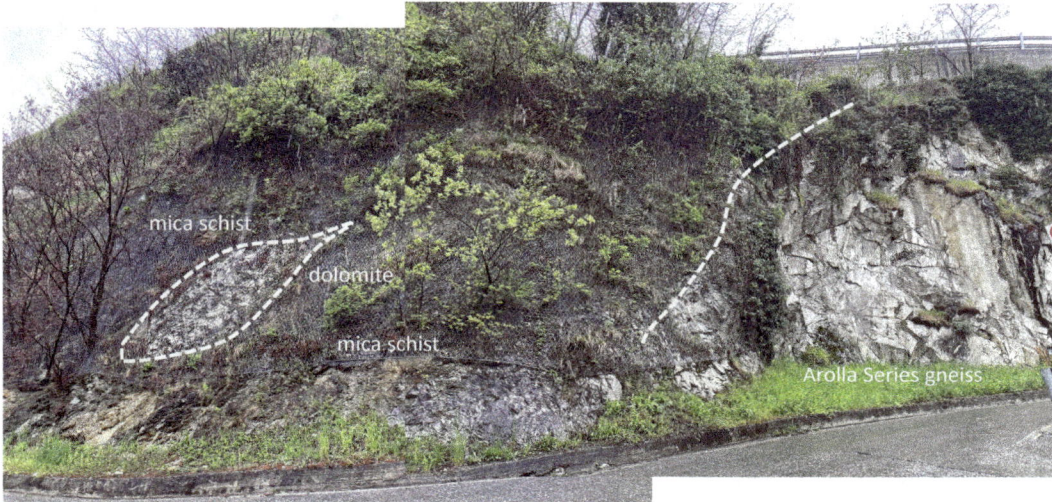

Mica schist and dolomite pod in the Roison Zone at Gignod. Geology derived from Manzotti and Ballèvre, 2016.

Side Trip 5, Gignod to Courmayeur/Ermitage (tolls): *Return south to Aosta; at the roundabout on the north side of Aosta take the 2nd exit (west) onto SS26; at the roundabout take the 2nd exit (south) onto SR47, then immediately take the E25/A5 toll road ramp to Mont Blanc; stay on E25 to the Courmayeur exit; exit north onto SS26dir; immediately after the sign for Courmayeur turn right (east); continue east through the roundabout and turn left (north) onto Via Circonvallazione; Via Circonvallazione turns slight left and becomes Via le Monte Bianco; at the roundabout take the 2nd exit (north) onto Str. Sergio Viotto; turn right (east) onto Str. Adolphe Rey; continue straight on Strada Grand Ru; turn left (northeast) on Strada del Villair; turn right (southeast) on Str. per Plan Gorret; turn left to stay on Str. per Plan Gorret and drive to the end of the road at parking for the Ermitage Restaurant on the left. This is* ***Stop 9.1, Mont Blanc Massif, Courmayeur/Ermitage*** *(45.796003, 6.982320) for a total of 43.9 km (27.3 mi; 45 min).*

STOP 9 MONT BLANC AND THE HELVETIC ZONE

Massif is the French word meaning "a large mountain mass or compact group of connected mountains forming an independent portion of a range." In geology a "massif" is a section of the crust, and the mountains associated with it, that is delineated by faults or flexures and retains its internal structure. Mont Blanc (White Mountain; Monte Bianco in Italy) meets both of these criteria.

We are now in the Dauphino-Helvetic Domain of the Western Alps. Within that domain, Mont Blanc is considered part of the Outer (or External) Crystalline Massifs that include, from southwest to northeast, the Argentera, Aiguilles Rouges (adjacent to the west), and Gastern-Aar massifs. These form a discontinuous series of elongate, domal features that follow the outer arc of the Western Alps (Egli and Mancktelow, 2013).

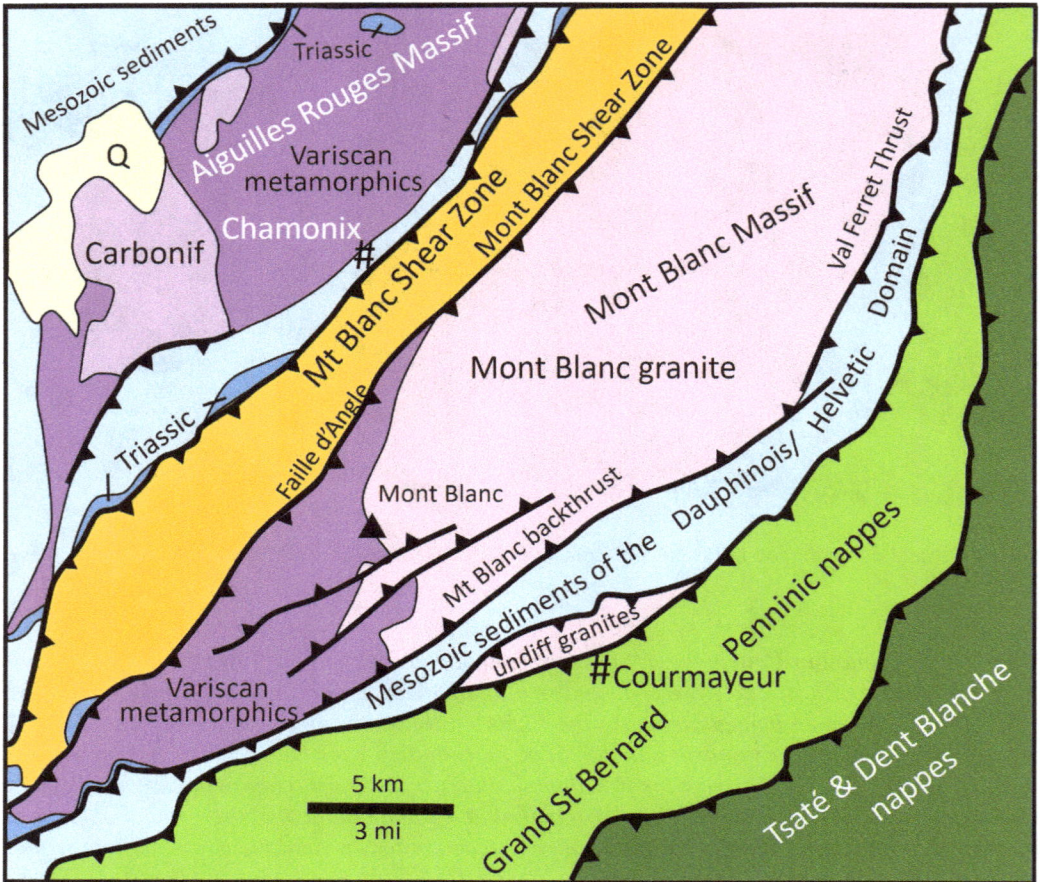

Geologic map of the Courmayeur, Mont Blanc, and Aiguilles Rouges area. Heavy lines with teeth are thrust faults (teeth on the thrusted sheet). Modified after Ravanel et al., 2010.

The Mont Blanc Massif, elongated southwest-northeast, is 46 km (29 mi) long and 20 km (12 mi) across at its widest point. It covers ~350 km^2 (135 mi^2) and reaches an elevation of 4,808 m (15,774 ft), the highest point in the Alps and in all of western Europe. Glaciers cover 170 km^2 (66 mi^2) of the massif. At 12 km (7.5 mi), the Mer de Glace is the longest glacier in the range as well as the longest in France and the second longest in the Alps (Wikipedia, Mont Blanc Massif). The massif straddles the border between Italy and France, and it forms the drainage divide between Rhone and Po basins along a 35 km (22 mi) long crest.

Mont Blanc is mainly a granitic batholith (303 ± 2 Ma) and associated rhyolites (307 ± 2 Ma) formed during the Late Carboniferous (Pennsylvanian) Variscan Orogeny. The magmas intruded Paleozoic gneiss and mica schist, part of the proto-European basement rock (Ravanel et al., 2010; Ribes et al., 2020). The sedimentary cover, preserved mainly along the margins, is Triassic to Miocene age. The Grès Singuliers/Lotharingien/Domérien sandstones comprise an Early Jurassic siliciclastic sequence deposited around the Mont Blanc massif. It was mainly derived from erosion of local basement and pre-rift sediments (Ribes et al., 2020). The cover rock also includes thick Late Jurassic and Early Cretaceous reef limestones.

The Mont Blanc massif is an elongated antiform. The folded and thrusted Dauphino-Helvetic cover verges to the northwest (Faure and Chen, 2018). The detailed structure of the massif and how it came to be is still a matter of debate.

One proposal is that the massif is a crustal scale recumbent antiform produced by southeast-directed back-folding simultaneous with northwest-directed ductile shear. Another idea is that the Penninic Thrust (the basal thrust of the Penninic nappes), reactivated as normal faults, may have led to significant erosion on the east side of the massif. Others have proposed a two-phase thrust/back-thrust model that has the northwest-verging Mont Blanc Massif thrust onto the Aiguilles Rouges Massif. This model suggests thrusting with 4-8 km (2.5 to 5 mi) of vertical displacement beginning around 12 Ma. Young uplift and erosion in this scenario are considered to be the result of displacement along a Mont Blanc back-thrust starting around 2.5 Ma. Other workers favor uplift during a combined compression and shearing setting during mid-Miocene (Egli and Mancktelow, 2013).

The Mont Blanc granite itself was intruded at depths of 4 to 14 km (2.5 to 8.7 mi). Early Jurassic rifting along the southern margin of Europe led to numerous faulted basins (half grabens) in the Dauphino-Helvetic Domain. A half-graben is a basin or valley bounded by a fault on one side, as opposed to a full graben bordered by faults on both sides. Recognition of Jurassic cataclasites suggests that the massif formed in response to crustal extension and thinning along the European margin of the Alpine Tethys rift system. Uplift of basement during crustal/lithospheric stretching produced the locally derived clastic sediments surrounding the massif. Early Jurassic sedimentation included a "Helvetic facies" (sandstone-siltstone-shale) and a marly "Dauphinois facies" (Ribes et al., 2020).

Triassic to Eocene sedimentation created a distinctive stratigraphy on the southern margin of the former European continent. The end of sedimentation and onset of deformation is marked by imbricate thrust stacking of the Penninic Nappes. Continued shortening due to collision between Adria and Europe created the Helvetic Fold-Thrust Belt, with ~100 km (60 mi) of north-south shortening. Continuing compression led to uplift and erosion of the thrust stack, finally exposing the external crystalline igneous basement in the Neogene.

Most Alpine uplift occurred in Miocene time. The geometry of the massif at that time strongly influenced later thrusting during continued Alpine compression.

The External Crystalline Massifs all share a similar Neogene uplift and erosion history. On the order of 10–14 km (33,000 to 46,000 ft) of material was eroded off the Gastern-Aar and Aiguilles Rouges-Mont Blanc massifs. The Mont Blanc Massif is characterized by young and rapid exhumation. None of the major faults or shear zones around the Mont Blanc Massif (i.e., Mont Blanc Shear Zone, Mont Blanc back-thrust, Penninic Thrust) are younger than Miocene (<5 Ma). Young, brittle faults are dominantly strike-slip and have no significant vertical offset. The young (<2 Ma) unroofing of Mont Blanc favors a scenario of enhanced uplift and erosion by regional upwarping, perhaps in part related to glacial melting and isostatic rebound of the crust (Egli and Mancktelow, 2013). Isostatic rebound occurs when a landmass, depressed by the immense weight of an ice sheet, rises back to its original elevation after the ice melts.

During the mid-19th century the granite of the Mont Blanc massif was an important source of stone for buildings. The massif has been a source of mineral specimens for over 250 years. However, in 2008 the Community of Chamonix banned all mineral-hunting activities and collection of specimens without a permit. Permitted collectors must offer the Chamonix Crystal Museum (Musée des Cristaux) first option to acquire any significant specimens for its collections. The museum opened to the public in 2006 (Wikipedia, Mont Blanc Massif).

The first ascent of Mont Blanc in 1786 by Jacques Balmat and Michel Paccard initiated the sport of alpine mountaineering.

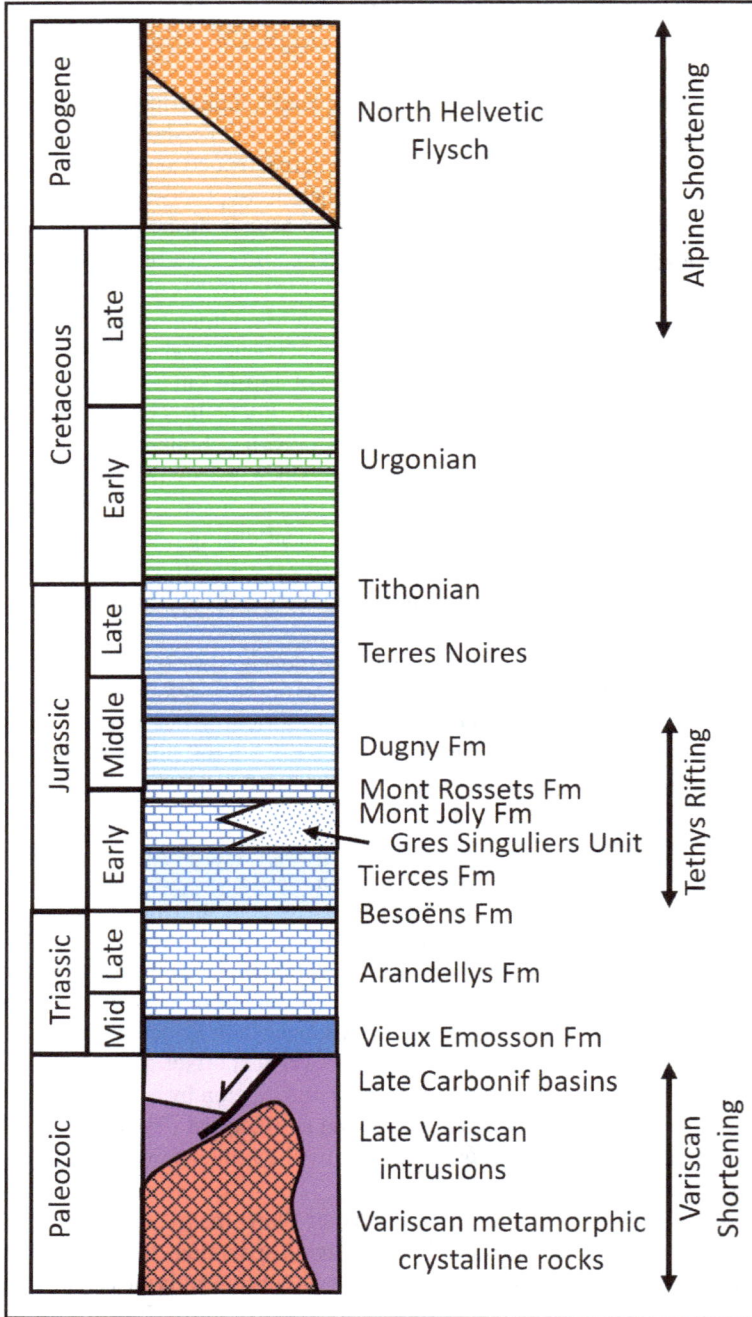

Stratigraphy of the Mont Blanc Massif and Dauphino-Helvetic Domain. Modified after Ribes et al., 2020.

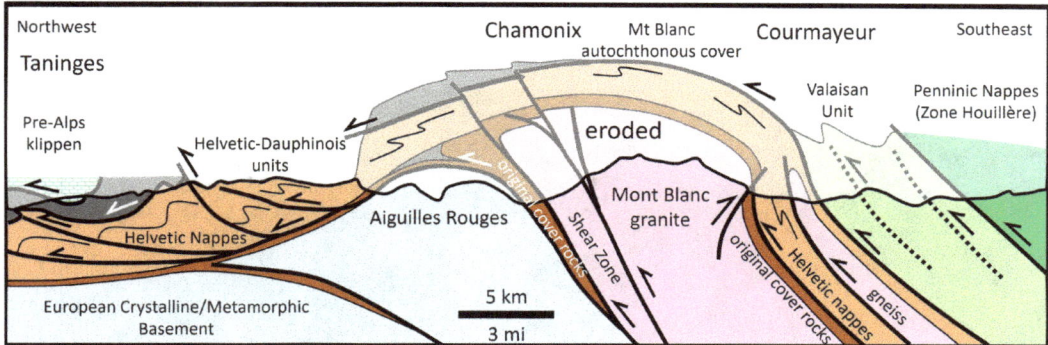

Cross section through the Mont Blanc and Aiguilles Rouges massifs. This shows the massif as a crustal scale recumbent antiform with northwest-directed ductile shear. Modified after Egli and Mancktelow, 2013.

STOP 9.1 MONT BLANC MASSIF, COURMAYEUR/ERMITAGE

According to the locals I spoke to, this stop provides the best panorama of the eastern side of the Mont Blanc massif. Whereas the town of Courmayeur sits on Triassic carbonates, all the peaks in the spectacular view to the north are Carboniferous granites. If you look southwest, you can see Mont Crammont capped by an amazingly flat dip slope and composed of southeast dipping Cretaceous marble.

Courmayeur/Mont Blanc to Mont Blanc Skyway: *Return to Courmayeur; at the Grand Hotel Courmayeur Mont Blanc turn right (north) onto Str. Sergio Viotto; continue straight on Via dei Bagni; turn right (east) onto Strada la Montoux; hard left (north) onto SS26dir; at the roundabout take the 1st exit (east) to the Skyway Monte Bianco parking area. This is **Stop 9.2, Mont Blanc Skyway and Tunnel** (45.815318, 6.956683) for a total of 5.5 km (3.4 mi; 10 min).*

STOP 9.2 MONT BLANC SKYWAY AND MONT BLANC TUNNEL

The Mt. Blanc tunnel connects Chamonix, France, to Courmayeur, Italy, by way of French Route Nationale 205 and Italian Traforo T1 (European route E25). The tunnel is a major commercial route, with Italy relying on this tunnel for as much as one-third of its freight shipped to northern Europe.

The idea to build the tunnel dates back to the building of railroads in the 1800s, but it was impractical at the time. The French and Italians got serious after World War II, and an agreement between France and Italy was signed in 1949. One Italian and one French operating company would each be responsible for their half of the tunnel. Drilling began in 1959. The French and Italian drilling teams met on 4 August 1962 with a deviation of less than 13 cm (5 in). The tunnel was inaugurated by the French president, Charles de Gaulle, and the Italian president, Giuseppe Saragat, on 16 July 1965, and it opened to traffic on 19 July.

The tunnel is 11.6 km (7.2 mi) long, 8.6 m (28 ft) wide, and 4.35 m (14.3 ft) high. It is a single gallery with a two-lane road. The tunnel passes almost directly under Aiguille du Midi, at which point it lies 2,480 m (8,140 ft) beneath the surface, making it the world's second deepest operational tunnel.

Following a fire in 1999 in which 39 people died, the tunnel was closed for three years for safety upgrades. For safety reasons, all operations were merged into a single entity, MBT-EEIG, controlled 50–50 by the French and Italian operating companies.

The toll for cars one-way is €52.30 entering on the Italian side and €51.50 entering on the French side (2025). Price also varies by vehicle type and whether you are going one-way or return.

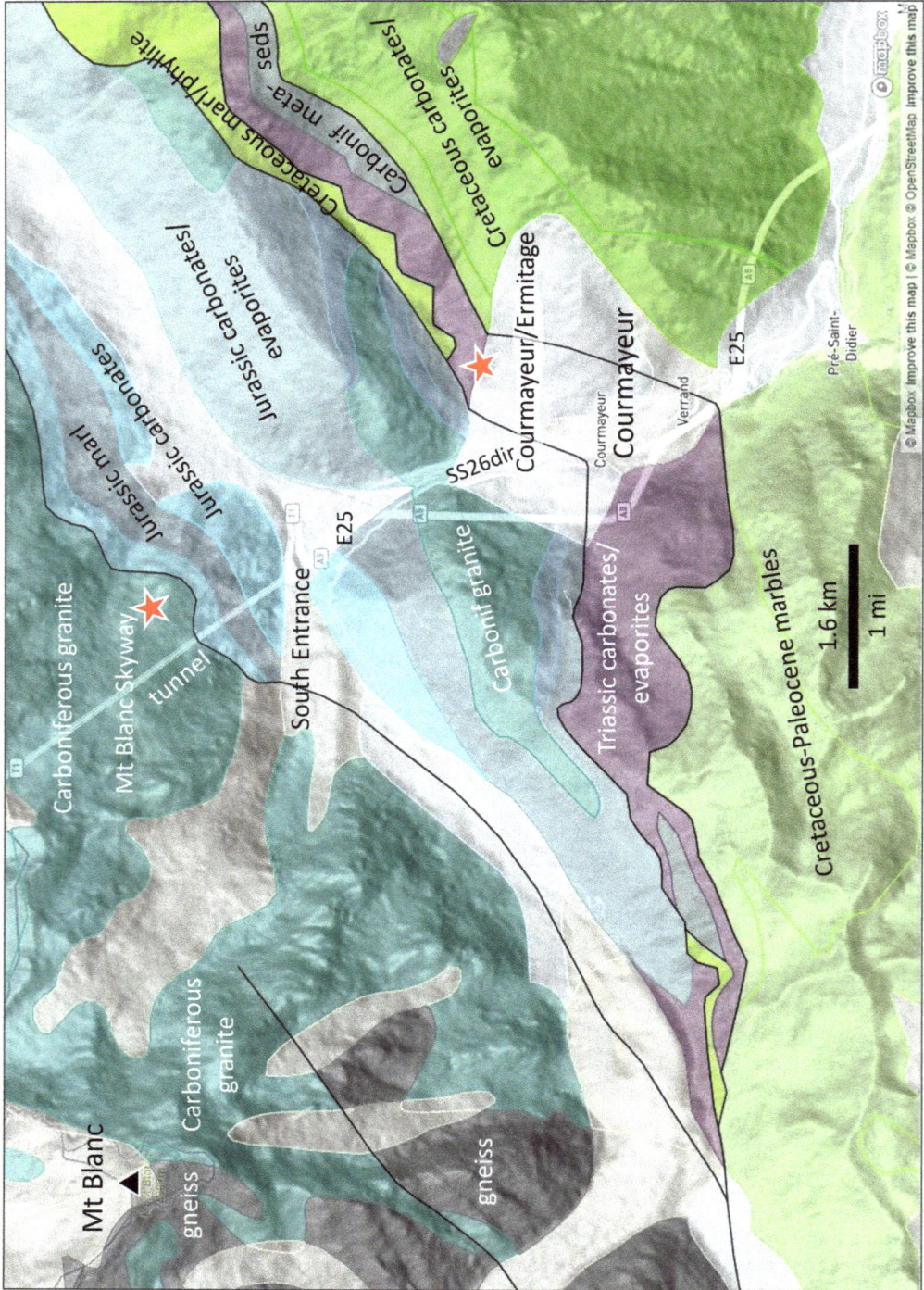

Geologic map of the Courmayeur–Mont Blanc area. Geologic base from MacroStrat, 2015.

Mont Blanc emerging from a snowstorm as seen from the Ermitage. View northwest.

Helbronner Peak (center), Mont Blanc Massif as seen from the Ermitage.

Mt. Crammont and southeast dipping Cretaceous marble layers. View southwest from the Ermitage.

The south entrance to the tunnel is in Jurassic carbonates. The Skyway terminates in the Mont Blanc granite.

View south over Courmayeur and Aosta Valley from Mont Blanc Skyway. Photo courtesy of Tiia Monto, https://commons.wikimedia.org/wiki/File:Courmayeur_and_mountains.jpg

Brenva side, Mont Blanc, from the Skyway. Photo courtesy of SteGrifo27, https://commons.wikimedia.org/wiki/File:Skyway_Mont_Blanc.jpg

Visit

Skyway Monte Bianco takes you to the highest point in Italy. It was inaugurated in 2015 and today includes three stations: Courmayeur/The Valley (1,300 m), Pavillon/The Mountain (2,173 m), and Punta Helbronner/The Sky (3,466 m).

Hours: Pavillon → Punta Helbronner, Monday–Friday 8:45 – 15:30 ; Weekend 8:30 – 16:30
Cost: see website for current prices
Address: Strada Statale 26 dir, 48, 11013 Courmayeur AO, Italy
Phone: +39016589196
Website: https://www.montebianco.com/en

Parking and facilities are available. Facilities include bars, restaurants, a cinema, museum, conference rooms, a botanical garden, a children's playground, and high-altitude adventure options.

Side Trip 6, Mont Blanc Skyway to Verney Lake: At the roundabout take the 2nd exit (south) onto SS26dir; turn right (southwest) toward SS26 della Valle d'Aosta; turn right (northwest) onto SS26; turn left (south) to stay on SS26 to La Thuile; drive to the Lago Verney parking area on the right. This is **Stop ST6.1, Petit Saint Bernard and Versoyen Complex, Verney Lake** *(45.695840, 6.870780) for a total of 29.3 km (18.2 mi; 55 min).*

If you prefer not to go on Side Trip 6:

Mont Blanc Skyway to Cascade Arpenaz: Continue north on E25/T1 throught the tunnel and into France. Within the tunnel the road becomes N205; continue northwest on N205; continue straight onto A40; take Exit 20 toward Annecy/Geneva; at the roundabout take the 1st exit (north) onto Rte de Luzier; turn left (north) onto Rte d'Oex; turn right (east) onto CR dit des Fours À Chaux and drive to the parking area on the left. This is **Stop 10, Alpine Folding, Cascade Arpenaz** *(45.974327, 6.639063) for a total of 45.1 km (28.0 mi; 45 min).*

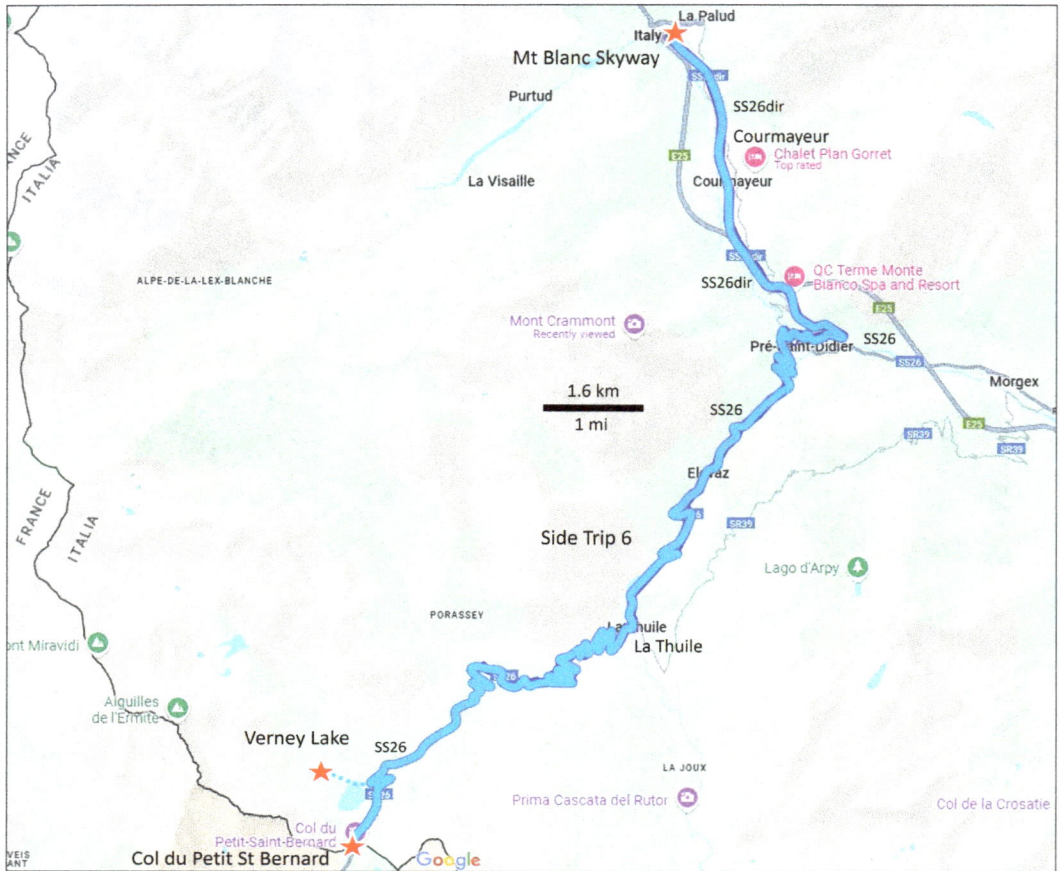

Side Trip 6 Penninic Domain, La Thuile Area

These stops are within the Penninic Domain near the boundary with the Helvetic Domain to the northwest (Kuderski, 2020). The units seen here help us understand the evolution of the European crust – oceanic crust transition from Jurassic rifting to the Alpine Orogeny.

The Petit St. Bernard Pass area preserves continental basement, serpentinized subcontinental mantle, and metasediments from the adjacent oceanic basin. It has been called one of the best examples of a "Zone of Exhumed Subcontinental Mantle," thought to be similar to present-day rifted margins where mantle is exposed by uplift and low-angle detachment faults. Remnant slivers of continental basement sit above the detachment surface as "extensional allochthons." Extensional allochthons are packages of rock that have moved significant distances from their original location and have been deformed by extensional tectonic forces. The distribution of serpentinized ultramafic units adjacent to continental basement and syn- to post-rift sediments may be inherited from the pre-subduction and Alpine deformation history.

Recent work in this area provides evidence of extensionally thinned continental crust that is directly onlapped by post-rift sediments. As the southern European continental crust thinned during Jurassic Tethys rifting, it graded into areas where mantle was exposed at the sea floor by uplift and sliding off of cover rock along shallow detachment faults (a process similar to metamorphic core complexes in the Basin-and-Range of North America). The extensional topography influenced the distribution of rift-related sediments, which were later covered by post-rift deep marine sediments. The stratigraphy of the Valaisan Domain near Petit St. Bernard Pass indicates an original position on the European continental margin between the European Plate and the Briançonnais rise (Beltrando et al., 2014).

Exposed continental mantle represents the Valaisan oceanic basin within the Penninic Nappes. Ophiolitic zones signify Piémont oceanic crust trapped between the European margin and the Briançonnais Microplate (Stampfli, 2001). The Briançonnais Domain is a segment of thinned European continental crust located between the partly oceanic Valaisan Basin to the northwest and the Piemont–Liguria Ocean to the southeast (Ballèvre et al., 2020). Thrust stacking of these domains produced tectonic packages known as the Valaisan (lower), Briançonnais (middle), and Piemontais (upper) units of the Penninic Domain (Aescher, 2001, in Stampfli).

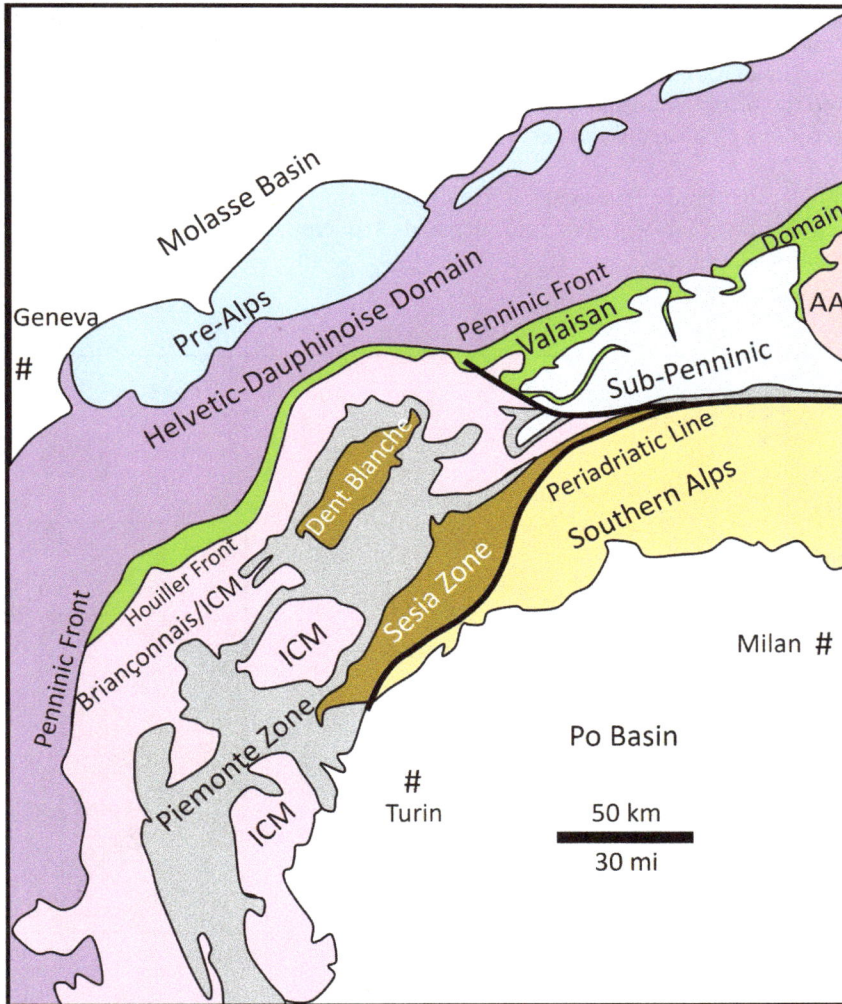

Tectonic map of the Western Alps. ICM = Inner Crystalline Massifs; AA = Austro-Alpine Domain. Modified after Beltrando et al., 2012.

The Valaisan Unit near the Petit St. Bernard Pass consists of serpentinized sub-continental mantle juxtaposed with Paleozoic granitic basement, Mesozoic metamorphosed pillow basalts, and Mesozoic to Tertiary metasediments, all of which underwent Alpine metamorphism and shortening. The different rock types appear to have been juxtaposed prior to the Alpine Orogeny: the unusual stratigraphic relationships are primarily a result of rift-related extension. Mantle peridotites were exposed by low-angle extensional faulting, with isolated remants of continental basement remaining above serpentinized mantle. Basement units were covered by locally sourced breccias prior to the deposition of deep marine shale and minor carbonates. These have been metamorphosed to micaschists and other basinal metasediments (Beltrando et al., 2014).

The Valaisan Domain is bounded on the northwest by the Penninic Front, which separates the Penninic nappes from the more external (northern and western) Helvetic-Dauphinois nappes. The Helvetic-Dauphinois Domain consists of Permo-Carboniferous basement overlain by a Late Permian to Lower Oligocene sedimentary cover. The Helvetic-Dauphinois Domain, 40 to 60 km (24–36 mi) wide, extends northeast-southwest through the Mont Blanc and Aiguilles Rouges massifs. The Valaisan Domain is bounded on the southeast by the Houillère Front that separates the Valaisan Unit from the overlying Houillère Zone. The Houillère Zone, part of the Briançonnais Domain, consists of Carboniferous and Permian continental sandstone, breccia, and shale, with minor Early Triassic dolomite and shale (Beltrando et al., 2014).

Units in the Petit St. Bernard Pass area include, from the base upward:

- Permo-Carboniferous Briançonnais (Houillère) Unit
- Triassic to Lower Jurassic Petit Saint Bernard–Arguerey Unit
- the Versoyen Complex
- the Valaisan Trilogy

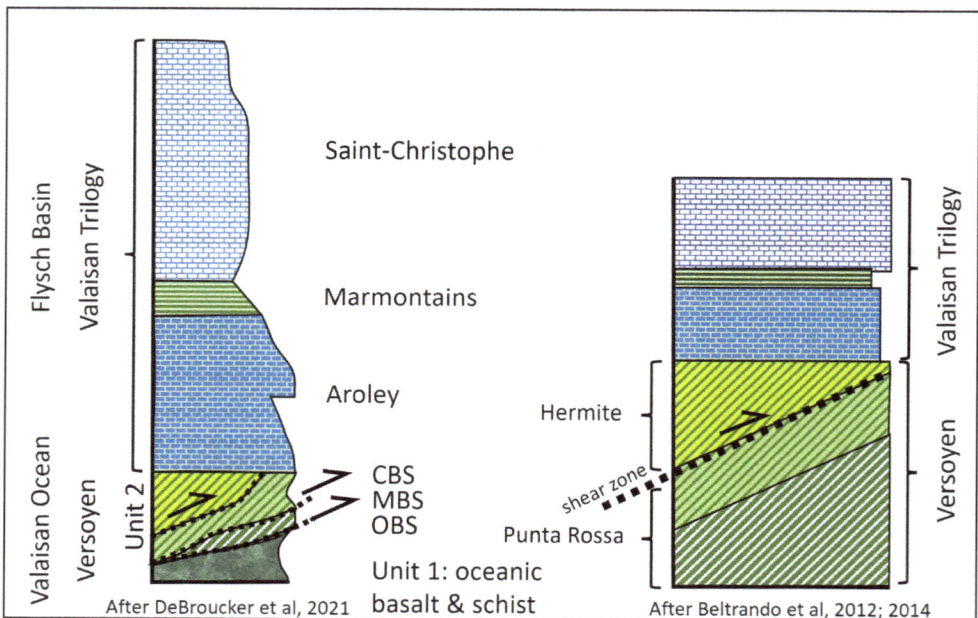

Stratigraphy of the Versoyen and Valaisan Trilogy groups, Petit Saint Bernard Pass area. OBS = Oceanic Blocky Schists; MBS = Mixed Blocky Schists; CBS = Continental Blocky Schists. Modified after Beltrando et al., 2012, 2014, and De Broucker et al., 2021.

The Briançonnais (Houillère) Unit comprises European continental margin sandstone, shale, and breccias.

The Triassic-Lower Jurassic Petit Saint Bernard–Arguerey Unit consists of calcschist, micaschist, and marble derived from Triassic dolomite and Lower Jurassic limestone.

The Versoyen Complex is a thrusted imbrication of four structural units. The first (lowest), corresponding to the Punta Rossa of Beltrando, et al. (2012, 2014), contains remnants of the Valaisan ocean floor. The overlying three units in the Complex, grouped together as "schistes à blocs" (blocky shales), correspond to the Hermite Unit of Beltrando (2012, 2014) and contain blocks of oceanic, continental, or mixed origin embedded in a matrix of gray mica schists. Paleozoic granitic continental basement rocks of the Versoyen Complex outcrop as a tectonic mélange, with imbricated blocks and sheared slivers in an Early Cretaceous mica schist matrix.

The Versoyen Complex is unconformably overlain by the Cretaceous Valaisan Trilogy (Aroley–Marmontains–Saint Christophe formations). Both the Versoyen Complex and Valaisan Trilogy were subjected to high-pressure Alpine metamorphism at around 40 Ma (Eocene).

The Trilogy were deposited during Valaisan rifting, which began in the Late Jurassic. The rifting appears to have followed preexisting Variscan structures along the Zone Houillère: this would explain the different Variscan basement types on either side of the Valaisan suture. Southward

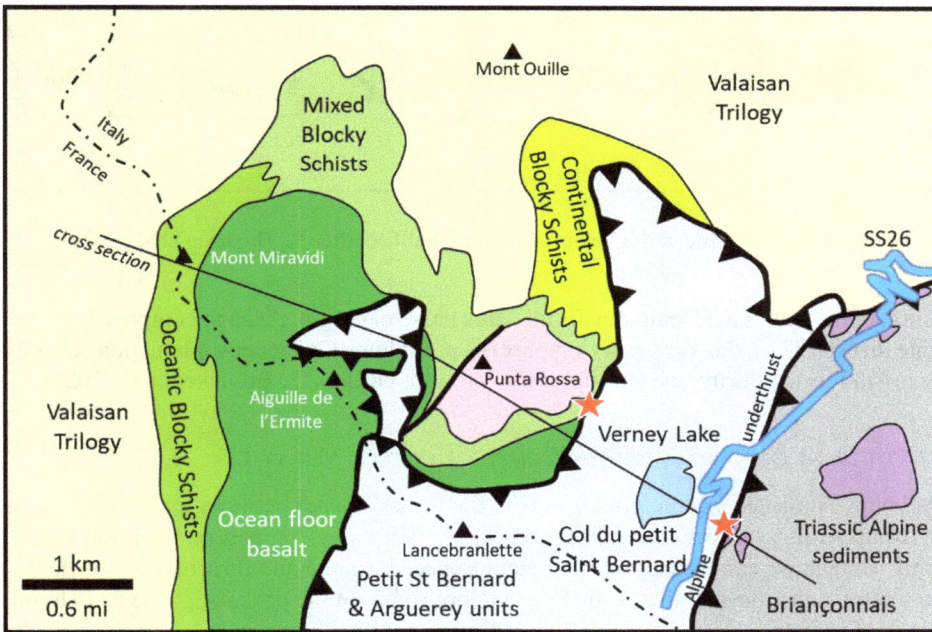

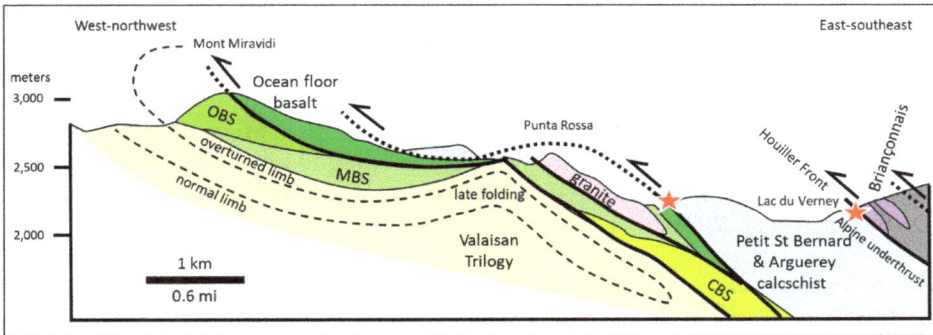

Above: Geologic map of the Petit Saint Bernard Pass area. Heavy lines with teeth are thrust faults. Below: Cross section from Mont Miravidi to Petit Saint Bernard Pass. OBS = Oceanic Blocky Schists; MBS = Mixed Blocky Schists; CBS = Continental Blocky Schists. See previous figure for location. Modified after De Broucker et al., 2021.

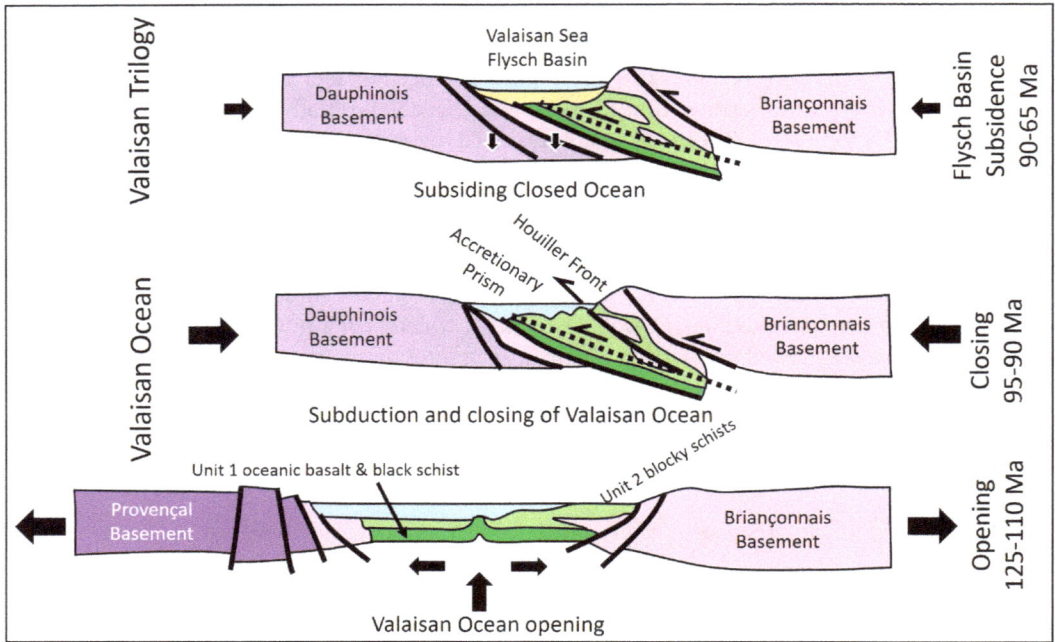

Progressive opening and closing of the Valaisan Ocean. Modified after De Broucker et al., 2021.

subduction of the Valaisan Ocean transformed this passive margin, through compression, to stacked imbricate thrusts. Thus, the Versoyen Complex is a pre-Alpine Cretaceous subduction-related accretionary prism formed during the closure of the Valaisan Ocean (De Broucker et al., 2021).

ST6.1 Petit Saint Bernard–Versoyen Complex Contact, Verney Lake

Besides being exquisitely scenic, this stop is at the thrust contact between the Petit Saint Bernard unit on the upper thrust sheet and Units 1 and 2 of the Versoyen Complex on the lower thrust sheet. From this vantage we can look up at the metamorphosed granite and thrust slivers of Punta Rossa Peak to the west and, beyond that, Unit 1 ocean floor rocks on the peaks Aiguille de l'Hermite and Aiguille de Beaupré.

The boundary of Unit 2/Hermite and Unit 1/Punta Rossa is southwest of here at the base of the grassy slope above the gray-green outcrops of the Unit 1 metagranite. The Versoyen Complex has a characteristic banded appearance due to the black schists and interbedded green-gray mafic sills (dark sheet-like layers of volcanic rock); it can be seen both in the cliffs from Mont Lancebranlette to the southwest to the Aiguille de l'Hermite to the west and to Mont Miravidi to the northwest. The Petit Saint Bernard–Arguerey calcschists (metamorphosed shaly limestone) outcrop in the grassy slope between here and Verney Lake to the southeast and all the way to the road (Beltrando et al., 2014; De Broucker et al., 2021).

Side Trip 6, Verney Lake to Col du Petit St. Bernard: *Return to the main road and turn right (south); drive a total of 1.4 km (0.9 mi; 2 min) to* ***Stop ST6.2, Houillère Front, Col du Petit St. Bernard*** *(45.684473, 6.887214) and pull over just north of chair lift north of Col du Petit St. Bernard.*

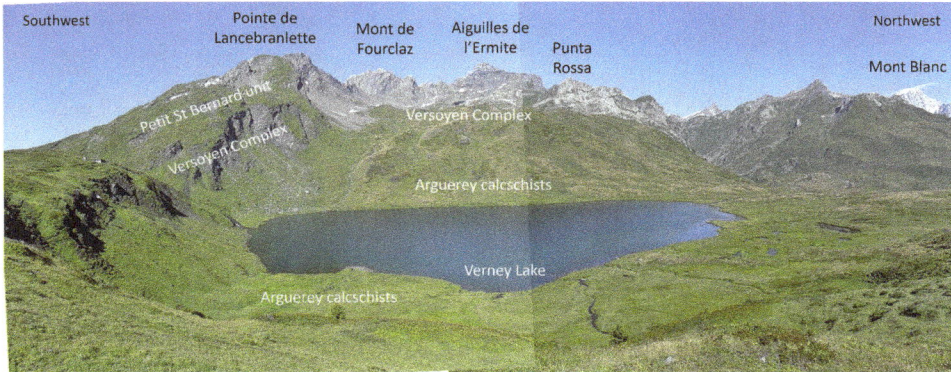

View west to northwest over Lake Verney. Geology derived from Beltrando et al., 2014. Photos courtesy of Florian Pepellin, https://commons.wikimedia.org/wiki/File:LacVerney_sousLancebranletteen_%C3%A9t%C3%A9_(2019).JPG and https://commons.wikimedia.org/wiki/File:LacVerneyetleMont-Blancen_ %C3%A9t%C3%A9_(2019).JPG

ST6.2 HOUILLÈRE FRONT/ALPINE UNDERTHRUST, COL DU PETIT ST. BERNARD

This stop is as much for the breathtaking views as for the geology. Geologically, Col du Petit St. Bernard is located along the trace of a west-directed thrust that puts the Carboniferous Briançonnais (Houillère) Unit sandstone-shale-breccia over the Petit Saint Bernard–Arguerey calcschist-micaschist-marble. So this is the fault zone that puts the Briançonnais microcontinent over the Valaisan Ocean.

We are near the southwest termination of a ~10 km (6 mi) northeast-southwest "multi-phase deformation zone" (ISPRA, 2011, Folio 89). De Broucker et al. (2021) calls this the "Alpine underthrust," which I take to mean a subduction-related thrust fault where the rocks beneath the fault are actively pushing downward beneath an essentially passive rock package. Beltrando et al. (2014) calls it the Houillère Front. The Houillère Zone is part of the Briançonnais Domain; it consists of tectonic units derived from a European crustal microplate separating the Valaisan Basin (north) from the

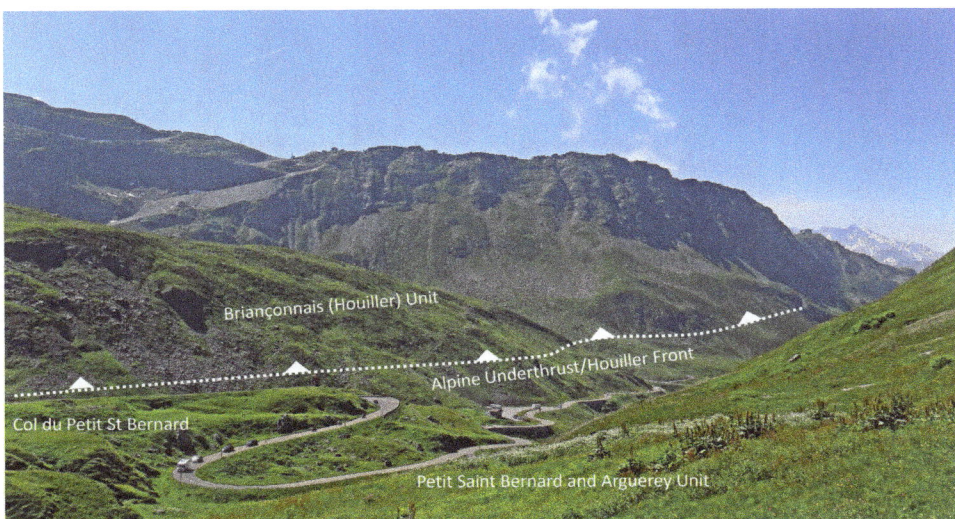

View south from Col du Petit Saint Bernard. Geology derived from De Broucker et al., 2021. Dotted line is a thrust fault with teeth on the thrust sheet. Photo courtesy of Rémih, https://commons.wikimedia.org/wiki/File :Mont_Valezan_@_Hospice_du_Petit_Saint-Bernard.jpg

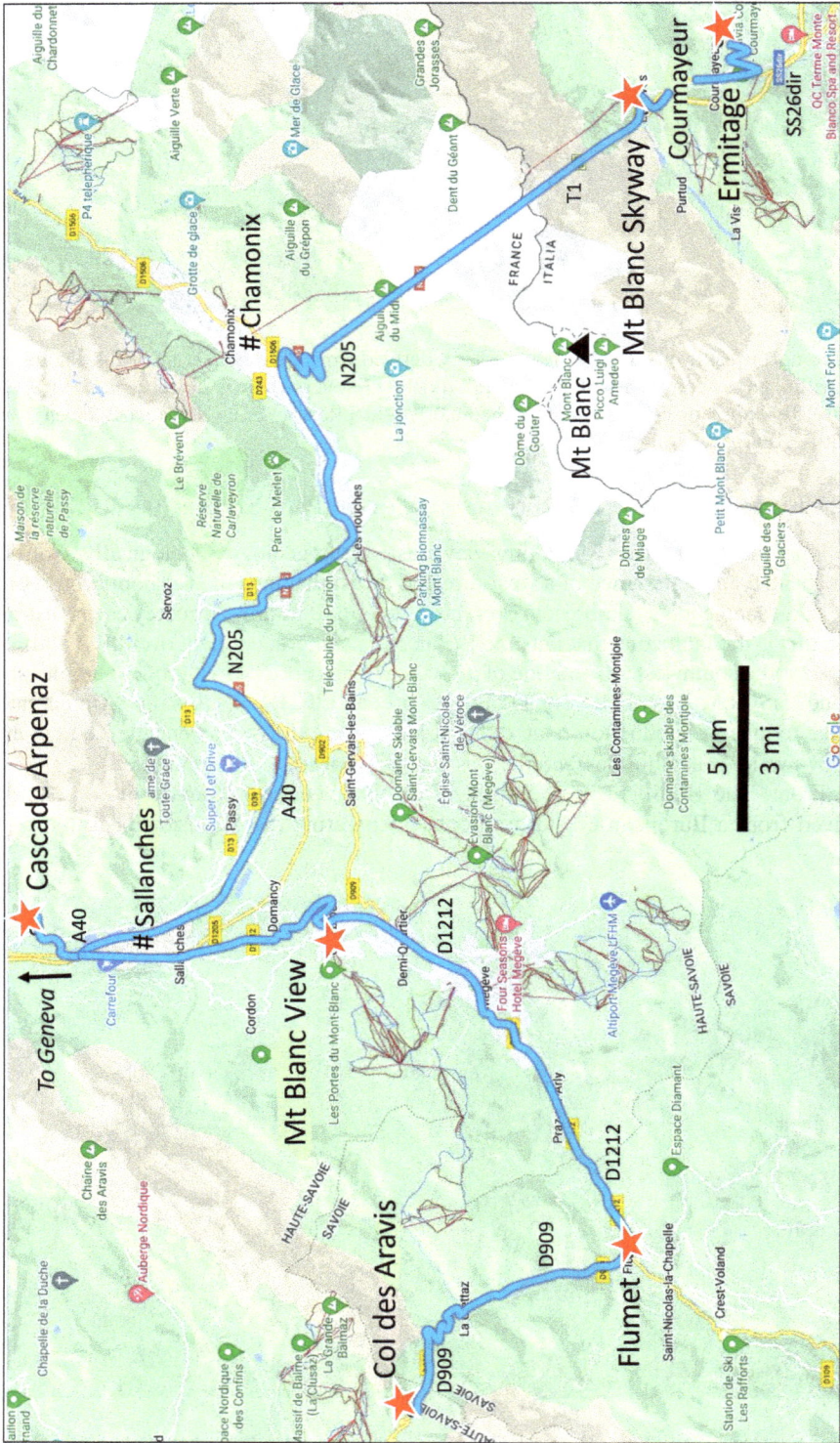

Penninic Piemonte–Liguria oceanic basin (south). The Briançonnais, or Houillère Unit consists of Carboniferous and Permian continental sandstone, breccia, and shale, with minor Early Triassic dolomite and shale. These formations underwent only minor Alpine metamorphism (Beltrando, 2014).

Side Trip 6, Col du Petit St. Bernard to Cascade Arpenaz: Return north on SS26 to SS26dir; turn left (north) onto SS26dir; at the tunnel entrance keep left to stay on E25/T1 north through the tunnel and into France. Within the tunnel the road becomes N205; continue northwest on N205; continue straight onto A40; take Exit 20 toward Annecy/Geneva; at the roundabout take the 1st exit (north) onto Rte de Luzier; turn left (north) onto Rte d'Oex; turn right (east) onto CR dit des Fours À Chaux and drive to the parking area on the left. This is **Stop 10, Alpine Folding, Cascade Arpenaz** *(45.974327, 6.639063) for a total of 73.8 km (45.8 mi; 1 hr 28 min).*

STOP 10 ALPINE FOLDING, CASCADE ARPENAZ

Cascade d'Arpenaz is both a beautiful waterfall and highlights an example of a recumbent fold. Horace Bénédict de Saussure, one of the original Alpine geologists, examined the area between 1760 and 1778, and in his *Voyages dans les Alpes* (1779) he gives what may be the first accurate description of Alpine folding. His description, translated into English, follows:

> A short distance from Maglan, a pretty waterfall, formed by a stream called the Nant d'Arpenaz, presents a spectacle as new as it is pleasant for those who have not been accustomed to this kind of pleasure from frequent journeys in the mountains. But a man curious about the structure of the mountains must, when approaching this waterfall, be mainly concerned with the rock from the top of which it falls.
>
> M. Bourrit drew before my eyes, with the greatest accuracy, the rock and the waterfall, seen from the middle of the meadows which are below the road….
>
> But we could not represent in this drawing a much higher mountain, which this rock hides entirely, because it is situated behind it. The layers of this mountain are the continuation of the upper layers of the rock of the waterfall, and form concentric arcs, turned in the opposite direction; so that the totality of these layers has the shape of an S, the upper part of which bends strongly backwards; the engraved plate represents only the lower part of this S. These large objects must be seen from afar and under different faces, so that we can grasp all of their shapes.

The fold being described is overturned to the northwest, indicating the direction of shear in the Helvetic Nappes is from southeast to northwest. The fold is developed in Late Jurassic (Malm) and Early Cretaceous (Urgonian) limestones. Defined in the Provence region of France, the rudist (bivalve)-bearing Urgonian Limestone is typical of a northern Tethyan shallow-marine carbonate rocks (Godet et al., 2024).

The scenic waterfall is about 279 m (915 ft) high.

Arpenaz to Mont Blanc View: Return south on Rte d'Oex; turn right (west) on Rte de Luzier; at the roundabout take the 2nd exit (south) onto D1205/Av de Genève; stay on D1205 to Sallanches; at the roundabout take the 1st exit (west) onto Quai de l'Hôtel de Ville; turn left (south) onto Rue Saint Eloy; continue straight onto D1212 for 0.9 km (0.57 mi) and pull into a parking area on the right. This is **Stop 11, Mont Blanc View** *(45.928635, 6.631625) for a total of 6.0 km (3.7 mi; 10 min).*

Photo looking east at the Arpenaz recumbent fold. From Lacassin, 2017.

Drawing of the Cascade d'Arpenaz by Jean-Antoine Linck. From the Swiss National Library, https://comm ons.wikimedia.org/wiki/File:CH-NB_-_Grenzgebiet,_Savoyen_Arpenas_-_Collection_Gugelmann_-_GS-GUGE-LINCK-B-16.tif?page=1

Geologic map of the Sallanches-Arpenaz area. Geologic base from MacroStrat, 2015.

Stop 11 Mont Blanc View

There are many great views of the Mont Blanc massif to the south. This is one that you can stop at. On a clear day you can see the glaciers filling the highest valleys.

View southeast to Mont Blanc.

Mt. Blanc View to Flumet: *Continue south on D1212 to Flumet; on the east side of Flumet pull into the parking area on the left. This is **Stop 12, Angular Unconformity, Flumet** (45.818678, 6.519515) for a total of 21.1 km (13.1 mi; 28 min). Carefully walk 115 m (375 ft) west to the small picnic area on the right. There is little to no shoulder on the road. Be mindful of traffic!*

Stop 12 Angular Unconformity, Flumet

There are lots of roadcuts with near-vertical Carboniferous schists in the canyon walls as you approach Flumet. At this stop there is an angular unconformity that places nearly flat-lying Triassic carbonates on steeply dipping Variscan (Carboniferous) Belledonne schists. The unconformity indicates the period of late Paleozoic erosion that accompanied Variscan uplift and deformation.

This unconformity is an erosion surface that represents a time gap of perhaps 40 million years when the mountains were raised up across Europe, then worn away by erosion.

Flumet to Aravis: *Continue southwest on D1212; in Flumet turn right (north) onto D909/Rue des Aravis; continue on D909 to Aravis Pass and pull over on the right at the hairpin turn. This is **Stop 13, Thônes Syncline, Col des Aravis** (45.871913, 6.465949), for a total of 11.8 km (7.4 mi; 19 min).*

Outcrop on the east side of Flumet. This angular unconformity has flat-lying Triassic carbonates above steeply inclined Variscan schists. Dashed lines follow bedding; dotted line is the unconformity.

STOP 13 THÔNES SYNCLINE FROM COL DES ARAVIS

We are still in the Helvetic-Dauphinois Domain, but this zone is called the Subalpine Chains, where Mesozoic and Cenozoic sedimentary cover rocks were folded and thrust over the European plate. A period of regional extension during the Triassic and Jurassic was related to opening of the Liguria–Piemontese Ocean. Alpine compression in Oligo–Miocene time reactivated and inverted many of the extensional normal faults as thrust faults (Marquer et al., 2006). Thrust sheets were folded during the compression and structural inversion.

A syncline is a fold where the rock layers dip (are inclined) toward the center of the fold to form a trough or basin. The Thônes Syncline is a large basin developed on a thrust sheet. It is said to be carried "piggyback" on the northwest-directed Montagne de Cotagne Thrust (Tardy and Doudoux, 1984).

Cretaceous limestones dip west into the Thônes Syncline at Aravis Pass. View is to the south. Photo courtesy of Guilhem Vellut, https://commons.wikimedia.org/wiki/File:L%27%C3%89tale_@_Col_des_Aravis_(51293667938).jpg

We are standing on the eastern limb of the syncline in gently northwest-dipping Cretaceous limestone. From Aravis Pass you get a view west to massive carbonates dipping west into the Thônes Syncline. Turn around and you get a stunning view back toward the Mont Blanc massif, about 30 km (18 mi) to the southeast.

View east to the Mont Blanc massif from Aravis Pass. Photo courtesy of chisloup, https://commons.wikimedia .org/wiki/File:La_massif_du_mont_blanc_vu_du_col_des_aravis_-_panoramio.jpg

Col des Aravis to St.-Jean-de-Sixt: *Continue north on D909 for 9.6 km (6.0 mi; 14 min) to* **Stop 14, Dauphinois Flysch, St.-Jean-de-Sixt** *(45.91929, 6.41549). Pull over on the left by the recycling bins just past the bend.*

STOP 14 DAUPHINOIS FLYSCH, ST.-JEAN-DE-SIXT

This stop exposes Eocene–Oligocene Dauphinois flysch near the axis of the Thônes Syncline. We are still on the southeast flank of the asymmetric syncline, which has a gentler west limb and steeper east limb, again consistent with compression directed to the northwest. The flysch sediments were deposited in a deep to shallow marine basin on the north flank of the rising Alpine mountain chain.

This roadcut consists of gently west-dipping Oligocene marl and calcareous shale interbedded with coarse, pebbly sandstone.

St.-Jean-de-Sixt to Les Villards-sur-Thones: *Continue west on D909 for 2.6 km (1.6 mi; 4 min) and pull over on the left. This is* **Stop 15, Bauges Massif, Les Villards-sur-Thones** *(45.919841, 6.387390).*

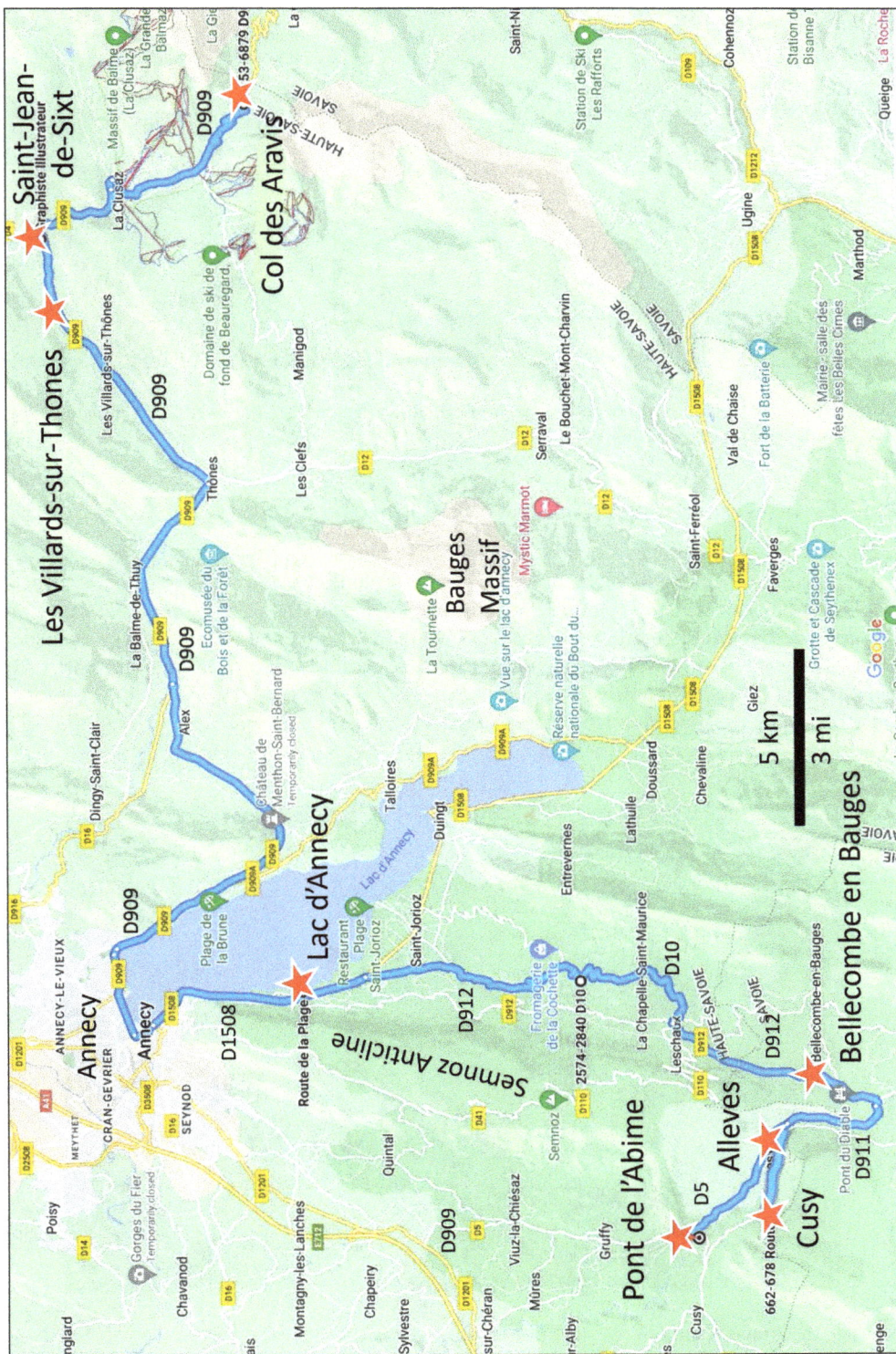

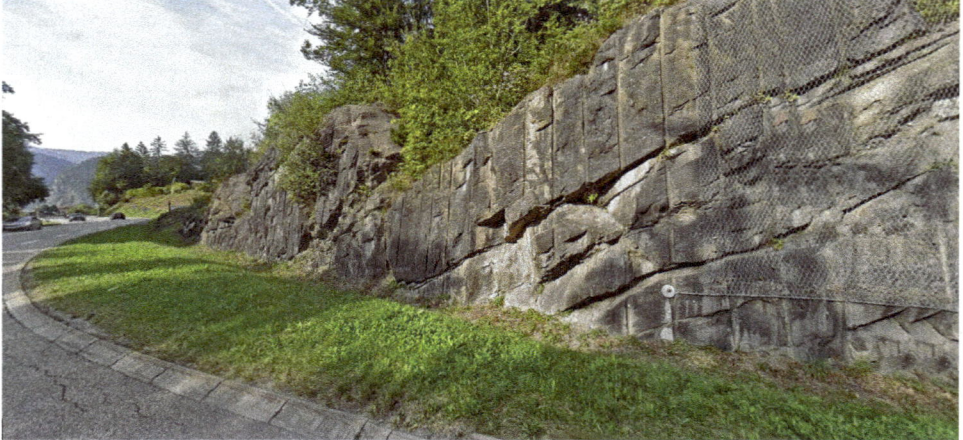

Shale-sandstone contact in the Eocene-Oligocene Dauphinois flysch near the axis of the Thônes Syncline. StreetView north, east side of St-Jean-de-Sixt.

STOP 15 PRE-ALPS AND BAUGES MASSIF FROM LES VILLARDS-SUR-THONES

This is a view stop. Looking east you see the Aravis Range of the Pre-Alps (or frontal Alps, sometimes written Pre-alps), largely made of Cretaceous limestone.

Looking southwest, the peaks in the first row are Rocher de Belchamp, Montagne de Cotagne, and Arguile de Serraval (from left to right), and behind in the second row are Roche Blanche and Dent du Cruet. All are Lower Cretaceous carbonates, mainly limestones, that are part of large regional folds such as the Semnoz Anticline.

Outcrops along the road here are Oligocene marl to calcareous shale.

View east to the Aravis Range.

View southwest to peaks of the Bauges Massif and Chaîne du Bargy.

MASSIF DES BAUGES UNESCO GLOBAL GEOPARK

Between Annecy and Chambéry we are in or skirting around the Massif des Bauges UNESCO Global Geopark. The Massif du Bauges Geopark extends from Annecy in the north to Aix-les-Bains and Chambéry on the west, to Albertville in the east, and to the Isère Valley in the south. The park encompasses over 856 km² (331 mi²) of gently folded and faulted limestones and marls deposited from the Jurassic to the Paleogene. The Alpine Orogeny thrust these units to the west, in the process forming parallel ridges and valleys, somewhat analogous to the Valley and Ridge Province in the Appalachian Mountains of North America. These are the Pre-Alps, the frontal zone of Alpine thrusting. Stark cliffs up to 1,000 m (3,280 ft) high and U-shaped glacial valleys surround the massif. The highest peaks are in the east, whereas the terrain gets gentler in the west. The park acts as a water recharge zone, with numerous karst and cavern networks and narrow canyons supplying Lac du Bourget and Lac d'Annecy, the two largest natural lakes in France. Karst is the term applied to limestone landscapes characterized by sinkholes, caverns, and fissures where the limestone has been slowly dissolved over eons by rainwater.

There are about 70,000 inhabitants living in 67 municipalities divided between the Departments of Savoie and Haute-Savoie.

The mandate of the Geopark is sustainable development of agriculture, forestry, and tourism, as well as public education and heritage preservation. Geotourism links geology not only to scenery, but also to agriculture. For example, the quality and uniqueness of the wines, apples, and pears of the region are a function of the area's soils and exposures (UNESCO, Massif des Bauges Global Geopark).

The Bauges Massif Regional Natural Park was created on 7 December 1995. The Heritage Paths system, initiated in 1997, aims to connect all the different sectors of the park by trails. In 2011 the park was designated a UNESCO Geopark, the third in France. Wildlife includes roe deer, ermine, chamois (antelope), and mouflon (mountain sheep). Popular activities include hiking, mountain biking, horseback riding, climbing, paragliding, caving, fishing, snowshoeing, and skiing.

Les Villards-sur-Thones to Lac d'Annecy: *Continue west on D909 to Annecy; at the roundabout east side of Annecy take the 3rd exit (straight); at the next roundabout take the 3rd exit (south)* *onto Chem. de Bellevue; next roundabout 2nd exit (straight); next roundabout take the 3rd exit (south) onto Rue de la Pesse; continue straight to Av. des Carrés; continue straight onto D5/Av. de Thônes; turn right (southwest) onto D5/Av. du Parmelan; at the roundabout take the 3rd exit (south) onto Rue Dupanloup; at the T intersection turn right (west) onto D909/Av. d'Abigny; continue left around the bend (south) on D1508; south of Sévrier turn left (east) on Rte du Port and left again (north) on Rte de la Plage; drive to parking lot at **Stop 16, Dauphinois Domain, Lac d'Annecy** (45.860902, 6.142935) for a total of 31.8 km (19.8 mi; 41 min).*

STOP 16 DAUPHINOIS DOMAIN, LAC D'ANNECY

We are now in what is called the Pre-Alps, or Alpine foothills. This stop is on the east limb of the Semnoz Anticline, one of the large folds developed in Jurassic and Cretaceous limestones within the Dauphinois Domain. From here there is a good view east to the mountains on the east side of Lake Annecy.

The thick Early Cretaceous platform carbonates (Urgonian Limestones) and detachment in Jurassic marls (a "sticky detachment") tend to generate box folds with abrupt hingelines. This can be seen in the folds developed in the Bornes Massif to the east, and will be seen again in the natural cross section through the Semnoz Anticline at Alleves.

Alpine shortening in the Helvetic-Dauphinois Domain led to northeast-oriented thrusted folds that involved Cretaceous limestones and sandstones, Eocene limestones and marls, and Oligocene Val d'Illiez Sandstones (flysch) in the Bornes Massif. In this case the massif does not expose basement crystalline (igneous) rocks, although the thrusts are thought to root in basement. The massif is an internally consistent structural package that has formed a large folded mountain group. The Bornes folds are thrust over the Bornes molasse along the Sub-alpine Frontal Thrust, located at the western front of the westernmost anticlines (Berio et al., 2021).

This region is scenic as well as rich in geology. In 1891 Mark Twain traveled from Aix-les-Bains to Lake Annecy and wrote:

> [T]he excursion which satisfied me best was a trip to Annecy and its neighborhood. You go to Annecy in an hour by rail, through a garden land that has not had its equal for beauty, perhaps, since Eden; and certainly Eden was not cultivated as this garden is. The charm and loveliness of the whole region are bewildering. Picturesque rocks, forest-clothed hills, slopes richly bright in the cleanest and greenest grass, fields of grain without fleck or flaw, dainty of color, and as shiny and shimmery as silk, old gray mansions and towers half buried in foliage and sunny eminences, deep chasms with precipitous wars, and a swift stream of pale blue water between, with now and then a tumbling cascade, and always noble mountains in view, with vagrant white clouds curling about their summits.
>
> Then at the end of an hour you come to Annecy and rattle through its old crooked lanes, built solidly up with curious old houses that are a dream of the middle ages, and presently you come to the main object of your trip-- Lake Annecy. It is a revelation, it is a miracle. It brings the tears to a body's eyes it is so enchanting. That is to say, it affects you just as all things that you instantly recognize as perfect affect you--perfect music, perfect eloquence, perfect art, perfect joy, perfect grief. It stretches itself out there in the caressing sunlight, and away towards its border of majestic mountains, a crisped and radiant plain of water of the divinest blue that can be imagined. All the blues are there, from the faintest shoal water suggestion of the color, detectable only in the shadow of some overhanging object, all the way through, a little blue and a little bluer still, and again a shade bluer till you strike the deep, rich Mediterranean splendor which breaks the heart in your bosom, it is so beautiful.

Lac d'Annecy to Bellecombe: *Return to D1508 and turn right (north); at the roundabout take the 2nd exit (west) onto D912/Rte du Col de Leschaux; continue south on D912 to D61A; turn left (east) onto D61A and immediately park in the parking area on the left. This is **Stop 17, Grand Colombier, Bellecombe en Bauges** (45.7309, 6.1093) for a total of 17.8 km (11.1 mi; 22 min).*

View east across Lac d'Annecy to large folds of the Bornes Massif. The anticline is overturned to the west (toward the viewer).

Panoramic view west from Lac d'Annecy to the east flank Semnoz Anticline. Dashed lines follow bedding.

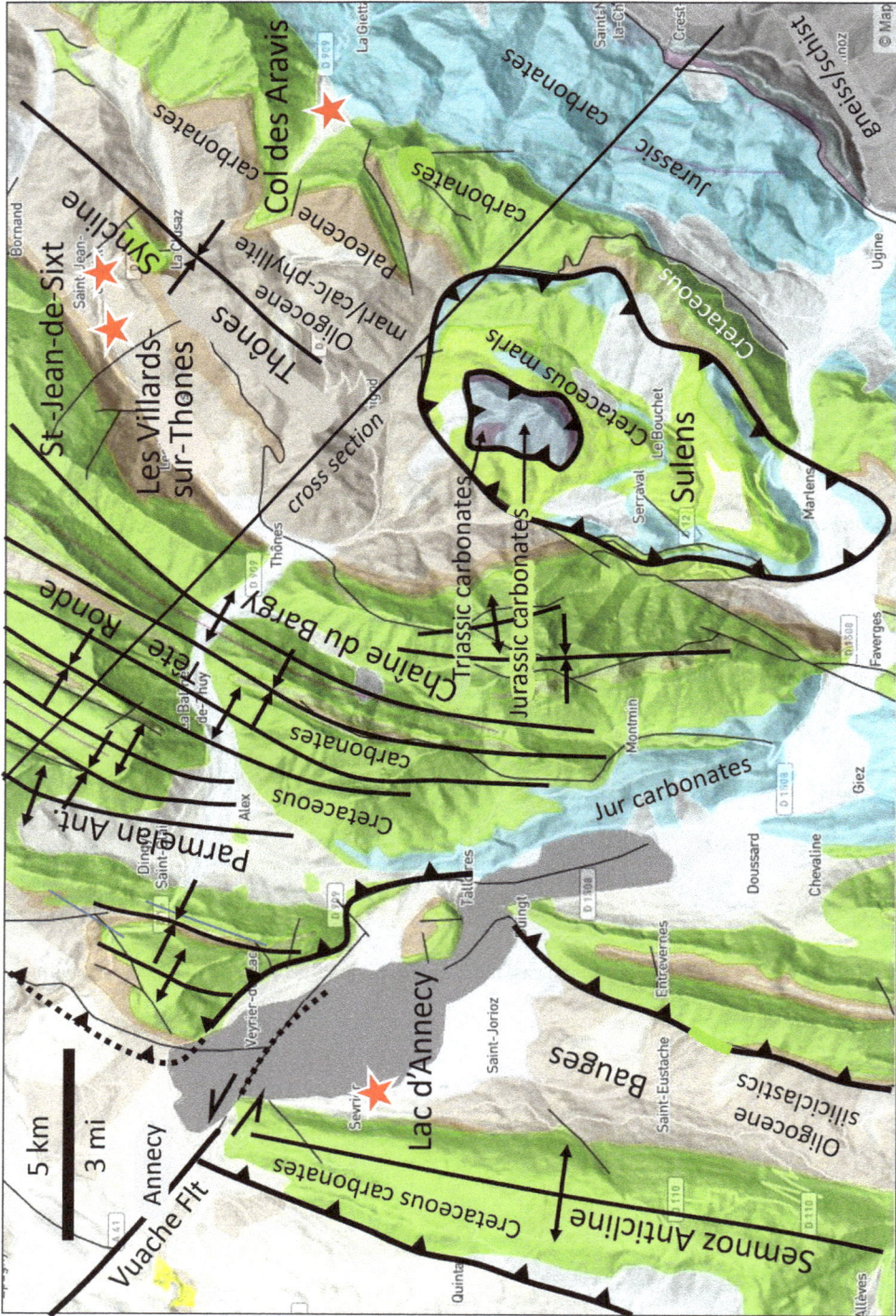

Geologic map of the Col des Aravis–St. Jean de Sixt–Lac d'Annecy area. Heavy lines are strike-slip faults; heavy lines with teeth are thrust faults. Folds with arrows pointing outward are anticlines; folds with arrows pointing inward are synclines. Geology base from MacroStrat, 2015.

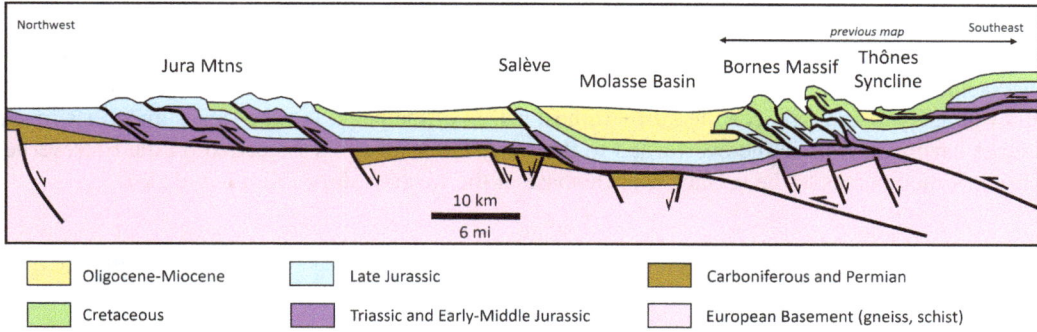

Northwest-southeast cross section from the Jura to the pre-Alps Bornes Massif to Aravis Pass. Modified after Berio et al., 2021. Line of section is shown on the previous map.

Stratigraphy of the pre-Alps in France. Stratigraphic information derived from Berio, et al., 2021.

STOP 17 GRAND COLOMBIER AND ENTREVERNES THRUST, BELLECOMBE EN BAUGES

The Grand Colombier (1,534 m, or 5,033 ft) is one of the southernmost peaks of the Jura Massif. Its spectacular relief is the result of Alpine thrusting that shaped the Jura into successive folds. From this vantage you can only see the northernmost ridges of the peak, but the ridges and cliffs reveal a large asymmetric anticline overturned to the west. The entire fold is carried on the Entrevernes Thrust, which comes to the surface near the base of the western slope (Kalifi et al., 2021).

View southwest to the Massif du Grand Colombier. Dashed lines follow bedding.

Bellecombe en Bauges to Cusy: *Continue south on D910; slight right onto D911 (sign for A41/ Cusy/Aix-les-Bains); at 7.3 km (4.6 mi; 8 min) turn into the parking area on the right. This is **Stop 18.1, Semnoz Anticline, Cusy** (45.7543, 6.0611).*

STOP 18 SEMNOZ ANTICLINE, ALLÈVES AREA

These stops are along a natural cross section through the Semnoz Anticline. Massive Cretaceous limestone defines the flank of the structure. Semnoz Anticline is a frontal fold of the Pre-Alps, and there is a thrust near the base of the west limb at the contact with molasse basin fill. This is the anticline we saw looking west from Lac d'Annecy. The entire fold is about 33 km (27 mi) long.

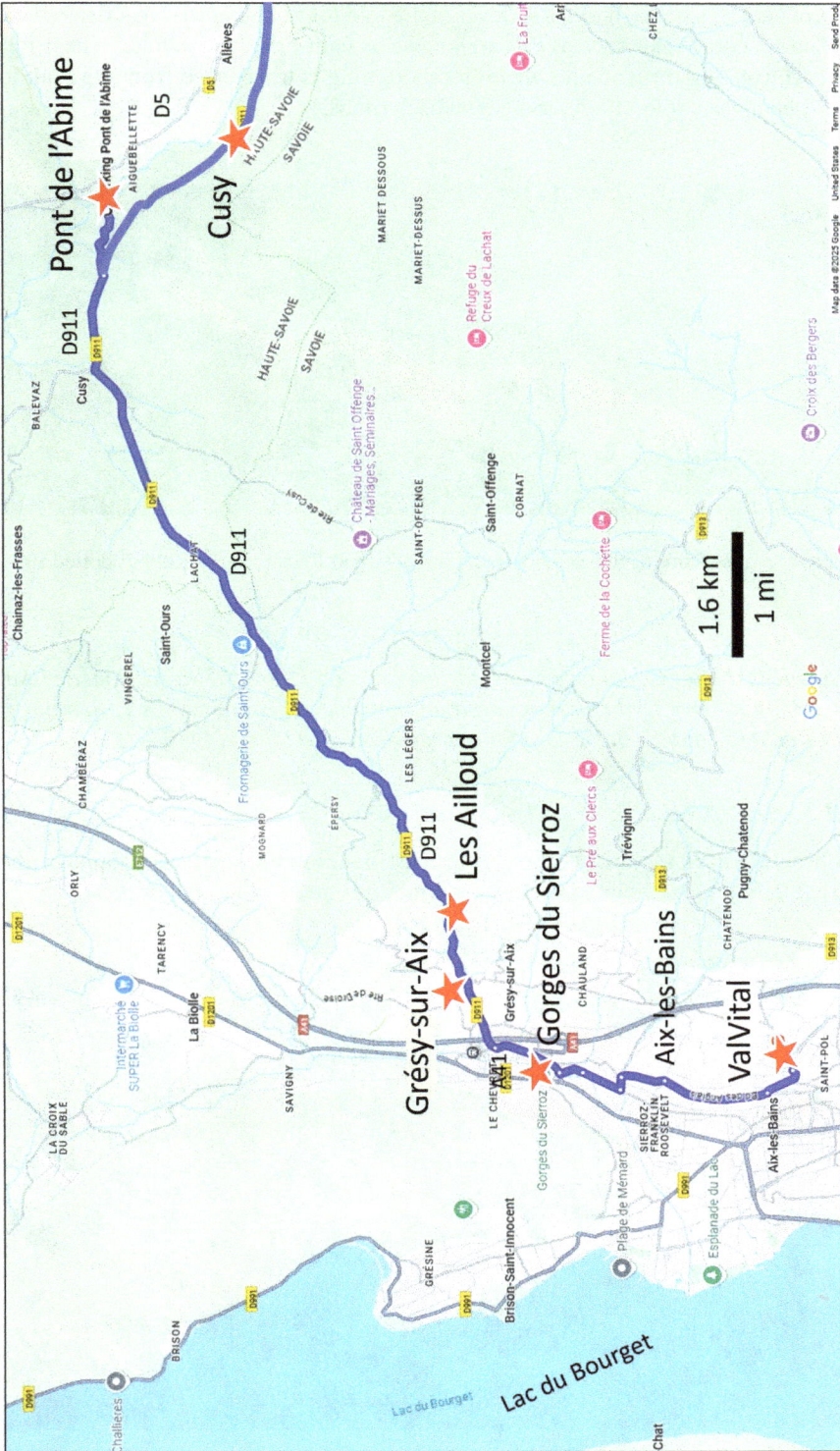

Stop 18.1 Semnoz Anticline, Cusy

This stop provides a view north to the rollover at the fold hinge and of the west flank of the Semnoz Anticline. Thrust-cored folds such as this are known as fault-propagation folds. These folds form at the tip of a propagating thrust fault, where the shortening is transferred from the fault to the fold, resulting in steep, narrow forelimbs and gentle back limbs.

View north to the west flank of the Semnoz Anticline. Dashed lines follow bedding; dotted line is a thrust fault.

*Cusy to Pont de l'Abîme: Continue driving northwest on D911 to D31/Rte du Pont de l'Abîme; hard right onto D31 and drive to the parking area on the right. This is **Stop 18.2, Semnoz Anticline, Pont de l'Abîme** (45.763691, 6.055487) for a total of 2.7 km (1.7 mi; 4 min).*

Stop 18.2 Semnoz Anticline, Pont de l'Abîme

This stop provides a view north at the gentler east flank and steeper west flank of the Semnoz Anticline, again highlighted by the Cretaceous limestone at the top of the ridge.

View south at the core of the Semnoz Anticline. Cretaceous limestone holds up the ridge. Dashed lines follow bedding; dotted line indicates a fault.

*Pont de l'Abîme to Les Ailloud: Continue west on D911/Rte des Bauges and drive to **Stop 19.1, Upper Marine Molasse, Les Ailloud** (45.727115, 5.942270) pullout on the left beside the river, for a total of 10.6 km (6.6 mi; 13 min).*

STOP 19 THE MOLASSE BASIN

We have just crossed into the Molasse Basin. The Molasse Basin lies north and west of the Helvetic-Dauphinois Domain. The basin is filled with sediments derived from the uplifting Alps to the south and east. In North America "molasse" would be called "syntectonic sediments" and the basin type would be a "foreland basin." Syntectonic just means these sediments were being deposited as the mountains were undergoing uplift and erosion. Foreland basin means a subsiding area in front of the fold-thrust belt. These sediments rest on Mesozoic European marine carbonate-platform limestones. The northwestern erosional margin of the basin is bordered by the Jura Fold-Thrust Belt. Isolated remnants of the Molasse Basin are preserved in synclines of the folded Jura (Schlunegger and Mosar, 2010).

The Molasse Basin is the type locality of "molasse," a sedimentary sequence of conglomerates and sandstones that were eroded from the rising mountain chain (Wikipedia, Molasse Basin). A type locality is where a rock sequence was originally defined.

Before the development of the Molasse Basin, during the Mesozoic era, the region was covered by a shallow sea that was part of the northern Tethys Ocean (the Valais Ocean in the west, the Tethys proper in the east). A major break in sedimentation (unconformity) occurred in this area from the Late Cretaceous to Eocene or Oligocene due to a foreland bulge, that is, a foreland or peripheral uplift in front of the Alpine orogenic stack (Stampfli, 2001). This uplift caused non-deposition instead of erosion and sedimentation.

The Molasse Basin formed when the actively uplifting Alps mountain range was pushed north over the European continental margin due to convergence of the European and Apulian plates during the Paleogene. The weight of the Alps bent the European Plate downward, forming a deep marine trough in which first deepwater flysch, and later molasse sediments were deposited. The huge amounts of sediments eroded from the forming mountain chain filled the basin during Oligocene and Miocene time (between 10 and 30 Ma).

The molasse deposits in the basin are up to 6 km (19,700 ft) thick. These rocks are divided into four formations based on whether they are continental or marine sediments (Wikipedia, Molasse Basin).

- The lowermost formation is the Lower Marine Molasse (early Oligocene, 34 to 28 Ma). It consists of shallow marine sand, clay, and marl.
- Above this is the Lower Freshwater Molasse (late Oligocen–early Miocene, 28 to 22 Ma) consisting of river sands and clays and alluvial fan conglomerates and breccias originating in the rising Alps.
- The Upper Marine Molasse was deposited in a shallow marine environment that extended from Lyon to Vienna. It consists of early Miocene (22 to 16 Ma) marine sands, clays, marls, and fan conglomerates.
- The Upper Freshwater Molasse consists of continental river sands, clays, and alluvial fan conglomerates deposited during late Alpine uplift that caused the sea to retreat one final time (late Miocene, 16 to 5 Ma).

Between 10 and 5 million years ago the Molasse Basin had itself become involved in the uplift and deformation, and sedimentation ended.

The southern and eastern parts of the Molasse Basin (the Subalpine Molasse) were strongly affected by Alpine deformation; the northern and western parts less so. The molasse deposits were overthrust about 10 km (6 mi) by the Helvetic Nappes, which caused deformation in the Subalpine

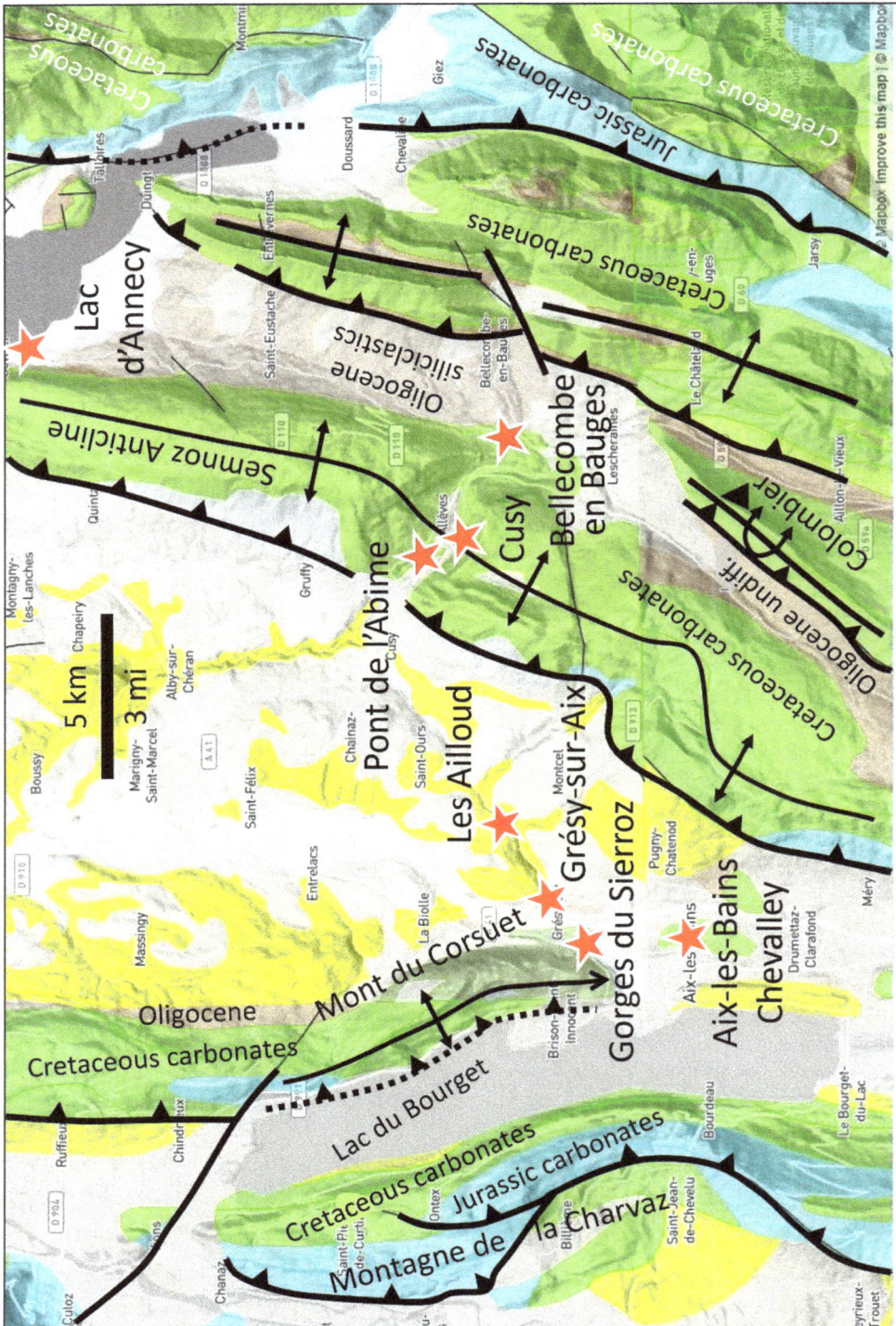

Geologic map of the Lac d'Annecy–Alleves–Aix-les-Bains area. Heavy lines are normal faults; heavy lines with teeth are thrust faults. Folds with arrows pointing outward are anticlines; folds with arrows pointing inward are synclines. Geology base from MacroStrat, 2015.

Molasse. The Swiss part of the Molasse Basin, between the Alps and the Jura, is a large piggyback basin carried on deep thrusts (Wikipedia, Molasse Basin). The Molasse Basin extended over a large part of the Jura before it was folded (Stampfli, 2001).

STOP 19.1 UPPER MARINE MOLASSE SANDSTONE, LES AILLOUD

The outcrop at this stop exposes east-dipping Miocene Upper Marine Molasse sandstone. At this location the unit is a massive, fine-to-coarse-grained and clean, so probably a lower shoreface sandstone (beach sand below the fair weather wave base).

Miocene sandstone of the Upper Molasse unit at Les Ailloud.

Les Ailloud to Grésy-sur-Aix: *Continue west on D911/Rte des Bauges for 0.85 km (0.5 mi; 1 min) and pull over on the gravel parking area on the right. Walk back east to outcrops on the south side of D911. This is **Stop 19.2, Lower Marine Molasse, Grésy-sur-Aix** (45.724920, 5.932591). This outcrop is rapidly being overgrown by vines.*

STOP 19.2 LOWER MARINE MOLASSE SANDSTONE, GRÉSY-SUR-AIX

We are on the east flank of the Mont du Corsuet Anticline, formed during the last stages of the Alpine Orogeny. This stop provides an example of east-dipping Oligocene Lower Marine Molasse sandstone. The shallow marine unit is mostly fine-to-medium-grained buff sandstone, perhaps a shoreface (beach) sand deposited in the foreland basin west of the uplifting Alps.

East-dipping Oligocene sandstone of the Lower Molasse at Grésy-sur-Aix.

Grésy-sur-Aix to Gorges du Sierroz: *Continue south on D911 1.8 km (1.1 mi; 3 min) to **Stop 20.1,** **Gorges du Sierroz** (45.715491, 5.918751) parking area on the right. Walk up the gorge for about 1 km.*

Stop 20 Aix-les-Bains Area

Aix-les-Bains is located on the eastern shore of the largest natural glacial lake in France, the Lac du Bourget. It is a spa town, with a population of just under 30,000 in 2018. Although the thermal baths are no longer the main attraction, they have compensated by developing other forms of tourism (Wikipedia, Aix-les-Bains).

Aix-les-Bains lies between the Jura Mountains and the Sub-alpine Bauges Massif and is at the southwest end of the Molasse Basin that extends from Chambéry (France) to Linz (Austria). The rocks exposed here are mostly Jurassic and Cretaceous limestone. The limestones are karsted (contain caverns) and act as groundwater aquifers, that is, the rock layers that contain groundwater. The limestone units are carried on thrust faults, some of which follow evaporite zones, and these fault zones act both as groundwater conduits and as sources of minerals in the groundwater. The thrust faults have generated a series of anticlines, including the Montagne de la Charvaz and Mont du Corsuet, that are thrust over synclines such as the Lake Bourget Syncline (Gallino et al., 2009). Tectonic activity continues today, as evidenced by field observations, geodetic surveys, and historical seismicity. As much as 1 to 2 mm/yr (0.04 to 0.08 in/yr) of vertical displacement has been measured on several structures (Chapron et al., 1996).

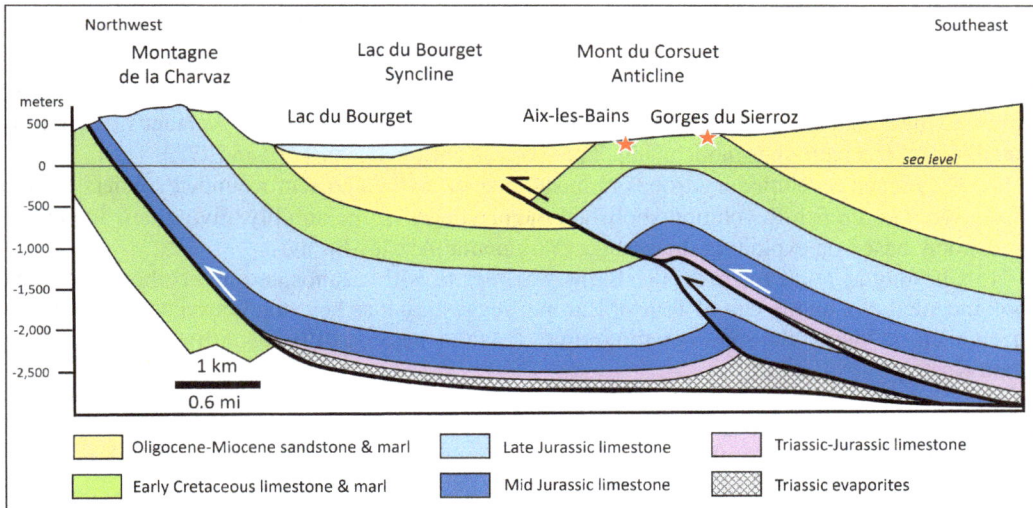

Cross section through Aix-les-Bains showing relative positions of the Gorges du Sierroz and Aix-les-Bains stops. Modified after Gallino et al., 2009.

Aix-les-Bains was and still is famous for the thermal springs that occur along the eastern shore of Lake Bourget. The recharge zone for the thermal springs is in Late Jurassic to Early Cretaceous limestones on the Montagne de la Charvaz west of Lake Bourget. Infiltrating water, locked in the limestone by overlying impermeable marls, is carried to depths of 2,200 m (7,200 ft) in the Lake Bourget Syncline. Warmed at this depth, the water then follows the thrust fault that carries the Aix-les-Bains Anticline and acquires chlorine, sulfate, sodium, and potassium ions. The mineralized water encounters vertical fractures below Aix-les-Bains and moves upward through the fractures to the surface where it then mixes with surface water (Gallino et al., 2009). The Grotte des Serpents is a horizontal cavern that formed at the water table. Grotte des Serpents contains Alun Spring, the main natural discharge of this hot water system.

The Urgonian Limestone is where thermal waters rising through the anticline mix with waters percolating down from the surface. Dye-tracing tests have shown that there is one main ascending hot water conduit. Circulation of 40°C (104°F), sulfur-rich water has dissolved limestone and led to the development of caverns along bedding planes, enlarged vertical fractures, and precipitation of gypsum and sulfur (Hoblea et al., 2010). The final emerging water temperature is a function of the rate of mixing of surface with subsurface waters (Gallino et al., 2009).

The Romans established the town of Aquae here in 131 BCE when they discovered a series of thermal springs gushing from the foothills (MacEacheran, 2019; Wikipedia, Aix-les-Bains). Archaeologists have found a large thermal complex close to its source. On a lower terrace, to the west, they built the Arch of Campanus, probably in the 1st century and, farther downstream, a second terrace held the Temple of Diana.

The Roman baths fell into ruin after being abandoned in the 5th century. Aix is a corruption of the Latin Aquae. Aix-les-Bains is mentioned in royal charters in 867 and in 1011. In the latter, King Rudolph III of Burgundy donated the villa of Aix, a royal seat, with its settlers and its slaves to his wife Ermengarde, who, in turn, passed them to the Bishopric of Grenoble. Bishop Hugues of Grenoble then donated Aix to the monastery of Saint-Martin de Miserere at the beginning of the 12th century.

The writings of the Dauphinois physician Jean Baptiste Cabias at the beginning of the 17th century promoted the medicinal value of the hot springs of Aix. The Duke of Chablais, son of King Victor Amadeus III, having experienced the benefit of the springs and finding them unattractive, suggested the construction of a thermal spa to the king. In 1776 King Victor Amadeus III commissioned plans for a bathing establishment. The spa was built between 1779 and 1783.

In 1792 French revolutionary troops requisitioned the baths and sent wounded soldiers there to convalesce. The French Revolution abolished the privileges of the nobility, giving new impetus to an economy based on exploiting the springs (Wikipedia, Aix-les-Bains).

In its heyday as a spa town, Aix-lesBains was replete with casinos and luxury hotels. Thomas Cook included Aix on his Grand Tour of Europe, suggesting it as key stop between Paris and Italy. Queen Victoria visited the spa town three times (MacEacheran, 2019). Regarding the baths, Mark Twain wrote

> I had the rheumatism and was advised to go to Aix, not so much because I had that disease as because I had the promise of certain others. What they were was not explained to me, but they are either in the following menu or I have been sent to the wrong place. Dr. Wakefield's book says:
> "We know that the class of maladies benefited by the water and baths at Aix are those due to defect of nutrition, debility of the nervous system, or to a gouty, rheumatic, herpetic, or scrofulous diathesis--all diseases extremely debilitating and requiring a tonic, and not a depressing action of the remedy. This it seems to find here, as recorded experience and daily action can testify.... According to the line of treatment, followed particularly with due regard to the temperature, the action of the Aix waters can be made sedative, exciting, derivative, or alterative and tonic."
> ... I began to take the baths, and found them most enjoyable; so enjoyable that if I hadn't had a disease I would have borrowed one, just to have a pretext for going on. They took me into a stone-floored basin about fourteen feet square, which had enough strange-looking pipes and things in it to make it look like a torture chamber. The two half-naked men seated me on a pine stool, and kept a couple of warm-water jets as thick as one's wrist playing upon me while they kneaded me, stroked me, twisted me, and applied all the other details of the scientific massage to me for seven or eight minutes. Then they stood me up and played a powerful jet upon me all around for another minute. The cool shower bath came next, and the thing was over. I came out of the bath-house a few minutes later feeling younger and fresher and finer than I have felt since I was a boy (Twain, 1891).

A multi-year project is underway to restore the city's ancient thermal baths. The city's historic ruins are soon to reopen to the public after having been closed for decades. Until then, check out the Le Soufre sulfur spring and the original Roman thermal baths (Vestiges Romains des Thermes d'Aix-les-Bains) in the official tourist office on Place Maurice Mollard (MacEacheran, 2019).

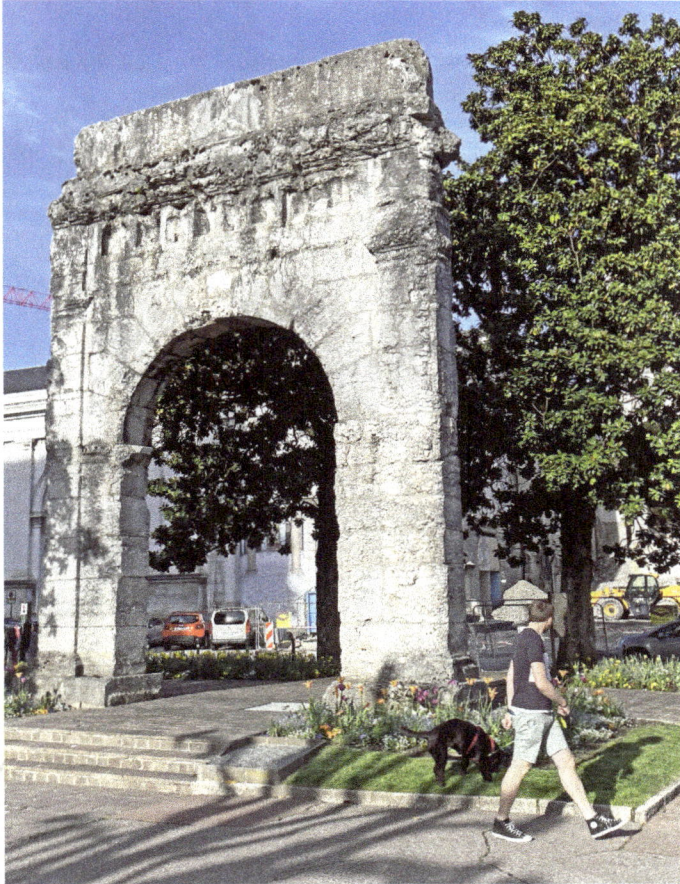

The 9.15 m (30 ft) high Arc de Campanus, Place Maurice Mollard, sits next to the Temple of Diana. It was built of limestone blocks (probably Cretaceous) placed without mortar.

Aix-les-Bains is nestled on the east shore of Lac du Bourget and has a spectacular backdrop of mountains on all sides. These are the Chartreuse Mountains to the south. The prominent peaks are all erosion-resistant Cretaceous limestone carried on west-directed thrust faults.

Stop 20.1 Gorges du Sierroz, Aix-les-Bains

This stop is in outcrops of light gray Early Cretaceous limestone on the east flank of the north-trending and south-plunging, thrust-cored Mont du Corsuet Anticline.

The question is often asked, "why is this gorge here?" A study of the "*Hautes Gorges du Verdon*" in the foreland of the southwestern Alps used Cosmic Ray Exposure dating to determine stream downcutting rates. The study found that downcutting is a function of tectonic or isostatic uplift during the Late Quaternary (Cardinal et al., 2024). In other words, the stream was probably already there, perhaps flowing over a gentle surface. The land gradually began to uplift, either by renewed mountain building or by rebound after the Ice Age glaciers melted, and downcutting by these "antecedent rivers" kept pace, causing deep and narrow gorges in the resistant limestone.

From the car park a path takes you along a narrow gorge eroded into Cretaceous limestone by the Sierroz River. A walk of roughly 200 m (660 ft) brings you to the Grésy Cascade. This park had been closed to the public for 40 years, but it reopened in 2021. It is a quiet, shady, cool, and lush green urban park, a respite from the hustle and bustle of the nearby city. It is not surprising that the park is a "classified natural site" intended to promote public access to nature as well as local history (Wikipedia, Gorges du Sierroz).

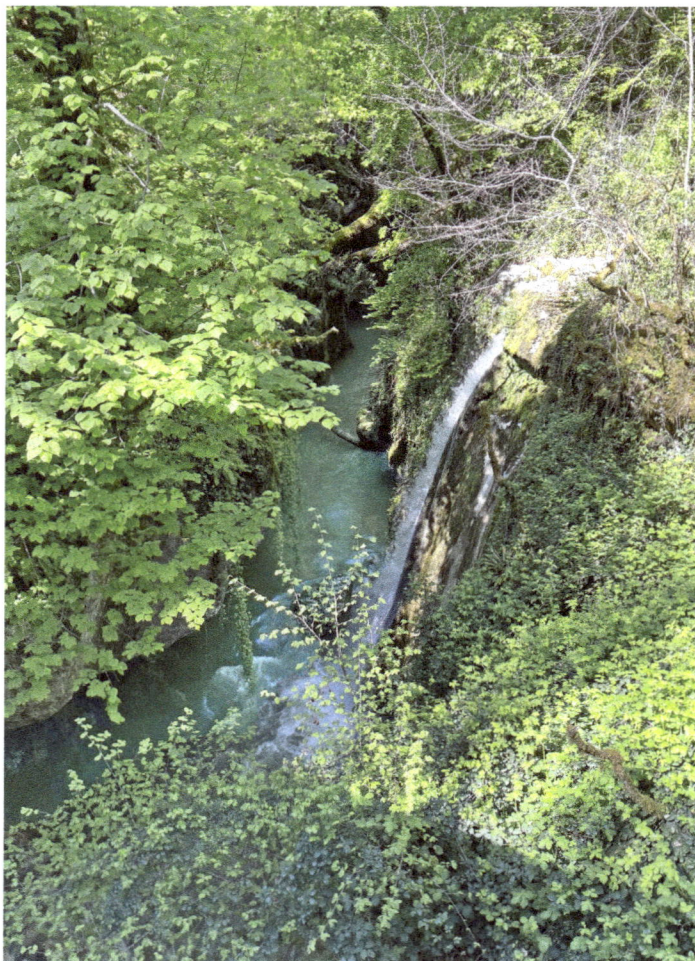

The Grésy cascade plunges 30 m (100 ft) into the Gorges du Sierroz.

*Gorges du Sierroz to Aix-les-Bains: Exit the roundabout onto the Rte de Pont Pierre (south); continue straight onto Av de St-Simond; turn left (east) onto Chem de la Baye; at the roundabout take the 1st exit (south) onto Chem des Prés de la Tour; at Mnt Rabut cross the intersection and bear slightly right onto Bd des Anglais; at the Y turn left (south) onto D913/Bd des Côtes; continue on D913 to **Stop 20.2, Chevalley Aix-les-Bains** (45.689966, 5.919348) parking on the right for a total of 3.4 km (2.1 mi; 7 min).*

STOP 20.2 CHEVALLEY AIX-LES-BAINS, VALVITAL THERMES AND SPA

An 1,800 m (5,900 ft) deep borehole taps into the source of the hot springs to provide the spa with water at 34°C (93°F). It is then provided to indoor and outdoor pools, saunas, and steam rooms, and it is used for swimming, hydrotherapy, and scrubs. Both French and English are spoken. The facility provides towels; you bring your bathing suit, bathing cap, and sandals.

VISIT

There is a large pay parking garage.

Spa Entry (2025) for adults: €24 to €26.
Hours: Daily 10:00 to 19:00
Address: 10 route du Revard73100 Aix-les-Bains
Phone: 04 79 35 68 66
Website: https://www.aixlesbains-rivieradesalpes.com/en/equipement/chevalley-thermal-baths
                 -aix-les-bains/

The Thermes Chevalley/ValVital Spa.

Side Trip 7, Aix-les-Bains to Vins de Savoie: *Continue east and south on D913; continue straight onto Chem. de la Croix du Chenoz; continue straight on Chem. des Bugnards; continue straight on Chem des Grands Champs; continue straight on Rte de Biolay; at the roundabout take the 3rd exit (south) onto D17A/Pl. de l'Horloge; stay on D17 to D127; at the roundabout take the 4th exit (southeast) onto D127/Rte de l'Échargeur; follow the signs for A41/Lyon/Grenoble and merge onto the A41/E712 south (toll road); take the exit to Grenoble/Albertville/Chambéry and merge onto N201 south; take Exit 20 to D9/Rte de Saint-Baldoph; at the roundabout take the 2nd exit (west) to D9/St Baldoph/Apremont; continue onto D201/Baldoph/Apremont/Myans; turn left (south) onto D201/Rte d'Apremont; continue south on D201; at the Apremont roundabout continue straight (south) onto D12; from Apremont drive south on D12 ~ 1.1 km (0.7 mi) to Domaine Adrien Veyron et Fils, Route de Saint-André, Lieu-dit La Ratte, 73190 Apremont on the left. This is **Side Trip 7, Vins de Savoie, Chambéry** (45.504269, 5.964658) for a total of 26.0 km (16.2 mi; 30 min) one-way.*

If you prefer not to go on Side Trip 7:

Aix-les-Bains to Belvédère du Col de l'Épine: *Continue east and south on D913; continue straight onto Chem. de la Croix du Chenoz; continue straight on Chem. des Bugnards; continue straight on Chem des Grands Champs; continue straight on Rte de Biolay; at the roundabout take the 3rd exit (south) onto D17A/Pl. de l'Horloge; stay on D17 to D127; at the roundabout take the 4th exit (southeast) onto D127/Rte de l'Échargeur; follow the signs for A41/Lyon/Grenoble and merge onto* *the A41/E712 south (toll road); take the E70/E712 exit toward A43/Turin/Grenoble; follow the signs to D16A/Rue du Nant Bruyant; take Exit 14 to La Motte-Servolex; merge onto D16A/Rue du Nant Bruyant; at the roundabout take the 1st exit (north) onto D1/Av Costa de Beauregard toward La Motte-Servolex Centre; at the roundabout take the 3rd exit (west) onto D1/Av Jean Marie Michellier; at the roundabout take the 4th exit (south) to D15/Av Jean Moulin; at the roundabout take the 1st exit (west) onto D3/Rte de l'Épine; continue on Rte de l'Épine to D916; take a hard right (north) on D916 to Col de l'Épine and drive to **Stop 21, Belvédère du Col de l'Épine** (45.581536, 5.825602); pull into the picnic area/viewpoint on the left for a total of 25.6 km (15.9 mi; 36 min).*

SIDE TRIP 7 VINS DE SAVOIE, CHAMBÉRY AREA

The French Department of Savoie was originally inhabited by the Gallic Allobroges, then became part of Gallia Transalpina, the first Roman province north of the Alps. The Allobroges had been cultivating *Vitis allobrogica* long before Roman occupation. The plant yielded grapes that matured before the frost and were able to withstand the alpine climate. Long after the fall of Rome, Savoie was still part of the Kingdom of Italy. The region became part of France following the Treaty of Turin in 1860 (Kuderski, 2020).

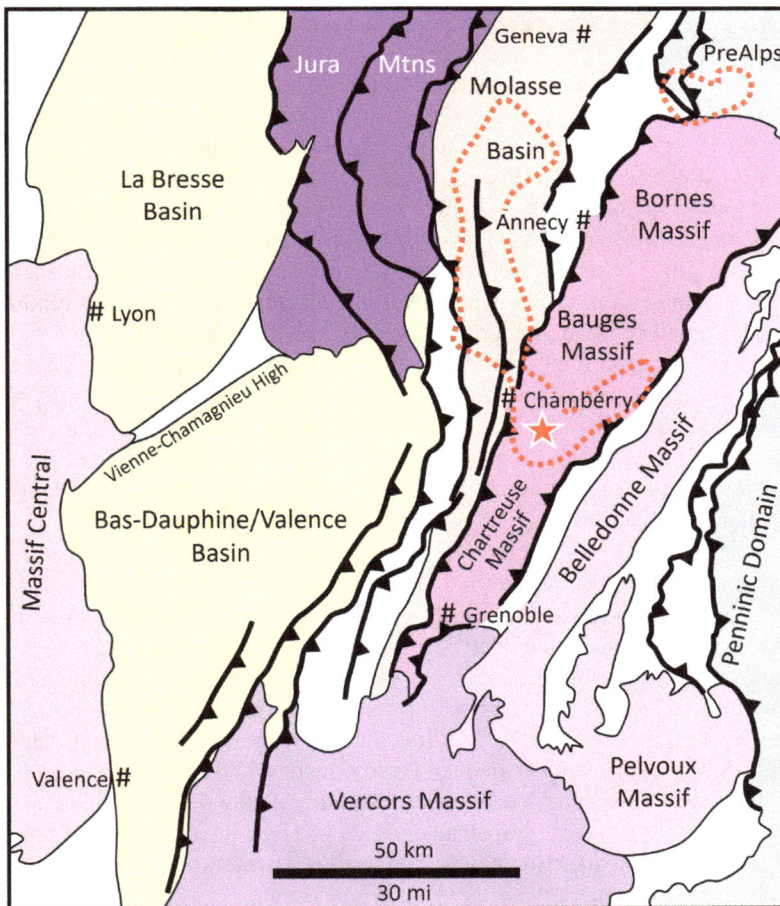

The primary geologic features in the Geneva to Valence area. Dotted line encloses the Savoie wine region; the star is Domaine Adrien Veyron et Fils. Modified after Kalifi et al., 2021.

Domaine Veyron is located at the base of Mont-Granier, an imposing Cretaceous limestone cliff at the north end of the Chartreuse Massif. Dominating the scenery at an altitude of 1,933 m (6,342 ft), Mont-Granier was the scene of a massive rock slide on the night of 24 November 1248. A huge section of the mountain broke loose and covered five towns, killing thousands. Much of the scree has since been cultivated, and some large scattered blocks are still visible today.

Wine Varieties

Most Savoie vineyards produce dry white wines. The wines are often described as being distinctly "alpine," with fresh, mineral flavors. The Vin de Savoie AOC (*Appellation d'Origine Contrôlée*) was created in 1973, and covers 18 km² (6.9 mi²) and 28 communities in Savoie, 20 in Haute-Savoie, 2 in Ain, and 1 in Isère. Wines labeled Savoie or Vin de Savoie may be white, red, rosé, and sparkling (WineTourism.com, Vin de Savoie; WineSearcher.com – Vins de Savoie, Terroir).

A Vin de Savoie label may include the name of one of 16 crus, or *dénominations géographiques* named for villages. For example, around Chambéry are:

- Les Abymes, Apremont, St-Jeoire-Prieuré: white only, at least 80% Jacquère grape
- Chignin: white, at least 80% Jacquère; red and rosé, at least 90% Gamay, Mondeuse, or Pinot Noir
- Chignin-Bergeron: white only, made entirely from Bergeron (the local name for the Roussanne grape; WineSearcher.com, Vins de Savoie, Terroir).

Savoie wines are special because many of them are made with old grape varieties native to Savoie. Although this wine-growing region is the second smallest in France (after the Jura), it boasts the greatest diversity of grape varieties. The main white wine varieties include Altesse, Jacquère, Chasselas, Gringet, Roussanne, and Chardonnay.

Altesse, also known as Roussette, is a white grape variety grown exclusively in Savoie and the Bugey region. Duke Amédée II of Savoy is said to have brought this grape back from Cyprus in 1432 for the wedding of his son to the daughter of the King of Cyprus, hence the name Altesse ("highness"), chosen in her honor. The vines cover ~250 ha (618 ac). The wines produced from this iconic grape are rich and complex, with delicate fruity and floral bouquets, and they can also have citrus overtones (Veyron, email, 16 April 2025). When young the flavors range from fresh almonds and bergamot to pineapple, peach, and quince. With age, the wines develop aromas of honey, toast, nuts, and white truffle. It is said that these wines should be set aside for at least three years to allow their flavors to develop (Kuderski, 2020). Its minerality can vary depending on the soil type. For example, the "Monthoux" Cru is grown on rocky soil; therefore, it is strongly mineralized; the same is true for the Jongieux area, the other birthplace of the Altesse grape.

"Altesse produces fresh and aromatic wines with notes of dried fruit, fresh and delicate in its youth," according to Franck Berkulès of the Interprofessional Committee of Savoie Wines. Unusual for white wines, "It has the advantage of aging well for over twenty years. On the other hand, it is a very fragile grape variety, sensitive to heat." Demanding to produce, Roussette de Savoie is one of the essential grapes of this mountain terroir, unique in the French wine-growing landscape.

Jacquère, another native grape, covers 50% of Savoyard vineyards (~1,100 ha; 2,720 ac) and is the most widespread grape variety in Savoie. It is known by many local names: Abymes, Chignin, Saint-Jeoire du Prieuré, Cruet, Saint-Jean de la Porte. The best known is Apremont. It produces low alcohol, lively dry wines. Flavors have been described as floral (hawthorn, jasmine, and honeysuckle) and fruity (apple, pear, white peach, grapefruit). Highly mineral, it can have notes of gunflint.

Jacquère is used in still or sparkling wines. The resulting wines are renowned for their lightness, freshness, and liveliness. Jacquère wines are best drunk young.

Chasselas is a native grape variety found only in Haute-Savoie. It covers only 100 ha (247 ac).

Gringet is a native variety covering 40 ha (100 ac) in the commune of Ayse.

Roussanne (called "Bergeron" in Savoie) is planted only in the communes of Chignin, Francin, and Montmélian. It covers an area of ~80 ha (198 ac).

Chardonnay covers an area of ~ 50 ha (124 ac).

The main red varieties include Gamay Noir, Pinot Noir, Persan, and Mondeuse Noire.

Gamay Noir covers ~ 300 ha (740 ac) and is a light-bodied red wine similar in taste to Pinot Noir. This variety grows primarily in the Beaujolais region just west of Haute-Savoie. Gamay wines are known for their delicate floral aromas and subtle earthy notes. Gamay makes everything from light and easy *nouveau* styles (wines released the same year the grapes are picked) to beautifully complex medium-bodied wines. High-quality Gamay wines are usually more affordable than Pinot Noirs.

Pinot Noir covers ~50 ha (124 ac) in Savoie. Pinot Noir is a red-wine grape grown mostly in cooler climates. This grape is also used to make champagne and sparkling white wines. Pinot noir is a difficult variety to cultivate and transform into wine, as the grapes grow in tightly packed clusters that make them susceptible to rot, thus requiring diligent canopy management. The thin skins and low levels of phenolic compounds cause pinot to produce mostly light-colored, medium-bodied, and low-tannin wines. When young, these wines tend to have aromas of cherries, raspberries, and straw-berries. As the wine ages, Pinot has the potential to develop more vegetal and "barnyard" aromas that can contribute to the complexity of the wine.

The native Persan grape produces a red wine. The grape, grown primarily in the Savoie region, may be varietal or blended with other local grapes such as Mondeuse Noire. The variety produces ageworthy red wines with a strong aromatic profile and attractive raspberry flavors.

Mondeuse Noire is a native grape variety that is becoming more popular. It covers ~300 ha (740 ac). Mondeuse is a red grape that produces wines that are full-bodied with noticeable but subtle tan-nins. They exhibit aromas of blackberry and blueberry, sometimes with spicy, floral, and slightly peppery notes (Veyron, email 16 April 2025).

In addition to its wine, Savoie is famous for vermouth. Vermouth became popular starting in Turin in 1786. Vermouth is an aromatic mix that includes wine, botanicals, some sugar (or grape juice), and spirits to fortify the wine. Botanicals, which include herbs, spices, and bitter roots, make vermouth unique (Kuderski, 2020). Vermouth was considered a health beverage for centuries due to several of its bitter ingredients. Today, vermouth is primarily an ingredient in cocktails. There are three varieties:

- Sweet vermouth: a sweet red vermouth
- Dry vermouth: a dry white vermouth
- Blanc vermouth: a sweet white vermouth.

Terroir

The Savoie wine region is divided roughly into three districts: the glacial terrain along the south shore of Lake Geneva, the hill country near the north end of Lac de Bourget, and the area border-ing the Massif de Bauges and Chartreuse Massif south and east of Chambéry (Wikipedia, Vin de Savoie AOC). It is claimed that the light, fresh style of the wine is directly related to the alpine *ter-roir* of Savoie. Most of the vineyards are planted on steep, south-facing slopes between 250 and 550 m (820 and 1804 ft) elevation, where favorable light and good drainage provide excellent ripening conditions despite the cold continental climate. The mountains offer protection from northwesterly weather systems, and rivers and lakes help moderate the climate. Most Vin de Savoie vineyards are planted on limestone-derived soils, scree, alluvium, and glacial moraines. These are good at storing heat during the day and emitting it back onto the vines at night. The loose soils tend to be thin and have little water-holding capacity, an important factor given the relatively high rainfall. Limited soil water leads to a reduction in yield but concentrates the flavor and increases the quality of the grapes (Kuderski, 2020; WineSearcher.com, Vins de Savoie, Terroir).

The local climate is continental, with mountainous and, increasingly, Mediterranean influences. The seasons are quite distinct, and rainfall is abundant due to the presence of mountains, which promote cloud formation and create a unique microclimate.

An important feature of the terroir is that vines located on hillsides generally mature earlier than those located on the plains, due to (1) the fact that these vineyards are located on stonier soils, which are therefore more easily drained, and which significantly improves the sugar content of the grapes (especially near Chignin, where this is particularly noticeable due to the numerous alluvial fans present in the Bauges Massif), and (2) sun exposure is greater due to the greater slope of the soil.

Wines from the plains, with more clayey soils, produce a more varietal wine profile, with herbaceous, fruity, and minty notes and a rounded feel, whereas wines grown on hillsides are stronger due to the soil's high mineral content, with more body than the others, with a fruitier and more floral profile. Because hillside vines are generally located in better drained soils, ripening is more advanced, with higher sugar content and lower yields than lowland vineyards. Wines grown on steeper slopes also have the advantage of being more suited to cellaring (Veyron, email, 16 April 2025).

Domaine Adrien Veyron et Fils

An example of Chambéry vineyards can be visited at Domaine Adrien Veyron et Fils. I chose this vineyard at random because of the scenic setting. I have no interest in promoting or advertising their wines. A quick Google search provides many vineyards that you can visit.

Domaine Veyron produces from two large geographical zones. The Apremont appellation (Jacquère grape) is produced in three villages and covers ~300 ha (740 ac): Apremont / Saint Baldoph / Les Marches (recently renamed "Porte de Savoie"). The Apremont region is located at an altitude of approximately 350 m (1,150 ft) and benefits from an east/northeast exposure. It rests largely on the limestone scree of Mont Granier and glacial moraines in areas not covered by scree. The soil is mainly clay-limestone with marl-limestone in some areas. The limestone is 80 to 120 million years old (Cretaceous). This limestone soil strongly influences soil drainage, water retention, and mineral content, thus contributing to the complexity of the wines.

The second region, the Combe de Savoie, is located in the eastern foothills of the Bauges Massif. It rests on a medium to deep clay-limestone soil and benefits from a south/southeast exposure. Five of the grape varieties grown by the estate, Altesse, Chardonnay, Pinot Noir, Gamay, and Mondeuse Noire, are found there.

In the 1860s Capuchin monks decided to live together in a small cellar at a place called "la ratte" in Apremont (a commune at the foot of Mont Granier south of Chambéry). A few years later Jean and Victorine Cartier bought this cellar and made it their home. They worked the land, raised some farm animals, and grew a few vines. In 1955, Jean-Baptiste Veyron married Victoria Chabert,

Domaine Adrien Veyron et Fils and Mont Granier. View southwest.

granddaughter of Jean and Victorine. They continued to grow vines, raise farm animals, and grow wheat. In the 1970s Jean-Baptiste decided to take advantage of a tourist boom to grow their operation. He planted the native Jacquère grapes to produce the traditional white wine demanded by tourists. By 1975 Adrien Veyron, son of Jean-Baptiste and Victoria, developed commercial ties to resort restaurants, wine merchants, and regional shops throughout Savoie. His son Thomas began studying viticulture and oenology to improve their winemaking. In 2009, he joined the family business and developed a new collection of original cuvées. He gradually moved toward more organic and sustainable vinticulture. Today 13 ha (32 ac) of vines are cultivated in a vineyard extending from Apremont to the Combe de Savoie (the valley of the Isère River between Albertville and Montmélian).

The Domaine Veyron winemaking process is based on traditional and ancestral knowledge updated with modern technology. They claim that they combine tradition and innovation. The use of an external oenologist and monitoring of wines by an external laboratory since the 1970s has allowed Domaine Veyron to stand out from the competition, earning the estate numerous medals in national competitions. The estate remains open to new ways of doing things. For example, they have developed an "orange wine" cuvée using the Altesse grape, as well as special vintages produced in egg-shaped vats.

The estate offers 14 different vintages based on six different grape varieties (including the three native grapes Jacquère, Altesse, and Mondeuse) across two large geographical areas in Savoie. The vineyard produces Apremont, Chardonnay, Gamay Rouge, Mondeuse, Pinot Noir, and Roussette de Savoie.

The vines have an average age of around 60 years, a key element of the terroir. The wines are thus recognized for their authenticity, quality, and consistency. Reviews have stated that the "wines reflect our mountain climate, the nature of our clay-limestone soil, but also the meticulous work carried out every year in the vineyard and the winery." "What makes this type of wine so successful is primarily its excellent value for money, palatability and relatively low alcohol content. Global warming, which is now unquestionable, promotes quality in the wines. The sugars in the grapes are incredibly rich and acidity levels are tending to drop" (Veyron, email, 16 April 2025).

White wine grapes are harvested and pressed using a horizontal half-membrane press. Depending on the year, grapes may macerate for several hours before being pressed, depending on the vintage. Maceration is the process where grape skins, seeds, and sometimes stems are left in contact with the grape juice during fermentation. This extended contact allows the juice to extract color, flavor, tannins, and aroma from the solids.

Grapes are harvested at full ripeness, followed by a cold skin maceration and strict settling for the whites. This obtains a juice of around 100 NTU (National Turbidity Units) before fermentation begins. The flotation settling technique allows fine-tuning of the turbidity, and thus the viscosity of the juice. This is crucial for obtaining the clearest possible juice, limits the appearance of vegetal notes, minimizes the risk of microbial deviations, and enables vinification with little or no sulfites (preservatives). It also avoids runaway fermentation. Once the fermentation is complete, the wine proceeds to rough racking, a natural and gentle pre-filtration of the wine. Aging follows and takes place in thermoregulated stainless steel vats at a temperature of around 14°C (54°F) until final filtration before bottling.

Eighty percent of the vintages are from single-plot vineyards. This constitutes a strength in wine blending. Apremont remains the estate's main specialty. It is available in four distinct vintages:

- Apremont
- The "Intense" vintage
- The "Révélation Vieilles-Vignes" vintage, a vintage bearing the image of Jean-Baptiste Veyron, the third generation of winemakers at the estate.
- "La Mystique" vintage (new in 2022). It is a single-plot selection, from a century-old vine located on the heights of the village of Apremont.

The Roussette de Savoie is the next most popular, with two different vintages:

- The Roussette de Savoie (no cuvée name)
- The Roussette cuvée "Plénitude," aged in egg-shaped vats.

For red wines, the winery specializes in the traditional Mondeuse with two cuvées:

- Mondeuse (no cuvée name)
- Mondeuse cuvée "Alchimia." This is the "cru Saint-Jean de la Porte." A single plot of vines, about 12 years old, is located at the foot of the hillside in the commune of Saint-Jean de la Porte. This *terroir* is known for producing dense wines with great depth that are ideal for aging.

Visit

Guided tours are available by appointment for groups of a minimum of 5 and a maximum of 20 people. Days and times are flexible and by arrangement. There is no charge for group visits at this time.
Wines are sold at the winery store. Prices per bottle range from €6.50 to €17.00.

Address: Route de Saint-André, Lieu-dit La Ratte, 73190 Apremont, France
Phone: +33 04 79 28 2020
Email: veyron.vins.savoie@wanadoo.fr
Website: https://www.vins-veyron.com/

The wine shop, Domaine Adrien Veyron, showing the selection of wines available.

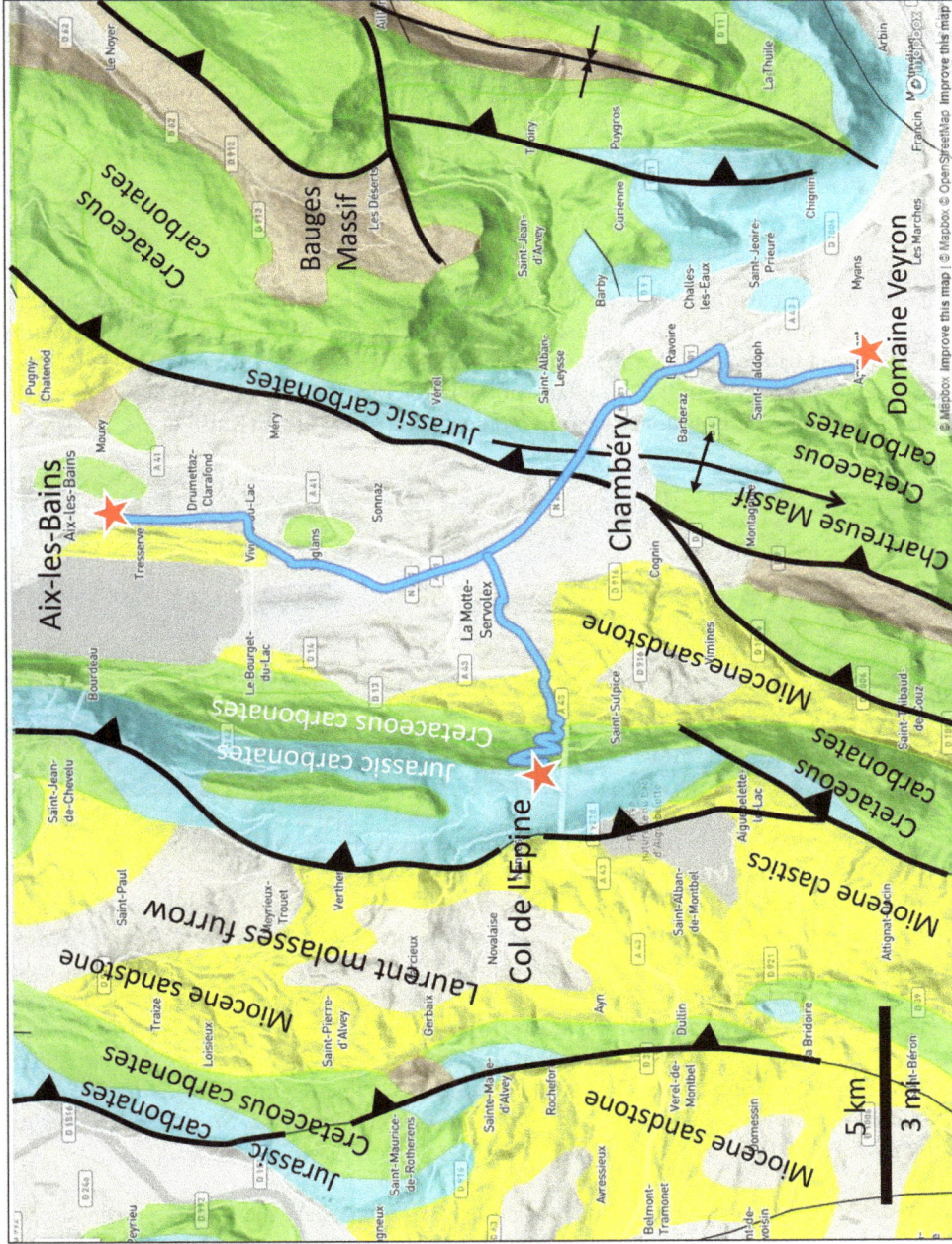

Geologic map of the Aix-les-Bains-Col de l'Épine area. Heavy lines with teeth are thrust faults. Folds with arrows pointing outward are anticlines; folds with arrows pointing inward are synclines. Geology base from MacroStrat, 2015.

Side Trip 7, Vins de Savoie to Belvédère du Col de l'Épine: *Return north through Apremont to D201/Rte d'Apremont; turn right (east) onto D9/Rte des Chevaliers Tireurs and follow the sign for Challes les Eaux; at the roundabout take the 2nd exit onto D9/Rte de Saint-Baldoph; at the round-about take the 2nd exit onto the A43 ramp north toward Chambéry/Lyon/Genève; merge onto A43 north; continue straight on N201; take Exit 14 to La Motte-Servolex; bear left to D16A/Rue du Nant Bruyant; at the roundabout take the 3rd exit (west) to D16A; at the roundabout take the 1st exit (north) onto D1/Av Costa de Beauregard toward La Motte-Servolex Centre; at the roundabout take the 3rd exit (west) onto D1/Av Jean Marie Michellier; at the roundabout take the 4th exit (south) to D15/Av Jean Moulin; at the roundabout take the 1st exit (west) onto D3/Rte de l'Épine; continue on Rte de l'Épine to D916; take a hard right (north) on D916 to Col de l'Épine and drive to **Stop 21, Belvédère du Col de l'Épine** (45.581536, 5.825602); pull into the picnic area/viewpoint on the left for a total of 24.8 km (15.4 mi; 33 min).*

STOP 21 BELVÉDÈRE DU COL DE L'ÉPINE

We are standing on the east-dipping flank of the north-south oriented l'Épine Anticline that has been carried westward on the l'Épine Thrust. The overlook has a display naming the mountains seen in the panorama to the east. To the southeast (south of Chambéry) are near-vertical Upper Jurassic–Cretaceous limestones of the west flank of the Outherans Anticline. To the east and north-east are east-dipping Jurassic limestones carried on the Chartreuse Thrust (Kalifi et al., 2021).

Panoramic view east and southeast to the Bauges Massif (left) and Chartreuse Massif (right).

Belvédère du Col de l'Épine to Cognin-les-Gorges: *Belvédère du Col de l'Épine to Gorges du Nan: Return south on D916; slight right to Rte du Freney; turn right (south) onto Rte du Freney; continue straight on Rte de St. Sulpice; turn right (south) on D47/Rte de St. Thibaud; turn right (south) on D1006; at roundabout take 2nd exit onto D520; continue straight on D520A; in Voreppe turn right onto D3A/Av Stalingrad; turn right (west) on D1075; at roundabout take 4th exit onto D3 (south); at roundabout take the 1st exit (north) to A48; stay right to continue on A49; take Exit 10 toward Vinay; at roundabout take 3rd exit (south) to D22; turn right (south) onto D1532; in Cognin-les-Gorges pull over on the right just before the Nant River. This is **Stop 22.1, Gorges du Nan** (45.169617, 5.410554) for a total of 81.3 km (50.5 mi; 1 hr 21 min).*

Here we leave the Molasse Basin proper and enter a transition zone where the Molasse Basin merges with the Central European Graben system, specifically the Bresse–Valence Graben.

THE CENTRAL EUROPEAN RIFTS

Europe began tearing itself apart by east-west extension during the Tertiary. A group of basins, the Central European Rift System (or European Cenozoic Rift System), formed to the north and the west of the Alps. The rift basins stretch from the North Sea to the Mediterranean. The elongated, roughly north-south basins that make up the southern and central branches include the Gulf of Lions, Valence, Bresse, and Limagne grabens (Schlunegger and Mosar, 2010). Rifting continues to the north as the Rhine Graben.

Cenozoic deposits in the Bresse–Valence Graben are up to 2,300 m (7,550 ft) thick. Sedimentation began with locally derived middle Eocene river and lake deposits. Late Eocene and early Oligocene units are hundreds of meters thick and include evaporites interbedded with fluvio-lacustrine limestones, marls, conglomerates, and breccias. By late Oligocene time a shallow marine incursion led to marginal marine sandstones and shales; as much as 1,200 m (3,940 ft) of evaporites, mainly salt, was deposited in the graben center during the Oligocene. Sandstone, siltstone, shale, and limestone units overlie the Oligocene evaporites (Sissingh, 1998). By early Miocene time freshwater limestones were being deposited. Another marine transgression led to shallow marine sandstones and conglomerates in mid-Miocene and lasted to Messinian (latest Miocene) time. Up to 300 m (1,000 ft) of Pliocene lake marls and river delta sands and shales indicate renewed subsidence in the Bresse Graben. Extension and subsidence of the Bresse Graben continued during the Early Pleistocene, finally ending around 1.4 Ma when the graben was uplifted and subjected to erosion (Schlunegger and Mosar, 2010).

The Bresse Graben is bounded on the north by the anticlinal Burgundy Sill, and on the south it is separated from the Valence Basin by the Lyon Sill. (A sill here is a narrow ledge or ridge.) Continued subsidence eventually breached the Lyon Sill, causing the Bresse Graben to merge with the Valence Basin. The Bresse segment of the Cenozoic Rift System is defined by subparallel basin-bounding faults (Sissingh, 1998).

The depositional history and tectonic development of the Central European Rifts appears to be somewhat connected with that of the Molasse (or North Alpine Foreland) Basin. The Rhine and Bresse grabens both began to subside during the middle Eocene. The Molasse Basin became a

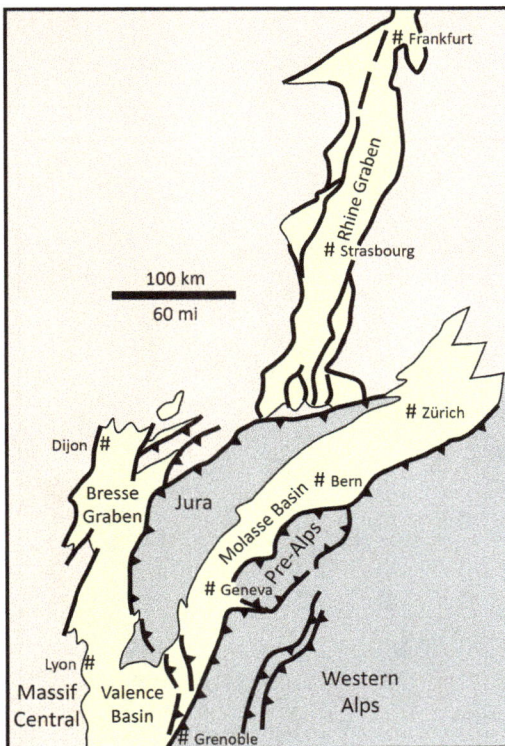

Left: Map of the Rhine, Bresse, and Valence grabens of the Central European Rift system, as well as the western Molasse foreland basin of the Alps. Heavy lines are normal faults; heavy lines with teeth are thrust faults. Right: Stratigraphy of the Bresse Graben. Modified after Sissingh, 1998.

flexural foredeep on the northern margin of the Alpine Orogeny in Eocene time. A flexural fore-deep is a depression or basin formed by the bending of a tectonic plate under the load of a mountain range, in this case a northwest-advancing fold-thrust belt.

STOP 22 COGNIN-LES-GORGES

The river Le Nant has cut a narrow gorge through impressive limestone cliffs. The Jurassic and Cretaceous limestone layers are carried on leading-edge thrusts of the Alps.

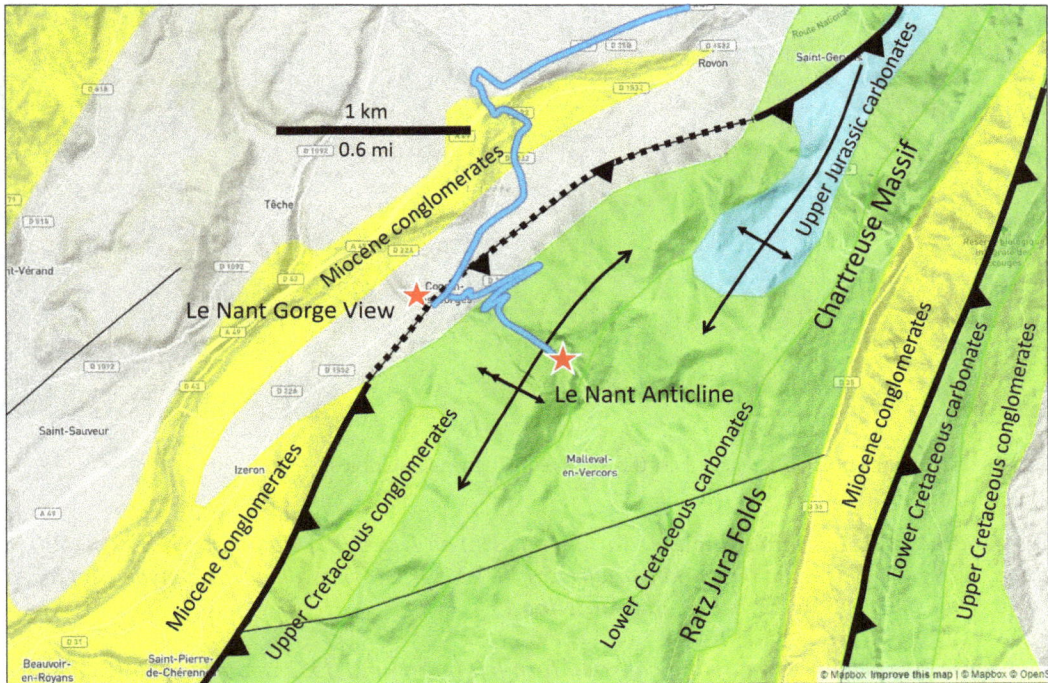

Geologic map of the Cognin-les-Gorges area. Heavy lines with teeth are thrust faults. Folds with arrows pointing outward are anticlines. Geology base from MacroStrat, 2015.

STOP 22.1 GORGES DU NAN

This scenic gorge is cut through Cretaceous limestone of what I'm calling the Le Nant Anticline. The anticline is on the leading edge of a west-directed thrust that originates below the Vercors Massif. The gorge slices across the Chartreuse Massif, in what have been called the Ratz Jura Folds.

As with the Gorges du Sierroz, the Le Nant River appears to have cut down as these structures were forming and uplifting. It is thus likely an antecedent stream.

The next stop allows us a view of a natural cross section through the center of the Le Nant Anticline.

Gorges du Nan viewpoint. View east past the village of Cognin-les-Gorges.

Gorges du Nan to Le Nant Anticline: *Turn left (north) on D1532/Rte Nationale; in 150 m (500 ft; 1 min) turn right (east) into the parking area (45.170773, 5.412431). Walk up the canyon trail for ~1 km (0.6 mi). This is* **Stop 22.2, Le Nant Anticline***.*

Alternatively, you can drive up the canyon: *Turn left (north on D1532 and drive 0.5 km (0.3 mi) and turn right (east) on Rue Saint Joseph; turn left (north) on D22/Rue de Malleval; continue east on D22 for ~5 km (3 mi) to La Frete (45.156684, 5.439629) where it is safe to turn around. There are multiple pullouts/viewpoints along the way. Note that the drive up D22 is wide enough for only one car, so if you meet someone going the other way one of you will have to back up until you come to a pullout. There are also hairpin curves and blind curves.*

STOP 22.2 LE NANT ANTICLINE

Walk (or drive) through a natural cross section across the Le Nant Anticline in Vercors National Park. The dips in the Lower Cretaceous limestone change from gentle on the east to steep on the west, revealing the vergence of the anticline: the asymmetric fold leans to the west, indicating it formed on a west-directed thrust. In addition, there is evidence of karsting, as caverns are exposed high on the canyon walls.

The walk up the canyon has moderate to steep sections and can be strenuous. There is also a spectacular road cut into the cliffs that leads from the north end of Cognin village to the crest of the anticline. The trail eventually emerges at the road, and the walk from town to the anticlinal axis takes at least an hour.

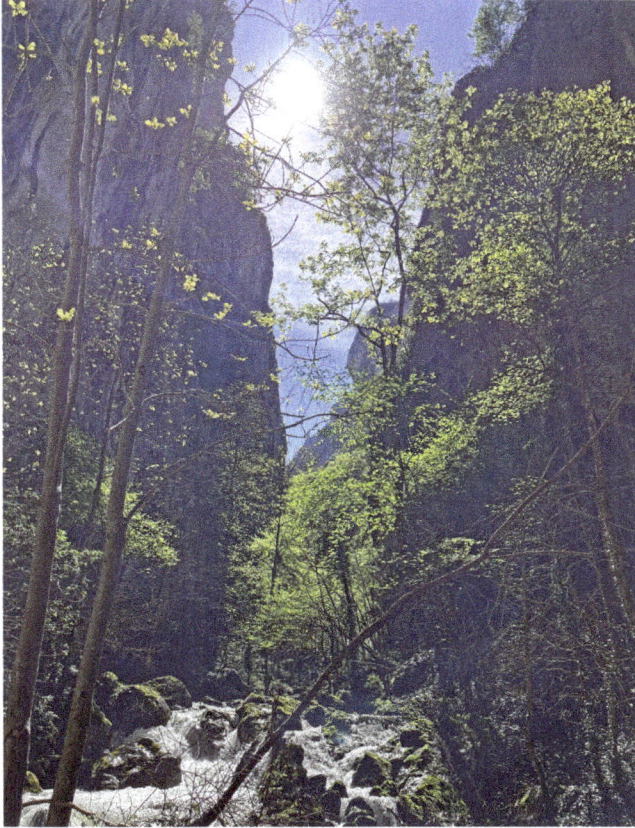

Looking up the impressive gorge cut by the Le Nant stream. In places the walls rise over 600 m (just under 2,000 ft).

View west toward the steeper west flank of the Le Nant Anticline. Note the road chisled out of the limestone cliff on the right (north) side of the canyon.

Side Trip 8, Le Nant Anticline to Châteauneuf-sur-Isère: Return northwest to D1532/Rte Nationale; turn right (northeast) and drive to Le Rivier; at the signs to Vinay/Grenoble-Isère turn left (west); at the roundabout take the 2nd exit and get on the A49 toll road ramp going southwest to Valence; continue straight on N532; take Exit 5 west toward Châteauneuf-sur-Isère; bear right toward Châteauneuf s I /Alixan; at the roundabout take the 2nd exit (west) onto D101; at the Salle des Fetes roundabout take the 3rd exit (west) onto D67/Av de Valence; at the next roundabout take the 3rd exit (south) onto Rue du Châtelard; from this roundabout drive ~130 m (0.1 mi) and pull over in the parking area on the left across from the school. This is to **Side Trip 8, Valence Graben, Châteauneuf-sur-Isère** *(45.012999, 4.934669) for a total of 52.6 km (32.7mi; 35 min).*

If you prefer to skip Side Trip 8:

Le Nant Anticline to **Pont-en-Royans***: Return northwest to D1532/Rte Nationale; turn left (southwest) and drive to Saint-Romans; at the sign for "Romans 28" turn left (south) and take D518 to Pont-en-Royans; in Pont-en-Royans turn left (east) onto D531 toward Choranche and Villard-de-Lans; continue another 700 m (0.4 mi) and pull into the parking area on the left. This is* **Stop 23, Vercors Massif, Pont-en-Royans** *(45.06469, 5.35253) for a total of 19.6 km (12.2 mi; 21 min).*

Side Trip 8 The Valence Graben, Châteauneuf-sur-Isère

This outcrop exposes middle Miocene upper shoreface sandstone near the base to possibly alluvial conglomerate near the top (Faure and Chen, 2018). It would be the equivalent of the contact between the Upper Marine Molasse and Upper Freshwater Molasse units in the Molasse Basin to the north-east. You may have noticed other spectacular roadcuts in this soft sandy material along the way. Here we can safely examine the outcrop.

The unit consists of a friable yellow sandstone capped by a pebble and cobble conglomerate. The tabular cross-beds are formed by the migration of dunes or ripples and have essentially planar bounding surfaces. These sandstones, shales, and conglomerates formed as shallow marine deposits in a foreland basin in front of the rising Alps. The sediments are derived from erosion of the uplift to the east.

The Valence Graben, part of the Central European Rift System, is deepest in the west. It is a tilted half graben bounded by normal faults on the west and rising gently to the east. The western edge of the Valence Basin is bounded by the Massif Central, a large uplifted plateau cored by Variscan-age granitic and metamorphic basement. The east side of the basin is overlapped slightly by thrusts and folds of the Vercors Massif Fold-Thrust Belt.

Side Trip 8, Châteauneuf-sur-Isère to **Pont-en-Royans***: Return north on Rue du Châtelard to the D67 roundabout; take the 1st exit (east) onto D67; at the next roundabout take the 1st exit (east) onto D101; pass under the N532 toll road and at the next roundabout take the 3rd exit (northeast) onto Bayanne/Rte des Poles toward N532/A49/Romans; merge onto N532; continue straight on A49; take Exit 8 toward Villard de Lans/Pont en Royans; at the roundabout take the 2nd exit (south) onto D325A; at the next roundabout take the 3rd exit (east) onto D532 and drive to Saint-Nazair-en-Royans; turn right (east) onto D1532; at the next roundabout take the 1st exit onto D531 toward Grenoble; continue straight onto D518 to Pont-en-Royans; in Pont-en-Royans turn left (east) onto D531 toward Choranche and Villard-de-Lans; continue another 700 m (0.4 mi) and pull into the parking area on the left. This is* **Stop 23, Vercors Massif, Pont-en-Royans** *(45.06469, 5.35253) for a total of 39.2 km (24.4 mi; 37 min).*

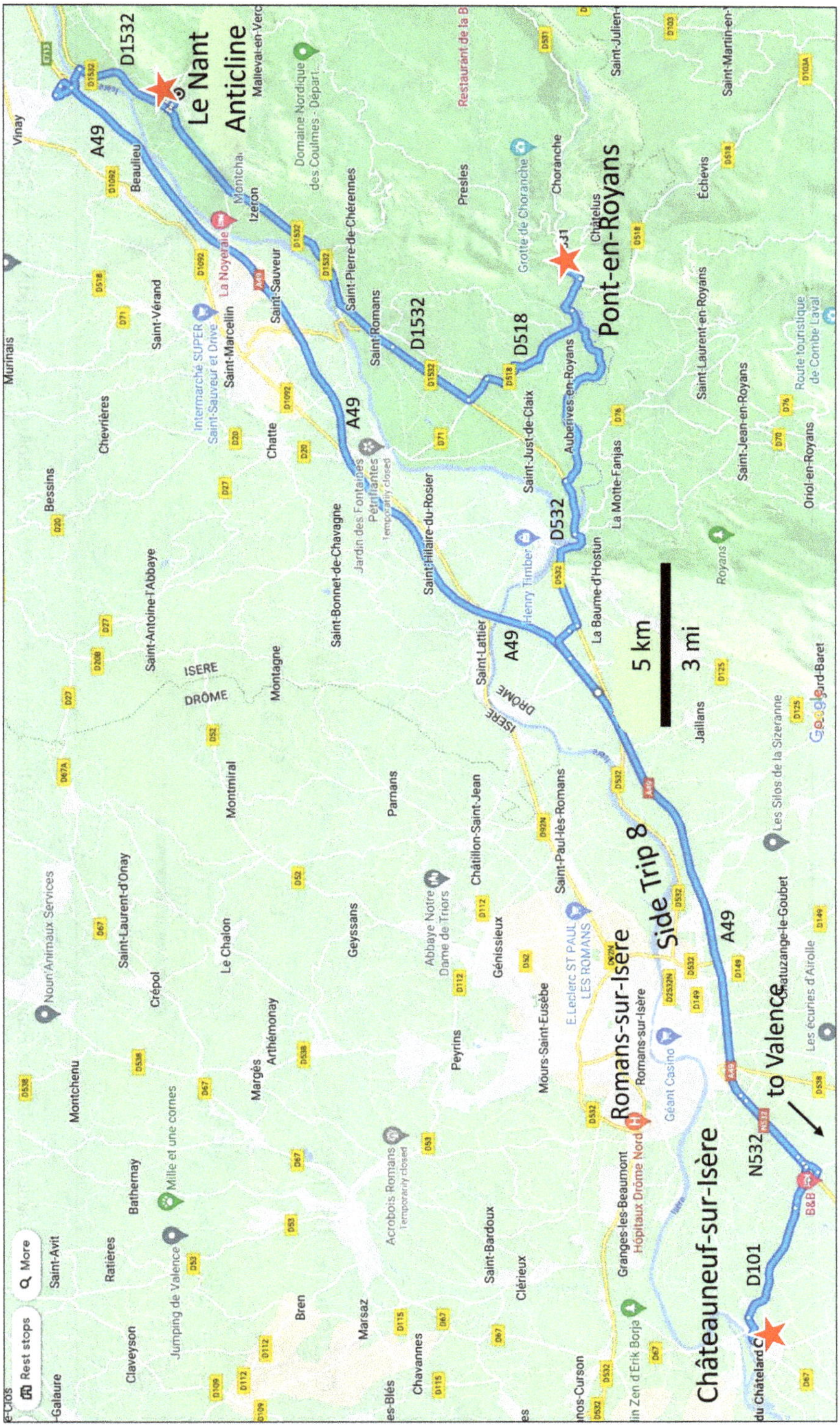
D1532
Le Nant
Anticline
A49
D1532
A49
Pont-en-Royans
D518
D532
A49
5 km
3 mi
Side Trip 8
A49
Romans-sur-Isère
to Valence
N532
Châteauneuf-sur-Isère
D101

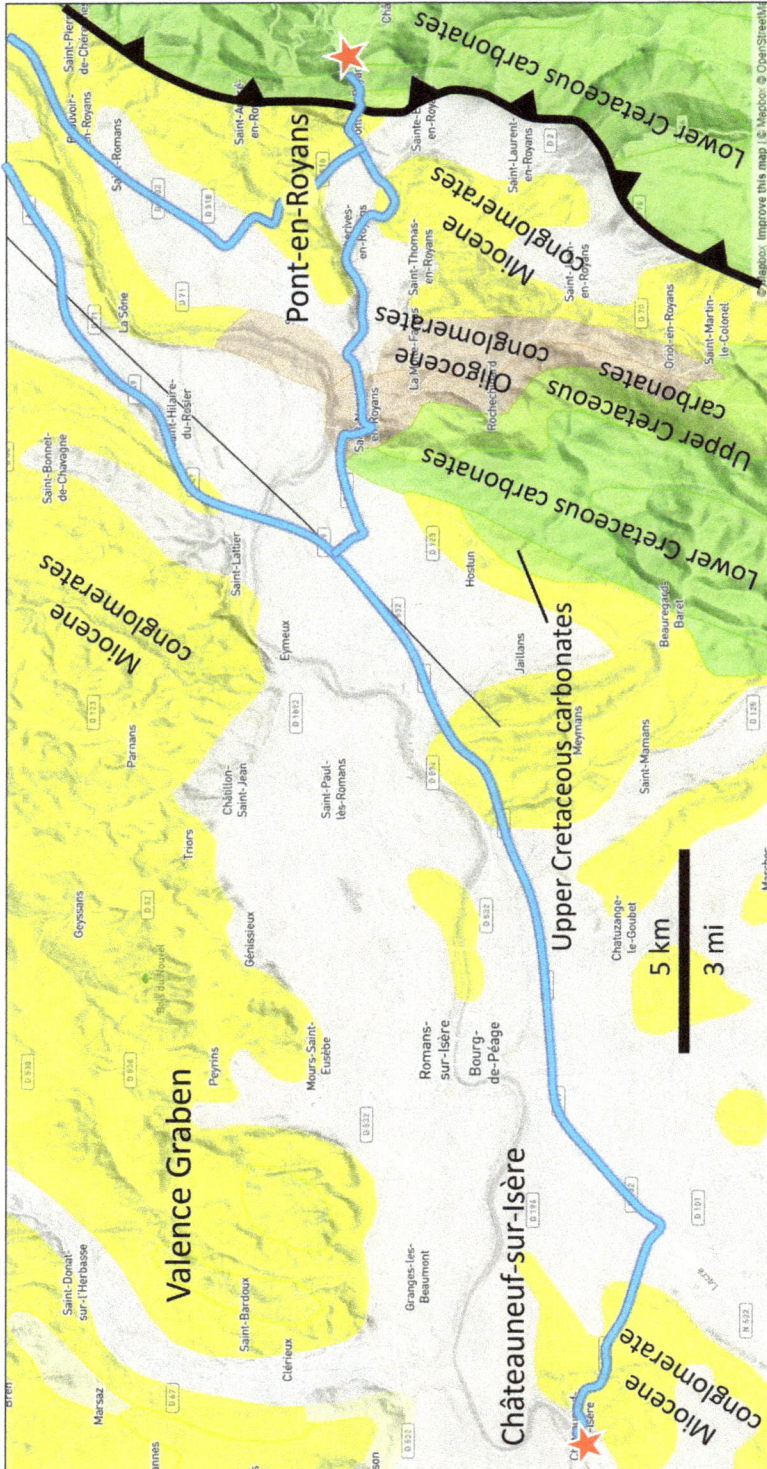

Geologic map of the Châteauneuf-sur-Isère to Pont-en-Royans area. Heavy lines with teeth are thrust faults. Geology base from MacroStrat, 2015.

Middle Miocene beach sandstone (molasse) of the Valence Basin, Châteauneuf-sur-Isère.

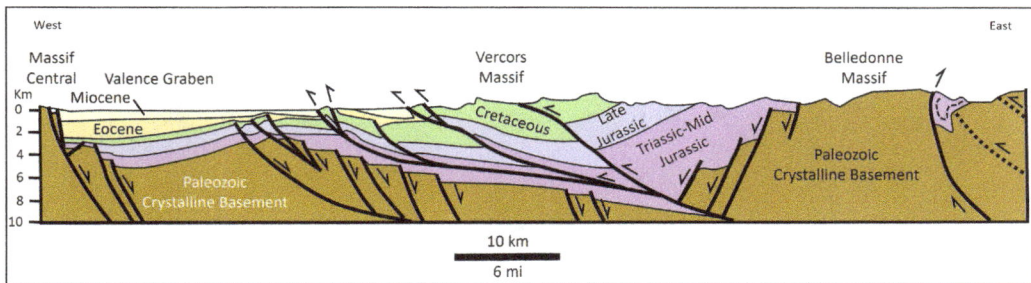

Cross section from the Belledonne Massif to the Massif Central and including the Valence Graben. Estimated west-east shortening in the Vercors Massif is ~20%. Modified after Bellahsen et al., 2014; Chapelet et al., 2019.

STOP 23 VERCORS MASSIF OF THE DAUPHINO-HELVETIC DOMAIN, PONT-EN-ROYANS

Here we reenter the Pre-Alps, in this case the Vercors Massif of the Dauphino-Helvetic Domain.

This is another place where you can traverse a natural cross section through a massive thrust-cored assymetric anticline near the western leading edge of Alpine thrusting. The Pont-en-Royans Anticline, developed in Lower Cretaceous Urgonian Limestone, has a steeper west flank and gentler east flank, again a consistent indicator that it was thrust toward the west. The thrust comes to the surface near this stop (Kalifi et al., 2021).

Pont-en-Royans to Choranche Overlook: Continue driving east on D531 for 7.0 km (4.3 mi; 9 min) and pull into the parking area on the right by the panels. This is **Stop 24, Urgonian Limestone, Choranche Overlook** *(45.06364, 5.42608).*

Front ranges of the southwestern Alps looking east from D101 in the Valence Graben.

Approaching Pont-en-Royans Anticline from the west. Dashed lines follow bedding.

View north into the core, or axis of the Pont-en-Royans Anticline. The west-dipping ridge is Lower Cretaceous Urgonian Limestone.

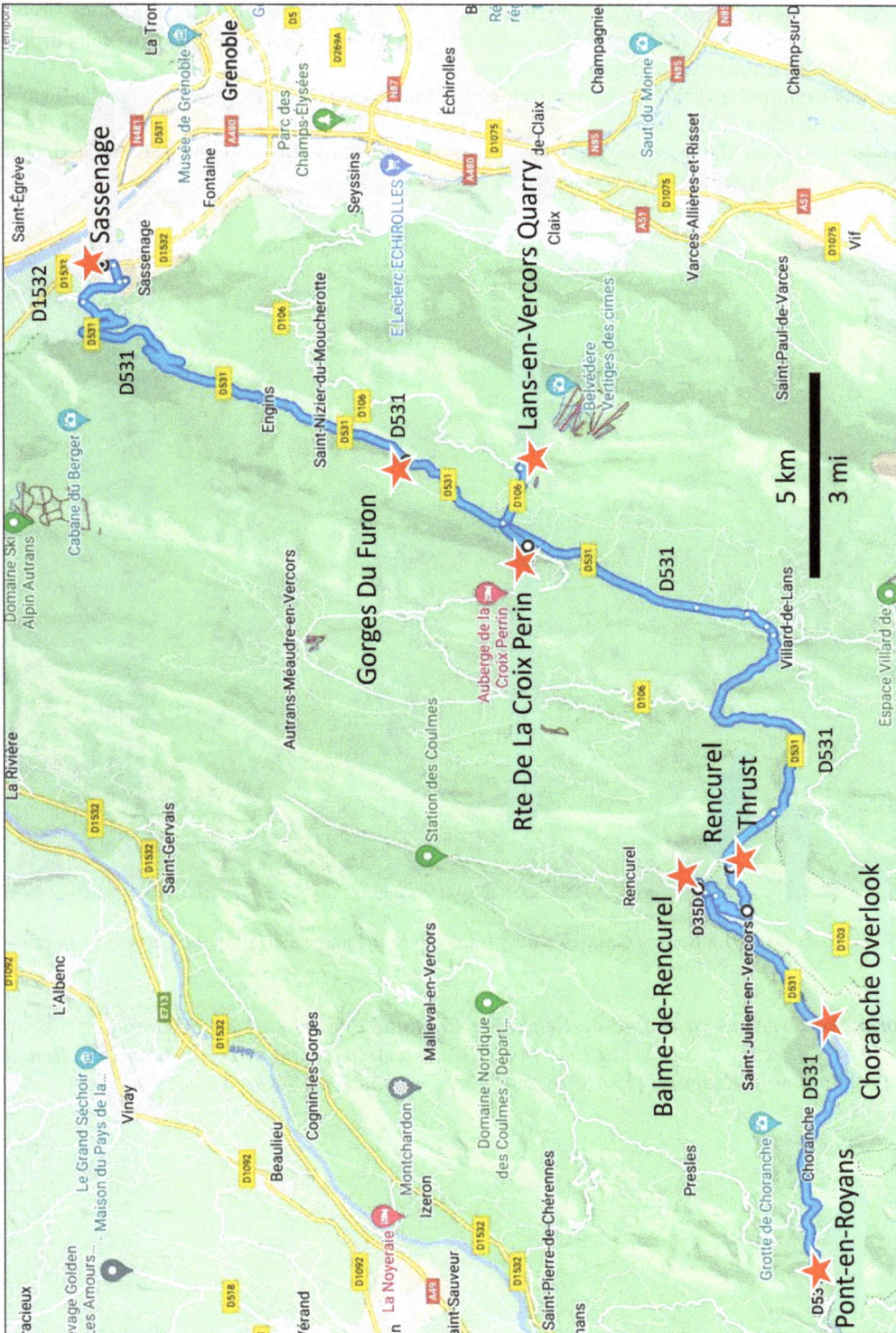

Stop 24 Urgonian Limestone, Choranche Overlook

This is a view stop. Looking south from this viewpoint you see massive Lower Cretaceous Urgonian Limestone cliffs as well as a spectacular waterfall in the Cirque de Bournillon. The falls drop 300 m (1,000 ft) over the cliffs and into the gorge.

Panels explain the various landmarks along the La Bourne River gorge as well as the history of dam construction in the gorge.

Behind you on the north side of the highway are gently east-dipping thin-bedded middle Cretaceous deep marine shales and marls.

View south across the La Bourne gorge to a truly impressive cliff and waterfall in the Urgonian Limestone.

Choranche Overlook to Balme-de-Rencurel: *Continue driving east on D531 to D35; turn left (north) on D35 and drive 192 m (0.1 mi) to the T-intersection and park. This is **Stop 25.1, Rencurel Thrust, Balme-de-Rencurel** (45.08978, 5.46903) for a total of 5.2 km (3.2 mi; 7 min).*

STOP 25 RENCUREL THRUST AND VERCORS MASSIF

The Vercors Massif is the southern extension of the Chartreuse Massif: it extends from Grenoble in the northeast to Crest in the southwest. The Vercors Massif comprises rugged plateaus and mountains in the departments of Isère and Drôme of the French Pre-Alps. At the start of the Miocene, the Alpine Orogeny involved westward thrusting of Jurassic and Cretaceous limestones. Sedimentary rocks were raised 2,000 m (6,600 ft) and faulted and folded. Subsequent Miocene marine transgressions and erosion from the uplifts were responsible for molasse sedimentation in basins to the west. Widening of the valleys and erosion of the mountains led to deposition of Miocene molasse in the Royans and Rencurel valleys, among others. The final receding of the sea occurred at the end of Miocene time. The massif was subject to severe erosion during the Ice Age, accentuating its relief (Wikipedia, Vercors Massif).

The Rencurel Thrust outcrops near La Balme-de-Rencurel and to the south. Early Cretaceous Urgonian Limestone is thrust above overturned Miocene molasse deposits. The Miocene strata are also limestones. The fault strikes roughly north-south and dips 40° to the east. The thrust sheet has been folded into two anticlines separated by a syncline. Farther east, two parallel thrusts with smaller offsets (0.8 and 0.9 km, or 0.5 and 0.6 mi) come to the surface. The Rencurel Thrust extends south for 6 km (3.7 mi), where it terminates against a northwest-southeast-oriented left-lateral fault. The thrust continues north to the Col de Montaud, always with Miocene in the footwall. At this location the thrust appears to turn from north-south to north-northeast and extends to Veurey–Voroize where Upper Jurassic sediments are thrust above the Miocene. The Rencurel thrust may be continuous with the Voreppe Thrust north of the Isère Valley. Multiple thrusts and strike-slip faults indicate west-northwest shortening and compression (Kalifi et al., 2021).

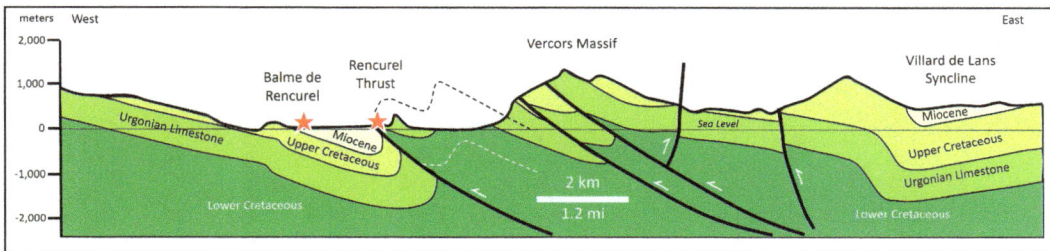

Cross section of the Vercors Massif showing the position of the Rencurel Thrust. Location is shown on the next figure. Modified after Kalifi et al., 2021.

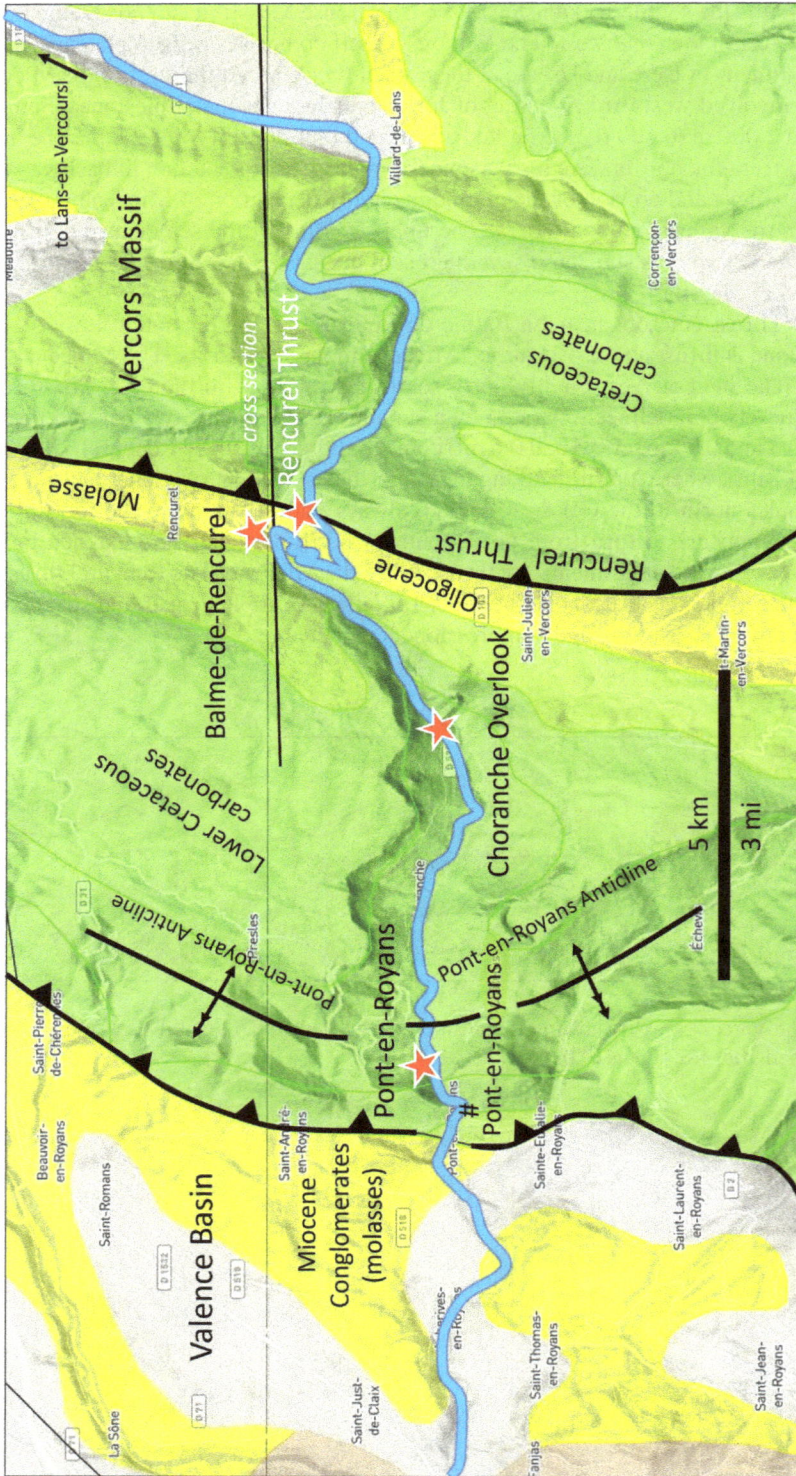

Geologic map of the Rencurel Thrust and Vercors Massif. Heavy lines with teeth are thrust faults. Folds with arrows pointing outward are anticlines. Geology base from MacroStrat, 2015.

STOP 25.1 RENCUREL THRUST, BALME-DE-RENCUREL

This stop provides a view east to the trace of the Rencurel Thrust where it is crossed by the La Bourne River. The thrust is at the base of the massive Lower Cretaceous limestone cliffs.

The rocks in the roadcut here are thin to thick beds of Oligocene molasse sandstone, pebbly sandstone, and shale.

View east to the trace of the Rencurel Thrust (dotted line) that breaches the surface at the base of the Urgonian Limestone cliffs. The thick limestone was carried west on this thrust.

Balme-de-Rencurel to Rencurel Thrust: *Return south on D35D; turn right (east) onto D531/Ham. de la Balme and drive southwest; turn left (south) onto D255/Ham. de la Balme toward St. Julien en Vercors; cross the Bourne River and drive to T intersection with D103; turn left (east) on D103 and drive 1.3 km (0.8 mi) to **Stop 25.2, Rencurel Thrust, D103 Roadcut** (45.082955, 5.473920) for a total of 4.2 km (2.6 mi; 7 min). Park on the right and walk ~220 m (730 ft) east to the base of the cliff. There is little or no shoulder; watch for traffic!*

STOP 25.2 RENCUREL THRUST, D103 ROADCUT

The Rencurel Thrust is exposed in this roadcut along road D103. A zone of black sheared fault gouge (broken fault zone rock) is just below the Urgonian Limestone cliff. The thrust is inclined about 40° to the east. Below the gouge is steeply east-dipping overturned Miocene molasse (Faure and Chen, 2018; Kalifi et al., 2021). This is one of the few places you can lay hands on a thrust fault.

The Rencurel Thrust, which outcrops near La Balme de Rencurel, is the main thrust in the Vercors Massif. Apparent offset for the top of the Urgonian Limestone is ~1.5 km (1 mi) to the west. The Miocene strata are thin-bedded limestones. Slickenside lines (scratches) on the fault surface indicate almost pure thrust faulting, that is, almost due west shortening.

Rencurel Thrust at base of cliff. Street View east.

Rencurel Thrust to Rte de la Croix Perin: *Continue east on D103; at the intersection with D531 bear to the right and merge onto D531 going east; at the roundabout in Lans-en-Vercors take the 3rd exit (southwest) onto D106/Rte de la Croix Perin and drive about 800 m (0.5 mi) to **Stop 26.1, Villard-de-Lans Syncline, Rte de la Croix Perin** (45.125739, 5.574364) and pull over on the right by a gravel track for a total of 17.7 km (11.0 mi; 21 min).*

STOP 26 LANS-EN-VERCORS AREA

South of the Isere Valley, the Moucherotte Thrust brings Cretaceous sediments over Miocene molasse in the Villard-de-Lans Syncline.

STOP 26.1 VILLARD-DE-LANS SYNCLINE, RTE DE LA CROIX PERIN

At this stop you can see outcrops and boulders of the Miocene molasse conglomerate in the core of the Villard-de-Lans Syncline. The core of a syncline always has the youngest rock layers in the fold. The conglomerate here is characterized by large angular fragments and poor sorting. This indicates that the components of the rock were not transported long distances but rather are an alluvial deposit close to the source of sediment.

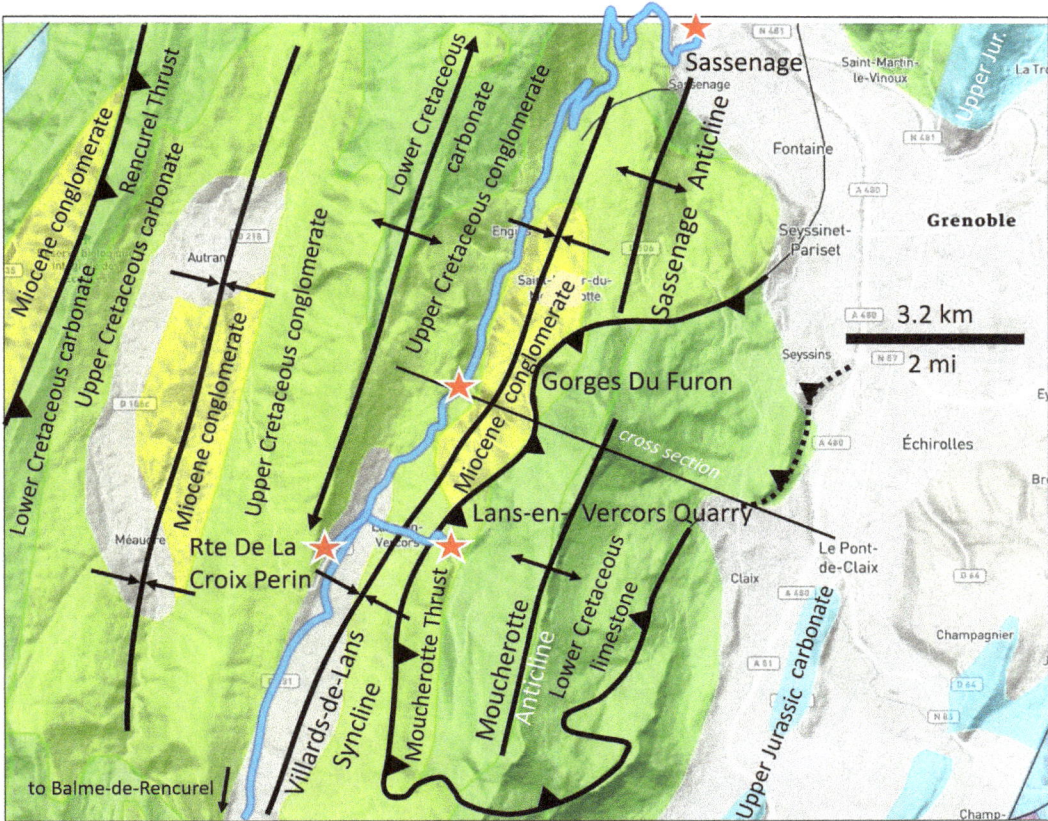

Geologic map of the Villard-de-Lans Syncline and Moucherotte and Sassenage anticlines, Vercors Massif. Heavy lines with teeth are thrust faults. Folds with arrows pointing outward are anticlines; folds with arrows pointing inward are synclines. Geology base from MacroStrat, 2015.

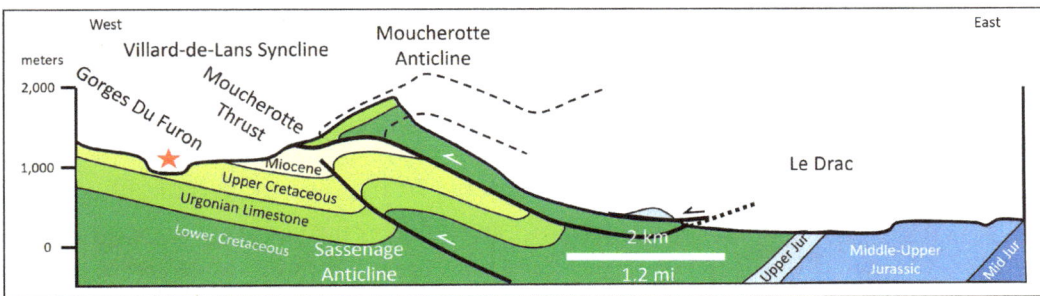

Cross section through Lans-en-Vercors. Location shown on previous figure. Modified after Kalifi et al., 2021.

*Rte de la Croix Perin to Lans-en-Vercors Quarry: Return north on D106 to Lans-en-Vercors; at the roundabout take the 2nd exit (east) onto D106/ Av. Léopold Fabre; continue southeast to D106L/Rte des Montagnes de Lans; turn right (southeast) on D106L and drive ~310 m (1,020 ft) to parking area in old quarry on the left. This is **Stop 26.2, Urgonian Limestone, Lans-en-Vercors Quarry** (45.125523, 5.602150) for a total of 2.7 km (1.7 mi; 4 min).*

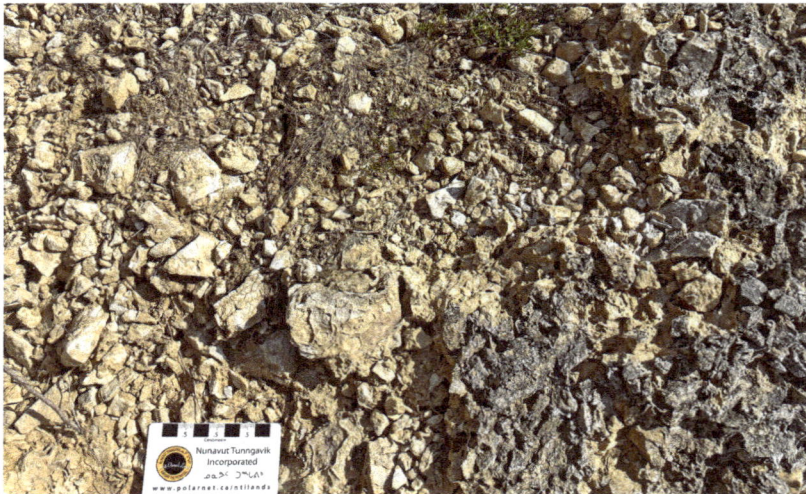

Miocene conglomerate outcrops. Top: Street View looking south. Bottom: angular fragments and poor sorting in the conglomerate indicate the source of the sediment is nearby.

STOP 26.2 URGONIAN LIMESTONE OF THE MOUCHEROTTE THRUST, LANS-EN-VERCORS QUARRY

We are in the hanging wall (the rock layers above the thrust) of the Moucherotte Thrust. This is the Lower Cretaceous Urgonian Limestone where it has been thrust west over Miocene molasse conglomerate in the Villard-de-Lans Syncline. Massive limestone with some thin beds and some fault breccia is exposed here. The trace of the thrust is buried beneath the valley fill just west of us.

This is a popular hiking and climbing area. The display panel shows a number of hiking trails that begin here.

Lans-en-Vercors Quarry to Gorges du Furon: *Return to D531 on the west side of Lans-en-Vercors; turn right onto D531 and drive to* **Stop 26.3, Upper Cretaceous Limestone, Gorges du Furon** *(45.14831, 5.60030) for a total of 4.1 km (2.5 mi; 5 min) and pull over on the right.*

Massive Urgonian Limestone, Lans-en-Vercors quarry, View north.

STOP 26.3 UPPER CRETACEOUS LIMESTONE, GORGES DU FURON

This stop exposes cliffs of Upper Cretaceous limestone carried on the Rencurel Thrust and in the footwall to (that is, just below) the Moucherotte Thrust. We are on the west flank of the Villard-de-Lans Syncline. Sheer cliffs of brilliant white limestone 76 m (250 ft) high make this gorge unforgettable.

*Gorges du Furon to Sassenage: Continue north on D531 to Sassenage; at Pl Jean Prevost take the 3rd exit (north) onto D1532/Av de Valence; turn right onto Rte de Valence and park. This is **Stop 27, Sassenage Anticline, Grenoble** (45.21554, 5.66144) for a total of 15.6 km (9.7 mi; 17 min).*

Gorges du Furon and Upper Cretaceous limestone. View north.

Stop 27 Sassenage Anticline, Grenoble

This is a view stop. From here you get a cross-sectional view of the overturned Sassenage Anticline where Lower Cretaceous Urgonian Limestone and Upper Cretaceous platy limestone are thrust westward over the Villard-de-Lans syncline containing Miocene Molasse. The bedding in the limestone is clear to see, as are the thrusts. This north-south trending fold is cored by an unnamed thrust fault that splits into two thrusts to the west.

Sassenage Anticline is outlined by the bedding (dashed lines). The thrust is indicated by the dotted lines. The fact that it is leaning to the west indicates it was pushed to the west. View south.

Sassenage is a suburb of Grenoble. Grenoble, the largest city of the Isère department in the Auvergne–Rhône–Alpes region, lies where the River Drac flows into the Isère. The population of Grenoble was 158,198 as of 2019, while the population of the greater metropolitan area was 714,799. This makes it the largest city in and, as it calls itself, the "Capital of the Alps."

Over 2,000 years ago Grenoble was the village of Cularo of the Allobroges Gallic tribe. With Rome's expansion it became part of Roman Gaul. In 381 Emperor Gratian visited Cularo and, touched by his welcome, made the village a Roman city. In honor of this, Cularo was renamed Gratianopolis ("city of Gratian"), which became Graignovolduring in the Middle Ages and, ultimately, Grenoble. It became the capital of the Dauphiné, a state within the Holy Roman Empire, in the 11th century: the local ruler, Count Guigues IV of Albon, had a dolphin on his coat of arms and was nicknamed 'le Dauphin' (the dolphin). The state took the name Dauphiné, which, in turn, was appropriated by geologists describing the Dauphinois Domain, the local region with a common geological framework. After annexation to France, Grenoble became a parliamentary and military city. The city's prominence grew starting with a booming glove industry in the 18th and 19th centuries, continuing with the development of hydropower in the late 19th and early 20th centuries, and ending with a post–World War II tourism boom symbolized by the Tenth Olympic Winter Games in 1968. Historically, Grenoble and the surrounding areas contained heavy industry and mines. Abandoned mills and factories can still be found in small towns and villages. The city has become one of Europe's most important research, technology, and innovation centers, and it earned the title of "European Green Capital" in 2022 (Wikipedia, Grenoble). I can attest that it is a charming and classic European city with wide, tree-lined boulevards and abundant parks, cafés, and boulangeries.

Grenoble is surrounded by mountains. To the north lies the Chartreuse Massif, to the south and west the Vercors Massif, and to the east the Belledonne Range. Except for a few houses on the slopes of the Chartreuse, Grenoble is built on the alluvial plain of the rivers Isère and Drac at an altitude of 214 m (702 ft). Beneath the valley are several hundred meters of post-glacial lake sediments (Chaljub et al., 2006).

Mountain sports are an important tourist attraction in summer and winter. Twenty large and small ski resorts surround the city, the nearest being about a 15 minute drive.

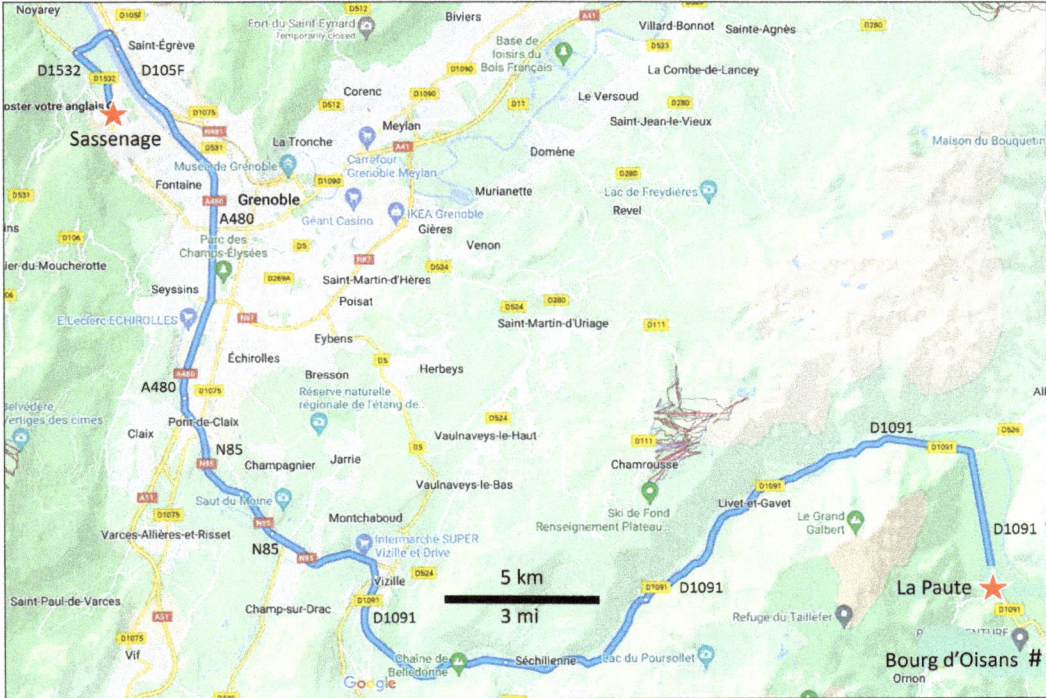

Sassenage to La Paute: Continue north on D1532; at the roundabout take the 1st exit (east) onto D105F; cross the Isère River and at the roundabout take the 1st exit (southeast) onto A48 to Grenoble; take Exit 8 for N85 toward Gap and Briançon; at Vizille stay left to continue onto D1091; just north of Oisans turn right (southwest) onto Mas du Vernay; turn left (south) toward Ham de Paute and pull over on the right. This is Stop 28, Alpine Folding, La Paute (45.07431, 6.01392) for a total of 56.0 km (34.8 mi; 46 min).

STOP 28 ALPINE FOLDING, LA PAUTE

This is a view stop. There are giant folds in the cliffs on all sides. The folds involve Early Jurassic limy mudstone and are a result of Miocene Alpine deformation. Regionally, most folds are west vergent, but the folds here verge east and the thrusts are steeply inclined to the west. The implication is that either (1) folds originally verged west and thrusts that were inclined to the east were rotated, or (2) there is backthrusting in this area.

View southwest at faulted folds in Jurassic units at La Paute. The folds verge east, and thrusts are steeply inclined to the west, suggesting either rotated imbricate thrusts or backthrusting. Dashed lines follow bedding; dotted lines are reverse faults.

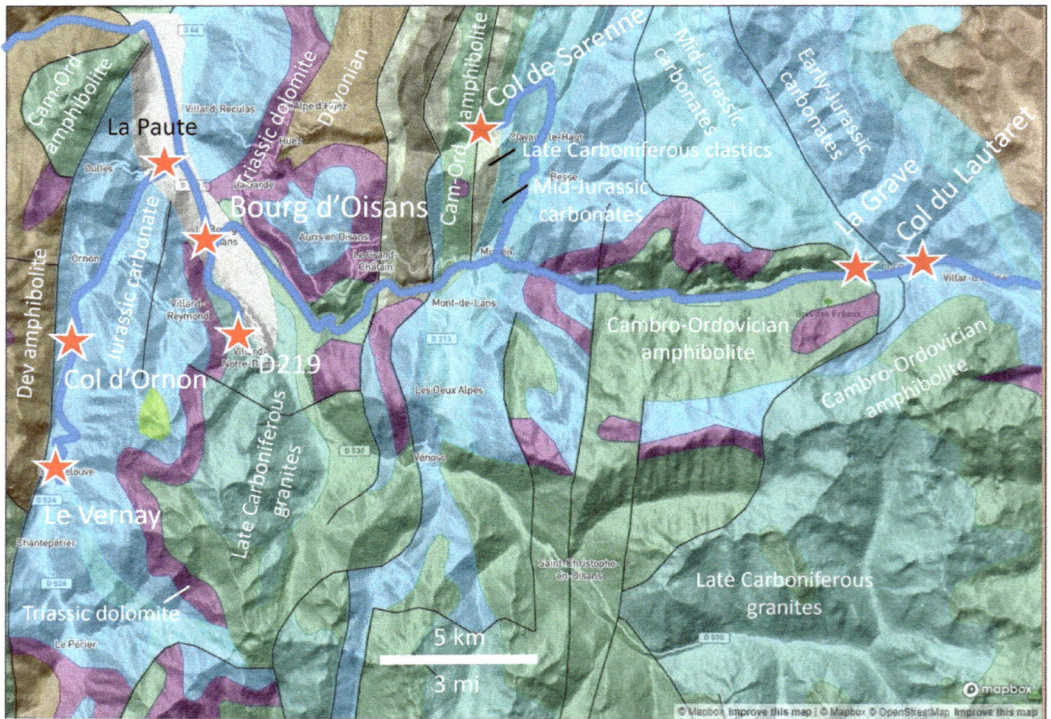

Geologic map of the Bourg d'Oisans area. Geology base from MacroStrat, 2015.

Side Trip 9, La Paute to Col d'Ornon: *Drive 30 m (100 ft) north to Ham de Paute; turn right (southeast) and drive to D526; turn right (southwest) on D526 and drive to* **ST 9.1, Triassic–Jurassic Normal Faulting, Col d'Ornon** *(45.0156, 5.9692) and pull over by the house on the right for a total of 10.3 km (6.4 mi; 14 min).*

If you prefer to skip Side Trip 9:

La Paute to Bourg d'Oisans*: Drive 30 m (100 ft) north to Ham de Paute; turn right (southeast) and drive to D526; turn left (south) on D526 and drive to D1091/Av. de la Gare; turn right (south) on D1091 and drive to parking area on the right just past Rue Gambetta. This is* **Stop 29.1, Alpine Folding, Bourg d'Oisans** *(45.055980, 6.029624) for a total of 2.5 km (1.6 mi; 4 min).*

SIDE TRIP 9 PANGEAN RIFTING, COL D'ORNON

These stops examine Upper Triassic–Lower Jurassic rift-related normal faults that occur in this area. This is the post-Variscan rifting episode that broke up Pangea, led to the separation of the European and African tectonic plates, and resulted in opening of the Piemont–Liguria/Tethys Ocean. Many of these ancient normal faults were reactivated as thrusts during the Alpine Orogeny many millions of years later. Alpine basement-involved folds and thrusts were strongly influenced by the orientation of the Triassic and Jurassic rift fault patterns (Dumont et al., 2008).

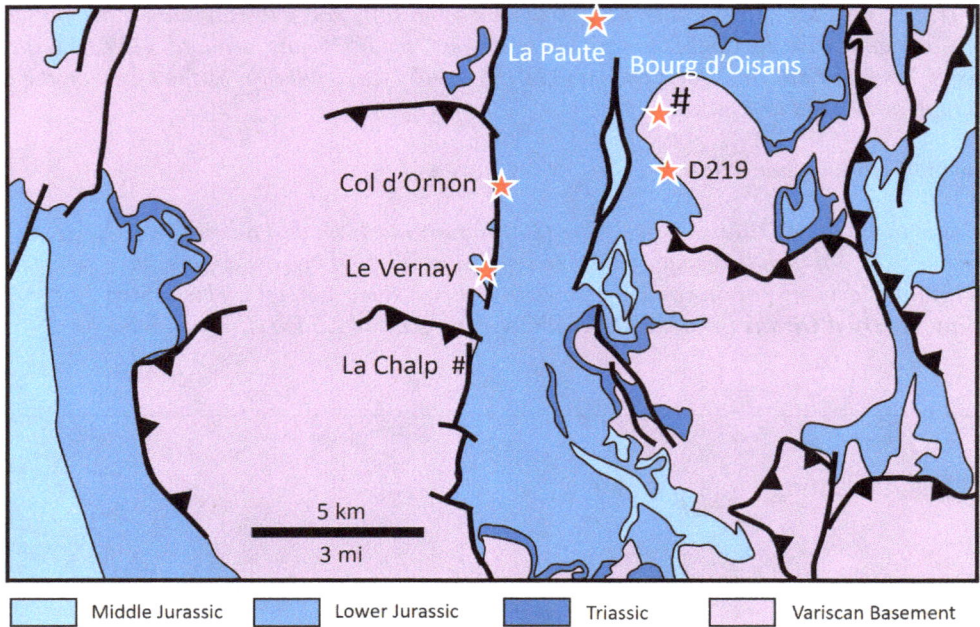

Triassic–Jurassic rifting, Dauphinois Domain. Heavy lines are normal faults; heavy lines with teeth are thrust faults. Modified after Dumont et al., 2008.

ST9.1 TRIASSIC–JURASSIC NORMAL FAULTING, COL D'ORNON

Roughly north-south oriented Jurassic normal faults separate the Variscan strongly foliated basement from Jurassic carbonates. The view north up this valley shows uplifted Devonian amphibolites known as the Taillefer Crystalline Basement forming the highlands to the west. Driving here from La Paute passes many roadcuts in the foliated bedrock, although there are few places to stop. A representative roadside measurement gives a foliation strike of N22°E and dip of 61° southeast.

View north to strongly foliated and faulted amphibolites bounding the west side of the Rif Garcin Valley.

Side Trip 9, Col d'Ornon to Le Vernay: *Continue driving south on D526 for 1.4 km (0.9 mi; 1 min) and pull over on the right. This is **Stop ST9.2, Triassic–Jurassic Rifting, Le Vernay** (45.0031, 5.9677).*

ST9.2 TRIASSIC–JURASSIC RIFTING, LE VERNAY

This stop is to observe one of the Upper Triassic–Lower Jurassic rift-related normal faults that are characteristic of this area. This is the rifting episode that broke up Pangea, led to the separation of the European and African tectonic plates, and resulted in opening of the Piemont–Liguria/Tethys Ocean.

You can see the fault contact between the Taillefer crystalline basement and the Early Jurassic (Lias) sediments.

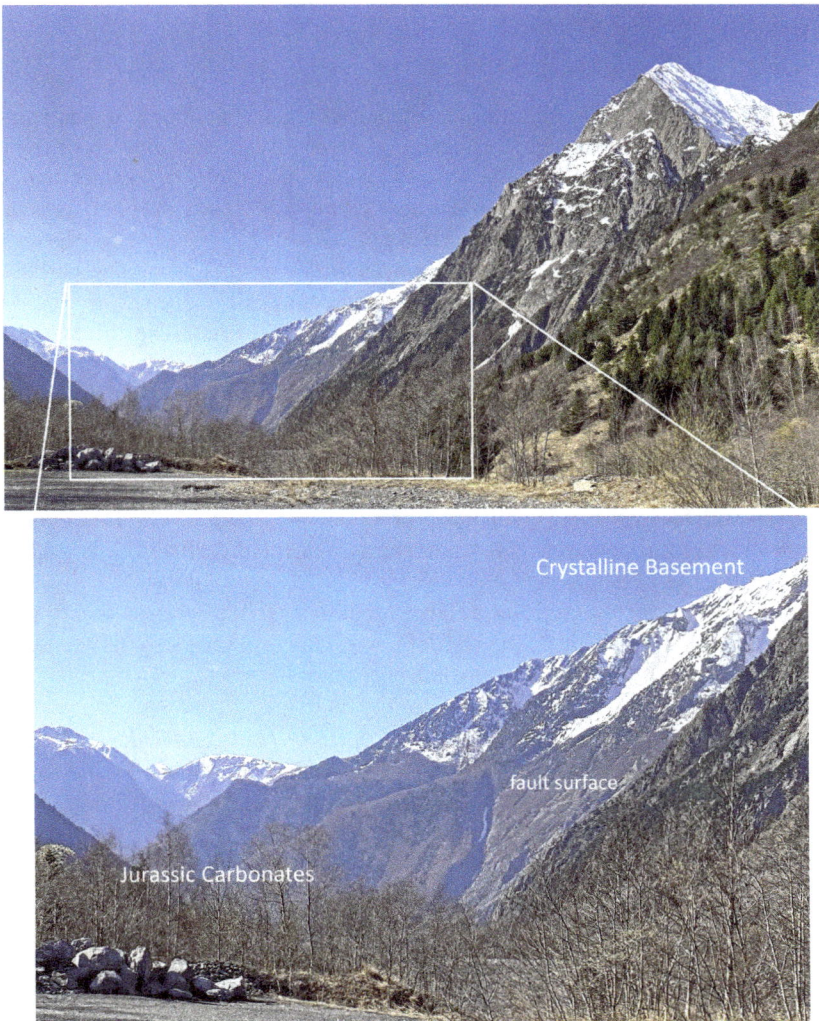

View south toward La Chalp d'Ornon. A Triassic normal fault separates Lower Jurassic carbonates on the east (left) from the Taillefer Crystalline Basement on the west (right).

*Side Trip 9, Le Vernay to Bourg d'Oisans: Return north on D526 and turn right (south) on D1091/Av de la Gare toward Bourg d'Oisans; turn right (west) into the parking area just past Rue Gambetta. This is **Stop 29.1, Alpine Folding, Bourg d'Oisans** (45.055980, 6.029624) for a total of 13.9 km (8.6 mi; 18 min).*

STOP 29 BOURG D'OISANS AREA

Bourg d'Oisans is a lovely small tourist and ski village nestled in the French Alps. The area around Bourg d'Oisans is part of the Dauphinois Domain. The northern Dauphiné region shows north-south to northeast-southwest-trending massifs composed of Cambro-Ordovician to Early Carboniferous metamorphic basement intruded by Variscan granites. The Palaeozoic basement and overlying sediments were truncated by a regional angular unconformity and overlain by widespread, thin Triassic shallow marine carbonates and deeper marine Jurassic limestones and marls. The Mesozoic section was altered to zeolite/greenschist (low temperature–low pressure) facies during Alpine metamorphism (Dumont et al., 2008).

STOP 29.1 ALPINE FOLDING, RUE GAMBETTA, BOURG D'OISANS

Jurassic rifting and later basement fold-and-thrust shortening were strongly dependent on the orientation of Tethyan extension and Alpine shortening, respectively. Reactivation of the preexisting Variscan tectonic fabric (faults) in response to oblique Jurassic extension produced an *en-échelon*, or staggered rift fault pattern. Mesozoic rift structures are still clearly visible, as seen at the Le Vernay stop, despite superimposed Alpine shortening.

Four post-rifting compressive events have been identified in this area. The earliest generated west-southwest-directed folds in the Mesozoic cover and could be Eocene age or older. The second deformation event created north-northwest-directed basement thrusts. The third, and main compressional event was west-northwest-directed and occurred before 24 Ma under a thick load of Penninic Nappes. The change in shortening direction from episode 2 to episode 3 appears related to a transition from northward propagation of the Alpine collision directly driven by Africa–Europe convergence, to onset of westward-driven thrusting in the western Alps. The last event is late Miocene and coincides with the final uplift of the Grandes Rousses and Belledonne external massifs and indicates an east-west to east-southeast-oriented maximum stress direction. It produced strike-slip faulting and local rotations that significantly deformed earlier Alpine folds and thrusts (Dumont et al., 2008).

From this stop you can see an impressive, large recumbent fold emphasized and highlighted by Early Jurassic thin bedded dolomites to the west of town. This fold verges to the southwest, suggesting it is the result of a phase 1 compressive event.

*Bourg d'Oisans to D219 Roadcut: Continue south on D1091B/Av. de la Republique; turn right (west) toward D219/Rue Humbert; turn left (south) onto D219/Rue Humbert and drive to pullout on left before the tunnel. This is **Stop 29.2, Variscan Granite and Unconformity, D219 Roadcut** (45.040935, 6.029686) for a total of 1.9 km (1.2 mi; 4 min).*

A large recumbent fold is outlined by Early Jurassic thin bedded dolomites west of Bourg d'Oisans.

STOP 29.2 VARISCAN GRANITE AND TOP PALEOZOIC UNCONFORMITY, D219 ROADCUT

At this stop we observe Triassic dolomite over Late Carboniferous (Pennsylvanian) granite. This is the regional angular unconformity at the top of the Paleozoic section. If you look north you can see the folded Early Jurassic (Lias) dolomite above the granite.

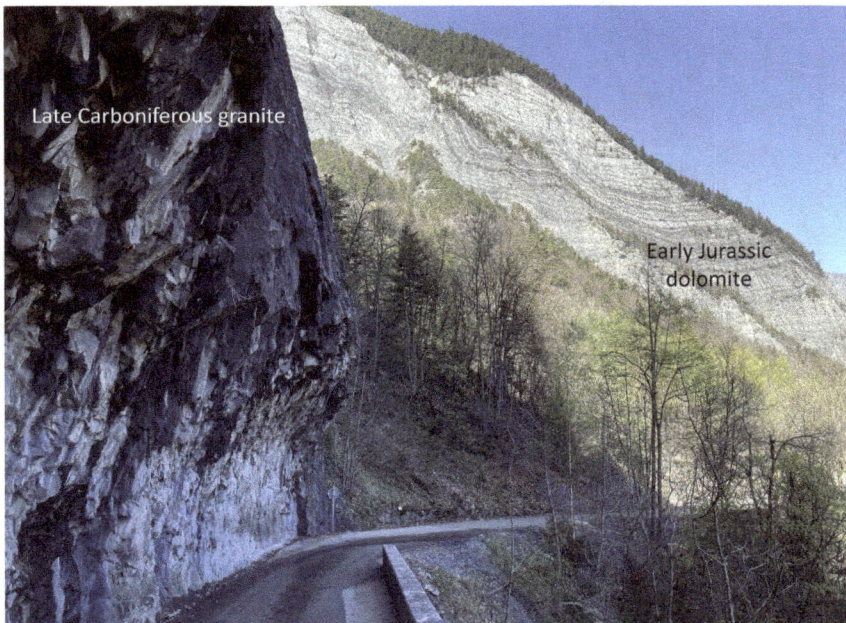

Variscan granite and folded Mesozoic sedimentary cover. The unconformity is in the trees. View to the northwest.

*Side Trip 10, D219 Roadcut to Col de Sarenne: Return north on D219 to Bourg d'Oisans; turn right (northeast) on Rue Gambetta; turn left (north) on D1091B/Av de la République; at the round-about take the 1st exit (east) onto D1091; at Mizoën turn left (north) onto D25; in about 3 km (1.9 mi) turn left (north) onto D25A; in 2.3 km (1.4 mi) turn left (southwest) onto Rte du Col de Sarenne; drive to the pass at **ST10, Alpine Ductile Shear Zone, Col de Sarenne** (45.087654, 6.149054) for a total of 30.6 km (19.0 mi; 45 min).*

The Col de Sarenne road is closed in winter due to snow. Check road conditions before you go.

If you prefer to skip Side Trip 10:

*D219 Roadcut to Meije: Return north on D219 to Bourg d'Oisans; turn right (northeast) on Rue Gambetta; turn left (north) on D1091B/Av de la République; at the roundabout take the 1st exit (east) onto D1091; drive 27.7 km (17.2 mi) on D1091 and pull over on the right. This is **Stop 30, Basement Thrust, Meije** (45.043381, 6.292373) for a total of 30.4 km (18.9 mi; 38 min).*

SIDE TRIP 10 ALPINE DUCTILE SHEAR ZONE, COL DE SARENNE

Walk to the outcrop on the east side of the pass to see a ductile shear zone, formed during the Alpine Orogeny, cutting through Paleozoic metasediments. A ductile shear zone is characterized by a zone of rock where deformation occurs primarily through ductile flow rather than brittle fracturing. This indicates that the zone formed at great depths. Such zones frequently contain mylonites, rocks with a distinct fabric (foliation and lineation) caused by crushing, shearing, and recrystallization of minerals.

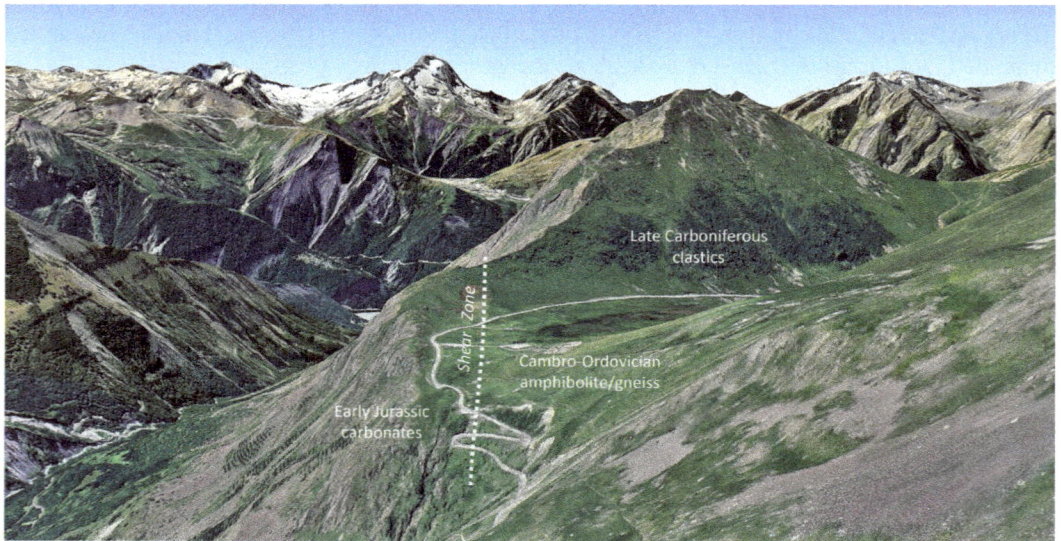

Google Earth oblique view south along the Col de Sarenne shear zone.

Google Street View south along the Col de Sarenne shear zone.

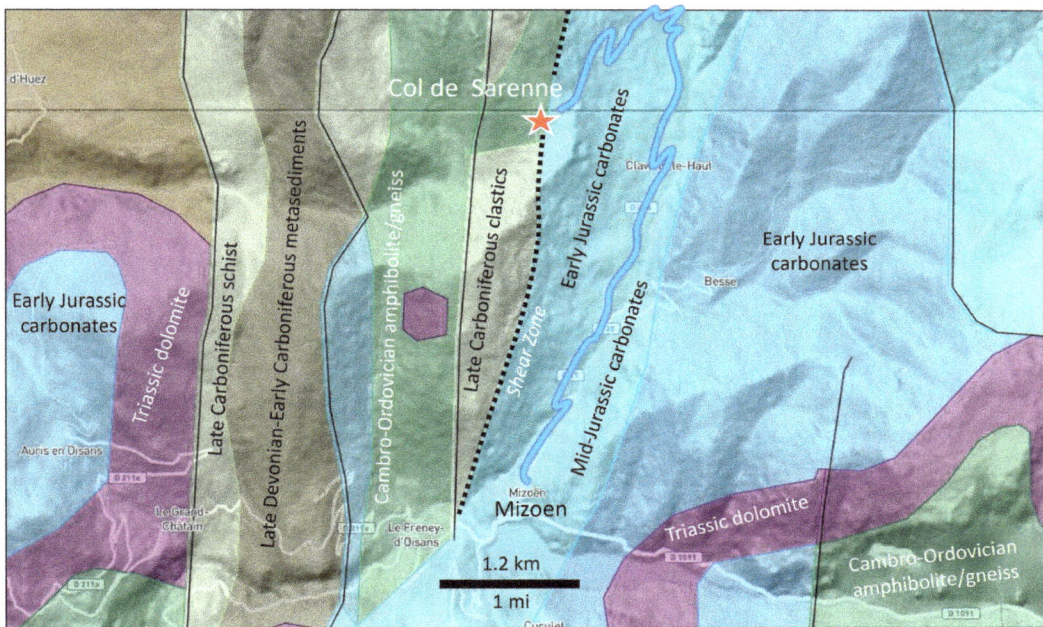

Geologic map of the Col de Sarenne area. The shear zone is indicated by the dotted line. Geology base from MacroStrat, 2015.

COTTIAN ALPS UNESCO GLOBAL GEOPARK

Starting just west of La Grave, we will be driving through the Cottian Alps Geopark (Alpes Cottianne Geoparc) until we emerge from the mountains at Pinerolo southwest of Turin. Spanning the French-Italian border, this park was established to promote the geologic heritage of the region, introduce the public to the history of the Alps, and encourage geotourism. The park has over 100 geologic sites and many museums and exhibits.

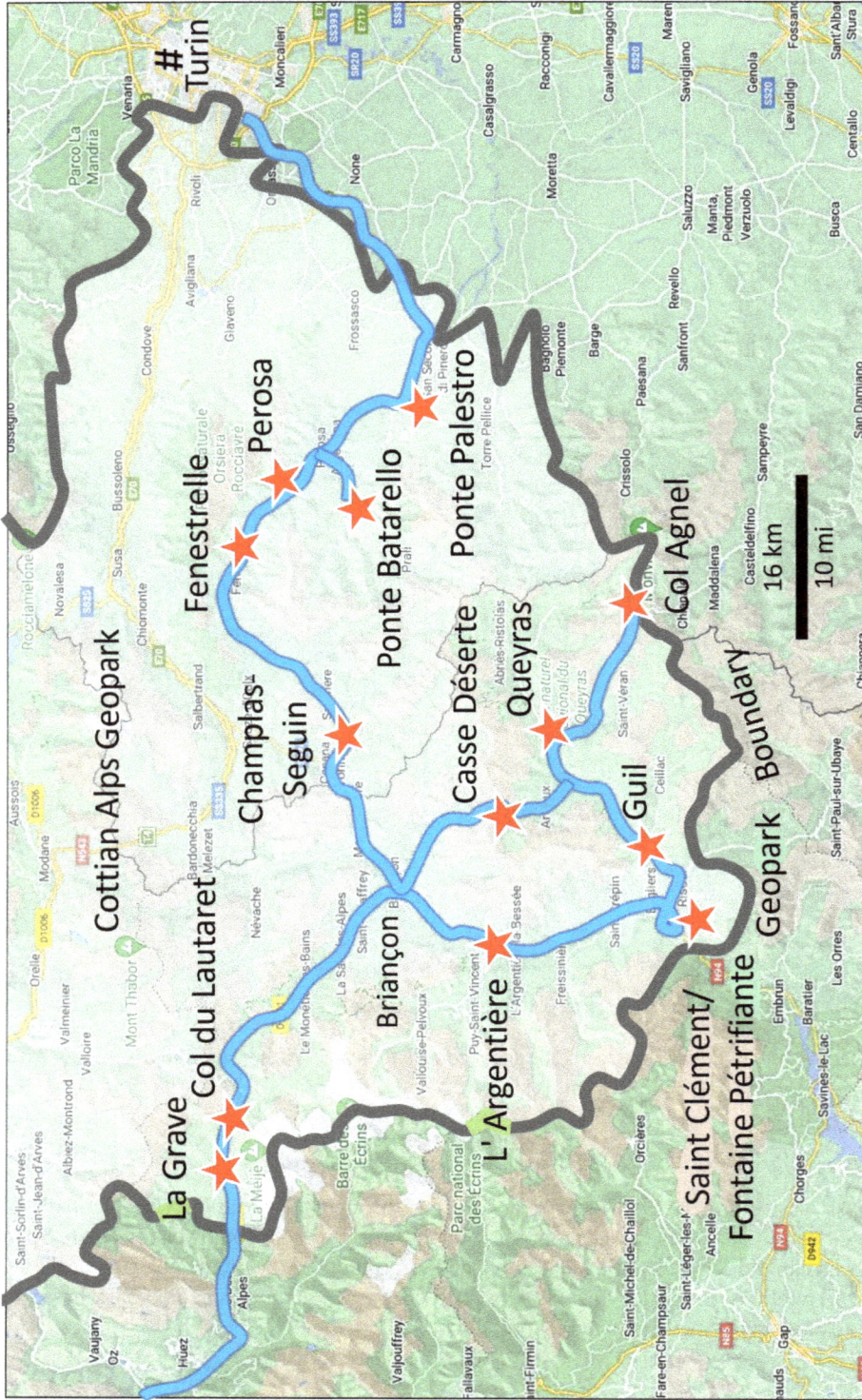

Map of our route and stops in the Cottian Alps Global Geopark.

Side Trip 10, Col de Sarenne to Meije: Return to Mizoën and turn left (east) onto D1091; drive to
Stop 30, Basement Uplift, Meije (45.043381, 6.292373) on the west side of La Grave; pull over on
the right for a total of 25.5 km (15.8 mi; 36 min).

STOP 30 BASEMENT UPLIFT, MEIJE

Looking northwest from this location you can see uplifted Meije crystalline basement along with
Penninic Nappes thrust east over Early Jurassic schist (metamudstone). Whereas most thrusting in
this region is west-directed, there appear to be small-scale east-directed thrusts along the eastern
margin of the Meije uplifted block.

The Meije block, containing uplifted Variscan basement, comprises mainly migmatite (partially
melted) gneiss and amphibolites overlain by a thin cover of Triassic quartzite and metadolomite
and Early Jurassic calcareous schists. This basement block has been affected by polyphase (that is,
multiple) deformation. Alpine deformation, superimposed on much older Variscan deformation, is
characterized by a 60°–70° east-dipping schistosity and shallow east-dipping shear zones (~25°).
The amphibolites were buried to depths between 10 and 15 km (6 to 9 mi); uplift and erosion began
at least at 27 million years ago (late Oligocene) for the Meije block based on zircon fission track
work (Bellanger et al., 2014; Lacombe and Bellahsen, 2016).

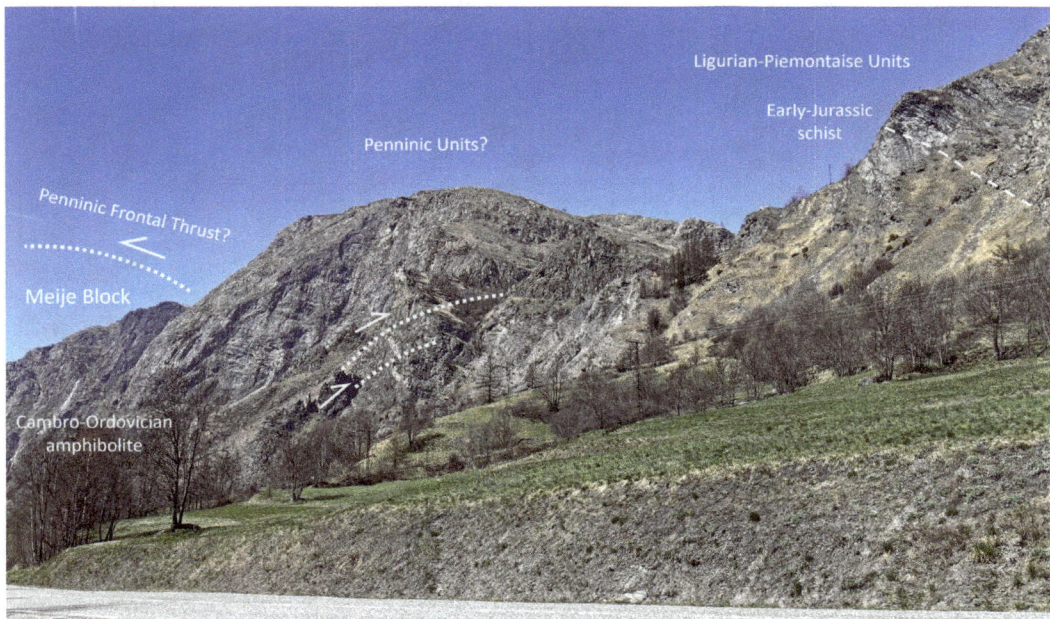

View northwest from Meije stop. Amphibolites are thrust over Jurassic schistose mudstone. The dashed line
follows bedding; faults are indicated by dotted lines. Structure is partly derived from mapping by Bellanger
et al., 2014.

Meije to La Grave: Continue east on D1091 for 2.1 km (1.3 mi; 3 min) to the first curve after the
*tunnel and pull over on the right. This is **Stop 31, Polyphase Deformation, La Grave** (45.048956,*
6.315516). Walk east ~40 m (130 ft) to see the roadcut. The shoulder is narrow: mind the traffic!

Stop 31 Polyphase Deformation, La Grave

The roadcut at this stop shows crenulation cleavage, an indicator of multiphase deformation of the Early Jurassic schist (Faure and Chen, 2018). Jurassic mudstones were overridden, metamorphosed, and deformed by Alpine thrusting and deep burial.

Crenulation cleavage provides insights into the complex deformation history of rocks, specifically those that have undergone multiple stages of metamorphism. It indicates that a rock has been folded and deformed across an earlier set of folds, a phenomenon that is common in metamorphic rocks that have undergone multiple episodes of deformation. This type of cleavage is characterized by small-scale wrinkles or rippling of an original foliation. Crenulation cleavage is significant because it reveals a history of multiple deformations, particularly those rich in phyllosilicate minerals like mica. This process can lead to the development of zones within the crenulated structure, for example, quartz and feldspar concentrated in the hinges of the folds, and phyllosilicates (micas, clays) concentrated in the limbs.

Roadcut in foliated Jurassic black schist. The foliation is inclined to the right (east) here.

The view to the south, besides being stunning, also shows large areas of highly eroded black schist in the middle distance. Outcrops along the highway are also very good, and there is another pullout on the right about 100 m farther east to see more outcrops of this black Jurassic schist.

La Grave to Col du Lautaret: *Continue driving east on D1091 for 9.9 km (6.1 mi; 13 min) to* ***Stop 32, Penninic Front, Col du Lautaret*** *(45.035752, 6.404863) parking area on the left.*

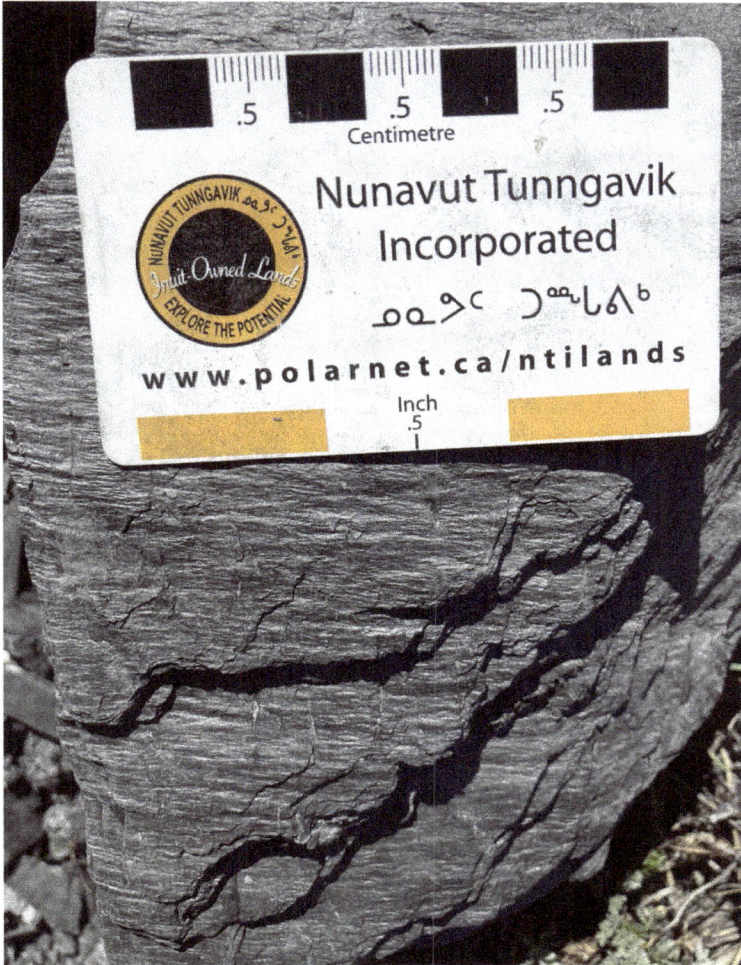

Crenulation cleavage in black schist, indicating multiple episodes of folding and metamorphism.

View south to outcrops of eroded black schist, center.

Stop 32 Penninic Front and Briançonnais Zone, Col du Lautaret

This stop provides a view east to the Penninic Front and Briançonnais Zone with its stack of "inner zone" nappes.

We are currently in the Ultra-Dauphinois zone, the easternmost part of the Dauphinois Domain, which is adjacent to and just west of the Briançonnais Zone of the Penninic Domain. The Ultra-Dauphinois zone is a nappe largely transported west-northwest over a thick gypsum decollement.

The Briançonnais Zone has been divided into a western Briançonnais "Zone Houillère" (the Paleozoic coal-bearing unit) and an eastern Briançonnais "Zone Vanoise," (where the Mesozoic series is folded and metamorphosed). The Mesozoic is a condensed section that spans Triassic through Cretaceous rocks. A condensed section consists of thin layers that cover a large span of time. The condensed section is topped by an unconformity. Above the unconformity are Early Eocene deposits of black and reddish shaley limestone and black mudstone. Sedimentation ended in middle Eocene time and was followed by uplift, thrusting, and erosion.

Structurally, the Briançonnais Zone is characterized by polyphase deformation. Briançonnais Nappes are characterized by east-verging folds. Overall, the Briançonnais Zone is described as a fan structure with rather more west-verging folds on the west and backfolded, strongly east-verging folds on the east (Faure and Chen, 2018).

Panoramic view east from Lautaret Pass showing the various nappes. Dashed lines follow bedding. The thrust is derived from Faure and Chen, 2018.

Col du Lautaret to La Mine d'Argent: Continue east on D1091 to to Briançon; at the roundabout take the 1st exit (south) onto N94; drive south on N94 to L'Argentière-la-Bessée; turn right (west) onto D994E/Rue des Giraudes; turn left (south) onto Av Charles de Gaulle; at the roundabout take the 2nd exit (south) onto Rue Saint-Jean; continue straight on Rue du Château; drive to **Stop 33, La Mine d'Argent, L'Argentière-la-Bessée** *(44.785178, 6.553222) parking area on the right for a total of 43.5 km (27.0 mi; 51 min).*

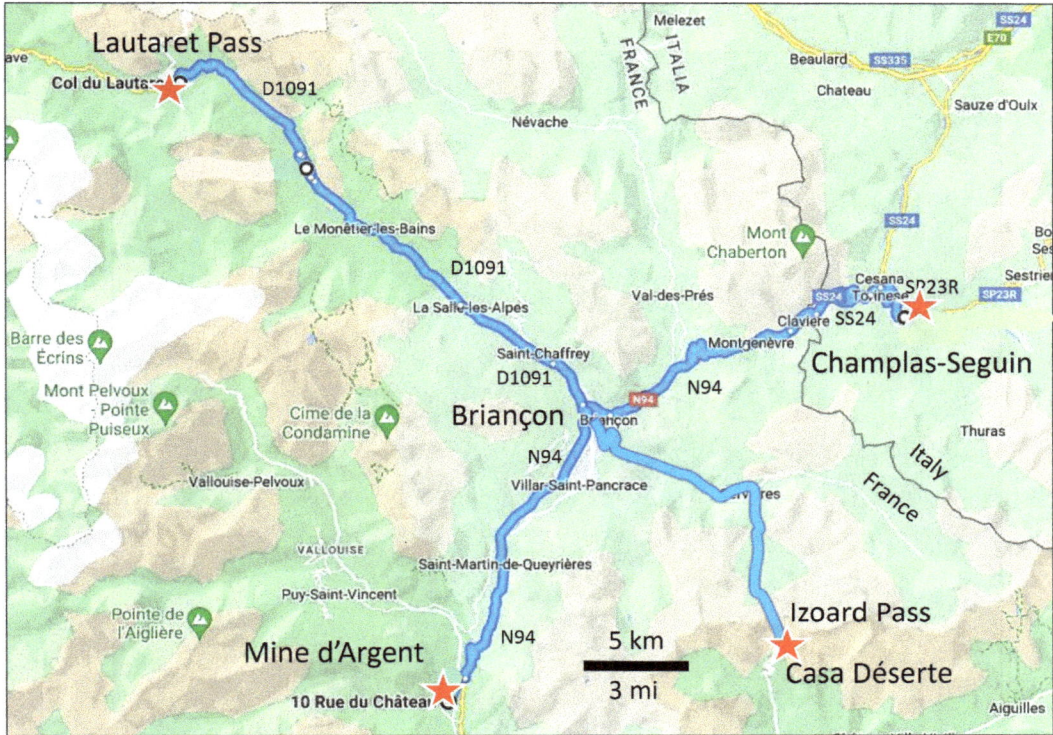

STOP 33 LA MINE D'ARGENT, L'ARGENTIÈRE-LA-BESSÉE

The Fournel deposit in the Hautes-Alpes (High Alps) is one of the main silver-lead deposits in the Alps. This Mississippi Valley–type deposit is hosted in Jurassic limestone. The source of the metal is unknown, but it is likely related to circulating low-temperature acidic hydrothermal (hot water) fluids reacting with the carbonates. The ore is almost exclusively galena, a silver-lead sulfide, with associated quartz and barite. At one time the ore yielded half a tonne of galena per m^3 (0.4 tons/yd^3). Extraction was difficult: hard country rock; a fragmented deposit; silver that was difficult to separate from the lead.

The mine was active between the 10th and the 13th centuries. The ore was mined until the surface deposits were depleted, some of the tunnels reaching to 150 m (490 ft) deep. Over 300 to 400 years, several thousand tons of lead and about 10 tons of silver were recovered. The silver was used mainly for local currencies.

The silver mining gave rise to the name of the village (Argentière, meaning silversmith). The mining village grew up around the castle of Argentière, eventually replaced the neighboring village of Rame, a former Roman outpost.

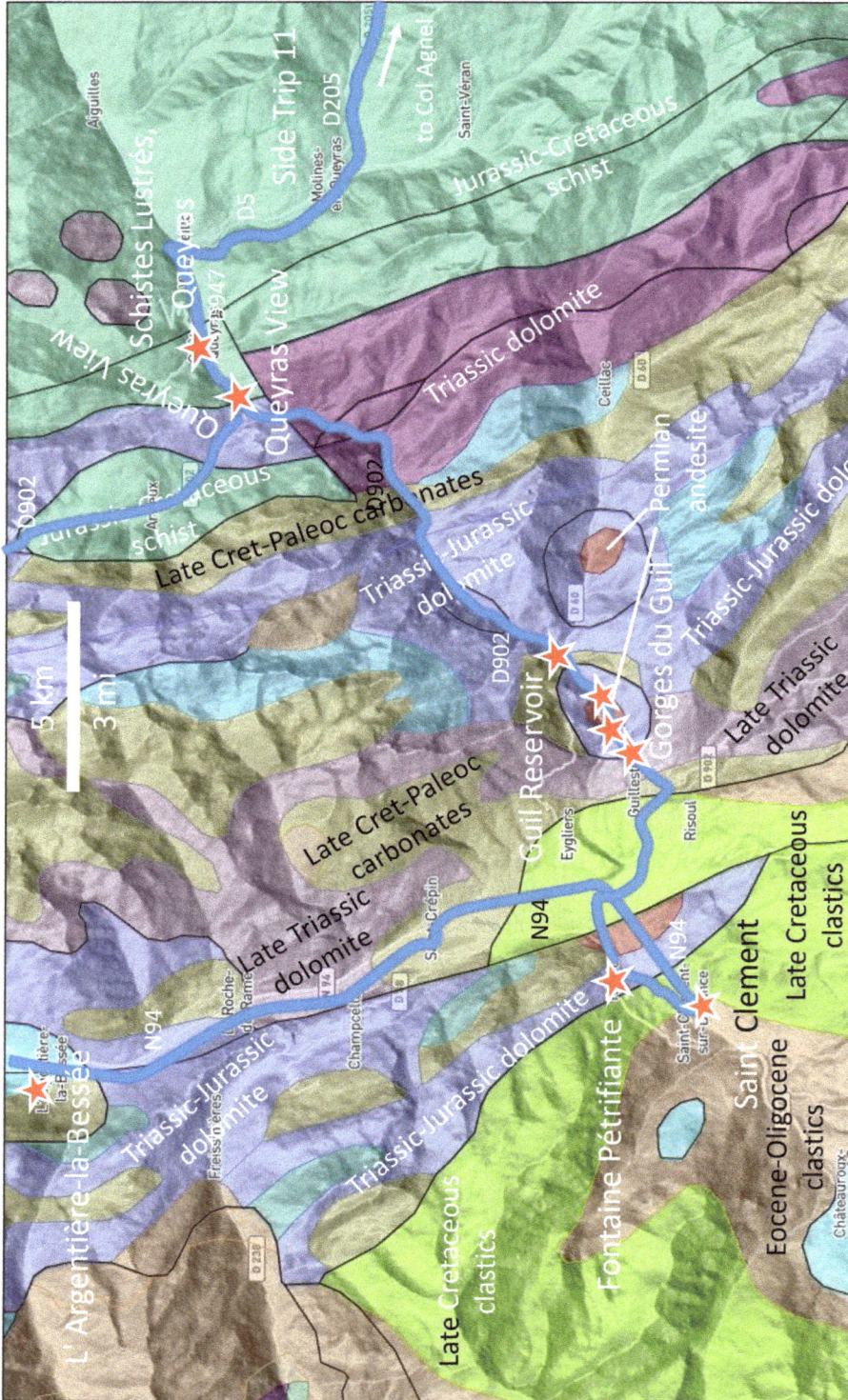

Geologic map of the L'Argentière-la-Bessée – Saint Clément – Gorges du Guil – Queyras area. Geology base from MacroStrat, 2015.

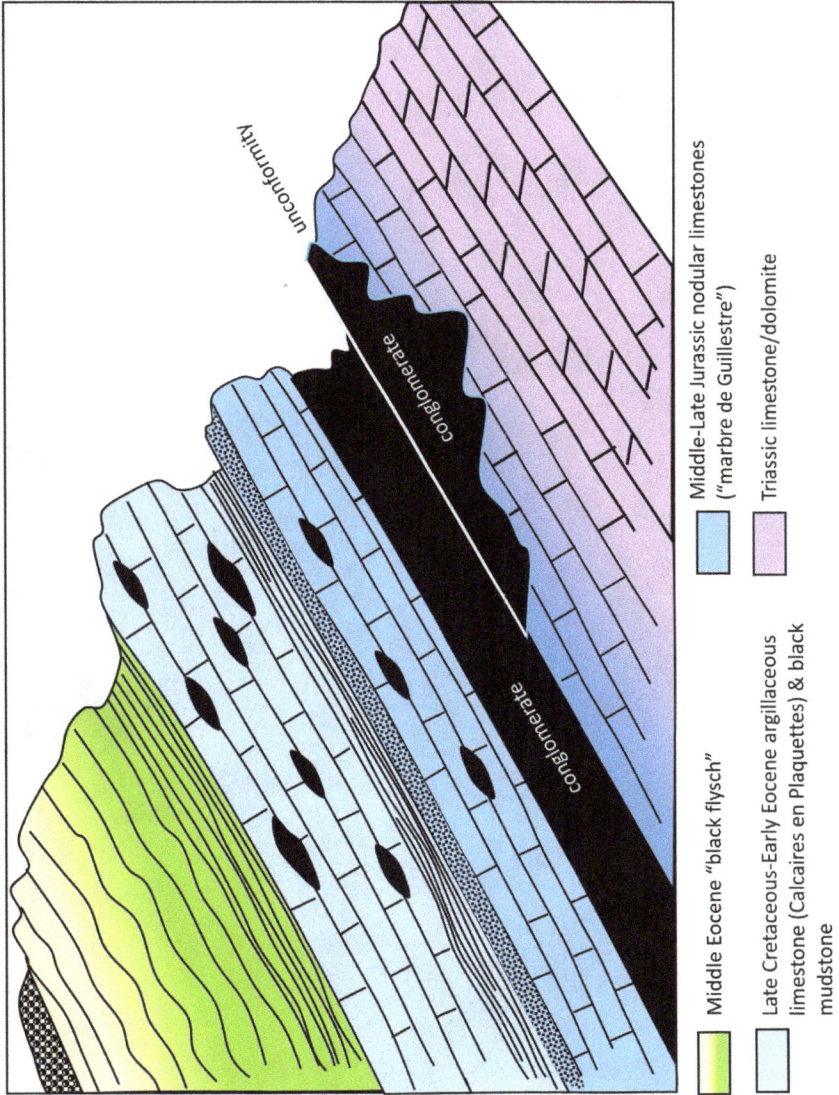

Condensed section in the Briançonnais zone. Modified after Faure and Chen, 2018.

Legend:

Middle Eocene "black flysch"

Late Cretaceous–Early Eocene argillaceous limestone (Calcaires en Plaquettes) & black mudstone

Middle–Late Jurassic nodular limestones ("marbre de Guillestre")

Triassic limestone/dolomite

unconformity

conglomerate

Nouveaux Etablissements d'Exploitation des Mines de l'Argentière
dans les Gorges du Fournel (Hautes-Alpes)
Lambert, édit., à Gap - Perrin, phot.

La Mine d'Argent, L'Argentière-la-Bessée.

After being abandoned for centuries, the mine reopened from 1785 to 1908. The Schlagberg company reopened it as the Fournel Mine, but neglected investment in the subsurface. This caused problems. A new manager, Duclos de Boussois, got the mine back on track, but his difficult personality led to his dismissal. Mining reached its peak under Pierre Suquet, a civil engineer who introduced pumps and a winch to lift ore from the mine. Production rose to 500 to 700 tonnes (550 to 770 tons) of ore concentrate per year. The concentrate was sent to the Luce and Rozan foundry in Marseille. During this period there were 300 to 400 workers at the mine, and the town grew to about 1,300 inhabitants.

After the banking crisis of 1870, the mine went into decline. Reserves were nearly exhausted. Two companies tried to restart the operation. The Argentiere Mineral Co. completed the Great Flow Gallery in 1879 but did not find economic amounts of ore. In 1892, the French Mine Ltd modernized the facility but failed to find reserves. By 1908 the deposit was abandoned for good.

Visit

The mine is open to the public from the spring (April) until the end of autumn (end of October) depending on the snow. Reservations are strongly advised. The silver mines in the Fournel valley have been open to the public since 1992 for guided tours.

The tour begins at the Mines' Museum (film and exhibition area). Visitors then travel to the Fournel Gorge by vehicle (10 min). During peak season (July and August) there is a shuttle bus due to a shortage of parking. The tour is not suitable for people in poor health or who have walking difficulties. Helmets are provided for the mine tour. There are informative signs explaining the underground workings and the surface buildings. The underground tour is ~600 m (1,970 ft) long and goes through two levels.

Address:	La Mine d'Argent (44.757108, 6.788484) 10 Rue du Château, 05120 L'Argentière-la-Bessée, France
Phone:	+33 4 92 23 02 94
Web site:	https://www.mines-argent-fournel.com/en
Hours:	9:00 to 17:00
Fees (2024):	
Full price:	Adults (+16 years old): €16, Children (-16 years old): €11.
Reduced price:	Adults (+16 years old): €13, Children (-16 years old): €10. Discounts for groups of 10 or more, job seekers, students, large families, on presentation of proof.

La Mine d'Argent to Saint Clément: *Continue south on Rue du Quartz; at the roundabout take the 3rd exit (east) onto D104; at N94 turn right (south) and drive to **Stop 34, Recumbent Fold, Saint Clément** (44.65048, 6.58427) and pull over on the right for a total of 18.4 km (11.4 mi; 19 min).*

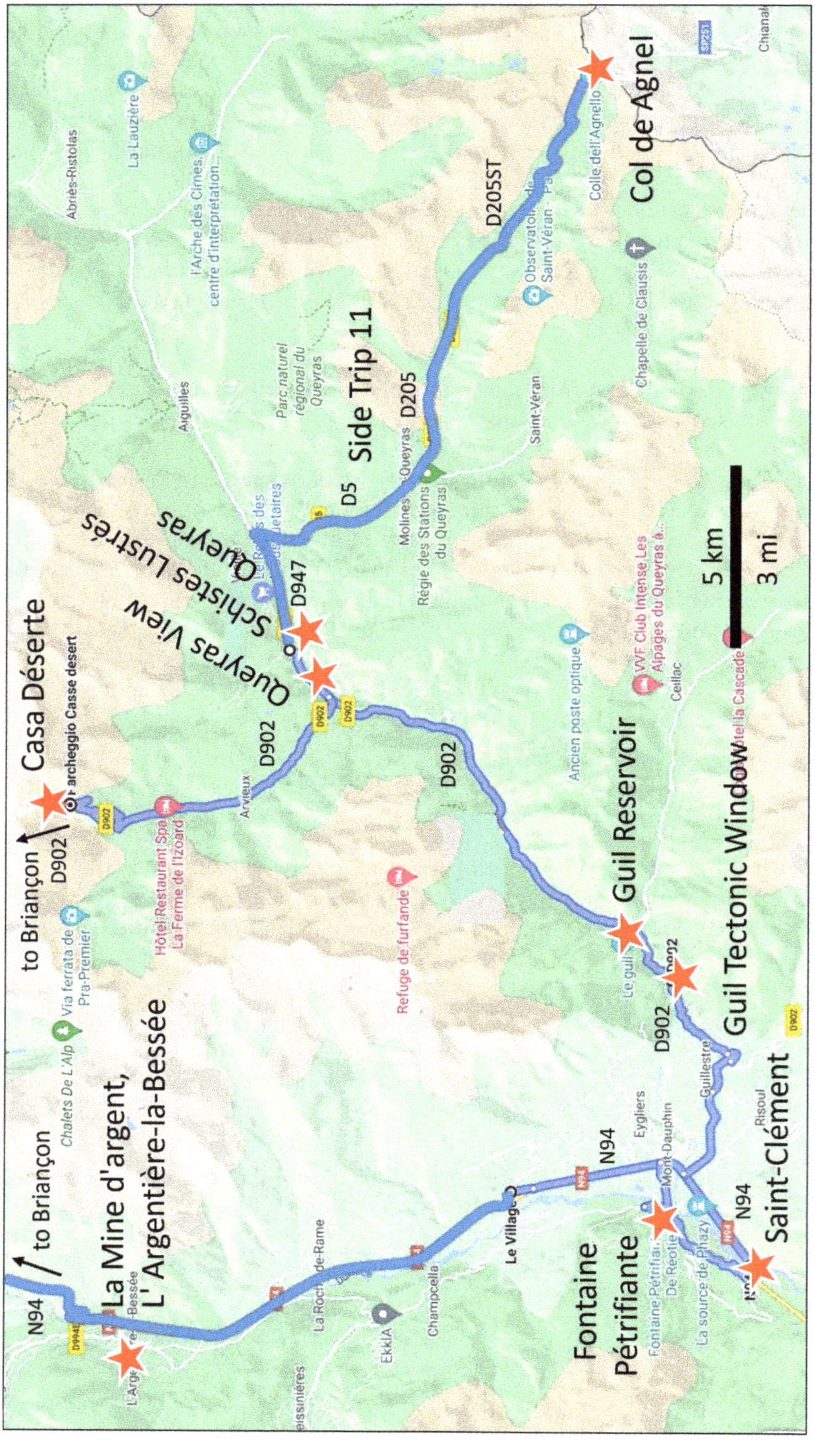

Col de Agnel
Side Trip 11
Queyras Schistes Lustrés
Queyras View
Casa Déserte
to Briançon
D902
Guil Reservoir
Guil Tectonic Window
Saint-Clément
Fontaine Pétrifiante
La Mine d'argent, L'Argentière-la-Bessée
to Briançon
N94
D947
D5
D205
D205ST
5 km
3 mi

STOP 34 RECUMBENT FOLD, SAINT CLÉMENT

This stop presents a spectacular example of a recumbent fold in the mountainside to the east.

The St. Clément fold is developed in Late Cretaceous to Eocene "Helminthoid Flysch." The mostly Late Cretaceous flysch consists of regionally widespread deep-water carbonate-rich turbidites deposited during early stages of the Alpine collision. The flysch is mainly basinal carbonate ooze mixed with clay and minor shale beds deposited in a subduction trench. But it was never subducted, and never metamorphosed.

This fold is considered to be part of a deformed thrust sheet.

The deformation of the thrust sheet is part of the fan folding in the Briançonnais Zone. In this case the fold is verging to the north or northwest (Faure and Chen, 2018).

Recumbent fold verging (inclinded to) northwest. View southeast.

*Saint Clément to Fontaine Pétrifiante: Continue south on N94 to D38; turn right (north) on D38 and drive to D37; turn right (east and south) onto D37 and drive to Fontaine Pétrifiante parking on the right. This is **Stop 35, Travertine Spring, Fontaine Pétrifiante de Réotier** (44.669765, 6.603084) for a total of 5.1 km (3.1 mi; 8 min). Follow the path south 340 m (1,120 ft) to the spring.*

STOP 35 TRAVERTINE SPRING, FONTAINE PÉTRIFIANTE DE RÉOTIER

Along the way to the travertine spring keep your eyes open for nice roadcuts through steeply east-inclined to near-vertical Late Cretaceous meta-shale and sandstone. The dip is likely related to down-to-the-east offset on the Durance Fault (Mathey et al., 2020).

A stop is often just as much about great scenery as it is about geology. This is one of those stops.

Fontaine Pétrifiante. The fantastic shapes of the travertine allow one's imagination to conjur all kinds of animals.

The "petrified fountain" is fed by waters emanating from the Durance Fault. The water enters the ground as rain or snowmelt, descends by gravity, acquires dissolved minerals as it percolates through limestone, dolomite, and gypsum, all the while warming 3°C (5.4°F) with every 100 m (330 ft) of depth. When the heated groundwater encounters the normal fault, it rises to the surface where it emerges at a temperature of 21°C (70°F). The water, saturated in calcium carbonate and carbon dioxide, precipitates calcite as travertine, a variety of limestone. Travertine precipitates because of the lower pressure and temperature at the surface (French Wikipedia, Fontaine Pétrifiante de Réotier).

Fontaine Pétrifiante to Gorges du Guil: *Continue east on D37 to N94; turn right (south) onto N94 and drive to the roundabout; take the 3rd exit (southeast) onto D902A; continue straight on D902 to* ***Stop 36.1, West Entrance, Gorges du Guil*** *(44.665898, 6.666564) and pull over on the left for a total of 8.0 km (4.9 mi; 10 min).*

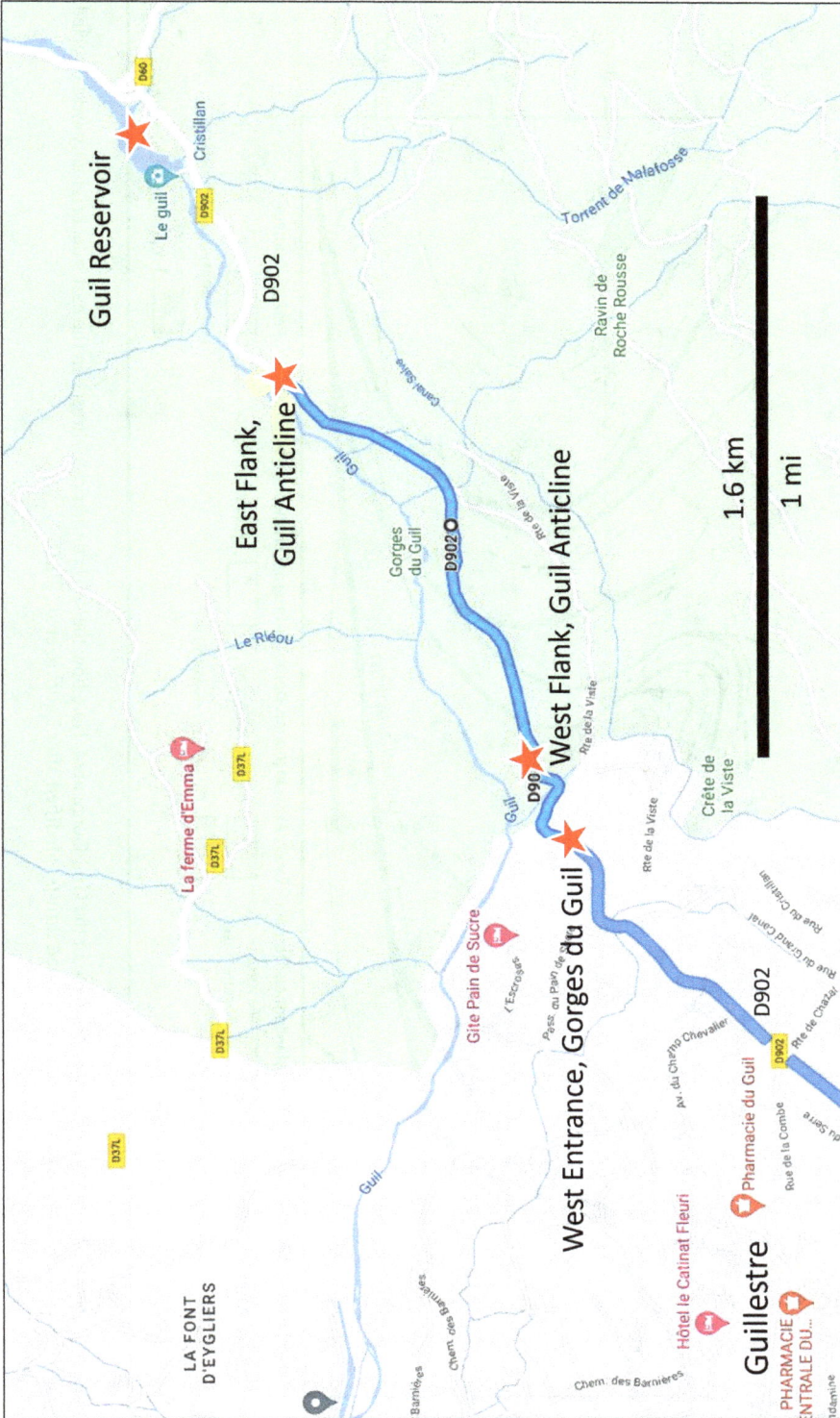

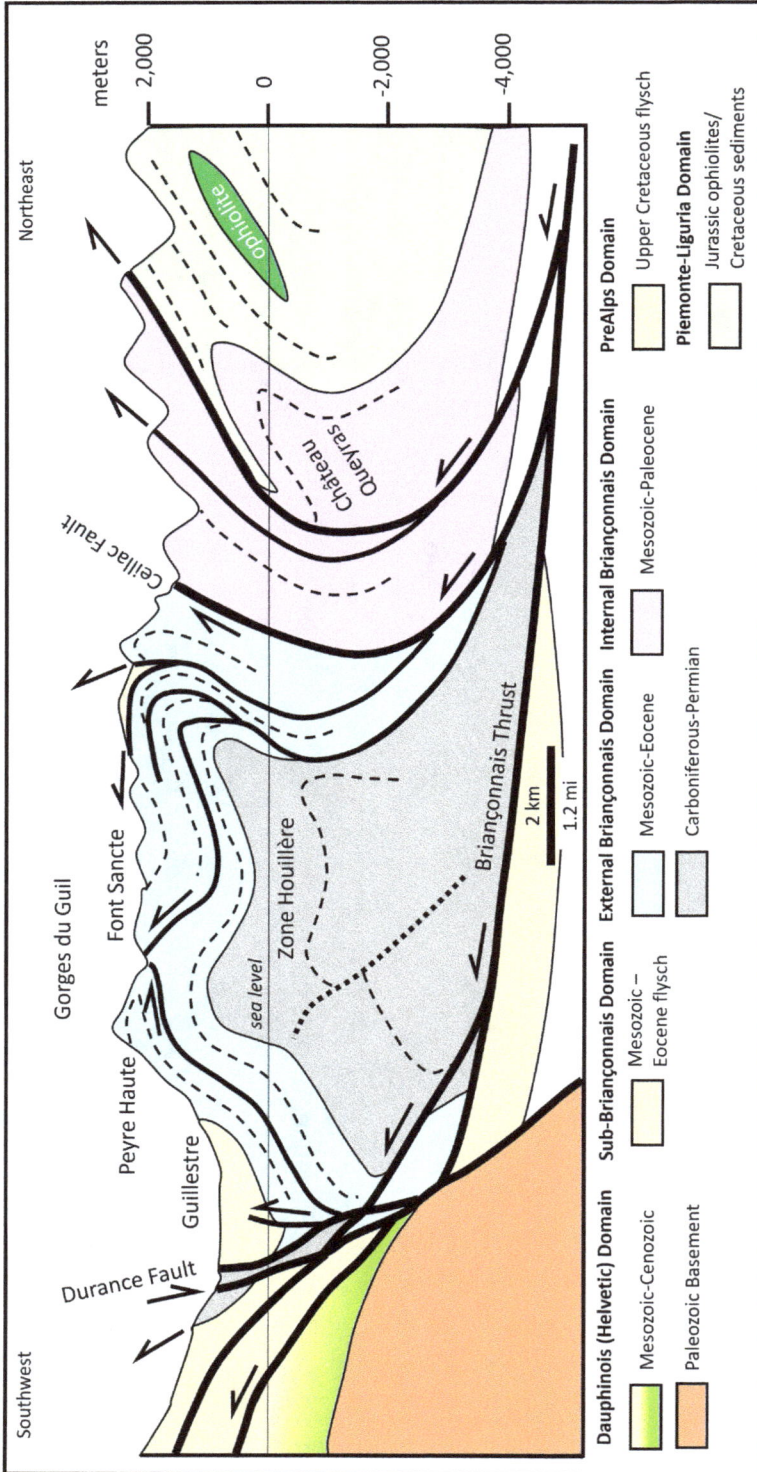

Regional cross section through the Briançonnais Zone in the Guil Gorge area. Note how the stacked imbricate thrusts on the east side, originally west-directed, have been rotated to overturned and appear to be east-directed thrusts. Modified after Ballèvre et al., 2020.

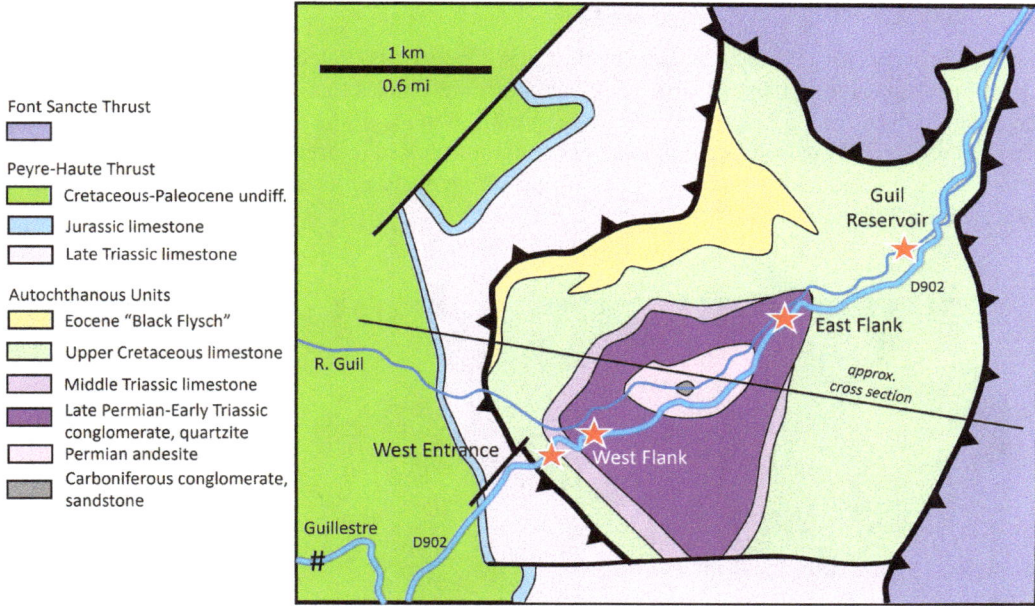

Geologic map of the Guil Tectonic Window. Heavy lines are normal faults; heavy lines with teeth are thrust faults. Modified after Ballèvre et al., 2020.

STOP 36 GUIL TECTONIC WINDOW

From here we get our first view of the Guil Anticlinorium, a complex southwest-verging fold that has been backfolded to the east. The anticlinorium uplifted the overlying thrust sheets, which have since been eroded off the uplift to form a fenster, or "tectonic window" where a hole has been eroded in young cover rocks and older rocks are exposed.

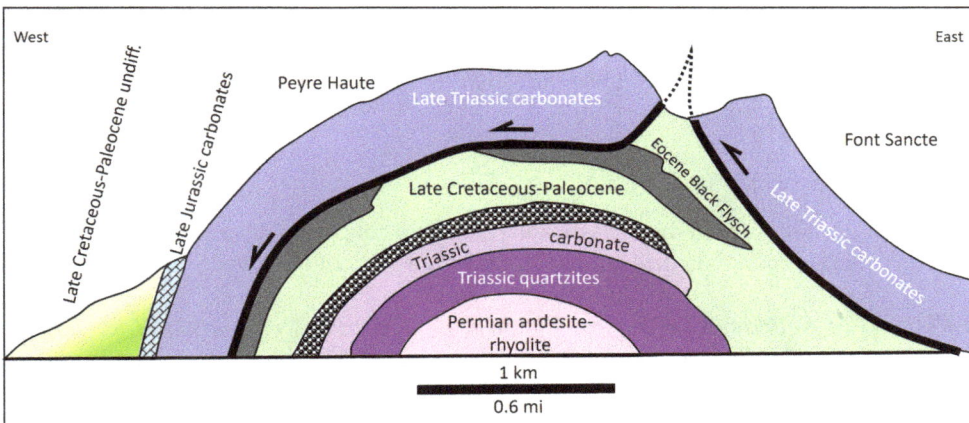

Cross section of the Guil Anticlinorium, Guil Gorge. Approximate location shown on previous figure. Modified after Faure and Chen, 2018.

STOP 36.1 WEST ENTRANCE GORGES DU GUIL

Beside the road are outcrops of deformed Upper Cretaceous–Paleocene schist. These Schistes Lustrés (glimmering schists) consists of organic-rich metashales that were originally deposited on the far margin of the European Plate or as deep marine muds in the Ligurian Ocean. These muds were deeply buried in a subduction trench and metamorphosed to greenschist and blueschist facies before being upifted and folded during the Alpine Orogeny. Rare foraminifera, a type of plankton, indicate a Late Cretaceous age (Faure and Chen, 2018).

West-inclined deformed slate, west side of Guil Anticlinorium. West entrance to Guil Gorge.

*Entrance, Gorges du Guil to West Flank Guil Anticlinorium: Continue east on D902 for 350 m (0.2 mi; 1 min) to **Stop 36.2, West Flank Guil Anticlinorium** (44.666859, 6.669740) and pull over on the right.*

STOP 36.2 WEST FLANK GUIL ANTICLINORIUM, GORGES DU GUIL

This stop provides a view of Permo-Triassic quartzites and conglomerates and the overlying Middle Triassic limestones exposed on the west flank of the Guil Anticlinorium.

*West Flank Guil Anticlinorium to East Flank: Continue driving east on D902 for 1.4 km (0.9 mi; 2 min) to **Stop 36.3, East Flank Guil Anticlinorium** (44.673218, 6.683462) and pull into the turnout on the left.*

Panoramic view north at the west flank and core of the Guil Anticlinorium. Dashed lines highlight bedding.

STOP 36.3 EAST FLANK GUIL ANTICLINORIUM

On the east flank of the structure are east-dipping Permo-Triassic quartzites and conglomerates. In the distance are north-inclined Upper Cretaceous–Paleocene marly limestones.

Panoramic view of the center and east flank of the Guil Anticlinorium and tectonic window. Dashed lines highlight bedding.

East Flank Guil Anticlinorium to Guil Reservoir: Continue driving east on D902 for 1.1 km (0.7 mi; 2 min) to **Stop 36.4, Guil Reservoir** *(44.676718, 6.693007) and pull into the parking area by the recycling bins on the left.*

STOP 36.4 VIVID COLOR, GUIL RESERVOIR

Guil Reservoir is a brilliant turquoise-colored lake. The color is every bit as vibrant as the world-famous aquamarine lakes of the Canadian Rockies. The colors are a result of dissolved calcium carbonate from the surrounding limestones scattering blue-green light. That, and perhaps some glacial-derived rock flour. It is worth a stop just to see the intense colors.

Side Trip 11, Guil Reservoir to Fort Queyras View: Continue driving northeast on D902 to D947; continue straight on D947 to **Stop ST11.1, Glacial Lock, Fort Queyras** *(44.7511, 6.7770) pullout on the right for a total of 12.2 km (7.6 mi; 12 min). Pull into the parking area on the right.*

Le Guil Reservoir, view north.

If you prefer to skip Side Trip 11:

Guil Reservoir to Casse Déserte: *Continue driving northeast on D902 to D947; turn left to stay on D902 to Briançon and Col d'izoard; drive another 11.8 km (7.3 mi) and pull over on the left. This is **Stop 37, Dolomite Hoodoos, Casse Deserte** (44.809051, 6.737722) for a total of 23.5 km (14.6 mi; 35 min).*

SIDE TRIP 11 SCHISTES LUSTRÉS AND COL AGNEL, QUEYRAS AREA

The side trip to Col Agnel provides a close-up view of the Schistes Lustrés and Monte Viso. Schistes Lustrés in the Queyras area are either European distal margin deposits or deep Ligurian Ocean deposits. The 10 km thick section consists mainly of metamorphosed ocean-bottom muds and shales that have also been deeply buried (to greenschist and blueschist stages), then uplifted and tightly folded during the Alpine Orogeny.

Encased within the Schistes are bits of ocean floor, including pillow basalts, ribbon chert, and oceanic crust, or ophiolites. Blocks of these rocks can be seen making up the bulk of Monte Viso. It is also a stunningly spectacular peak.

ST11.1 GLACIAL LOCK, FORT QUEYRAS

This stop provides a nice view of the castle at Fort Queyras and the jagged peaks beyond. While Fort Queyras does not sit on a classic roche moutonnée, it is perched upon a rocky promontory shaped by glaciers. The glacier that carved the Guil Valley at Queyras encountered a hard, resistant rock that was difficult to erode. This resistant rock became a "glacial lock," a dam-like feature that constricted glacial erosion and the subsequent valley. Fort Queyras was built on this naturally defensible, glacially shaped rock to take advantage of its commanding position in the narrow valley. A good example of how geology can shape military history.

Fort Queyras dates to the 13th century when it was originally built by the Dauphins of Viennois, then part of an independent principality. The fortress was built to protect the Guil Valley from

Fort Queyras castle and the "glacial lock" constricting the Guil Valley. View northeast.

invasions and threats from neighboring rulers, particularly from Italy. The original castle was far more modest than the fort that exists today: it had a defensive tower and several residential buildings surrounded by a wall. In the 17th century, under the reign of Louis XIV, the fort was upgraded and modernized. Renovations included new bastions, reinforcement of the walls, and expansion of the defensive system. Fort Queyras played a crucial role in several conflicts, including during the French Wars of Religion and invasions by Italian forces. The fort served as a refuge for the local population during armed conflicts. Although it was attacked multiple times, its strong walls and well-designed fortifications ensured it was never captured. In the 19th century, due to changing military strategies and technological advancements, Queyras gradually lost its significance as a defensive fortress. Today it is a historical landmark. It is open to visitors who want to explore the impressive walls, bastions, and breathtaking views of the surrounding Alps. In recent years, the fort has become a popular venue for cultural events, including historical reenactments, exhibitions, and festivals.

Side Trip 11, Fort Queyras View to Schistes Lustrés, Queyras: *Continue driving northeast on D947 for 1.3 km (0.8 mi; 2 min) and pull into parking area on the right. This is **Stop ST11.2, Schistes Lustrés, Queyras** (44.756808, 6.788843). Walk back 20 m (65 ft) to the roadcut at the curve. The shoulders are narrow: mind the traffic!*

ST11.2 SCHISTES LUSTRÉS, QUEYRAS

We are back in the Schistes Lustrés (Glimmering Shales), which we saw last in the Sesia-Lanzo Zone around Andrate. There the outcrops are Late Cretaceous high-pressure micaschists subducted beneath the Adriatic Microplate and metamorphosed to blueschist-eclogite facies during the Alpine Orogeny.

In the Queyras area the Schistes Lustrés is known for monotonous exposures of metamorphosed marl-rich sediments. Bodies of gabbro, basalt, ophiolite, and dolomite up to a few km in size are randomly distributed in these metasediments. This *mélange* (rock mixture) appears to be partly sedimentary and partly tectonic in origin (Tricart and Schwartz, 2006).

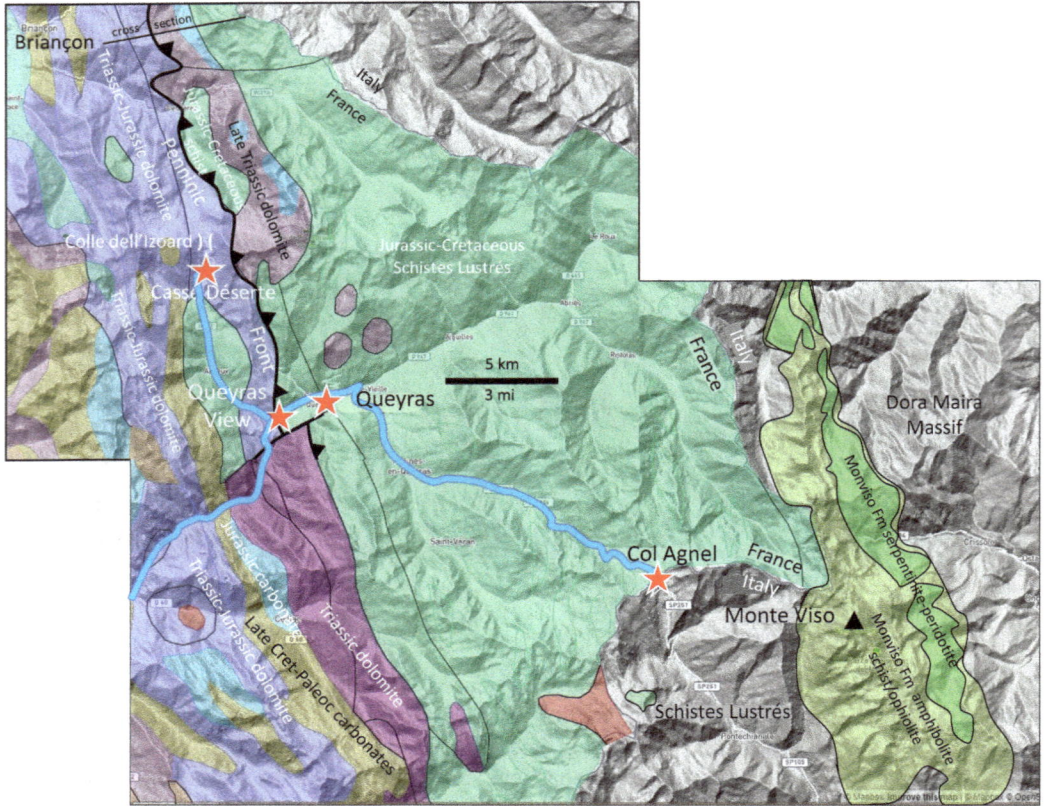

Geologic map of the Queyras, Col Agnel, and Casse Déserte area. Heavy lines are normal faults; heavy lines with teeth are thrust faults. Geology base from MacroStrat, 2015.

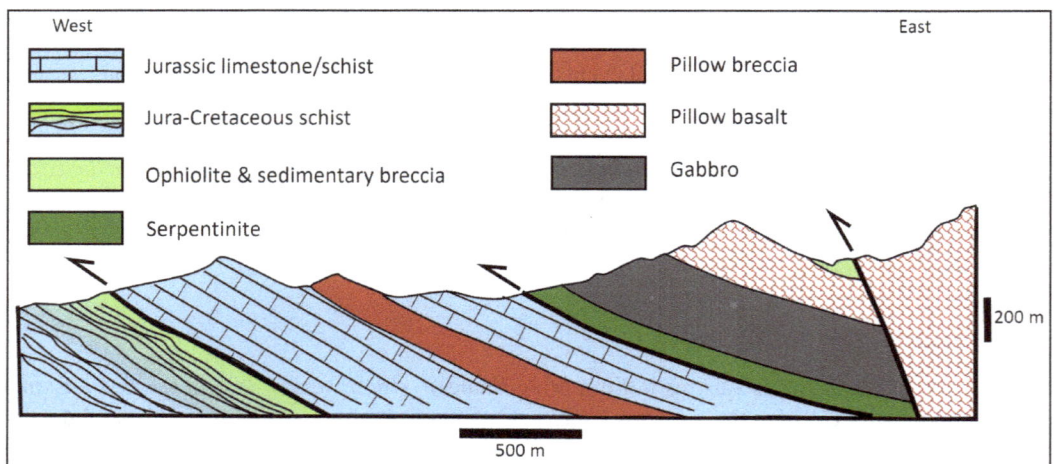

Cross section of the Chenaillet Ophiolite and Penninic Front. See previous figure for approximate location. Modified after Faure and Chen, 2018.

Schistes Lustrés roadcut at Queyras. View east.

Close-up of the Schistes Lustrés at Quayras.

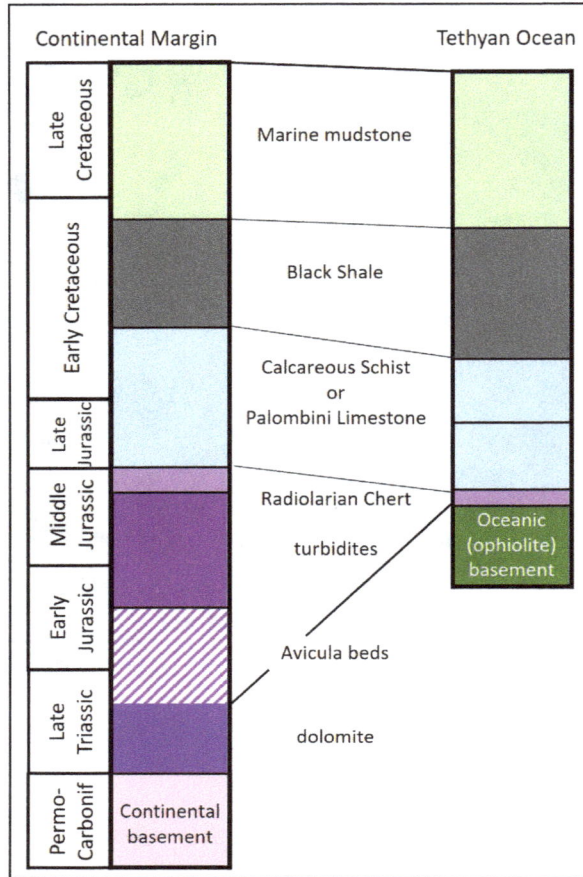

Stratigraphy of the Schistes Lustrés near Queyras. Stratigraphic information derived from Tricart and Schwartz, 2006.

There are two main types of Schistes Lustrés in the Queyras area. The first type consists of thrust sheets containing deposits derived from continental rifting and the distal European margin. These nappes are detached on Late Triassic evaporites and contain latest Triassic to latest Cretaceous units. The second type of Schistes Lustrés is derived from the Ligurian Ocean. This unit begins with reworked ophiolites. These breccias are locally associated with pillow basalts and grade upward into Jurassic radiolarian cherts (a silica-rich rock derived from the shells of radiolarian plankton). The old ocean floor is overlain by widespread Late Jurassic marine limestone (the "Malm Marble"). These are overlain by alternating Cretaceous limestones, black shales, and marls occasionally containing km-scale ophiolite blocks. The section has been tightly folded and comprises a large part of the Schistes Lustrés in the Queyras area (Tricart and Schwartz, 2006).

The 10 km (33,000 ft) thick Schistes Lustrés section in the Queyras area underwent a complex deformation history that began with the building of a subduction wedge (the sediments piled up in a subduction zone trench). This wedge consisted of oceanic and distal European margin rocks. Early development of the tectonic wedge is associated with blueschist (high pressure–low temperature) metamorphism in Paleocene–Eocene time. During the Alpine collision, mainly in Oligocene time, the wedge continued to grow and was subject to greenschist facies (low temperature–low pressure) metamorphism.

Alpine compression was followed by two distinct uplift and erosional episodes. First, a deep, pervasive east-vergent ductile event was responsible for most of the uplift of the Schistes Lustrés complex from 50–40 Ma. Second, a shallower west-vergent ductile-to-brittle regional uplift and extension event occurred from 40–35 Ma (Agard et al., 2001).

Side Trip 11, Queyras to Col Agnel: *Continue northeast on D947 to D5; turn right (south) onto D5 and drive to D205 in Molines-en-Queyras; turn left (east) onto D205; continue straight onto D205ST and drive to* **Stop ST11.3, Col Agnel and Monte Viso** *(44.684014, 6.979660) for a total of 25.1 km (15.6mi; 50 min).*

Col Agnel road is closed in winter and spring due to snow. Check road conditions before you go.

ST11.3 Col Agnel and Monte Viso Ophiolites

From Agnel Pass there is a magnificent view of Monte Viso (also known as Monviso) and ophiolites that originated at the bottom of the Liguro-Piemontais Ocean.

The ophiolites lie within the Schistes Lustrés nappes discussed at the previous stop. Planktonic foraminifera indicate a Late Cretaceous age for these rocks. Ophiolites indicate a true oceanic domain. Ophiolites occur as meter- to kilometer-sized blocks. Monte Viso is known for ophiolites that have undergone high-pressure metamorphism to blueschist and eclogite facies.

It is worth noting that, in the Alps, a complete ophiolitic sequence is never observed. A typical ophiolite sequence, from top to bottom, includes sediments, pillow lavas, sheeted dikes, gabbro, and peridotite. Gabbro, peridotite, pillow lavas, and siliceous sedimentary rocks are scattered within the Schistes Lustrés. In some places cherts, mudstone, or limestone directly cover serpentinite (hydrated peridotite) or peridotite. Most workers believe that the Liguro-Piemontais Ocean originated as a rift basin that evolved into an ocean with a mid-oceanic ridge (Faure and Chen, 2018).

Closeup of the west face of Monte Viso as seen from Col Agnel. Photo courtesy of Luca Bergamasco, https://commons.wikimedia.org/wiki/File:VisoVisolottoVisovallanta.jpg?uselang=uk

Side Trip 11, Col Agnel to Casse Déserte: *Return north on D205 to D5; turn right (north) on D5 and drive to D947; turn left (west) onto D947 and drive to D902; turn right (north) on D902 to Briançon and Col d'izoard; drive another 11.8 km (7.3 mi) and pull over on the left. This is **Stop 37, Dolomite Hoodoos, Casse Déserte** (44.809051, 6.737722) for a total of 37.2 km (23.1 mi; 61 min).*

The road across Izoard Pass is usually closed in winter and spring due to snow. Check road conditions before you go.

STOP 37 DOLOMITE HOODOOS, CASSE DÉSERTE

From this stop you get a view of the Piemonte Thrust front to the east and eroded dolomite pillars and spires in a glacial landscape. These hoodoos are bits of more resistant dolomite that weather into fantastic shapes, a geologic fairyland. It could be a smaller version of Bryce Canyon National Park in the American Southwest (in that case, eroded pinnacles of pink limestone).

A 1.3 km (0.8 mi) trail has been described by Climbing7.com (https://climbing7.com/2021/08/09 /la-casse-deserte-from-col-dizoard-arvieux-queyras/) as

> From the Col d'Izoard, at an altitude of almost 2400m, the short Casse Déserte hike offers a spectacular, almost effortless walk. The traverse of the giant scree is impressive, and the geological structures that emerge from the scenery are unusual. Strange orange-hued shapes rise up in places, forming tooth-like formations, oddly shaped needles and even natural arches. The views over the Arvieux Valley, the Clot la Cime summit opposite and the Col d'Izoard are superb.

Dolomite outcrop and scree at trailhead. View north, Casse Déserte.

Casse Déserte to Champlas-Seguin: *Continue north on D902 to Briançon; at the roundabout take the 1st exit (northeast) onto D2; turn right (east) onto N94/Baldenberger; drive northeast on N94 to the Italian border; continue straight on SS24; at the roundabout in Cesana Torinese take the 1st exit (southeast) onto SP23R; continue east on SP23R to **Stop 38, Ophiolite and Chert, Champlas-Seguin** (44.94112, 6.80767) and pull onto the paved track on the left by the fenced outcrop for a total of 47.8 km (29.7 mi; 1 hr 20 min).*

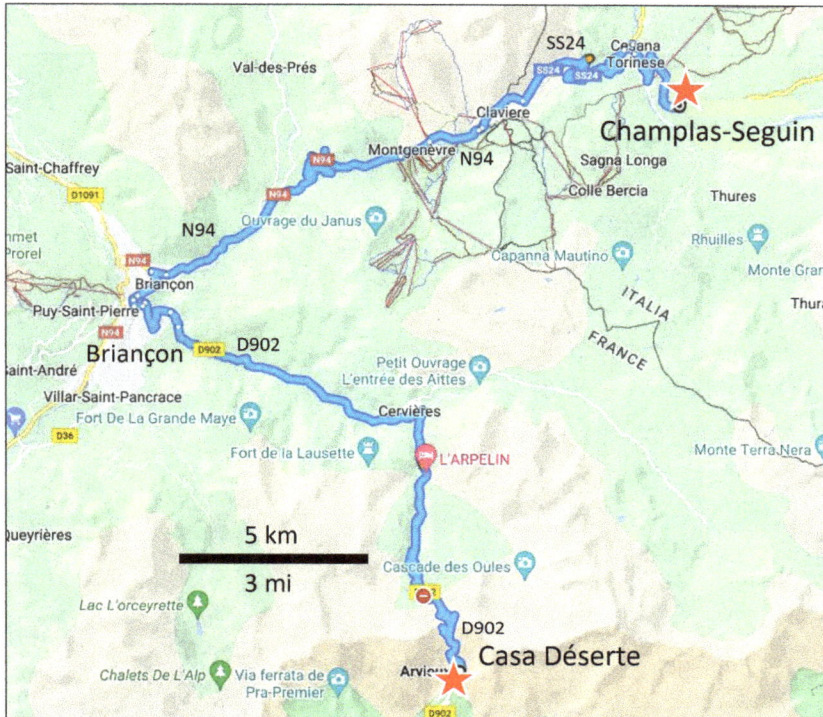

Besides the fenced serpentinite, walk up the jeep track 150 m (500 ft) to the display panel and banded radiolarian chert outcrops.

STOP 38 OCEAN FLOOR TO THE SKY: OPHIOLITE AND RADIOLARIAN CHERT AT CHAMPLAS-SEGUIN

In the Western Alps, remnants of the Liguro-Piemontaise Ocean are preserved in the Schistes Lustrés, also known as Piemontese or Ligurian-Piedmont units. These include slices of the thinned continental margin of the Europe plate. The Piemontaise Unit has been divided in two main groups based on their stratigraphic and metamorphic characteristics. The first group, including the Tsaté, Charbonnel, and Mirabouc-Bouchet units, is characterized by greenschist and blueschist (low to moderate grade) metamorphism of ultramafic bodies with subordinate basalts and gabbros enclosed in thick calcschistes. The second group includes the Zermatt-Saas Fee, Rocciavré, and Monviso units and consists of large and strongly deformed ophiolitic sequences and minor metasediments (quartzites and calcschists). These ophiolites were affected by high-pressure metamorphism. The Lago Nero Unit falls in the second group of the Schistes Lustrés complex (Burroni et al., 2003).

At Col Agnel we looked out over the Schistes Lustrés terrane to see the ophiolites of Monte Viso that formed on the bottom of the Liguro-Piemontaise Ocean. At this stop we can actually touch both an ophiolite block and red-and-green radiolarian cherts encased within the Schistes Lustrés.

Champlas–Seguin to Mount Ciantiplagna: Continue east on SP23R for 26.0 km (16.1 mi) to the sign for Laux; take a sharp right onto Via al Lago and drive another 0.8 km (0.6 mi) to a parking area on the left. This is Stop 39, Mount Ciantiplagna Rock Avalanche (45.041719, 7.021434) for a total of 26.8 km (16.7 mi; 34 min).

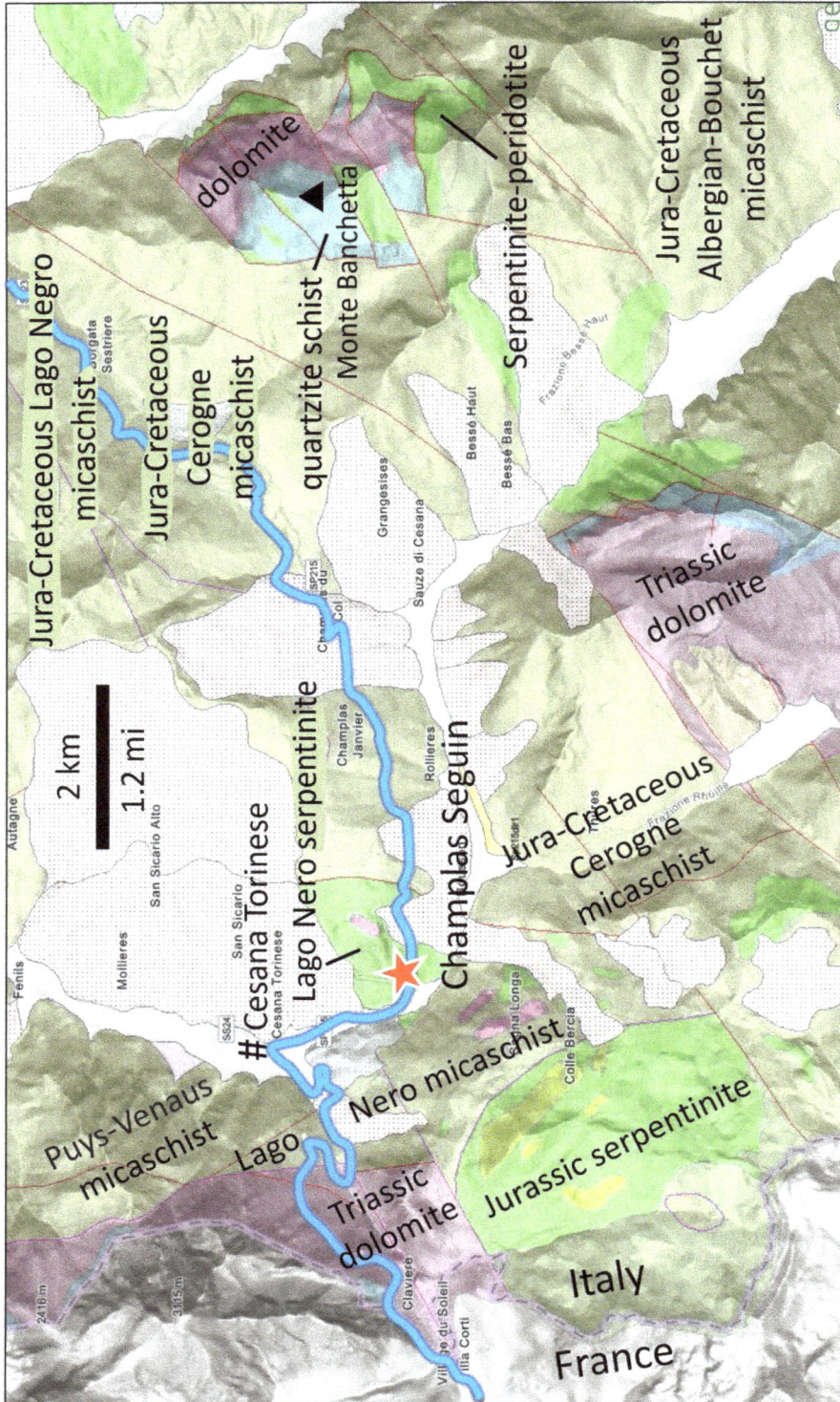

Geologic map of the Champlas–Seguin area. Geology base from Arpa GeoPiemonte, 2021, https://webgis.arpa.piemonte.it/Geoviewer2D/index.html?config=other-con-figs/geologia250k_config.json

Outcrop of serpentinized ophiolite within the Schistes Lustrés, Champlas–Seguin.

Oucrop of red and green banded radiolarian chert, Champlas–Seguin.

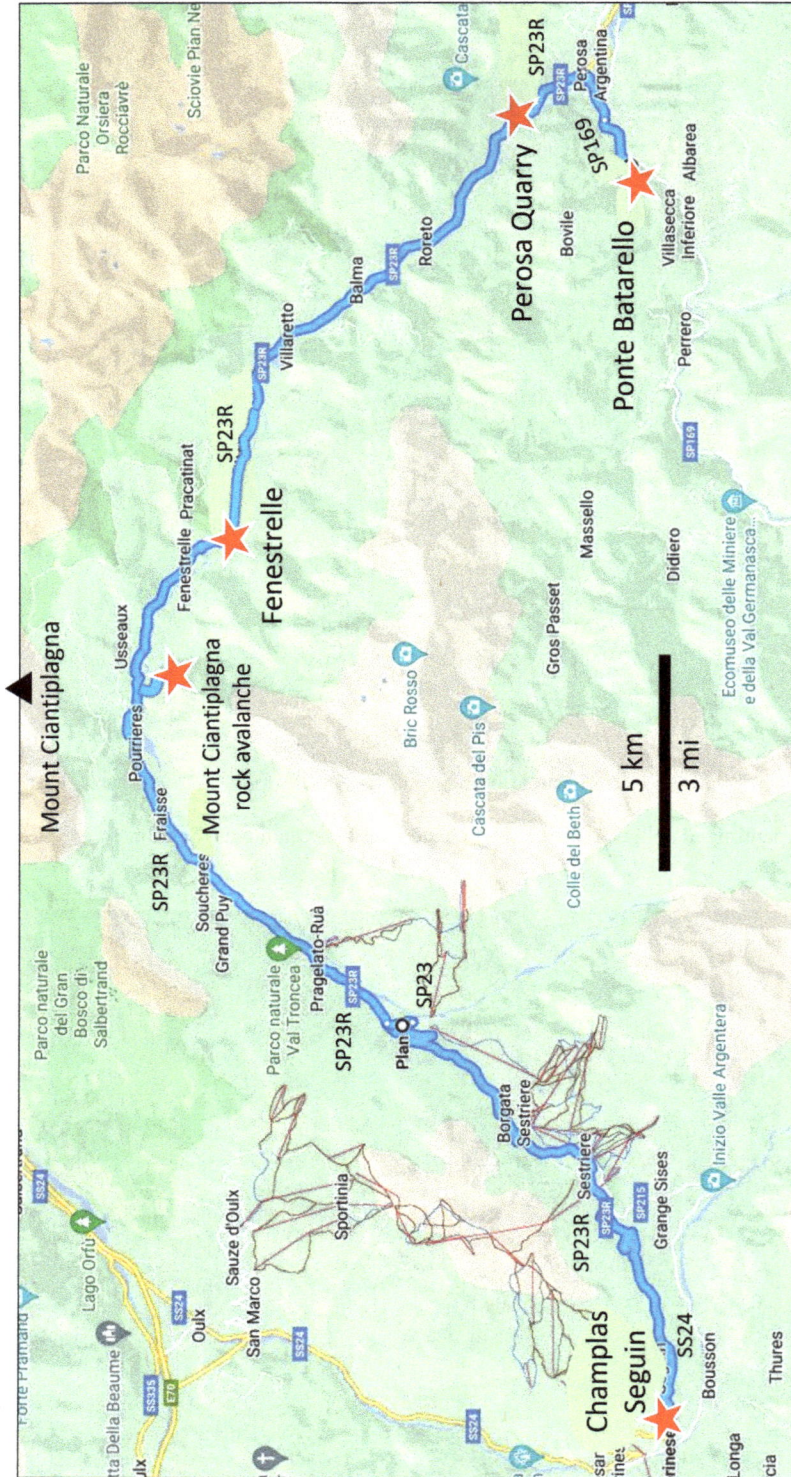

Stop 39 Mount Ciantiplagna Rock Avalanche

Driving between Pourriers and Laux, you cross the largest rock slide in the Western Alps of Italy, the Mount Ciantiplagna (locally Cima Ciantiplagna) rock avalanche. I wasn't sure what to look for, but then looking northeast from the highway, I saw a peak that looks like a giant ice cream scooper had excavated the front of the mountain. The landslide, on the north side of the highway, is 2.9 × 3.1 km (1.8 by 1.9 mi) and involved about 157 million m³ (205 million yd³), of rock. The rock avalanche traveled a horizontal distance of 4.6 km (2.8 mi) over a vertical drop of 1,455 m (4,770 ft). While the exact age is unknown, it is considered early post-glacial, perhaps as much as 10,000 to 12,000 years old. The rock involved is massive calcschist of the Cerogne–Ciantiplagna Unit, with foliation that dips 20°–50° downslope. The downslope dip was likely a factor in triggering the slide. The slide is thought to have occurred because the retreat of glacial ice removed an ice buttress to the oversteepened rock slope. The slide may have been caused by heavy rain, rapid snowmelt, seismic activity, or some combination of these (Fioraso and Baggio, 2013).

The Mount Ciantiplagna rock slide as seen from the south. Not only has the mountain been scooped out, but the landslide forms the hummocky terrain in the middle foreground.

Mount Ciantiplagna to Fenestrelle: *Continue east on SP23R for 5.1 km (3.2 mi; 7 min) and pull over on the right. This is **Stop 40.1, Dora–Maira Ophiolite, Fenestrelle** (45.027449, 7.056598). Carefully cross the highway and walk around the curve. The shoulder is narrow, so beware of traffic!*

STOP 40 THE DORA–MAIRA MASSIF

The remaining stops between here and Turin are in the Dora–Maira Massif, one of the Inner Crystalline Massifs (the others are the Gran Paradiso and Monte Rosa massifs).

As mentioned previously, the Western Alps orogenic belt is the result of the collision between the Adriatic Microplate and Europe during and after closure of the Liguro-Piemontais Ocean. The orogenic belt evolved from Late Cretaceous–middle Eocene subduction, to late Eocene–early Oligocene continental collision, and late Oligocene–Neogene deep crust/mantle uplift and thrusting. The first of these stages involved high and ultra-high-pressure metamorphism; the last stage involved retrograde greenschist metamorphism. (Retrograde metamorphism is the alteration of rocks that occurs as they are uplifted and cooled.) Thus, the Dora–Maira Massif is a slab of the paleo-European continental margin that suffered intense subduction and exhumation. It is now part of the the Penninic Domain, and it is overlain by thrusted metaophiolite complexes of the Liguro-Piemontais Ocean floor.

The Dora–Maira Massif comprises multiple tectonic units ("ensembles," or lithostratigraphic units). These units are characterized by different stratigraphies, peak metamorphism, and structural positions. The lowermost tectonic zone, the Pinerolo Unit, lies in a tectonic window in the central Dora–Maira Massif. It was metamorphosed to blueschist facies (high pressure–low to moderate temperature) metamorphism, and consists of Upper Carboniferous (Pennsylvanian) metasediments intruded by Lower Permian diorite and granite/orthogneiss.

The Pinerolo Unit is tectonically overlain in the southern Dora–Maira by more intensely metamorphosed eclogite facies units consisting of a composite metamorphic basement intruded by Lower Permian granites, and successions of Permian to Triassic metasediments. These include the San Chiaffredo Unit, the Brossasco–Isasca Unit, and the Rocca Solei Unit. These eclogite facies (high pressure–high temperature) rocks are, in turn, tectonically overlain by the less metamorphosed blueschist facies Dronero and Sampeyre units consisting of mica schists, metaintrusives, interbedded Permian metasediments and metavolcanics, and Lower Triassic metasediments. Between the eclogite facies units and the blueschist facies Dronero and Sampeyre units are scattered metaophiolites containing serpentinite, metamorphosed mafic igneous rocks, and calcareous schist (Balestro et al., 2020).

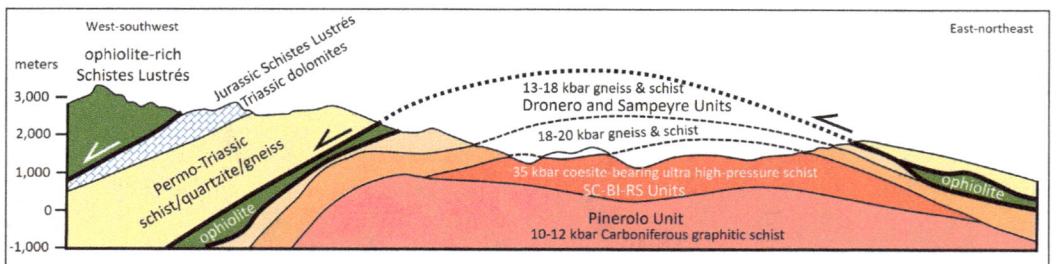

Schematic cross section of the Dora–Maira Massif. SC-BI-RS are the San Chiaffredo, Brossasco–Isasca, and the Rocca Solei units, respectively. Modified after Faure and Chen, 2018.

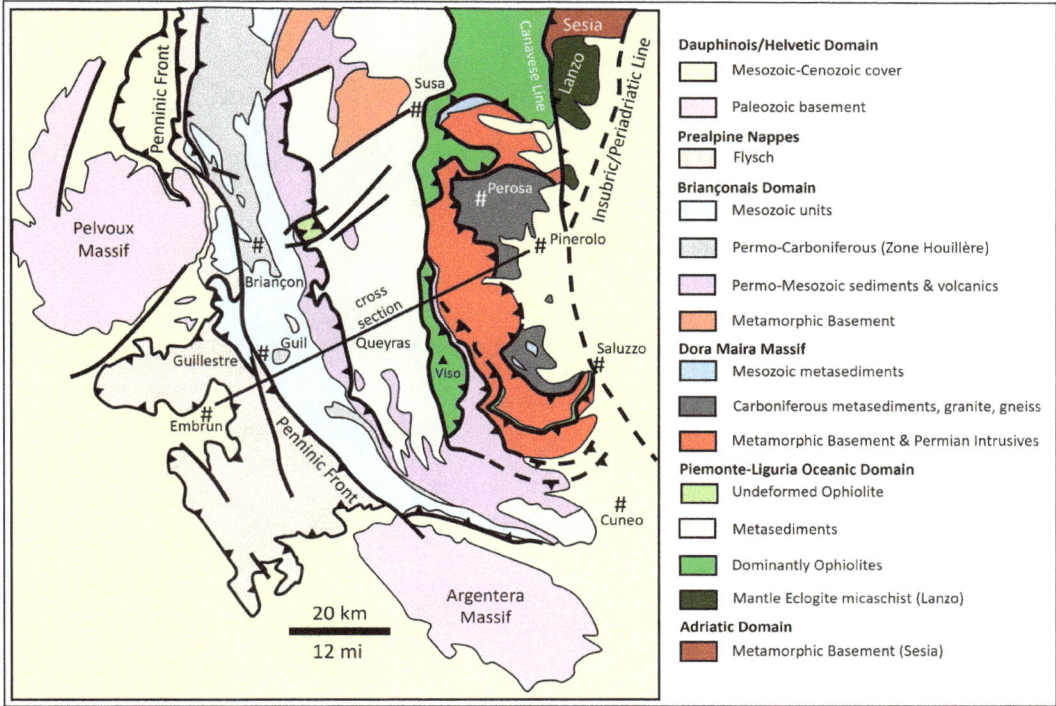

Geologic map of the Dora–Maira Massif. Heavy lines are normal faults; heavy lines with teeth are thrust faults. Cross section refers to the following figure. Modified after Ballèvre et al., 2020.

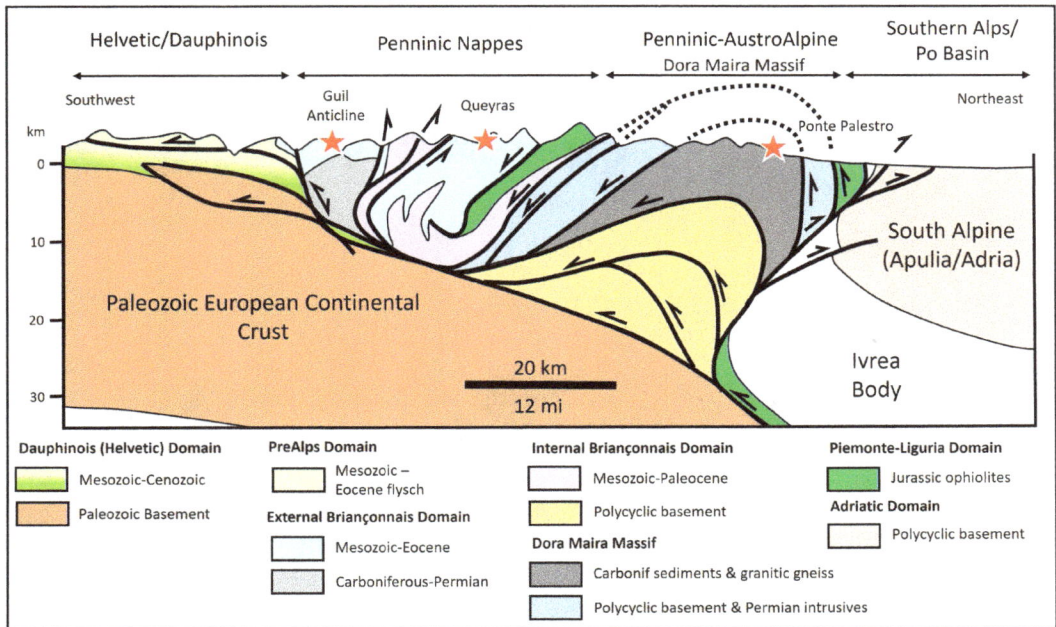

Cross section of the Western Alps through the Dora Maira Massif with stops shown as stars. Section location is shown on previous figure. Modified after Ballèvre et al., 2020.

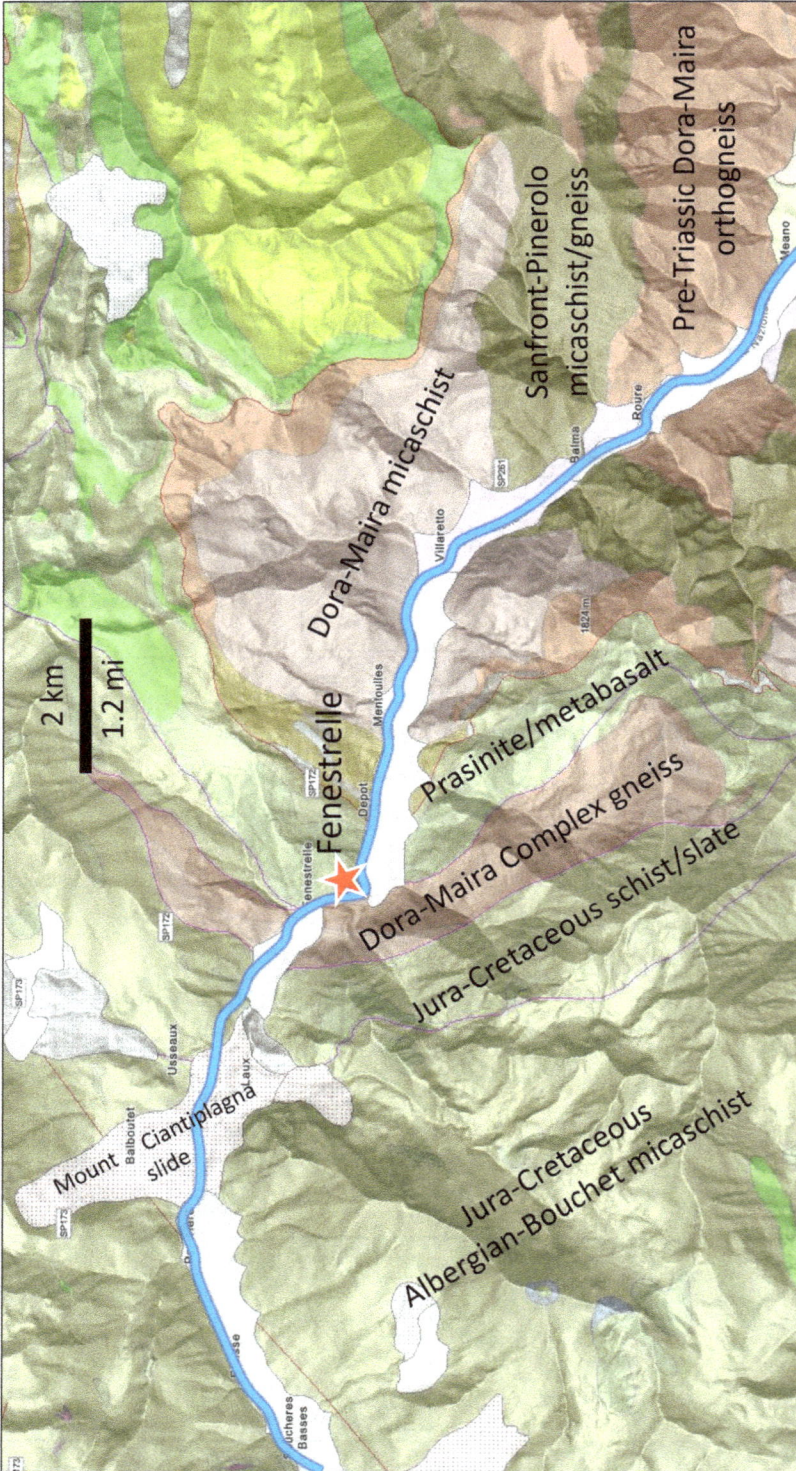

Geologic map of the Fenestrelle area. Geology base from Arpa GeoPiemonte, 2021.

STOP 40.1 DORA–MAIRA OPHIOLITE, FENESTRELLE

Although much of the roadcut is covered in chain link fencing, there are accessible outcrops of very green mica-rich ophiolite. These rocks are part of the Complesso Monometamorfico Unita Dora–Maira, a pre-Triassic gneiss and mica schist (Alpa Piemonte). They have also been described as high-pressure metaophiolites within the Schistes Lustrés (Faure and Chen, 2018; Ballèvre et al., 2020).

Roadcut in the Dora–Maira Complex prasinite (metabasalt) ophiolite at Fenestrelle.

Fenestrelle to Perosa Quarry: *Continue driving southeast on SP23R to **Stop 40.2, Dora–Maira Orthogneiss, Perosa Quarry** (44.970545, 7.177557) for a total of 12.9 km (8.0 mi; 14 min). Pull over on the right (west) on the road to the quarry.*

Stop 40.2 Dora–Maira Orthogneiss, Perosa Quarry

The Palmero Carlo e Figlio Quarry has been providing custom stone for over 150 years. The quarry is developed in a granodiorite/orthogneiss called "Perosa Stone" (Faure and Chen, 2018). This Carboniferous to Permian continental basement rock, the Paleozoic continental margin of Europe, is part of the Pinerolo Unit of the Dora–Maira Massif (Borghi et al., 2016; Arpa GeoPiemonte, 2021). Beautiful cut slabs of this rock are available to examine along the road leading into the quarry.

The Sanfront–Pinerolo Gneiss, a subset of the Pinerolo Unit of the Dora–Maira Massif, exhibits a distinct mineral fabric reflecting its metamorphic history. Intense deformation resulted in a consistent mylonitic foliation due to recrystallization and shear. The mineral fabric provides evidence of high-pressure conditions during the metamorphic evolution of the Sanfront–Pinerolo Unit.

Although not necessary to see the rock, if you would like to tour the quarry you should request permission well ahead of visiting.

Visit

Address: PALMERO CARLO & FIGLIO snc, Frazione Brandoneugna - 10063 Perosa Argentina (TO) – Italy, PI 02596990016
Phone: +39 0121 81306
Email: palmero.pietradiperosa@fastwebnet.it
Website: https://www.palmeropietradiperosa.com/

Perosa to Ponte Batarello: *Continue southeast on SP23R to Perosa Argentina; turn right (west) onto SP169/Via S. Giovanni Bosco and drive to* **Stop 40.3, Pinerolo Gneiss, Ponte Batarello** *(44.951162, 7.166883) for a total of 4.0 km (2.5 mi; 5min). Make a U-turn at the bridge and pull over on the right. Walk west past the bridge and take the fisherman's path about 100 m to the streambed.*

Stop 40.3 Pinerolo Gneiss, Deep Dora–Maira Massif, Ponte Batarello

We are still in the Pinerolo Unit, the lowest tectonic zone of the Dora–Maira Massif. Outcrops in the bed of the Germanasca River are metaconglomerates metamorphosed to blueschist facies gneiss. There is a clear, aligned metamorphic fabric. The conglomerate pebbles have all been elongated perpendicular to the maximum compression direction.

Three sequences have been described in Pinerolo Unit. The *Graphite-rich conglomerate-sandstone succession* contains gravel to pebble conglomerates consisting of quartz, gneiss, and schist. The *Graphite-rich siltstone succession* consists of metasiltstones with interbedded metasandstones and coal layers that have been metamorphosed to graphite. The *Conglomerate-sandstone succession with thin interbedded graphite layers* contains very fine grained, black, graphite-rich schists, interpreted as former organic-rich mudstones; metasandstone layers; and interbedded metaconglomerates and metasandstones. The original units were probably deposited alluvial to lagoonal environments and are considered Carboniferous. The maximum age of deposition, based on detrital zircon geochronology, is Early Carboniferous (Mississippian, around 340–330 Ma; Nosenzo et al., 2024).

The Pinerolo Unit is known for its metaconglomerates, best seen along the Germanasca River near Ponte Batterello.

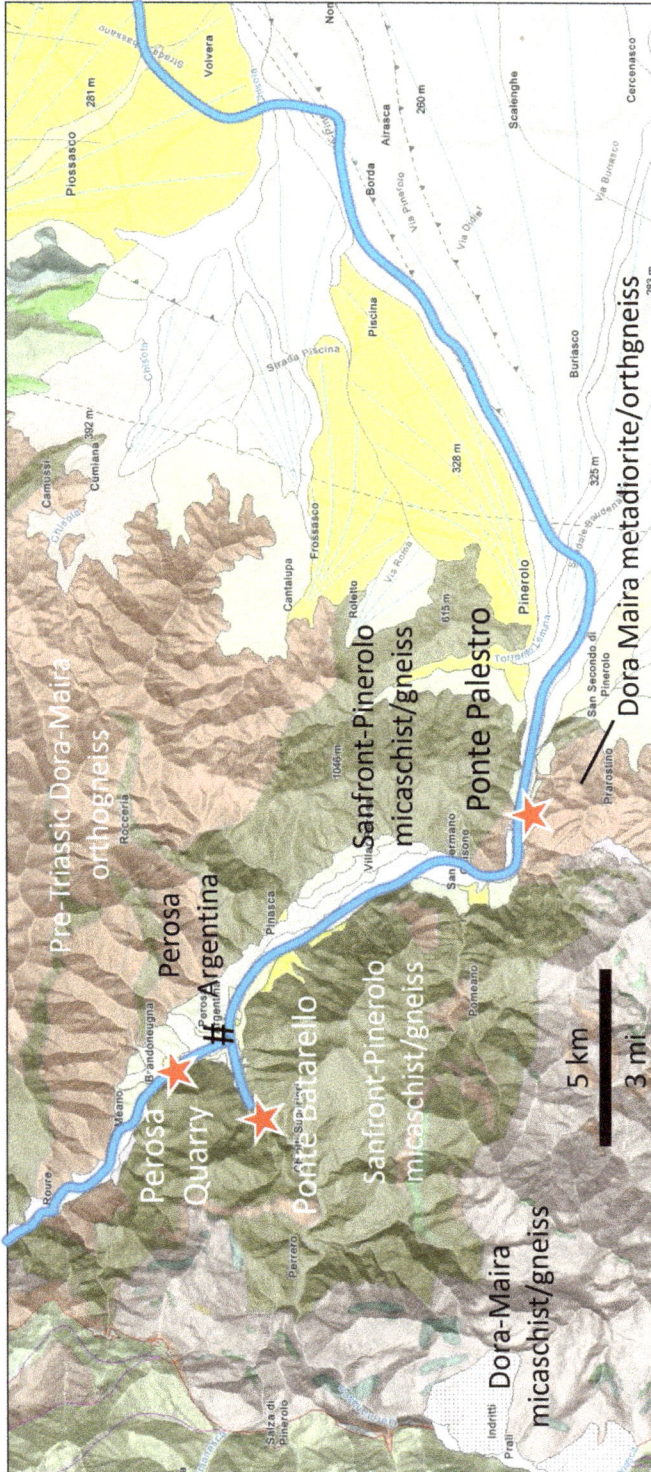

Geologic map of the Perosa Quarry–Ponte Batarello–Ponte Palestro area. Geology base from Arpa GeoPiemonte, 2021.

Entrance to the Palmero Carlo e Figlio Quarry at Perosa. View south.

Cut slabs of Sanfront–Pinerolo Gneiss from the Perosa Quarry. There is an obvious mineral fabric, or foliation.

Pinerolo conglomeratic gneiss in the Germanasca riverbed.

Detail of Pinerolo Gneiss texture showing the stretched and elongated pebbles (light blobs) of the original conglomerate.

Ponte Batarello to Ponte Palestro: *Return east on SP169; at the roundabout take the 1st exit (south) across the river; at the next roundabout take the 2nd exit (east) onto SP166; at roundabout continue straight on SP166dir; at the roundabout after crossing the Chisone River continue straight (south) onto SP23R; at the roundabout take the 2nd exit onto SR23/Via Nazionale; turn right (south) to Via Ponte Palestro; turn right (west) on Via Ponte Palestro and drive 260 m (850 ft) and pull over on the right. This is **Stop 40.4, Pinerolo Diorite, Ponte Palestro** (44.89008, 7.256491) for a total of 13.3 km (8.3 mi; 16 min). Take a fisherman's path down to the river for fresh outcrops.*

Stop 40.4 Pinerolo Diorite, Deepest Dora–Maira Massif, Ponte Palestro

Outcrops in the Chisone River at Ponte Palestro near Pinerolo expose the Carboniferous Pinerolo Unit. In contrast to the Pinerolo Gneiss seen at Ponte Batarello, here the unit is a diorite that was metamorphosed to blueschist facies during Late Eocene (Arpa GeoPiemonte, 2021; Ballèvre et al., 2020). There is a platy fabric that appears unrelated to the mineral fabric in the rock. Perhaps it is unloading joints causing exfoliation due to pressure release after uplift and erosion of the overlying rock.

These rocks are from the tectonically lowest part of the Dora–Maira Massif. The metamorphic facies indicates burial to between 15 and 40 km (9 to 24 mi), 10 to 12 kbar (145,000 to 174,000 psi) and 200° to 550°C (390° to 1,000°F). This is considered high pressure–low temperature metamorphism that is typical of subduction zones.

Detail of Pinerolo diorite at Ponte Palestro.

Pinerolo diorite at Ponte Palestro. View west up the Chisone River.

Return to Turin Airport and End: *Return to SR23 and turn right (east); at the roundabout on the west side of Pinerolo take the 1st exit (southeast) onto Via Giustetto; at the roundabout take the 3rd exit (east) to stay on Via Giustetto; at the roundabout take the 2nd exit (south) onto Via Battitore; bear right to stay on Via Battitore; at the roundabout take the 2nd exit onto SP23 and the ramp to Torino; bear left to continue on A55; take the E70 toll road north to Torino; bear right to continue north on A55 toward the airport/Torino; take the Caselle exit toward Aeroporto Caselle; drive north on RA10 to the airport for a total of 68.1 km (42.3 mi; 46 min).*

Back in Turin, we end our loop trip through the Western Alps. We have traveled from ancient Apulia to Europe, traversing the Alps twice, once from Turin to Aix-les-Bains, and again from Valence to Grenoble and back to Turin. We have seen the locations where fold-thrust belts were first described and explained; where the concept of Ice Ages and glaciation was conceived. This trek has taken us through three Global Geoparks, several World Heritage Sites, and past multiple quarries, mines, hot springs, and iconic peaks. This is well and truly one of the great geologic landscapes of the world.

REFERENCES

Aescher, Arthur. 2001. The overall tectonic framework of the Penninic Domain. In Stampfli, G.M. (ed.), *Geology of the western Swiss Alps, a guide-book.* Mémoires de Géologie, Lausanne, No. 36, p. 115–117.

Agard, P., L. Jolivet, and B. Goffe. 2001. Tectonometamorphic evolution of the Schistes Lustres Complex; Implications for the exhumation of HP and UHP rocks in the Western Alps. *Bulletin de la Société Géologique de France* v. 172 no. 5, p. 617–636. https://doi.org/10.2113/172.5.617

Alpes Cottianne Geoparc. Tourist Map. Accessed 6 March 2022. https://geoparc-alpescottiennes.eu/uploads/cartes/carte_geoparc_en.pdf

Arpa GeoPiemonte. 2021. Geologic Map of the Piemonte Region, Institute of Geosciences and Earth Resources – National Research Council of Italy, 1:250,000. https://webgis.arpa.piemonte.it/agportal/apps/webapp viewer/index.html?id=6ea1e38603d6469298333c2efbc76c72

Balestro, G., F. Nosenzo, P. Cadoppi, G. Fioraso, C. Groppo and A. Festa. 2020. Geology of the southern Dora-Maira Massif: Insights from a sector with mixed ophiolitic and continental rocks. *Journal of Maps* v. 16 no. 2, p. 736–744. https://doi.org/10.1080/17445647.2020.1824825

Ballèvre, M., A. Camonin, P. Manzotti, and M. Poujol. 2020. A step towards unraveling the paleogeographic attribution of pre-Mesozoic basement complexes in the Western Alps based on U–Pb geochronology of Permian magmatism. *Swiss Journal of Geosciences* v. 113 no. 12, 28 p. https://doi.org/10.1186/s00015-020-00367-1

Barale, L., A. Borghi, A. d'Atri, F. Gambino, and F. Piana. 2020. Ornamental stones of Piemonte (NW Italy): An updated geo-lithological map. *Journal of Maps* v. 16 no. 2, p. 867–878. https://doi.org/10.1080/17 445647.2020.1837685

Bellahsen, N., F. Mouthereau, A. Boutoux, M. Bellanger, O. Lacombe, L. Jolivet, and Y. Rolland. 2014. Collision kinematics in the western external Alps. *Tectonics* v. 33, p. 1055–1088. https://doi.org/10.1002 /2013TC003453

Bellanger, M., N. Bellahsen, L. Jolivet, T. Baudin, R. Augier, and A. Boutoux. 2014. Basement shear zones development and shortening kinematics in the Ecrins Massif, Western Alps. *Tectonics* v. 33, p. 84–111, doi:10.1002/2013TC003294.

Beltrando, M., G. Frasca, R. Compagnoni, and A. Vitale-Brovarone. 2012. The Valaisan controversy revisited: Multi-stage folding of a Mesozoic hyper-extended margin in the Petit St. Bernard pass area (Western Alps). *Tectonophysics* v. 579, p. 17–36.

Beltrando, M., R. Compagnoni, J. Barnes, M. Luce Frezzotti, D. Regis, G. Frasca, M. Forster, G. Loster. 2014. *From passive margins to orogens: The link between Zones of Exhumed Subcontinental Mantle and (U) HP metamorphism.* 10th International Eclogite Conference - Courmayeur (Aosta, Italy). Geological Field Trips v. 6 (1.1). ISPRA. https://doi.org/10.3301/GFT.2014.01

Berio, L.R., F. Storti, F. Balsamo, S. Mittempergher, A. Bistacchi, and M. Meda. 2021. Structural evolution of the Parmelan anticline (Bornes Massif, France): Recording the role of structural inheritance and stress field changes on the finite deformation pattern. *Tectonics* v. 40, 34 p. e2021TC006913. https://doi.org/10 .1029/2021TC006913

Borghi, A., P. Cadoppi, and G.A. Dino. 2016. Heritage stone 2. The Dora-Maira Unit (Italian Cottian Alps): A reservoir of ornamental stones since roman times. *Geoscience Canada* v. 43 no. 1, p. 13–30.

Buckley, J.J. 2020. Your guide to piemonte wine. Accessed 7 March 2022. https://www.jjbuckley.com/wine-knowledge/blog/your-guide-to-piemonte-wine/1086

Burroni, A., N. Levi, M. Marroni, and L. Pandolfi. 2003. Lithostratigraphy and structure of the lago nero unit (Chenaillet Massif, Western Alps: Comparison with internal liguride units of Northern Apennines. *Ofioliti* v. 28 no. 1, p. 1–11.

Butler, R. Thrusts. Accessed 7 March 2022. http://www.see.leeds.ac.uk/structure/tectonics/thrust_tectonics/

Cairngorms Nat Park, Roche Moutonnée, Dulnain Bridge. Accessed 22 March 2022. https://cairngorms.co.uk/photo-posts/photoposts/13/moreinfo/

Cardinal, T., C. Petit, Y. Rolland, S. Schwartz, P. G. Valla, B. Scalabrino, L. Audin, M. Bertauts, S. Zerathe, D. Thiéblemont, R. Braucher, and ASTER Team. 2024. The interplay of geology, climate and tectonics on river incision: the example of the High Verdon Gorges, Southwestern French Alps. *Comptes Rendus Géoscience, Sciences de la Planète* v. 356, Special Issue S2, p. 265–287.

Chaljub, E., C. Cornou, and P.-Y Bard. 2006. *Numerical benchmark of 3D ground motion simulation in the valley of Grenoble, French Alps.* 3rd Intl. Symp. On the Effects of Surface Geology on Seismic Motion, Grenoble, France, 30 Aug-1 Sep 2006, paper SB1, p. 1–11.

Chapelet, M., C. Dezayes, and T. Dewez. 2019. *Characterization of fracture network based on photogrammetry.* Proceedings World Geothermal Congress 2020, Reykjavik, Iceland, April 26–May 2, 2020, 10 p.

Chapron, E., P. Van Rensbergen, C. Beck, M. De Batist, and A. Paillet. 1996. Lacustrine sedimentary records of brutal events in Lake Le Bourget (Northwestern Alps-Southern Jura) [Enregistrements d'événements brutaux dans la sédimentation lacustre du Lac du Bourget (Alpes occidentales-Jura méridional)]. *Quaternaire* v. 7, no. 2–3, p. 155–168. https://doi.org/10.3406/quate.1996.2068

Cloetingh S., P.A. Ziegler, F. Beekman, E.B. Burov, D. Garcia-Castellanos, and L. Matenco. 2015. Tectonic models for the evolution of sedimentary basins. In: Schubert, G. (ed), *Treatise on geophysics*, 2nd edition, v. 6, Oxford: Elsevier, p. 513–592.

Compagnoni, R., M. Engi, and D. Regis. 2014. *Valle d'Aosta section of the Sesia Zone: Multi-stage HP metamorphism and assembly of a rifted continental margin.* 10th International Eclogite Conference, Courmayeur (Aosta, Italy) - Pre-conference excursions: September 5, 2013. Periodico semestrale del Servizio Geologico d'Italia - ISPRA e della Società Geologica Italiana, Geologic Field Trips, v. 6 no. 1.2, 44 p. https://doi.org/10.3301/GFT.2014.02

De Broucker, G., Y. Siméon, G.M. Stampfi, D. Thiéblemont, P. Lach, and M. Marthaler. 2021. Early Cretaceous Accretionary Complex of the Valaisan Ocean, Western Alps? *Ofoliti* v. 46 no. 2, p. 203–227. https://doi.org/10.4454/ofoliti.v46i2.547

Dela Pierre, F., M. Natalicchio, F. Lozar, S. Bonetto, G. Carnevale, S. Cavagna, S. Colombero, M. Sabino, and D. Violanti. 2016. The northernmost record of the Messinian salinity crisis (Piedmont basin, Italy). *Periodico semestrale del Dipartimento per il Servizio Geologico d'Italia - ISPRA e della Società Geologica Italiana Geologic Field Trips* v. 8 no. 2.1, 58 p. https://doi.org/10.3301/GFT.2016.03

DiBiase, D., and E. Mutti. 2002. Part 1 – The "Proto Adriatic Basin." In E. Mutti, F.R. Lucchi, and M. Roveri (eds.), Revisiting turbidites of the Marnoso-arenacea Formation and their basin-margin equivalents: problems with classic models - Excursion Guidebook. 64th EAGE Conference and Exhibition, Florence, Italy, 27 - 30 May 2002. p. I–1 to I--4.

Dumont, T., J-D Champagnac, C. Crouzet, and P. Rochat. 2008. Multistage shortening in the Dauphiné zone (French Alps): The record of Alpine collision and implications for pre-Alpine restoration. *Swiss Journal Geoscience* v. 101, Supplement 1, p. S89–S110. https://doi.org/10.1007/s00015-008-1280-2

Egli, D., and N. Mancktelow. 2013. The structural history of the Mont Blanc massif with regard to models for its recent exhumation. *Swiss Journal Geoscience* v. 106, p. 469–489. https://doi.org/10.1007/s00015-013-0153-5

Faure, M., and Y. Chen. 2018. *A geological excursion in the Western Alps, Field Guide Book 2018.* Orléans University-Institute of Geology and Geophysics Cooperation program, Orléans, 36 p.

Festa, A., G. Balestro, A. Borghi, S. De Caroli, and A. Succo. 2020. The role of structural inheritance in continental break-up and exhumation of Alpine Tethyan mantle (Canavese Zone, Western Alps). *China University of Geosciences (Beijing)* v. 11 no. 1, p. 167–188.

Fioraso, G., and P. Baggio. 2013. Geological map of the Mount Ciantiplagna rock avalanche (Chisone Valley, Italian Western Alps). *Journal of Maps* v. 9 no. 3, p. 336–342. https://doi.org/10.1080/17445647.2013.781967

Froitzheim, N. 2012. *Geology of the Alps Part 1: General remarks; Austroalpine nappes.* Lecture notes, Steinmann Institute, Universität Bonn, 73 p.

Gallino, S., J.-Y. Josnin, M. Dzikowski, F. Cornaton, and D. Gasquet. 2009. The influence of paleoclimatic events on the functioning of an alpine thermal system (France): The contribution of hydrodynamic–thermal modeling. *Hydrogeology Journal* v. 17, no. 8, p. 1887–1900.

Gambino, F., A. Borghi, A. d'Atri, L.M. Gallo, L. Ghiraldi, M. Giardino, L. Martire, M. Palomba, and L. Perotti. 2017. TourInStones: Application for mobile on ornamental stones of the city of Torino. *Società Geologica Italiana, Roma* v. 42, p. 81–84. https://doi.org/10.3301/ROL.2017.20

Giraudi, C. 2016. The evolution of the Northernmost Apennine front (Piedmont, Italy): Plio-pleistocene sedimentation and deformation in the po basin and monferrato hills. *Alpine and Mediterranean Quaternary* v. 29 no. 1, p. 45–65.

Giuntoli, F., and M. Engi. 2016. Internal geometry of the central Sesia Zone (Aosta Valley, Italy): HP tectonic assembly of continental slices. *Swiss Journal of Geoscience* v. 109, p. 445–471. https://doi.org/10.1007/s00015-016-0225-4

Godet, A., T. Adatte, A. Arnaud-Vanneau, L. Bonvallet, E. De Kaenel, and P-O Mojon. 2024. Rise and demise of the Urgonian platform in Switzerland. In Hart, M.B., Batenburg, S.J., Huber, B.T., Price, G.D., Thibault, N., Wagreich, M. and Walaszczyk, I. (eds), *Cretaceous project 200 volume 2: Regional studies.* Geological Society, London, Special Publications, 545, 117–135. First published online January 11, 2024. https://doi.org/10.1144/SP545-2023-103

Goodson, D.C. 2018. Turin history. Life in Italy.com. Accessed 7 March 2022. https://lifeinitaly.com/turin-history/

Halama, R., J. Glodny, M. Konrad-Schmolke, and M. Sudo. 2018. Rb-Sr and in situ 40Ar/39Ar dating of exhumation-related shearing and fluid-induced recrystallization in the Sesia zone (Western Alps, Italy). *Geosphere* v. 14 no. 4, p. 1425–1450. Accessed 7 March 2022. https://doi.org/10.1130/GES01521.1

Hoblea, F., S. Gallino-Josnin, and P. Audra. 2010. Genesis and functioning of the Aix-les-Bains hydrothermal karst (Savoie, France): Past research and recent advances. *Bulletin de la Société Géologique de France* v. 181 no. 4, p. 315–326. Accessed 7 March 2022. https://doi.org/10.2113/gssgfbull.181.4.315

IGoToWorld.com. Ponte Vittorio Emanuele I, Turin. Accessed 7 March 2022. https://it.igotoworld.com/en/poi_object/316983_bridge-vittorio-emanuele.htm

International Granites, Sienita della Balma Granite from Italy. Accessed 5 July 2025. https://igranites.com/sienite_della_balma.html

ISPRA, 2011, Folio 89 – Geologic Map, Courmayeur, 1:50,000. Accessed 7 March 2022. https://www.isprambiente.gov.it/Media/carg/89_COURMAYEUR/Foglio.html

ISPRA, 2012a, Folio 70, Geologic Map, Monte Cervino, 1:50,000. Accessed 10 March 2022. https://www.isprambiente.gov.it/Media/carg/70_MONTE_CERVINO/Foglio.html

ISPRA, 2012b, Folio 90, Geologic Map, Aosta, 1:50,000. Accessed 7 March 2022. https://www.isprambiente.gov.it/Media/carg/90_AOSTA/Foglio.html

James, T. 2002. *A study of the Geological Structure of the Massiccio dei Laghi (Northern Italy).* University of Manchester unpublished PhD Thesis, online preview, 24 p.

Kalifi, A., P.H. Leloup, P. Sorrel, A. Galy, F. Demory, V. Spina, B. Huet, F. Quillévéré, F. Ricciardi, D. Michoux, K. Lecacheur, R. Grime, B. Pittet, and J.-L. Rubino. 2021. Chronology of thrust propagation from an updated tectono-sedimentary framework of the Miocene molasse (western Alps). *Solid Earth* v. 12, p. 2735–2771. Accessed 7 March 2022. https://doi.org/10.5194/se-12-2735-2021

Kindler, P. 2014. Geological features between Chamonix and Geneva: Facts and controversies. *Bulletin fuer Angewandte Geologie* v. 19 no. 1, p. 41–51.

Kuderski, A. 2020. SavoieWine: Unusual Finds From the French Alps. Wine Folly. Accessed 7 March 2022. https://winefolly.com/deep-dive/savoie-wine-guide/

Lacassin, R. 2017. First description of Alpin Folding by H.B. De Saussure in 1778. Accessed 13 May 2025. https://tectoldies.mystrikingly.com/blog/first-description-of-alpine-folding-by-h-b-de-saussure-in-1778?fbclid=IwY2xjawIgFYNleHRuA2FlbQIxMAABHbmxEiLLq3iNggzXxGIMg86-YDAYY2h9SfsW1EefV4jQ4-dbtWgpXSg9cQ_aem_JEBsgNtzZJFOxCsPFopzCQ

Lacombe, O., and N. Bellahsen. 2016. Thick-skinned tectonics and basement-involved fold–thrust belts: insights from selected Cenozoic orogens. Geology Magazine v. 153, p. 763-810.

Laura, P., B. Attilio, G. Evelina. 2005. Sr, Nd, and Pb Isotope Investigation of the Serie dei Laghi (Southern Alps, Italy): The Behaviour of Different Isotope Systems During Crustal Metamorphism, Magmatism, and Deformation. Accessed 7 March 2022. https://www.researchgate.net/publication/326232238_Sr_Nd_and_Pb_isotope_investigation_of_the_Serie_dei_Laghi_Southern_Alps_Italy_the_behaviour_of_different_isotope_systems_during_crustal_metamorphism_magmatism_and_deformation

Lindquist, S.J. 1999. Petroleum Systems of the Po Basin Province of Northern Italy and the Northern Adriatic Sea: Porto Garibaldi (Biogenic), Meride/Riva di Solto (Thermal), and Marnoso Arenacea (Thermal). U.S. Geological Survey Open-File Report 99-50-M, 34 p.

Lonely Planet, Chiesa della Gran Madre di Dio. Accessed 8 March 2022. https://www.lonelyplanet.com/italy/liguria-piedmont-and-valle-daosta/turin/attractions/chiesa-della-gran-madre-di-dio/a/poi-sig/424663/359924

MacEacheran, M. 2019. Chilling Outin Aix-les-Bains, France's Ultimate Spa Town. Accessed 8 March 2022. https://www.roughguides.com/articles/chilling-out-in-frances-ultimate-spa-town/

MacroStrat. 2015. Accessed multiple times between 2022 and 2025. https://macrostrat.org/map/#/z=7.4/x=7.3387/y=45.5845/bedrock/lines/

Manzotti, P. 2011. Petro-structural map of the Dent Blanche tectonic system between Valpelline and Valtournenche valleys, Western Italian Alps. *Journal of Maps* v. 7 no. 1, p. 340–352. https://doi.org/10.4113/jom.2011.1179

Manzotti, P., and M. Ballèvre. 2016. Tectonic history of the Dent Blanche. *Periodico semestrale del Servizio Geologico d'Italia - ISPRA e della Società Geologica Italiana, Geologic Field Trips* v. 9 no. 2.1, 73 p. https://www.isprambiente.gov.it/it/pubblicazioni/periodici-tecnici/geological-field-trips

Marquer, D., P. Calcagno, J.-C. Barfety, and T. Baudin. 2006. 3D Modeling and Kinematics of the External Zone of the French Western Alps (Belledonne and Grand Châtelard Massifs, Maurienne Valley, Savoie). Eclogae Geologicae Helvetiae v. 99 no. 2. 13 p.

Mathey, M., A. Walpersdorf, C. Sue, S. Baize, and A. Deprez. 2020. Seismogenic potential of the High Durance Fault constrained by 20 yr of GNSS measurements in the Western European Alps. *Geophysical Journal International* v. 222, p. 2136–2146. https://doi.org/10.1093/gji/ggaa292

Mazzucchelli, M., J.E. Quick, S. Sinigoi, A. Zanetti, T. Giovanardi. 2014. Igneous evolutions across the Ivrea crustal section: the Permian Sesia Magmatic System and the Triassic Finero intrusion and mantle. *Periodico semestrale del Servizio Geologico d'Italia - ISPRA e della Società Geologica Italiana Geologic Field Trips* v. 6 no. 2.2, 98 p. https://doi.org/10.3301/GFT.2014.05

Mowery, L. 2019. A Beginner's Guide to Wines from Italy's Piedmont. Accessed 8 March 2022. https://www.winemag.com/2019/02/12/beginners-guide-wines-from-italys-piedmont/

Mutti, E., F.R. Lucchi, and M. Roveri (eds.). 2002. Revisiting turbidites of the Marnoso-arenacea Formation and their basin-margin equivalents: Problems with classic models. *Excursion Guidebook* (preliminary version). Parma, Italy 21–22 May 2002. Accessed 8 March 2022. https://www.researchgate.net/publication/284814160_Revisiting_turbidites_of_the_Marnoso-Arenacea_Fomation_and_their_basin-margin_equivalents_Problems_with_classic_models

Norman, J. 1991a. Origins of the Theory of the Ice Age; Global Cooling and Warming. Accessed 8 March 2022. https://www.historyofinformation.com/detail.php?id=2384

Norman, J. 1991b. Origins of the Theory of the Ice Age; Global Cooling and Warming, 1834 to 1841. The Haskell F. Norman Library of Science and Medicine nos. 17 & 462. Accessed 7 March 2022. https://www.historyofinformation.com/detail.php?id=2384

Nosenzo, F., P. Manzotti, M. Krona, M. Ballèvre, and M. Poujol. 2024. Tectonic architecture of the northern Dora-Maira Massif (Western Alps, Italy): Field and geochronological data. *Swiss Journal of Geosciences* v. 117 no. 6, 33 p.

Piana, F., G. Fioraso, A. Irace, P. Mosca, A. d'Atri, L. Barale, P. Falletti, G. Monegato, M. Morelli, S. Tallone, and G.B. Vigna. 2017. Geology of Piemonte region (NW Italy, Alps–Apennines interference zone). *Journal of Maps* v. 13 no. 2, p. 395–405. https://doi.org/10.1080/17445647.2017.1316218

Piemonte The History of Turin From Its Origins to the Present Day. Accessed 8 March 8, 2022. https://piemonte.italiaguida.it/en/torino/default.asp

Prost, G.L., and B.P. Prost. 2017. *The Geology Companion - Essentials for Understanding the Earth.* CRC Press, Boca Raton, FL, 467 p.

Puckette, M. 2021. Essential Guide to Piedmont Wine. Wine Folly, Accessed 8 March 2022. https://winefolly.com/deep-dive/piedmont-wine-guide/

Ravanel, L., F. Allignol, P. Deline, S. Gruber, and M. Ravello. 2010. Rock falls in the Mont Blanc Massif in 2007 and 2008. Landslides. Accessed 8 March 8, 2022. https://www.researchgate.net/publication/226540399_Rock_falls_in_the_Mont_Blanc_Massif_in_2007_and_2008

Regis, D., G. Venturini, and M. Engi. 2015. Geology of the Scalaro valley – Sesia Zone (Italian Western Alps), *Journal of Maps* v. 12 no. 4, p. 621–629. Accessed 8 March 8, 2022. http://dx.doi.org/10.1080/17445647.2015.1060182

Ribes, C., J.-F. Ghienne, G. Manatschal, N. Dall'asta, D. Stockli, F. Galster, M. Gillard, and G. Karner. 2020. The Grès Singuliers of the Mont Blanc region (France and Switzerland): stratigraphic response to rifting and crustal necking in the Alpine Tethys. *International Journal of Earth Sciences, Springer Verlag* v. 109 no. 7, p. 2325–2352. http://doi.org/10.1007/s00531-020-01902-z

Schlunegger, F., and J. Mosar. 2010. The last erosional stage of the Molasse Basin and the Alps. *International Journal of Earth Sciences*, 16 p. Accessed 8 March 8, 2022. https://www.researchgate.net/publication /49251296_The_last_erosional_stage_of_the_Molasse_Basin_and_the_Alps

Sinigoi, S., J.E. Quick, G. Demarchi, and G. Peressini. 2010. The Sesia magmatic system. In Beltrando, M., A. Peccerillo, M. Mattei, S. Conticelli, and C. Doglioni (eds.), *Journal of the Virtual Explorer* v. 36, paper 5. https://doi.org/10.3809/jvirtex.2010.00218

Sissingh, W. 1998. Comparative tertiary stratigraphy of the Rhine Graben, Bresse Graben and Molasse Basin: Correlation of Alpine foreland events. *Tectonophysics* v. 300, p. 249–284.

Stampfli, G.M. (ed.). 2001. Geology of the western Swiss Alps, a guide-book. *Mémoires de Géologie (Lausanne)* No. 36, 194 p.

Statista.com. 2019. Accessed 4 July 2020. https://www.statista.com/statistics/800371/crude-oil-production-by -region-in-italy/

Statista.com. 2024. Accessed 7 July 2025. https://www.statista.com/statistics/800396/natural-gas-production -by-region-in-italy/

Storta, E., L. Barale, A. Borghi, A. d'Atri, G.A. Dino, F. Gambino, L. Martire, L. Perotti, F. Piana, A. Acquarone, P. Sassone, M. Senesi, L. Mallen, M. Morelli, and G. Nicolò. 2024. The Database of the Ornamental Stones of Piemonte (NW Italy). Geoheritage. Accessed 8 October 2025, https://www. researchgate.net/publication/382014112_The_Database_of_the_Ornamental_Stones_of_Piemonte_ NW_Italy_Hosted_on_a_WebGIS_Service

Tardy, M., and B. Doudoux. 1984. Un trait nouveau du synclinal de nappes de Thônes (Massif des Bornes, Haute-Savoie): le chevauchement cisaillant de la Montagne de Cotagne. *Géologie Alpine* v. 60, p. 77–84.

Tavazzani, L., S. Peres, S. Sinigoi, G. Demarchi, R.C. Economos, and J.E. Quick. 2020. Timescales and mechanisms of crystal-mush rejuvenation and melt extraction recorded in permian plutonic and volca- nic rocks of the sesia magmatic system (Southern Alps, Italy). *Journal of Petrology* v. 61 no. 5, 36 p. https://doi.org/10.1093/petrology/egaa049

Tricart, P., and S. Schwartz. 2006. A north-south section across the Queyras Schistes Lustrés (Piedmont zone, Western Alps): Syn-collision refolding of a subduction wedge. *Eclogae Geologicae Helvetiae* v. 99, p. 429–442. https://doi.org/10.1007/s00015-006-1197-6

Turrini, C., O. Lacombe, and F. Roure. 2014. Present-day 3D structural model of the Po Valley basin, Northern Italy. *Marine and Petroleum Geology* v. 56, p. 266–289.

Twain, M. 1891. Mark Twain at Aix-les-Bains. Mark Twain's Travel Letters from 1891–92 with original illus- trations by Dan Beard and Harold R. Heaton. Chicago Daily Tribune, November 8, 1891.

UNESCO, Massif des Bauges Global Geopark. Accessed 8 March 2022. https://en.unesco.org/global-geopar ks/massif-des-bauges

UNESCO, Sesia Val Grande Geopark. Accessed 8 March 2022. https://en.unesco.org/global-geoparks/sesia- val-grande

Wikipedia, Aix-les-Bains. Accessed 8 March 2022. https://en.wikipedia.org/wiki/Aix-les-Bains

Wikipedia, Bornes Massif. Accessed 8 March 2022. https://en.wikipedia.org/wiki/Bornes_Massif

Wikipedia (French), Fontaine Pétrifiante De Réotier [Petrified Fountain of Réotier]. Accessed 8 March 2022. https://fr.wikipedia.org/wiki/Fontaine_p%C3%A9trifiante_de_R%C3%A9otier

Wikipedia, Geology of the Alps. Accessed 8 March 2022. https://en.wikipedia.org/wiki/Geology_of_the_ Alps

Wikipedia (French), Gorges du Sierroz. Accessed 8 March 2022. https://fr.wikipedia.org/wiki/Gorges_du_ Sierroz

Wikipedia, Grenoble. Accessed 29 May 2025. https://en.wikipedia.org/wiki/Grenoble

Wikipedia, Gran Madre di Dio. Accessed 8 March 2022. https://en.wikipedia.org/wiki/Gran_Madre_di_Dio, _Turin

Wikipedia, Ice Age. Accessed 8 March 2022. https://en.wikipedia.org/wiki/Ice_age

Wikipedia, Matterhorn. Accessed 8 March 2022. https://en.wikipedia.org/wiki/Matterhorn

Wikipedia, Messinian Salinity Crisis. Accessed 8 March 2022. https://en.wikipedia.org/wiki/Messinian_ salinity_crisis

Wikipedia, Metropolitan City of Turin. Accessed 8 March 2022. https://en.wikipedia.org/wiki/Metropolitan _City_of_Turin

Wikipedia, Molasse Basin. Accessed 8 March 2022. https://en.wikipedia.org/wiki/Molasse_basin

Wikipedia, Mont Blanc Massif. Accessed 8 March 2022. https://en.wikipedia.org/wiki/Mont_Blanc_massif

Wikipedia, Palazzo Carignano. Accessed 8 March 2022. https://en.wikipedia.org/wiki/Palazzo_Carignano

Wikipedia, Piemonte Wine. Accessed 8 March 2022. https://en.wikipedia.org/wiki/Piemonte_(wine)

Wikipedia, Po River. Accessed 8 March 2022. https://en.wikipedia.org/wiki/Po_(river)

Wikipedia, Po Valley. Accessed 8 March 2022. https://en.wikipedia.org/wiki/Po_Valley

Wikipedia (Italian), Ponte Vittorio Emanuele I. Accessed 8 March 2022. https://it.wikipedia.org/wiki/Ponte_Vittorio_Emanuele_I

Wikipedia, Thrust Fault. Accessed 8 March 2022. https://en.wikipedia.org/wiki/Thrust_fault

Wikipedia, Turin. Accessed 8 March 2022. https://en.wikipedia.org/wiki/Turin

Wikipedia, Vercors Massif. Accessed 8 March 2022. https://en.wikipedia.org/wiki/Vercors_Massif

Wikipedia, Vin de Savoie AOC. Accessed 8 March 2022. https://en.wikipedia.org/wiki/Vin_de_Savoie_AOC

WineSearcher.com, Vins de Savoie, *Terroir*. Accessed 8 March 2022. https://www.wine-searcher.com/regions-vin+de+savoie

WineTourism.com – Vin de Savoie. Accessed 8 March 2022. https://www.winetourism.com/wine-appellation/vin-de-savoie/

World Guides, Turin History Facts and Timeline. Accessed 8 March 2022. http://www.world-guides.com/europe/italy/piedmont/turin/turin_history.html

Woudloper Accessed 8 March 2022. https://commons.wikimedia.org/wiki/File:Hercynides_EN.svg

Index

For Product Safety Concerns and Information please contact our EU
representative GPSR@taylorandfrancis.com
Taylor & Francis Verlag GmbH, Kaufingerstraße 24, 80331 München, Germany

www.ingramcontent.com/pod-product-compliance
Lightning Source LLC
Chambersburg PA
CBHW060943210326
41598CB00031B/4707